P9-AGA-683

System Integration

System Integration

From Transistor Design to Large Scale Integrated Circuits

Kurt Hoffmann
University of the Bundeswehr Munich, Germany

John Wiley & Sons, Ltd

Telephone (+44) 1243 779777

Email (for orders and customer service enquiries): cs-books@wiley.co.uk
Visit our Home Page on www.wileyeurope.com or www.wiley.com

Other Wiley Editorial Offices

John Wiley & Sons Inc., 111 River Street, Hoboken, NJ 07030, USA

Jossey-Bass, 989 Market Street, San Francisco, CA 94103-1741, USA

Wiley-VCH Verlag GmbH, Boschstr. 12, D-69469 Weinheim, Germany

John Wiley & Sons Australia Ltd, 33 Park Road, Milton, Queensland 4064, Australia

John Wiley & Sons (Asia) Pte Ltd, 2 Clementi Loop #02-01, Jin Xing Distripark, Singapore 129809

John Wiley & Sons Canada Ltd, 22 Worcester Road, Etobicoke, Ontario, Canada M9W 1L1

Wiley also publishes its books in a variety of electronic formats. Some content that appears in print may not be available in electronic books.

British Library Cataloguing in Publication Data

A catalogue record for this book is available from the British Library

ISBN 0-470-85407-3

Typeset in 10/12pt Times by Integra Software Services Pvt. Ltd, Pondicherry, India
Printed and bound in Great Britain by Antony Rowe Ltd, Chippenham, Wiltshire
This book is printed on acid-free paper responsibly manufactured from sustainable forestry
in which at least two trees are planted for each one used for paper production.

Contents

Preface

This book is based on one with the title *VLSI Design*, published by Oldenbourg for the first time in 1990 and followed by four further editions. A substantially revised version with the title *System Integration* was published by Oldenbourg in 2003 and is the basis for this book. It includes lectures which the author teaches at the Bundeswehr University in Munich for graduate students and topics for adult education of professional engineers and physicists.

In order to introduce the book it is appropriate to consider more closely the history of silicon integrated circuits and how the design environment has changed over the years.

The development began with the integration of a couple of bipolar transistors on a piece of silicon. A similar development with field effect (MOS) transistors started later because of several technical problems. The number of integrated MOS circuits produced increased over the years enormously, compared to the number of bipolar circuits. The better chip area utilization together with simplified processes and circuits with reduced power consumption are some reasons for this development.

Due to a dramatic reduction in geometries and the availability of n-channel and p-channel transistors simultaneously on the same chip in so-called CMOS technologies, it is now possible to integrate millions of MOS transistors to a system on one chip.

This trend to ever-increasing chip density is going to continue in the foreseeable future, whereby physical constraints are probably not a limiting factor. It is more likely that the increase in complexity, in conjunction with production costs, will slow down this development.

CMOS processes are today and will be in the future mainstream technologies for integrating complex systems on chips. Despite this, bipolar technologies with a lower level of integration offer attractive solutions to the designer, particular in the area of analog and high frequency designs. These options are especially appealing when a BICMOS technology is available. The advantages of using bipolar and MOS transistors simultaneously in innovative designs of large scale integrated circuits are manifold.

The development of large scale integrated systems on a chip has changed the environment of circuit designers dramatically over the years. Initially hand calculations were sufficient to design a circuit. This became extremely troublesome with the ever-increasing transistor count. The situation was eased by the introduction of circuit simulation programs. But the number of transistors per chip continued to increase. The testing of the design, e.g. for electrical parameters, electrical rules, or the layout for design rule violations prior to production, became a problem. This conflict was eased with the introduction of appropriate CAD tools. Unfortunately other handicaps unfolded. The technologies became more complex due to the continued scaling of transistor geometries. New effects had to be taken into account in the design. This meant further that the accuracy of the transistor models used in the circuit simulation programs was no longer sufficient. As a consequence, many improved transistor models were developed and will be developed and adopted to each new technology. A further implication of the scaled transistor geometries is that quality and lifetime questions of a product gain more and more significance, not to mention the testing of these properties.

It is obvious from this discussion, that not one designer but a whole team of specialists has to be available to cope with the challenges. If something is overlooked in the design phase, the IC might not work or work only at a particular voltage, temperature, and timing condition – an undesirable situation, as the system has to be analyzed and a redesign started.

To cope with the mentioned or similar problems it is necessary to understand thoroughly the physical interrelationship between the function of integrated circuit components like transistors and their impact on circuit performance in an integrated system.

The reader is thus introduced in the first four chapters to the basic behavior and design of modern semiconductor components of integrated circuits.

With this background, the design of digital and analog circuits is presented in the six following chapters, with the emphasis on CMOS implementations. One goal of these chapters is to derive simple equations for an estimate of transistor performance, geometry sizing, and circuit behavior. If one starts a design without this information the possibility exists that an inappropriate circuit will be selected, innovations be suppressed or unnecessary simulation runs performed without the desired success. Or if one starts a failure analysis and in due cause a redesign without this knowledge, a wrong conclusion may be drawn.

Acknowledgments

The author would like to thank all those who contributed to the manuscript and its correction. These are the members of the Bundeswehr University, Munich, Dr Kowarik, Dr Kraus, and Dr Pfeiffer, and the graduate students, who were an invaluable help in providing ideas for ease of complex presentations. The artwork was done excellently by Mr W. Barth. The author is also deeply grateful to many colleagues at Siemens and Infineon for their contribution.

Finally, I would like to express my extreme gratitude to my wife Gisela for doing excellent proofreading and for enduring the long months while this book was being written. She has been a constant support and help during the writing of this manuscript.

Kurt Hoffmann
Munich, Germany

Physical Constants and Conversion Factors

Conversion factors

$1\,\text{eV} = 1.602 \cdot 10^{-19}\,\text{J}\,[\text{Ws}]$

$1\,\text{m} = 10^3\,\text{mm} = 10^6\,\mu\text{m} = 10^9\,\text{nm}$

$1\,\text{F} = 10^6\,\mu\text{F} = 10^9\,\text{nF} = 10^{12}\,\text{pF} = 10^{15}\,\text{fF}$

Physical constants

Symbol	Parameter	Value
Q	Magnitude of electronic charge	$1.602 \cdot 10^{-19}\,\text{C}\,[\text{As}]$
k	Boltzmann constant	$1.38 \cdot 10^{-23}\,\text{JK}^{-1}\,[\text{Ws K}^{-1}]$
$kT/q = \phi_t$	Temperature voltage	0.026 V at 300 K
ε_0	Permittivity of free space	$8.854 \cdot 10^{-14}\,\text{Fcm}^{-1}$
ε_{ox}	Dielectric constant of silicon dioxide	3.9

Important electrical properties of semiconductors

	Ge	Si	GaAs	Unit
Band-gap E_G	0.66	1.12	1.42	eV
Dielectric constant ε_r	16	11.9	13.1	
Intrinsic carrier concentration n_i	$2.4 \cdot 10^{13}$	$1.45 \cdot 10^{10}$	$1.79 \cdot 10^6$	cm^{-3}
Effective density of states				
Conduction band N_C	$1.04 \cdot 10^{19}$	$2.8 \cdot 10^{19}$	$4.7 \cdot 10^{17}$	cm^{-3}
Valence band N_V	$6.0 \cdot 10^{18}$	$1.04 \cdot 10^{19}$	$7.0 \cdot 10^{18}$	cm^{-3}

Symbols

Symbol	Description	Units
General		
C	Capacitance	F
C'	Capacitance per area	Fm^{-2}
C^*	Capacitance per perimeter	Fm^{-1}
Q	Charge	C
ρ	Charge per volume	Cm^{-3}
σ	Charge per area	Cm^{-2}
ϕ	Semiconductor voltage	V
V	Applied voltage	V
Detail		
A	Area	m^2
a, (a_o)	Small-signal gain v_o/v_i (at $\omega \rightarrow 0$)	
B_F, B_R	Current gain; forward, reverse	
BV	Breakdown voltage	V
b_E	Emitter width	m
C_d	Diffusion capacitance	F
C_{gdo}, C_{gso}	Overlap capacitance: gate drain, gate source	F
C_j, (C_{jo})	Depletion capacitance (at $V_{PN} = 0\,V$)	F
C_{be}, C_{bc}	BE- and BC-capacitance	F
C_{je}, C_{jc}	BE- and BC-depletion capacitance	F
C_{jeo}, C_{jco}	BE- and BC-depletion capacitance at $V = 0\,V$	F
C'_{ox}	Oxide capacitance per area	Fm^{-2}
D	Electrical displacement	Cm^{-2}
D_n, D_p	Diffusion coefficient: electrons, holes	m^2/s
d_{ox}	Oxide thickness	m
$\mathcal{E}$	Electrical field	Vm^{-1}
E_F, E_C, E_V	Energy: Fermi level, conduction and valance band edge	eV
E_i	Electron energy at intrinsic Fermi level	eV
E_G	Band-gap	eV
E_{ox}, E_{Si}	Electrical field: oxide, silicon	Vm^{-1}
F	Probability of occupation by an electron	
f, f_T	Frequency, transit frequency	1/s
G	Rate of generation of electron–hole pairs	$1/m^3\,s$
g_o	Conductance	Ω^{-1}
g_m	Transconductance (gate)	Ω^{-1}

Symbol	Description	Units
g_{mb}	Transconductance (substrate)	Ω^{-1}
g_{π}	Input conductance	Ω^{-1}
I	Current	A
I_C, I_E, I_B	Current: collector, emitter, base	A
I_{Co}	Collector current at $V_{BC} = 0\,V$	A
I_{KF}, I_{KR}	Knee current: forward, reverse	A
I_S	Saturation current	A
I_{SS}, (I_{SSo})	Transport current ($V_{BC} = 0\,V$)	A
I_{DS}	Drain source current	A
J_n, J_p	Current density: electrons, holes	Am^{-2}
k	Boltzmann constant	$1.38 \cdot 10^{-23}\,JK^{-1}$
k_n, k_p	Gain factor of the process: n-channel, p-channel	AV^{-2}
L	Length, channel length (drawn)	m
l_E	Emitter length	m
l	Effective channel length	m
M	Grading coefficient	
N	Emission coefficient	
N_A, N_D	Acceptor and donor density	m^{-3}
N_C, N_V	Effective density of states: conduction band, valence band	m^{-3}
n_o, p_o	Density at equilibrium: electrons, holes	m^{-3}
n_n, p_n	Density in n-doped region: electrons, holes	m^{-3}
n_{no}, p_{no}	Density at equilibrium in n-doped region: electrons, holes	m^{-3}
n_p, p_p	Density in p-doped region: electrons, holes	m^{-3}
n_{po}, p_{po}	Density at equilibrium in p-doped region: electrons, holes	m^{-3}
n_i	Intrinsic density of electrons and holes	m^{-3}
n_{iB}, n_{iE}	Intrinsic density: base, emitter	m^{-3}
n'_p	Excess electron density in p-doped region	m^{-3}
p'_n	Excess hole density in n-doped region	m^{-3}
P	Power dissipation	W
Q	Magnitude of electronic charge	$1.602 \cdot 10^{-19}\,C$
Q_p, Q_n	Charge: holes, electrons	C
Q_B, (Q_{Bo})	Base majority charge ($V_{BC} = 0\,V$)	C
R	Recombination factor	$1/m^3\,s$
R_E	Emitter resistor	Ω
R_B	Base resistor	Ω
R_C	Collector resistor	Ω
R_S	Sheet resistance	$\Omega/\square$
T	Temperature	K (°C)
t	Time	s
t_d	Delay time	s
t_r	Rise time	s
t_f	Fall time	s
t_S	Storage time, switching time	s
U	Netto generation rate	$1/m^3\,s$
V_{AF}, V_{AR}	Early voltage: forward, reverse	V
V_{BC}	Voltage between base and collector	V
V_{BE}	Voltage between base and emitter	V
V_{CC}, V_{DD}	Positive power-supply voltages	V

V_{CE}	Voltage between collector and emitter	V
V_{DS}	Voltage between drain and source	V
V_{FB}	Flat band voltage	V
V_{GB}	Voltage between gate and bulk	V
V_{GS}	Voltage between gate and source	V
V_I	Input voltage	V
V_{PN}	Voltage between p- and n-doped region	V
V_Q	Output voltage	V
V_{SB}	Voltage between source and bulk	V
V_{SS}	Negative power-supply voltage	V
V_{Ton}, V_{Top}	Threshold voltage: n- and p-channel transistor ($V_{SB} = 0\,V$)	V
V_{Tn}, V_{Tp}	Threshold voltage: n- and p-channel transistor	V
v_n, v_p	Average drift velocity: electrons, holes	m/s
v_{sat}	Saturation velocity	m/s
w	Effective MOS transistor width	m
w_E	Effective emitter length	m
x_d	Depletion region width of MOS transistor	m
x_j	Depth of source and drain diffusion	m
x_p, x_n	Width of depletion region: p-doped and n-doped region	m
x_B, (x_{Bo})	Base width ($V_{BC} = 0\,V$)	m
Z	Inverter sizing parameter	
β	Small-signal current gain i_o/i_g or i_o/i_b	
β_n, β_p	Gain factor: n- and p-channel transistor	AV^{-2}
γ	Body-effect parameter	$V^{1/2}$
ε_0	Permittivity of free space	$8.854 \cdot 10^{-12}\,Fm^{-1}$
ε_{ox}	Dielectric constant of silicon dioxide	3.9
ε_{Si}	Dielectric constant of silicon	11.9
λ	Channel length modulation factor	V^{-1}
μ_n, μ_p	Mobility: electrons, holes	m^2/Vs
ρ_d	Depletion charge per volume	Cm^{-3}
ρ_B	Base resistivity	Ωm
σ_g	Gate charge per area	Cm^{-2}
σ_n	Inversion layer charge per area	Cm^{-2}
σ_d	Depletion region charge per area	Cm^{-2}
σ_{SS}	Interface charge per silicon/silicon-dioxide area	Cm^{-2}
σ	Conductance	$(\Omega m)^{-1}$
τ_T	Effective transit time	s
τ_n, τ_p	Effective transit time, lifetime: electrons, holes	s
τ_F, τ_R	Effective transit time: forward, reverse	s
ϕ_C	Channel voltage, contact voltage	V
ϕ_F	Fermi potential, Fermi voltage	V
ϕ_i	Built-in voltage	V
ϕ_{ox}	Voltage across oxide layer	V
ϕ_t	Thermal voltage kT/q	V
ϕ_S	Surface potential, surface voltage	V
ω	(Angular) frequency	1/s
ω_T	Unit-gain (angular) frequency	1/s
ω_p	$-3\,dB$ bandwidth, pole-(angular) frequency	1/s
ω_z	Zero-(angular) frequency	1/s

1

Semiconductor Physics

In order to understand the behavior of semiconductor components, some basic understanding of semiconductor physics is absolutely necessary. The starting point of the explanation is the band theory of solids and the origination of free electrons and holes, as well as the carrier transport caused by drift and diffusion. The chapter ends with an important theoretical experiment. In this experiment the equilibrium condition of a semiconductor is disturbed, causing an injection or an extraction of minority carriers. By solving the diffusion equation, one finds the local carrier distribution and the minority carrier current. The equations derived are applied directly to the semiconductor devices described in the following chapters. A substantially simplified treatment results. For the interested reader an extended view of the experiment is included, which treats generation and recombination processes. When reading this book, this section may be omitted, without losing continuity.

1.1 BAND THEORY OF SOLIDS

One very important result of the application of quantum mechanics to the description of an atom is that the negatively charged electrons assume certain allowed energies (Figure 1.1(a)). The electrons fill the allowed energy levels, starting with the lowest one. When two atoms e.g. approach one another (Figure 1.1(b)), the Pauli exclusion principle demands that each allowed electron energy level has a slightly different energy, due to their mutual influence. In this case each of the original energy levels of the isolated atom differs slightly.

When N atoms are included, as e.g. in a crystal, the original energy levels will split into N different allowed energy states forming an energy band. Each state may contain two electrons at most with different spin. Since the number of atoms in a crystal is very large – in the order of $10^{22}\,\text{cm}^{-3}$ – the separation between the different energy levels within the bands is extremely small and electrons may easily move between levels. Therefore, one can speak of a continuous band of allowed states. Due to the Coulomb force between the positively charged atomic core and the negatively charged electrons, the latter have the highest energy the further they are from the core. The electrons can be considered to be free, when they approach the vacuum. This is described by arbitrary energy levels the electrons may assume. The energy levels are usually expressed in electron volts for convenience ($1\,\text{eV} = 1\,\text{V} \cdot 1.6 \cdot 10^{-19}\,\text{As}$).

System Integration: From Transistor Design to Large Scale Integrated Circuits. Kurt Hoffmann.
 ISBN: 0-470-85407-3

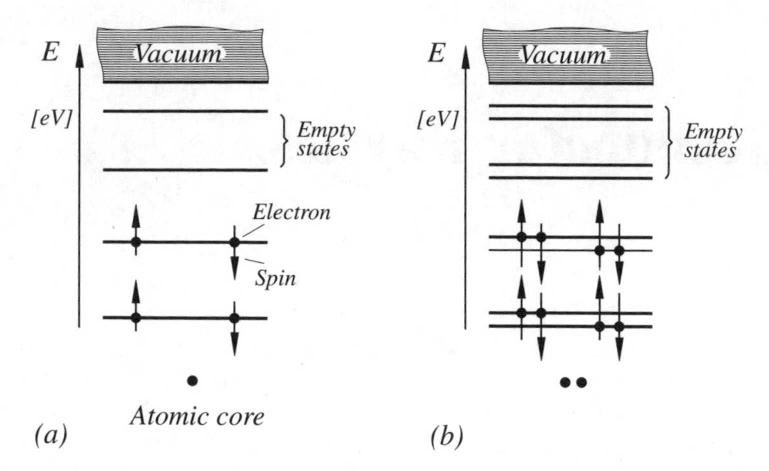

Figure 1.1 Schematic presentation of energy levels: (a) single atom and (b) two closely spaced atoms

Energy band diagram of a semiconductor

Of interest are only those energy bands with the highest energy, the conduction band and valence band (Figure 1.2). All other energy bands are completely filled by electrons. The electrons in the completely filled bands cannot assume kinetic energy and are thus unable to contribute to an electrical current. In order to distinguish between the electrons in the conduction band and valence band they are sometimes called conduction electrons and valence electrons, respectively. The allowed bands are bound by the conduction band edge E_C and the valence band edge E_V. The edges are separated by the band-gap E_G, where no energy state exists. Therefore this energy region is sometimes called the forbidden gap. At very low temperature almost all electrons of the conduction band are in the valence band. With increasing temperature, electrons are able to overcome the band-gap and reach the conduction band, leaving behind empty states in the valence band. The electrons in the conduction

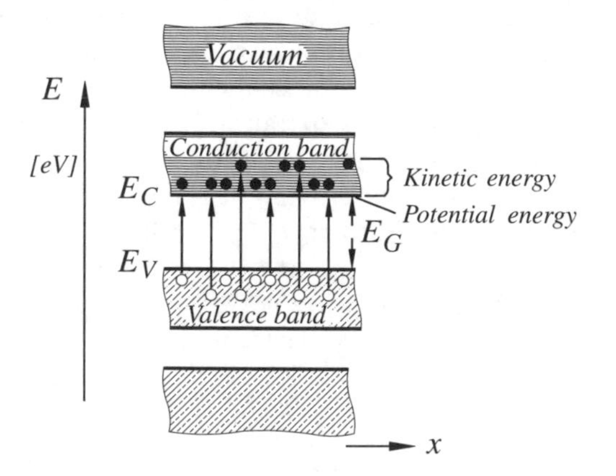

Figure 1.2 Schematic energy band diagram of a crystal

band can move freely within the band and occupy energetic higher or lower energy states, by picking up or giving up kinetic energy. The electrons have the lowest energy at the conduction band edge E_C. This energy can be considered to be the potential energy of the electrons in the conduction band. The described process, whereby electrons overcome the band-gap and reach the conduction band, can also be viewed in a different way. The vibration energy – hence the temperature – of the crystal lattice is responsible for the generation of free electrons by breaking crystal bonds.

The electrons in the conduction band are able to respond to an applied electrical field causing a current, described by

$$J_n = \rho v_n \tag{1.1}$$

where ρ is the charge per volume and v_n the net velocity of the charge. With a population of n conduction electrons per volume and $(-q)$ the electron charge ($q = 1.6 \cdot 10^{-19}$ As) a current density of

$$J_n = -qnv_n \tag{1.2}$$

results.

Hole concept

The applied electrical field not only has an influence on the conduction electrons but also on the valence electrons. But the charge transport is quite different. When electrons excite into the conduction band, empty states are left behind in the valence band. Since there are many more electrons in the valence band than empty states, only nearby valence electrons are able to contribute to a current. Therefore, it is easier to describe the motion resulting from valence electrons interacting with these empty states, than to describe the movement of all valence electrons. How this can be viewed is shown in Figure 1.3.

An electrical field applied to the semiconductor causes a nearby electron to move into an empty state. This electron leaves behind an empty state, into which another electron can jump, and so on. According to this model, empty states in the valence band move in the direction opposite to that of the negative valence electrons. They behave as if they

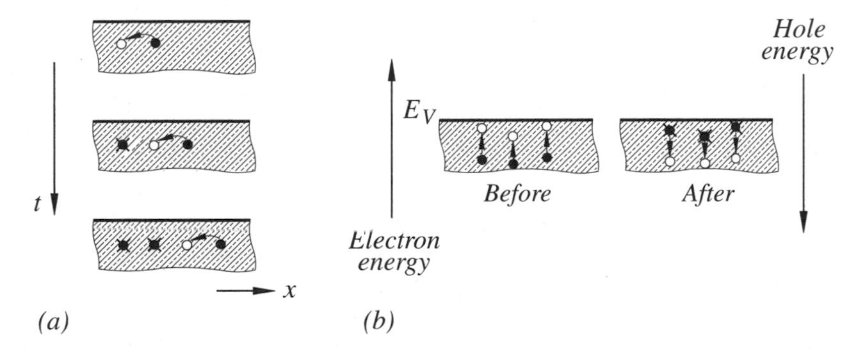

Figure 1.3 (a) Sketch of valence electron movement and (b) energy of electrons and holes

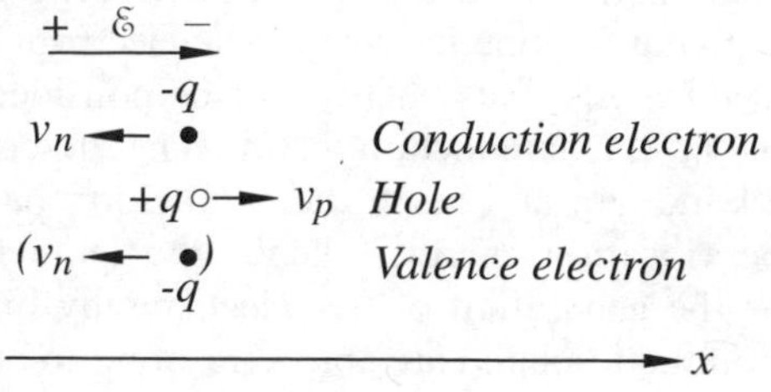

+ $\mathcal{E}$ −

-q

v_n ← • *Conduction electron*

+q ○→ v_p *Hole*

(v_n ← •) *Valence electron*

-q

→ x

Figure 1.4 Particle movement under the influence of an electrical field

had a charge of positive sign. These particles are called holes. The resulting current density can therefore be described by

$$J_p = +qpv_p \tag{1.3}$$

where $+q$ is the hole charge, p the hole density per volume, and v_p the net hole velocity. According to the previous discussion, the one-dimensional particle movement illustrated in Figure 1.4 results, under the influence of an electrical field $\mathcal{E}$.

Of further interest is how the hole energy is represented in the band diagram. To start with, empty states are assumed to be present in the valence band (Figure 1.3(b)) and valence electrons are able to jump into these states if, e.g., thermal energy is applied to the semiconductor. Empty states with higher energy are left behind. In other words, energy has to be applied in order to create these new empty states. The increase in hole energy can therefore be visualized as being opposite to that of the valence electrons. The hole energy is lowest at the valence band edge E_V corresponding to the potential energy. Within the band the holes move freely by picking up or giving up kinetic energy.

Band model comparison

A markedly different electrical behavior occurs in metals with overlapping bands (Figure 1.5). There is no forbidden band-gap. As a consequence, electrons can move freely when an electrical field is applied. In the case of the insulator, such as silicon dioxide, the situation is completely different. The band-gap energy is very large, about 8 eV, and no electron can overcome this barrier under normal circumstances. All energy levels in the conduction band are empty and those in the valence band are occupied by electrons. Therefore, no charged particle can pick up or give up kinetic energy and conduction is not possible. In comparison the band-gap of a semiconductor is small, e.g., 1.1 eV for silicon. At room temperature $1.45 \cdot 10^{10}$ electron–hole pairs per cm^{-3} are generated.

Intrinsic carrier concentration

In a semiconductor electrons will jump from the valence band over the forbidden gap into the conduction band, leaving behind empty states, when sufficient thermal energy is

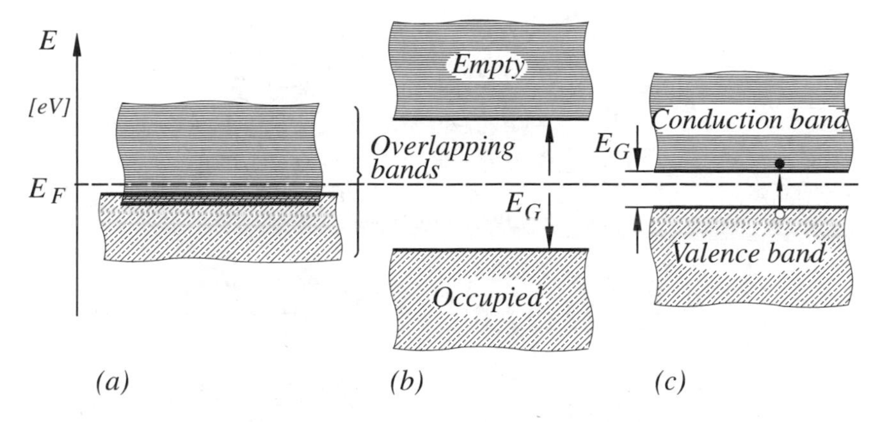

Figure 1.5 Band models: (a) metal, (b) insulator and (c) semiconductor

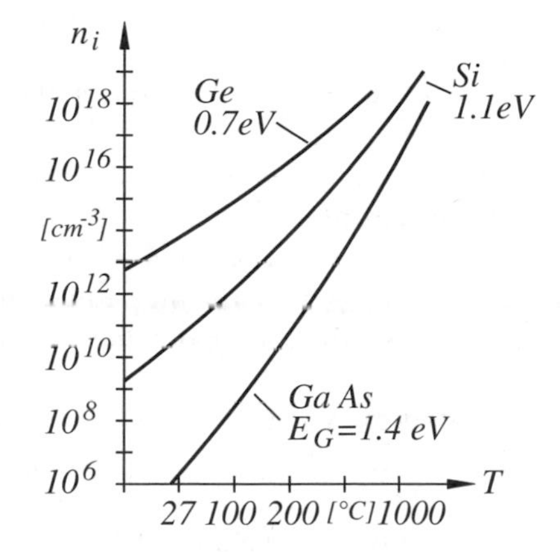

Figure 1.6 Intrinsic carrier concentration of Ge, Si and GaAs as function of temperature

applied. Electron–hole pairs are generated. Simultaneously, a reverse process takes place. Electrons lose energy and move over the forbidden gap into the valence band and vanish by recombining with holes. When the generation rate equals the recombination rate, thermal equilibrium exists. In a pure semiconductor the concentration of electrons n is equal to that of holes p. These concentrations are called intrinsic carrier concentration n_i. One expects that the intrinsic concentration is a function of the temperature and of the amount of energy required to break crystal bonds, which corresponds to the energy needed to overcome the forbidden gap (Thurmond 1975). This is shown in Figure 1.6. As expected, the largest intrinsic density results at elevated temperature and at materials with a small band-gap.

1.2 DOPED SEMICONDUCTOR

When an impurity, much larger than the intrinsic density, is incorporated into a semiconductor, its electrical behavior can be altered substantially. Figure 1.7(a) shows

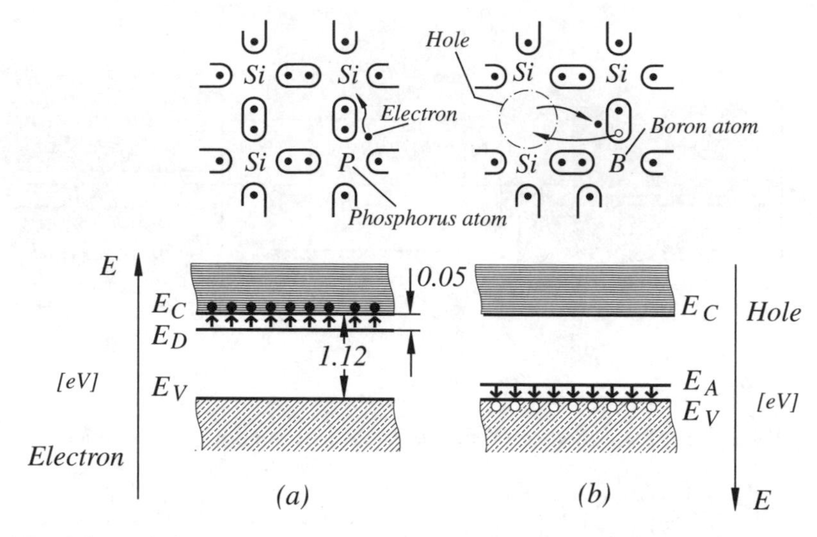

Figure 1.7 Schematic band diagram: (a) n-type semiconductor and (b) p-type semiconductor

the case when an impurity with five valence electrons, such as phosphorus, is added to silicon, which has four valence electrons. The extra electron of the phosphorus atom cannot be accommodated in the bonding arrangement of the silicon lattice. Therefore, it requires only a small amount of energy, the so-called ionization energy, to excite the electron from the phosphorus atom into the conduction band, while a much greater energy is required to make an electron jump from the valence band to the conduction band. The density of donors is designated by N_D. The corresponding donor state E_D is about 0.05 eV below the conduction band edge E_C. A thermal energy at a temperature higher than approximately 100 K is generally sufficient for the excitation. Once the electron of the donor atom is in the conduction band a fixed positively charged atom is left behind. One speaks of complete ionization, when all phosphorus atoms N_D have donated an electron. In this case the concentration of electrons n is equal to that of the donor atoms N_D. Since the electron density has been increased, the semiconductor is called n-type. An analogous situation (Figure 1.7(b)) exists when an impurity is introduced into the silicon lattice which has three valence electrons, such as boron. As boron has one electron less than silicon, one can consider it to carry a hole. This hole can easily be exited from the boron atom into the valence band by an ionization energy of about $E_A - E_V = 0.05\,\text{eV}$. This, of course, is equivalent to a valence electron jumping from the valence band to the boron state E_A, leaving behind an empty state. Once the hole is in the valence band, a fixed negatively charged atom is left behind. Complete ionization exists when all boron atoms have accepted an electron. In this case the concentration of holes p is equal to that of the acceptor atoms N_A. The altered semiconductor is called p-type.

Extrinsic carrier density

Figure 1.8 shows measured data of the electron concentration in an n-type semiconductor as a function of temperature. At temperatures below 100 K the thermal energy is

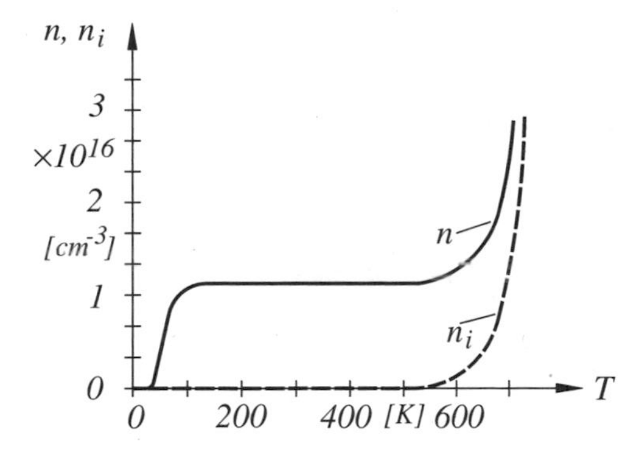

Figure 1.8 Electron concentration in an n-type semiconductor as a function of temperature

not sufficient to ionize all donor atoms. When the temperature is increased, the condition of complete ionization is approached, where the electron concentration equals the donor concentration. As the temperature is increased further, the electron concentration n remains basically constant. Finally a condition is reached where the intrinsic carrier concentration n_i becomes dominating, due to the increased breakage of silicon bonds. The region where the electron concentration is constant is the most important one for the application of semiconductor devices.

1.3 SEMICONDUCTOR IN EQUILIBRIUM

In this section the carrier density under equilibrium condition is derived. This condition is reached when there is no net transfer of electrons at any energy. This does not imply that all processes cease, rather it means that an occurring process and its inverse compensate each other. An example is the generation and recombination of electron–hole pairs. Equilibrium exists when the net rate between these processes is zero.

1.3.1 Fermi–Dirac Distribution Function

To start with the derivation of the density equation, one has to determine the probability that a given energy state is occupied by an electron first. The distribution of electrons in a semiconductor is governed by the laws of Fermi–Dirac statistics. The result is a distribution function which yields the probability F that an energy state E is occupied by an electron

$$F(E) = \frac{1}{1 + e^{(E - E_F)/kT}} \tag{1.4}$$

In this equation k is the Bolzmann constant and E_F the Fermi level. The Fermi level

$$F(E = E_F) = \frac{1}{1 + e^{(E_F - E_F)/kT}} = \frac{1}{2}$$

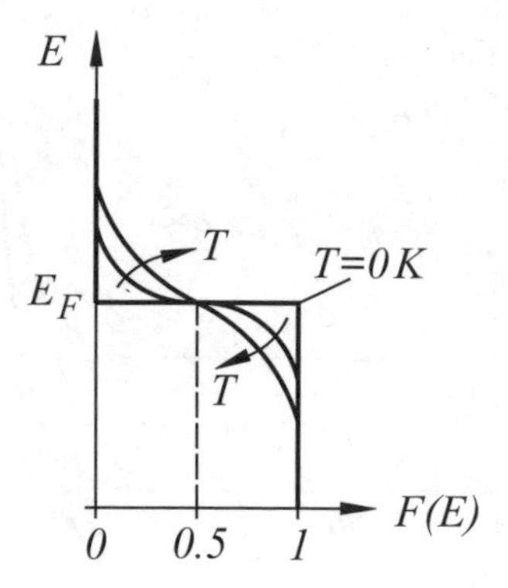

Figure 1.9 Fermi–Dirac distribution function

can be interpreted as the energy at which the probability of occupation by an electron is one-half or 50%.

The distribution function (Equation 1.4) is shown in Figure 1.9. To apply the distribution function more easily to the band diagram, ordinate and abscissa are exchanged. According to Equation (1.4) a rectangular distribution results when the temperature T approaches 0 K. In this case with $E > E_F$ the occupation probability is $F(E) = 0$ and with $E < E_F$ the probability is $F(E) = 1$. In other words, energy states below E_F are all occupied by electrons, whereas those above E_F are empty. By increasing the temperature, the occupation probability changes continuously. The Fermi–Dirac equation can be replaced by the Boltzmann distribution function for cases where the energy E is at least 0.1 eV above or below the Fermi level.

$$F(E) \approx e^{-(E-E_F)/kT} \text{ with } E > E_F \tag{1.5}$$

$$F(E) \approx 1 - e^{-(E_F-E)/kT} \text{ with } E < E_F \tag{1.6}$$

The distribution functions do not contain any information about the states available for occupancy. But quantum physics reveals information about the density of states. This is the number $N(E)$ of available states per energy and volume. The actual distribution of electrons can then be found by multiplying the density of available states with the probability that these states are occupied (Figure 1.10).

According to the foregoing discussion the following distribution of electrons and holes results:

$$n(E) = N(E)F(E) \tag{1.7}$$

$$p(E) = N(E)[1 - F(E)] \tag{1.8}$$

The term $[1 - F(E)]$ in Equation (1.8) describes the probability that electrons are absent, which is nothing else than the occupation probability by holes. The density of states in the conduction band is very large. But since the occupation probability is very low, only a small amount of electrons are available in the conduction band. Contrary to this is the

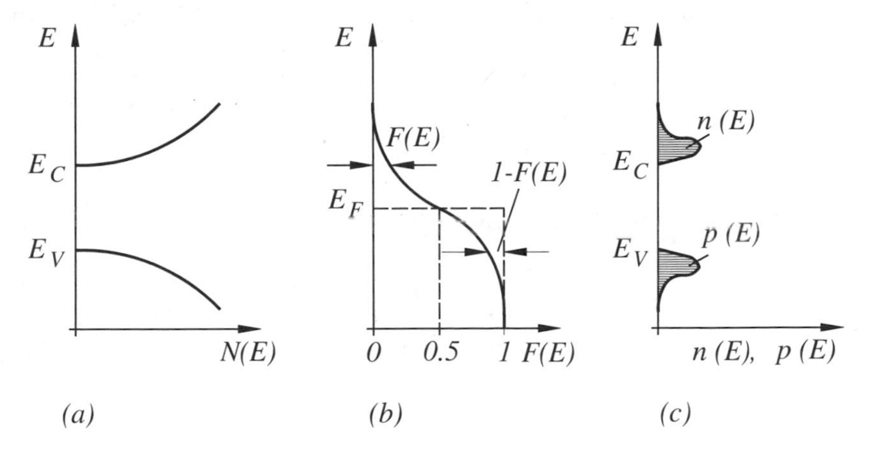

Figure 1.10 Intrinsic semiconductor: (a) density of states, (b) distribution function and (c) electron and hole distribution (not to scale)

situation in the valence band, where almost all states are occupied. This means only a small amount of holes are available.

In an n-type semiconductor the density distribution of electrons and holes is quite different (Figure 1.11). Additional to the density of states in the conduction band and valence band, a density of states exists generated by the donor atoms close to the conduction band edge. These states are represented by a Dirac function with the density N_D. Electrons are able to move at relatively low temperature from the donor states into the conduction band and occupy empty states. This is also shown in Figure 1.8, where from a temperature of about 100 K the thermal energy is sufficient for the ionization of almost all donor atoms. The increase in electron density in Figure 1.11 is taken into account by adjusting the Fermi–Dirac distribution function. In the case of the shown n-type semiconductor the function shifts toward higher energies, whereas in the case of a p-type semiconductor the function shifts toward lower energies. At the end of this

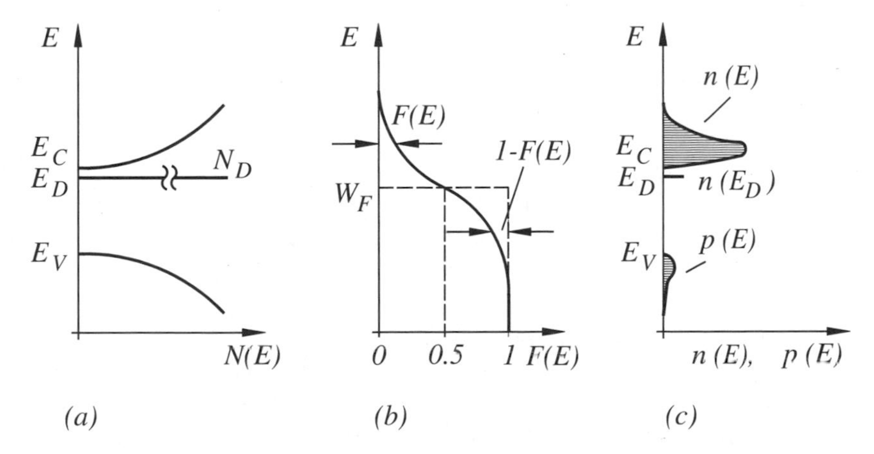

Figure 1.11 n-Type semiconductor: (a) density of states, (b) distribution function and (c) electron and hole distribution (not to scale)

section is described how the distribution function can be found by assuming that the semiconductor in equilibrium is electrically neutral.

If no electrons exist at the donor state E_D (Figure 1.11(c)) one speaks of 100% ionization. This is the case up to a density of about 10^{18} impurity atoms/cm^3 (Problem 1.2).

1.3.2 Carrier Concentration at Equilibrium

The distribution of electrons in the conduction band can be found, as already mentioned, by building the product of the density of allowed states $N(E)$ and the probability $F(E)$ that these states are occupied. Integrating over the conduction band from lower band edge E_C to the end of the band E_E leads to the total number of electrons per volume in the conduction band

$$n_o = \int_{E_C}^{E_E} N(E)F(E)\mathrm{d}E \tag{1.9}$$

The index o at the electron density, and later on at the hole density, is used throughout the book to indicate equilibrium condition. The result of the integration yields an electron density of

$$\boxed{n_o = N_C e^{-(E_C - E_F)/kT}} \tag{1.10}$$

and by analogy a hole density of

$$\boxed{p_o = N_V e^{-(E_F - E_V)/kT}} \tag{1.11}$$

where the Bolzmann distribution function, Equations (1.5) and (1.6), is used. In these equations N_C and N_V are effective density of states in the conduction and valence band, respectively. These quantities can be visualized as effective energy states, concentrated at the band edges E_C and E_V. Their value for silicon at room temperature (300 K) is $N_C = 2.8 \cdot 10^{19}\,\mathrm{cm}^{-3}$ and $N_V = 1.04 \cdot 10^{19}\,\mathrm{cm}^{-3}$. The difference between the two results from the different effective masses of electrons and holes.

Alternative carrier density equations

In the previous density equations the effective densities of states N_C and N_V are used to determine the carrier densities. Alternative equations with the intrinsic density n_i as parameter are very convenient to use, and are derived in the following. In an intrinsic semiconductor the density of electrons equals that of holes

$$n_o = p_o = n_i$$

$$N_C e^{-(E_C-E_F)/kT} = N_V e^{-(E_F-E_V)/kT} \tag{1.12}$$

This yields a Fermi level of

$$E_F = E_i = \frac{1}{2}(E_C + E_V) + \frac{1}{2}kT\ln\frac{N_V}{N_C} \tag{1.13}$$

where E_i is called intrinsic (Fermi) level. This level is in the middle of the band-gap, displaced from it by a term which can usually by neglected. Dividing Equation (1.10)

$$n_o = N_C e^{-(E_C-E_F)/kT}$$

by Equation (1.12) where $E_F = E_i$

$$n_i = N_C e^{-(E_C-E_i)/kT}$$

leads to an electron density of

$$\boxed{n_o = n_i e^{(E_F-E_i)/kT}} \tag{1.14}$$

and by analogy to a hole density of

$$\boxed{p_o = n_i e^{(E_i-E_F)/kT}} \tag{1.15}$$

In these alternative equations the intrinsic level E_i can be regarded as a reference level.

1.3.3 Density Product at Equilibrium

A very important relationship can be derived by multiplying the electron and hole density, Equations (1.10) and (1.11) as well as Equations (1.14) and (1.15)

$$p_o n_o = N_V N_C e^{-E_G/kT} \tag{1.16}$$

$$\boxed{p_o n_o = n_i^2}$$

where $E_G = E_C - E_V$ describes the band-gap.

From this equation is evident that the density product at equilibrium is independent of the Fermi level and therefore of the doping concentration. This can be explained in the following way. The rate of electron–hole pair generation G depends on the temperature

T and the property of the material and is almost independent of the number of carriers already present. The rate of recombination, on the other hand, depends on the concentration of electrons and holes, since both carriers must interact for recombination to occur. If one carrier type, e.g., is zero no recombination occurs. The recombination can therefore be represented by the product of the carrier concentrations

$$R(T) = n_o p_o r(T) \tag{1.17}$$

and a function $r(T)$ which describes the recombination mechanism in the crystal as a function of the temperature. At equilibrium condition $G(T) = R(T)$, which yields a density product (Equation 1.17) of

$$n_o p_o = G(T)/r(T) = n_i^2(T) \tag{1.18}$$

which is a function of the temperature only. This behavior is sketched in Figure 1.12 for intrinsic and n-type semiconductors. In the n-type semiconductor the increased electron density causes an increased recombination rate, and thus a reduction in the hole concentration, compared to that of the intrinsic semiconductor.

Temperature dependence of the intrinsic density

The temperature dependence of the intrinsic density n_i has a profound effect on semiconductor devices. It can be found directly from Equation (1.16).

$$n_i = \sqrt{N_C N_V} e^{-E_G(T)/2kT} \tag{1.19}$$

The density is exponentially dependent on the band-gap, since thermal energy is responsible for the excitation of electrons across the forbidden gap. If the temperature dependence of the effective density of states is taken into consideration (Sze 1981), the result is

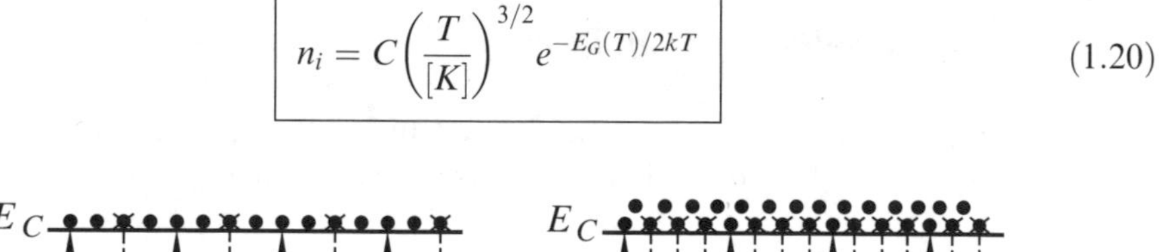

$$\boxed{n_i = C\left(\frac{T}{[K]}\right)^{3/2} e^{-E_G(T)/2kT}} \tag{1.20}$$

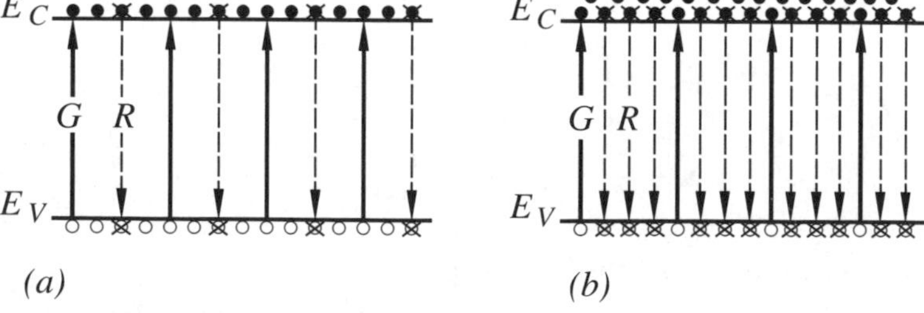

Figure 1.12 Schematic representation of generation G and recombination R: (a) intrinsic semiconductor and (b) n-type semiconductor

where C is a temperature-independent constant and $[K]$ indicates in Kelvin. This density function is discussed at the beginning of this chapter, and is shown in Figure 1.6 as a function of the temperature and different semiconductor materials.

Fermi level as function of doping concentration

This information can be gained by considering, e.g., a homogeneously doped semiconductor with donor and acceptor atoms. At equilibrium condition the net electrical charge density of the semiconductor is zero. Thus, if one adds up all positive and negative charges and assumes 100% ionization (Problem 1.2), the result is

$$q(p_o - n_o + N_D - N_A) = 0 \tag{1.21}$$

This equation and the density product equation (1.16) yield with $N_D > N_A$ an electron density of

$$n_{no} = \frac{1}{2}\left[N_D - N_A + \sqrt{(N_D - N_A)^2 + 4n_i^2}\right]$$

$$\boxed{n_{no} \approx N_D - N_A} \tag{1.22}$$

where the first index at the electron density indicates the type of semiconductor and the second one equilibrium condition. In all practical cases the magnitude of the net doping $N_D - N_A$ is much larger than the intrinsic carrier density (Figure 1.8), leading to the above shown simplification. The charge carriers n_{no} are called majority carriers. The minority carriers, in this case the holes, can be found by combining Equations (1.22) and (1.16)

$$\boxed{p_{no} = \frac{n_i^2}{n_{no}} \approx \frac{n_i^2}{N_D - N_A}} \tag{1.23}$$

With the charge carrier known, the band energies can be found by using, e.g., Equation (1.14)

$$\begin{aligned} E_F - E_i &= kT \ln \frac{n_{no}}{n_i} \\ &\approx kT \ln \frac{N_D - N_A}{n_i} \end{aligned} \tag{1.24}$$

An analogous situation in a p-type semiconductor with $N_A > N_D$ exists, yielding a majority and minority carrier density of

$$p_{po} = \frac{1}{2}\left[N_A - N_D + \sqrt{(N_A - N_D)^2 + 4n_i^2}\right]$$

$$\boxed{p_{po} \approx N_A - N_D} \tag{1.25}$$

$$\boxed{n_{po} = \frac{n_i^2}{p_{po}} \approx \frac{n_i^2}{N_A - N_D}} \tag{1.26}$$

The relationship between the majority carriers, in this case holes, and the band energies is therefore

$$E_i - E_F \approx kT \ln \frac{N_A - N_D}{n_i} \tag{1.27}$$

The following example gives an idea of the magnitude of the carrier densities involved.

Example

A homogeneously doped semiconductor contains the doping concentrations 10^{17} phosphorus atoms/cm^3 and 10^{16} boron atoms/cm^3. Calculate: The majority and minority density and the band energies (Figure 1.13).

Majority carriers:

$$n_{no} \approx N_D - N_A = 9 \cdot 10^{16}\,\text{cm}^{-3}$$

Minority carriers:

$$p_{no} \approx \frac{n_i^2}{N_D - N_A} = \frac{(1.45 \cdot 10^{10}\,\text{cm}^{-3})^2}{9 \cdot 10^{16}\,\text{cm}^{-3}} = 2.3 \cdot 10^3\,\text{cm}^{-3}$$

Band energies:

$$E_F - E_i = kT \ln \frac{n_{no}}{n_i} \approx 0.026\,\text{eV} \ln \frac{9 \cdot 10^{16}\,\text{cm}^{-3}}{1.45 \cdot 10^{10}\,\text{cm}^{-3}} = 0.41\,\text{eV}$$

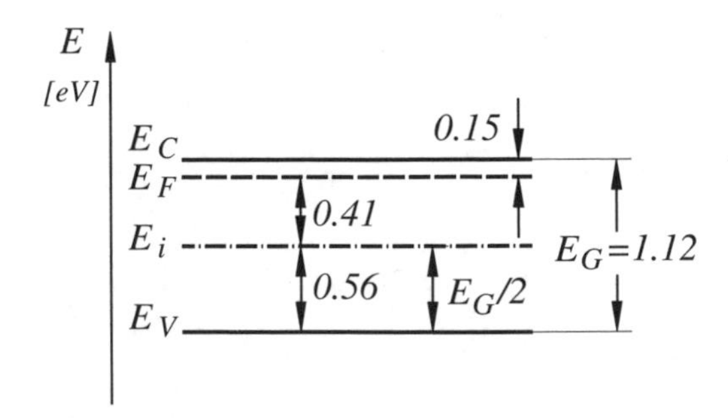

Figure 1.13 Band diagram of the n-type semiconductor example

1.3.4 Relationship between Energy, Voltage, and Electrical Field

In order to avoid problems with sign assignments or energy and voltage references in the following chapters, a few basic considerations are helpful. The band diagram displays the energy levels an electron can gain in a semiconductor. The electrons have the lowest energy, that is the potential energy, at the band edge E_C of the conduction band. To this energy an electrical potential can be assigned. This is the energy with respect to the electron charge of $-q$

$$\boxed{\psi_C = \frac{E_C}{-q}} \tag{1.28}$$

In the band diagram usually energy differences are of interest. This leads, according to the above definition, to differences in electrical potentials, namely voltages

$$\phi_C = \psi_C - \psi_{\text{ref}} = \frac{E_C - E_{\text{ref}}}{-q} \tag{1.29}$$

where ψ_{ref} and E_{ref} are arbitrary reference potentials and reference energies. In order to distinguish between externally applied voltages and internal voltages of the semiconductor, the following symbols are used: externally applied voltages V, internal voltages ϕ.

An inhomogeneously doped n-type semiconductor with bent band edges (Figure 1.14) is used to demonstrate the effect of different voltage references. Why these band edges are bent will be discussed in Chapter 2.

According to the made definition the voltage between region (2) and (1) – with region (1) used as reference point – has a value of

$$\phi_{2,1} = \frac{E_C(2) - E_C(1)}{-q} = \frac{-0.2V \cdot 1.6 \cdot 10^{-19}\,\text{As}}{-1.6 \cdot 10^{-19}\,\text{As}} = 0.2\,V$$

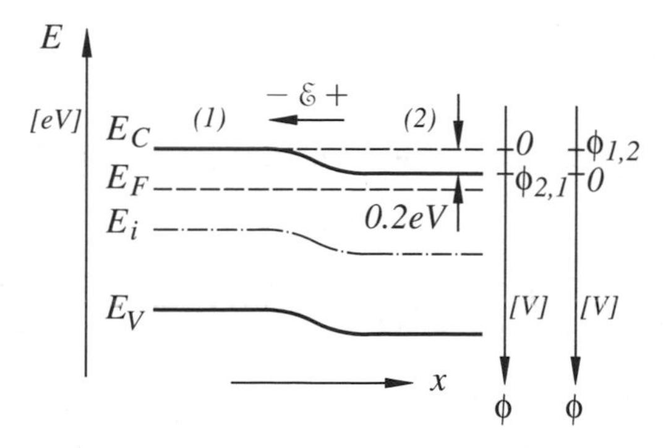

Figure 1.14 Band diagram of an inhomogeneously doped n-type semiconductor

or if region (2) is used as reference point a voltage of

$$\phi_{1,2} = \frac{E_C(1) - E_C(2)}{-q} = \frac{+0.2\,V \cdot 1.6 \cdot 10^{-19}\,\text{As}}{-1.6 \cdot 10^{-19}\,\text{As}} = -0.2\,V$$

results. In the band diagram the increase in electron energy and voltage are opposite to each other. This is caused by the negative sign of the electron charge in Equation (1.28). This leads in either case to the observation: *the band edge with the lower energy has a positive voltage with respect to the higher band edge*. Other voltage references are possible, e.g.

$$\phi_{CF} = \frac{E_C - E_F}{-q} \tag{1.30}$$

or

$$\phi_{FC} = \frac{E_F - E_C}{-q} \tag{1.31}$$

Since the intrinsic energy E_i always runs in parallel to the conduction band edge E_C the definition

$$\boxed{\phi_F = \frac{E_F - E_i}{-q}} \tag{1.32}$$

is also possible. ϕ_F is called Fermi potential or, better, Fermi voltage. According to this definition $\phi_F < 0$ in an n-type semiconductor and $\phi_F > 0$ in a p-type semiconductor. Sometimes one finds in the literature the definition $\phi_F = (E_i - E_F)/-q$. In this case a sign change occurs. To avoid ambiguities Equation (1.32) is used exclusively in this book.

Is a band bending present, and therefore a voltage across the semiconductor, as shown in Figure 1.14, an electrical field

$$\mathcal{E} = -\frac{\mathrm{d}\psi}{\mathrm{d}x} = -\frac{\mathrm{d}\phi}{\mathrm{d}x} \tag{1.33}$$

must exist. The negative sign is used because the electrical field direction is defined from the higher to the lower electrical potential. This leads to the relationship (Equation 1.29)

$$\boxed{\mathcal{E} = \frac{1}{q}\frac{\mathrm{d}E}{\mathrm{d}x}} \tag{1.34}$$

between the electrical field $\mathcal{E}$ and the electron energy (Figure 1.14). *The band bending is therefore a direct measure for magnitude and direction of the electrical field.*

1.4 CHARGE TRANSPORT

In the previous sections the generation of free electrons and holes were discussed and their densities determined. From the movement of these charge carriers the electrical current can be derived. This movement may be caused by an electrical field, or by a concentration gradient. Drift and diffusion currents result. To start with the derivation of these currents, the drift velocity has to be determined first.

1.4.1 Drift Velocity

The electrons in the semiconductor undergo a continuous random thermal motion interrupted by collisions. The two most important collision mechanisms are impurity and lattice scattering. The first occurs when an electron travels, e.g., past a fixed charge such as an ionized donor or acceptor, where its path will be deflected. The second, lattice scattering, occurs due to the thermal vibration of the crystal lattice atoms. The disrupted periodicity of the periodic lattice slows the motion of the electrons. The very simplified picture in Figure 1.15 demonstrates the situation.

The electrons in the uniform semiconductor perform a continuous random thermal motion, interrupted by the mentioned collisions. If an electrical field is applied an additional velocity component, in the direction against the electrical field $\mathcal{E}$, will add to the thermal movement. This additional movement is called drift velocity. Under the influence of the electrical field the electron is accelerated from $v_n = 0$ until a collision occurs. A new acceleration process starts and so on. The magnitude of the acceleration is given by Newton's second law

$$b = \frac{-q\mathcal{E}}{m_n} \tag{1.35}$$

where m_n is the effective mass of the electron in the semiconductor. This can be viewed as a quantity which takes the place of the mass of a free electron. The resulting average drift velocity can therefore be expressed as

$$v_n = \frac{v_e}{2} = \frac{1}{2} b\tau_{cn} = \frac{-q\mathcal{E}}{2m_n}\tau_{cn}$$

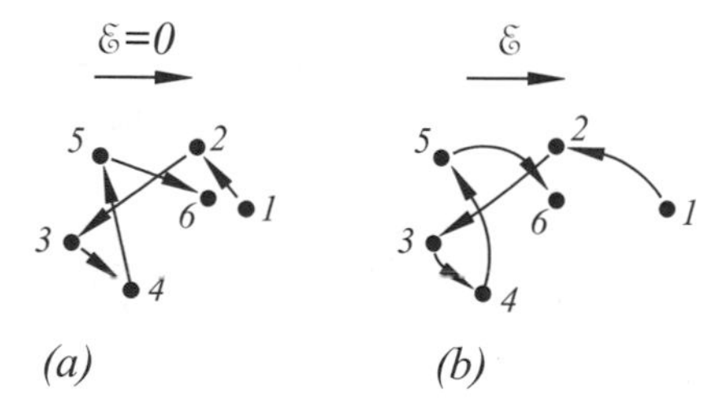

Figure 1.15 Illustration of the electron motion in a crystal: (a) without and (b) with an electrical field

$$\boxed{v_n = -\mu_n \mathcal{E}} \tag{1.36}$$

where v_e is the electron speed just before the collision and τ_{cn} the time between collisions. In analogy to the previous consideration, the drift velocity of the holes is

$$\boxed{v_p = \mu_p \mathcal{E}} \tag{1.37}$$

In the drift velocity equations μ_n and μ_p are called mobility, and the units are m^2/Vs.

In the simplified analysis the time between two collisions is assumed to be constant. At larger field strength this is no longer the case. Additional scattering mechanisms occur, whereby more energy is transferred from the electrons to the crystal lattice. The

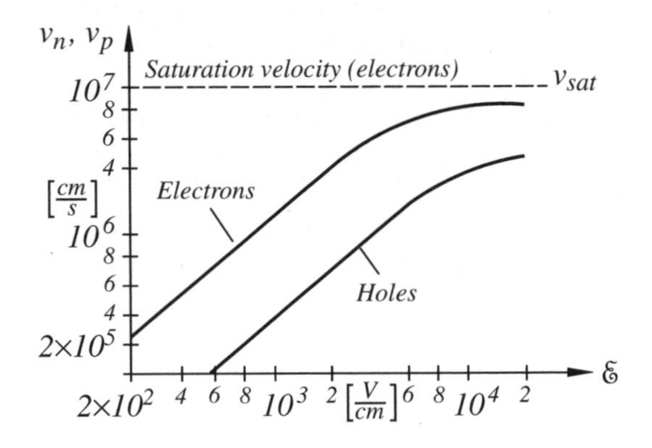

Figure 1.16 Drift velocity as a function of the electrical field for electrons and holes

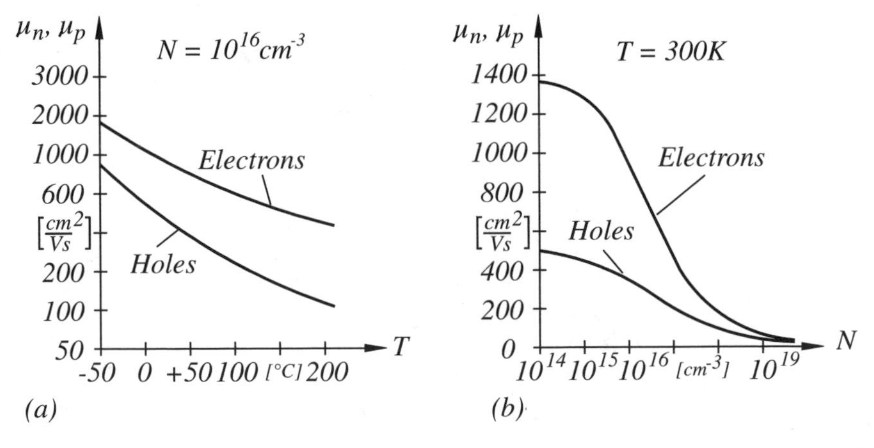

Figure 1.17 The effect of electron and hole scattering: (a) as a function of temperature and (b) as a function of the doping concentration ($N = N_A + N_D$)

drift velocity of the electrons approaches a saturation velocity v_{sat} of about 10^7 cm/s (Figure 1.16) (Ryder 1953).

The charge carriers suffer, as mentioned, collisions by impurity and lattice scattering. These effects are shown in Figure 1.17 as a function of temperature and doping concentration.

1.4.2 Drift Current

The conduction process caused by an electrical field is called drift current. According to Equation (1.2) the drift current density of electrons is

$$J_n = \rho v_n = -qnv_n = qn\mu_n\mathcal{E} \tag{1.38}$$

and that of holes is

$$J_p = \rho v_p = qpv_p = qp\mu_p\mathcal{E} \tag{1.39}$$

If both carrier types are involved in the conduction process a current density of

$$J = J_n + J_p = q(\mu_n n + \mu_p p)\mathcal{E} \tag{1.40}$$

results. This leads to a conductance of

$$\boxed{\sigma = q(\mu_n n + \mu_p p)} \tag{1.41}$$

The carrier movement and current flow are vector quantities and are defined by the unit vector $\vec{i}$ parallel to the x-axis (Figure 1.18).

Since in this book only one-dimensional presentations are considered the vector symbol is omitted. Therefore it is agreed: $J = \vec{J} \cdot \vec{i}$, $v = \vec{v} \cdot \vec{i}$ and $\mathcal{E} = \vec{\mathcal{E}} \cdot \vec{i}$. According to the nomenclature used, the electron movement v_n is opposite to that of the current density J_n, due to the multiplication by the minus sign of the electron charge. Both current densities must add up, since both carriers contribute to the total current density J. This is useful to remember, when considering the bipolar transistor of Chapter 3.

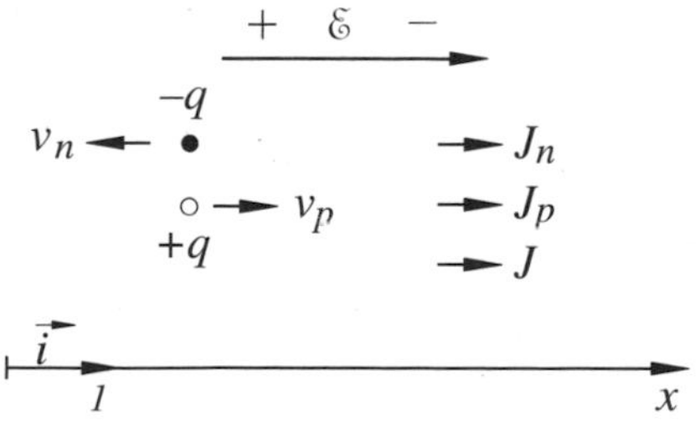

Figure 1.18 Relationship between carrier movement and current flow direction

A sketch of the conduction process in terms of the band diagram for a homogenously doped n-type semiconductor with short dimensions is shown in Figure 1.19.

Between the terminals N and M a voltage V_{NM} is applied. This voltage causes a decrease or an increase of the energy qV_{NM} at $x = 0$ with respect to the energy reference point E_i, chosen at $x = L$. During the movement the electrons suffer collisions. They are losing potential energy and increase their kinetic energy at the expense of the loss of potential energy, which is converted to heat. As shown in Figure 1.19 the electron trajectory moves the electron above the conduction band edge E_C, which represents the potential energy of the electrons. Similar is the energy loss of the holes. A hole moving from the metal contact into the semiconductor is equivalent to the movement of a valence electron into the metal. The opposite case, when a hole moves into the metal, can be viewed as the movement of a valence electron into the semiconductor.

The resistance of the semiconductor bar with length L, cross-section A, and conductance σ is

$$\boxed{R = \frac{1}{\sigma}\frac{L}{A} = \frac{1}{q(\mu_n n + \mu_p p)}\frac{L}{A}} \tag{1.42}$$

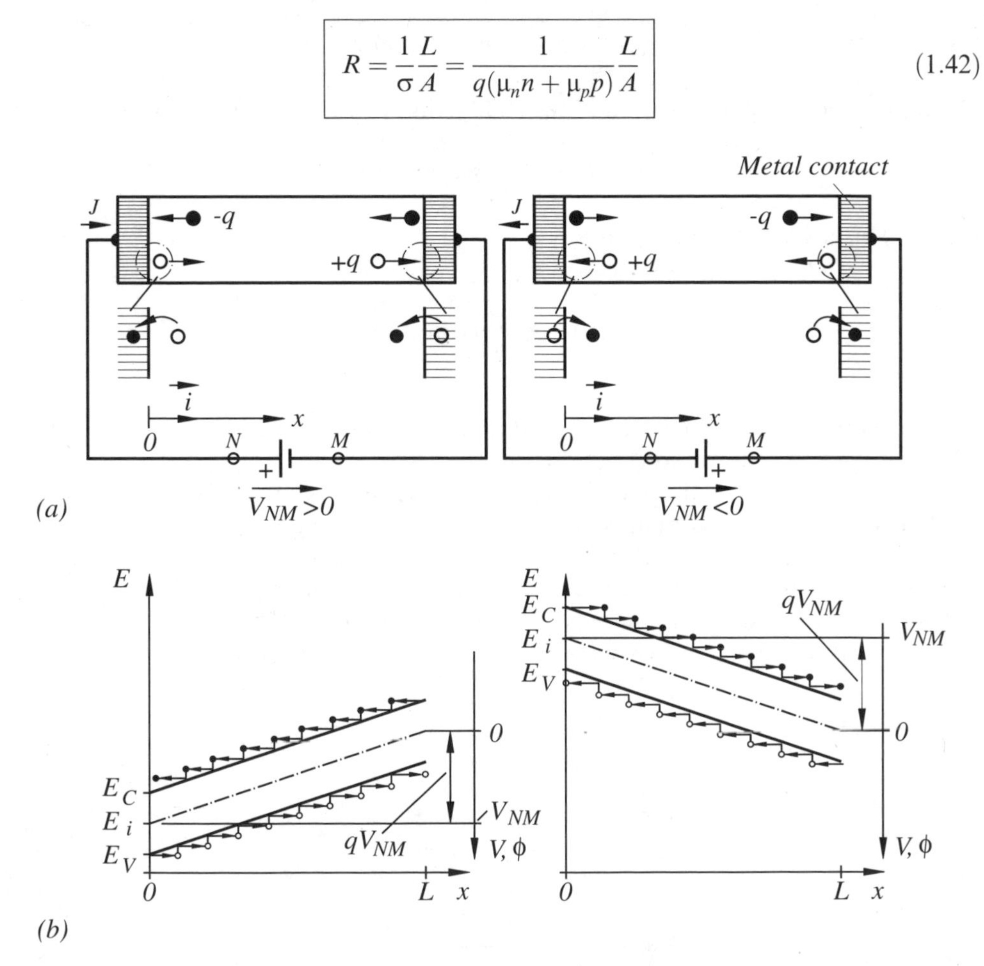

Figure 1.19 (a) Conduction process in an n-type semiconductor with $V_{NM} > 0$ and $V_{NM} < 0$ and (b) band diagram illustration of electron and hole movement

1.4.3 Diffusion Current

Another important current component, called diffusion current, exists when there is a spatial variation of a hole or electron density within the semiconductor. In this case the current arises from the random thermal motion of charge carriers. In order to demonstrate the effect, one-dimensional local hole and electron variations are assumed (Figure 1.20).

In this case the number of charge carriers crossing the plane at $x = 0$ starting at their mean-free path l is considered. Mean-free path is the distance the charge carrier travels without suffering a collision. The current density caused by the holes crossing the plane at $x = 0$ starting at l_l and l_r is therefore

$$\begin{aligned} J_p &= \underbrace{\frac{1}{2}qp(l_l)v_p}_{l_l \to 0} - \underbrace{\frac{1}{2}qp(l_r)v_p}_{0 \leftarrow l_r} \\ &= \frac{1}{2}q\left[p(0) - \frac{dp}{dx}l\right]v_p - \frac{1}{2}q\left[p(0) + \frac{dp}{dx}l\right]v_p \\ &= -qv_p l\frac{dp}{dx} \end{aligned} \tag{1.43}$$

The factor ½ is used since, in the case of a homogeneous carrier distribution, the probability that holes travel to the right or left in the one-dimensional case is 50%. If v_p and l are combined a hole current density of

$$\boxed{J_p = -qD_p\frac{dp}{dx}} \tag{1.44}$$

and similarly an electron density of

$$\boxed{J_n = qD_n\frac{dn}{dx}} \tag{1.45}$$

result.

The factors D_n and D_p are called diffusion constant (m^2/s). By use of the Einstein relationship these constants can be expressed with respect to the carrier mobility

$$\boxed{D_p = \frac{kT}{q}\mu_p(T) = \phi_t(T)\mu_p(T)} \tag{1.46}$$

$$\boxed{D_n = \frac{kT}{q}\mu_n(T) = \phi_t(T)\mu_n(T)} \tag{1.47}$$

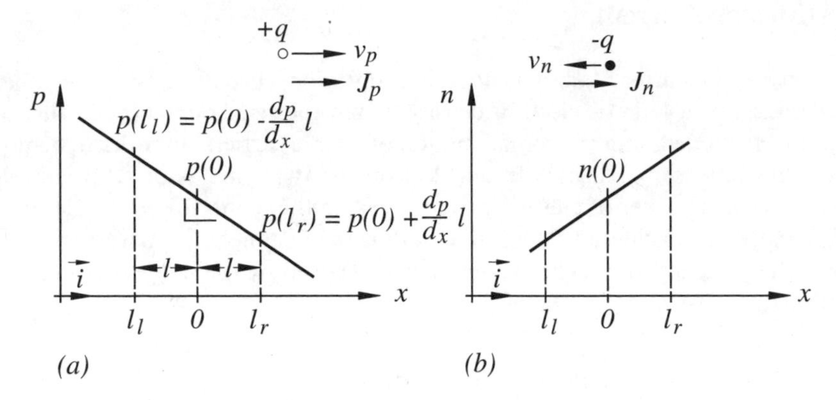

Figure 1.20 Spatial carrier distribution: (a) holes and (b) electrons

where ϕ_t is named temperature voltage. It is obvious that the diffusion constants are proportional to the temperature, since the diffusion mechanism is caused by the thermal motion of charge carriers.

In a semiconductor with a spatial hole and electron distribution, as shown in Figure 1.20, the diffusion currents of both carriers add up to a total current. To make this more clear, one has to keep in mind that the hole movement is nothing other than the movement of valence electrons in the opposite direction. If drift and diffusion processes are present simultaneously, the total current density of electrons and holes in a semiconductor is the sum of each contribution

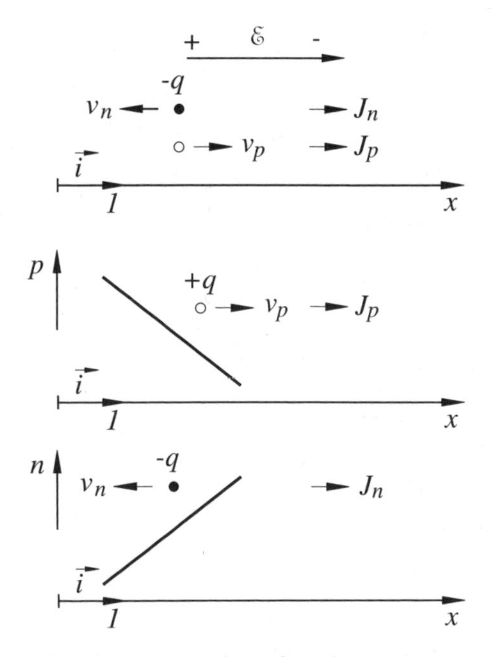

Figure 1.21 One-dimensional representation of carrier movement and current directions

$$J_n = q\mu_n n\mathcal{E} + qD_n \frac{dn}{dx} \tag{1.48}$$

$$J_p = q\mu_p p\mathcal{E} - qD_p \frac{dp}{dx} \tag{1.49}$$

This situation is illustrated in Figure 1.21 for the one-dimensional case.

This is a useful summary in order to avoid difficulties in the following chapters between current directions and charge carrier movements. Again as a reminder, the movement of holes is nothing other than the movement of valence electrons in the opposite direction.

1.4.4 Continuity Equation

With the described equations the drift and diffusion currents can be calculated, but not the switching performance of a semiconductor. This is possible with the continuity equation, accounting for the flux of charge carriers into and out of a semiconductor volume (Figure 1.22).

To derive the one-dimensional equation an infinitesimal slice of semiconductor with a thickness dx is considered. The number of electrons in the slice may increase or decrease due to a net flux of electrons per area into $J_n(x)/-q$ and out of $J_n(x + dx)/-q$ the slice and due to a difference between generated G and recombined R electrons within the slice.

The rate of change in the number of electrons per time in the infinitesimal slice is thus

$$\frac{\partial n}{\partial t} dx = \left[\frac{J_n(x)}{-q} - \frac{J_n(x + dx)}{-q}\right] + (G - R)\, dx \tag{1.50}$$

Using the derivative

$$\lim_{dx \to 0} \frac{J_n(x + dx) - J_n(x)}{dx} = \frac{dJ_n}{dx} \tag{1.51}$$

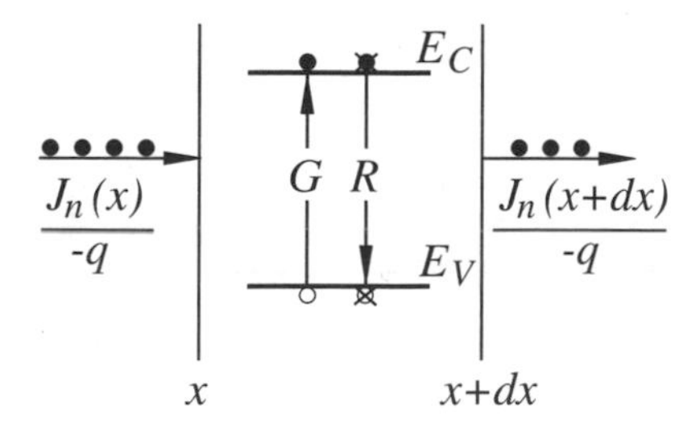

Figure 1.22 Semiconductor slice of infinitesimal thickness

the continuity equation for electrons can be simplified to

$$\boxed{\frac{\partial n}{\partial t} = \frac{1}{q}\frac{\partial J_n}{\partial x} + G - R} \tag{1.52}$$

By analogy the ones for holes

$$\boxed{\frac{\partial p}{\partial t} = -\frac{1}{q}\frac{\partial J_p}{\partial x} + G - R} \tag{1.53}$$

can be derived.

1.5 NON-EQUILIBRIUM CONDITIONS

Semiconductor devices do not operate at thermal equilibrium. Disturbances, for example caused by applying an external voltage, change the carrier densities. This implies that the pn-density product is different from n_i^2. A theoretical experiment is used to explain injection and extraction of charge carriers under non-equilibrium conditions. The first case, where $pn > n_i^2$, is called injection and the second, with $pn < n_i^2$, is called extraction of carriers from the semiconductor (Grove 1967).

The analysis of the experiment can be simplified substantially if a distinction is made between low-level and high-level injection of minority and majority carriers. This is illustrated with an n-type semiconductor in Figure 1.23.

In the low-level case the minority carrier density p_n is very small compared to the majority carrier density n_n. As a result the change in majority carrier density can be neglected. Contrary to this case is the high-level injection one, where the minority carrier density p_n is of the same magnitude as the majority carrier density n_n. The

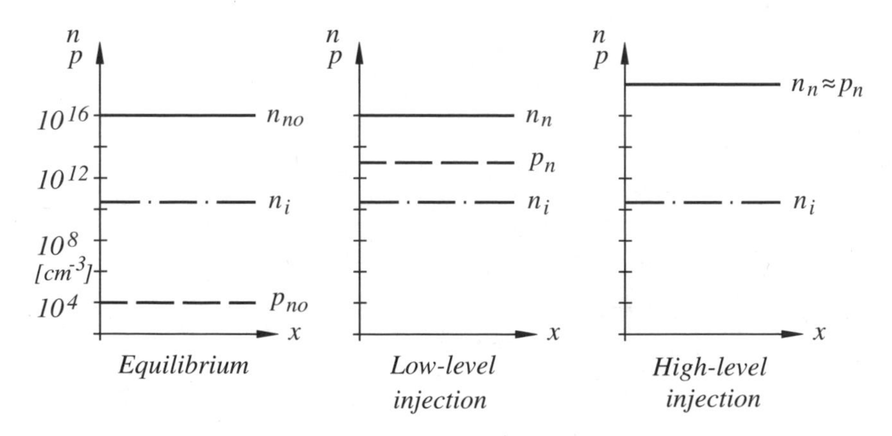

Figure 1.23 Simultaneous injection of electrons and holes in an n-type semiconductor

distinction between the two injection levels has the advantage that in the case of low-level injection – the proof follows – the influence of the electrical field on the minority carrier distribution can be neglected. This is usually the case at modern integrated semiconductor components. As a result the derivation of analytical equations can be simplified substantially.

Experiment

The major goal of this theoretical experiment is to determine the local carrier distribution and the current caused by an injection or an extraction of minority carriers into a semiconductor sample. The result leads to an equation which can be applied directly to a forward and reverse biased pn-junction and therefore to a bipolar transistor. Substantial mathematical simplifications result. In order to derive the equation an n-type semiconductor sample with a one-sided injection or an extraction of minority carriers is considered (Figure 1.24).

Minority carriers, in this case holes, move into the semiconductor during the one-sided injection. A hole meeting the metal contact is equivalent to the movement of a valence electron from the metal into the valence band. The situation is reversed in the case of hole extraction. A hole originates at the metal contact, when a valence electron moves into the metal and leaves behind an empty state.

The current density, caused by the injection or extraction of minority carriers, can be found by considering first the local carrier distribution (Figure 1.24(b)). The starting point is the continuity equation (Equation 1.53) with $dp_n/dt = 0$, since a stationary condition exists. Furthermore, it is assumed that generation and recombination are

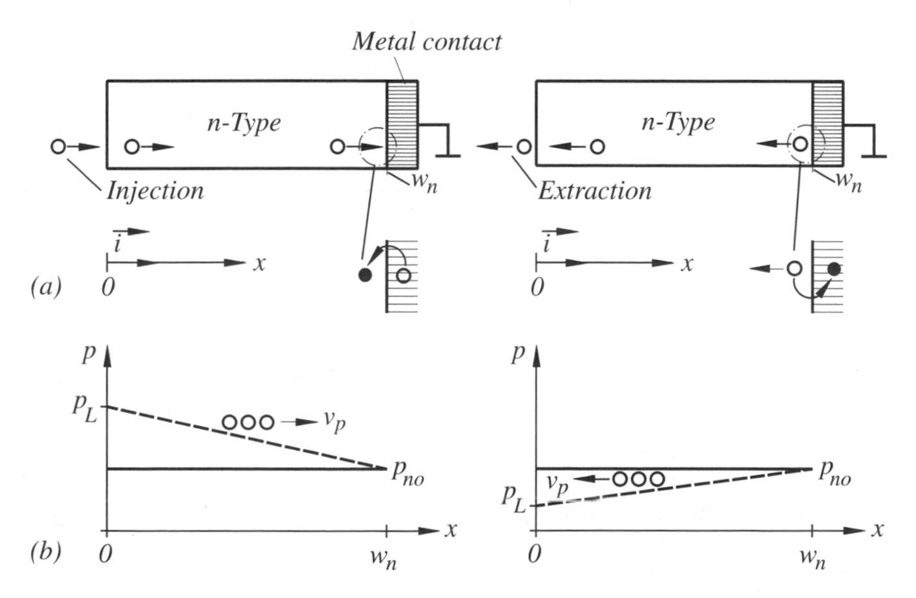

Figure 1.24 (a) One-sided injection and extraction of minority carriers and (b) minority carrier distributions

negligible, because either their rate is very small or the volume of the sample (see 'Extended view of the experiment'). This leads to the equation

$$0 = -\frac{1}{q}\frac{\mathrm{d}J_p}{\mathrm{d}x} \tag{1.54}$$

The hole current density can be determined by Equation (1.49)

$$J_p = \underbrace{q\mu_p p_n \mathcal{E}}_{\text{Drift}} - \underbrace{qD_p \frac{\mathrm{d}p_n}{\mathrm{d}x}}_{\text{Diffusion}}$$

An analytical solution can most easily be derived when the drift component of the current density equation can be neglected. This is possible, as will be shown in the following, as long as a low-level injection exists. With this assumption and an excess hole density of

$$p_n' = p_n - p_{no} \tag{1.55}$$

the following differential equation results

$$D_p \frac{\mathrm{d}^2 p_n'}{\mathrm{d}x^2} = 0 \tag{1.56}$$

With the border conditions $p_n'(w_n) = 0$ and $p_n'(x = 0) = p_L' = p_L - p_{no}$ the equation yields a hole distribution of

$$\boxed{p_n'(x) = p_L'\left(1 - \frac{x}{w_n}\right)} \tag{1.57}$$

which is shown in Figure 1.24(b). The border condition at w_n with $p_n' = 0$ is caused by the metal contact, which is assumed to reduce the excess hole density to zero.

Majority carrier behavior

The one-sided injection or extraction of minority carriers cannot happen without an influence on the majority carriers. How these majority carriers behave is important to know, when considering the function and operation of semiconductor devices. As shown in Figure 1.25, the majority carriers, the electrons, move in the opposite direction to the minority carriers, the holes. Again the hole movement is nothing but the movement of a valence electron in the opposite direction.

Thus the disturbance causes conduction electrons and valence electrons to move in and out of the semiconductor in the same direction. This movement of charge carriers produces a majority and minority current. How large the minority current

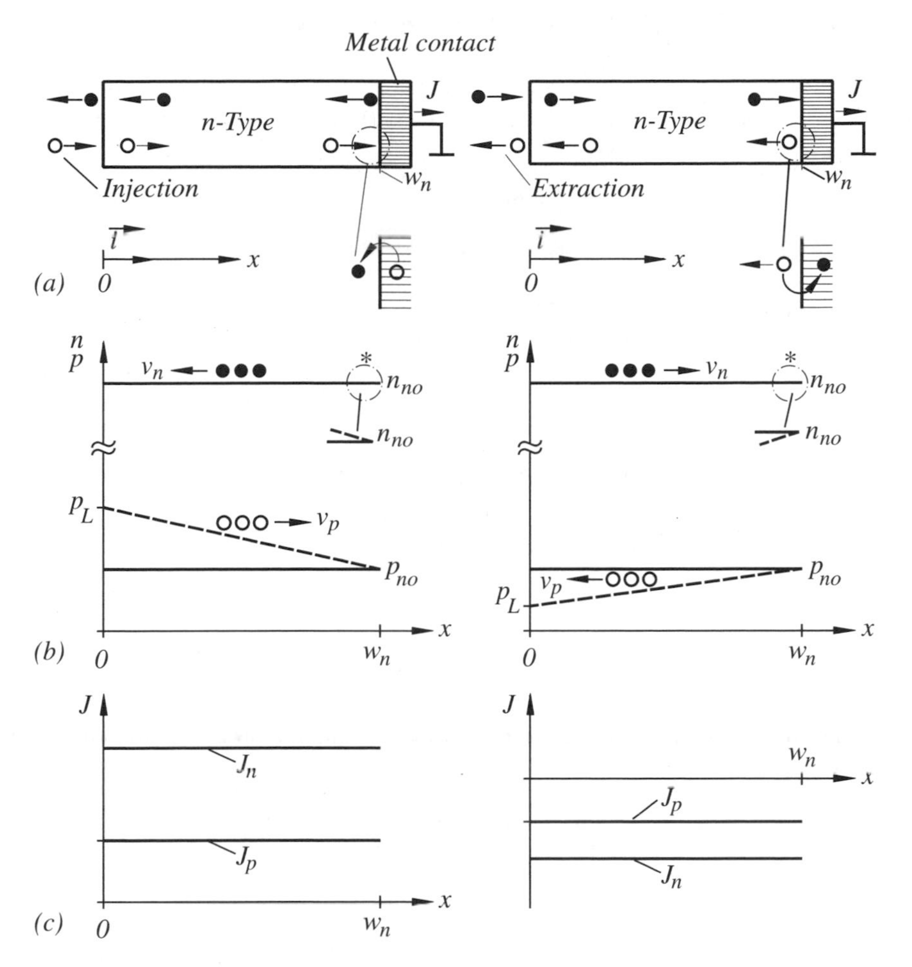

Figure 1.25 (a) One-sided injection and extraction of minority carriers, (b) carrier density distributions and (c) current density distributions (* negligible change in majority carrier density)

is, compared to the majority current, depends on the experiment. The excess majority carriers, the electrons, assume a distribution of about 10^{-12} s, which correlates to that of the excess minority carriers. This process is called dielectric relaxation and can be viewed in the following way. At a particular location in the semiconductor sample an excess small hole density is assumed to exist suddenly. This disturbance causes the negatively charged electrons to move from the large reservoir of majority carriers to this location and quasi-neutralize the excess positive minority carrier charge. The evolved change in majority carrier density can be neglected as long as low-level injection exists.

In the experiment the metal contact is assumed to produce an equilibrium between electrons and holes due to the overlapping conduction and valence bands. In reality, such a contact is a device which can have resistive or diode character (Sze 1981). The latter is undesirable, and is usually avoided during manufacturing. For simplicity reasons an ideal contact with no excess charge carriers is assumed throughout the book.

Influence of the electrical field on majority and minority carriers

The differential equation (1.56) was derived under the condition that the drift term can be disregarded compared to the diffusion term in the minority carrier current equation

$$J_p = \underbrace{q\mu_p p_n \mathcal{E}}_{\text{Drift}} - \underbrace{qD_p \frac{\mathrm{d}p_n}{\mathrm{d}x}}_{\text{Diffusion}}$$

if low-level injection is assumed. The starting point for the proof of this statement is the total current density (Equations 1.48 and 1.49)

$$\begin{aligned} J &= J_p + J_n \\ J &= q\mu_p p_n \mathcal{E} - qD_p \frac{\mathrm{d}p_n}{\mathrm{d}x} + q\mu_n n_n \mathcal{E} + qD_n \frac{\mathrm{d}n_n}{\mathrm{d}x} \end{aligned} \tag{1.58}$$

This equation can be solved for the electrical field

$$\mathcal{E} = \frac{J + qD_p \dfrac{\mathrm{d}p_n}{\mathrm{d}x} - qD_n \dfrac{\mathrm{d}n_n}{\mathrm{d}x}}{q\mu_p p_n + q\mu_n n_n} \tag{1.59}$$

Substituting Equation (1.59) into Equation (1.49) results in a minority current density of

$$J_p = \frac{J + qD_p \dfrac{\mathrm{d}p_n}{\mathrm{d}x} - qD_n \dfrac{\mathrm{d}n_n}{\mathrm{d}x}}{1 + \dfrac{\mu_n n_n}{\mu_p p_n}} - qD_p \frac{\mathrm{d}p_n}{\mathrm{d}x} \tag{1.60}$$

With the assumption that $D_p \mathrm{d}p_n/\mathrm{d}x \approx D_n\, \mathrm{d}n_n/\mathrm{d}x$, the following simplification can be made

$$J_p \approx \frac{\mu_p p_n}{\mu_n n_n} J - qD_p \frac{\mathrm{d}p_n}{\mathrm{d}x} \tag{1.61}$$

This equation indicates that at low-level injection, with $p_n \ll n_n$, the influence of the field term on the minority carrier current can be neglected, compared to that of the diffusion term.

The situation is completely different for the majority carriers. A similar derivation reveals a majority carrier current of

$$J_n \approx J + qD_n \frac{\mathrm{d}n}{\mathrm{d}x} \tag{1.62}$$

where no simplification is possible.

The minority carrier current can therefore be considered to be a diffusion current only, as long as low-level injection is guaranteed. This is a substantial mathematical simplification, and is the reason why minority carriers are usually considered. In the presented example the minority carrier current can therefore be derived directly from the minority carrier gradient (Figure 1.25(b)).

$$J_p = -qD_p \frac{\mathrm{d}p_n}{\mathrm{d}x} = -qD_p \frac{p_n(w_n) - p_L}{w_n} \tag{1.63}$$

Extended view of the experiment

In the presented experiment the generation and recombination of carriers are neglected, since it is assumed that either the rates are very low or the volume of the semiconductor sample is very small. In this section a proof is presented, to justify this assumption.

Imperfections within the semiconductor can disrupt the periodicity of the crystal lattice, causing energy states to be introduced into the forbidden gap. These energy states function as stepping-stones for the generation and recombination of electrons and holes. This is different from the assumption, which is made in the preceding sections, where a band-to-band generation and recombination is assumed for reasons of simplicity (Figure 1.26).

An electron captures, with a particular probability, in a first step the energy state *1a* and, in a second one an empty state in the valence band *1b*. This process demonstrates recombination, whereas steps *2a* and *2b* show a generation process via an intermediate energy state.

The net rate of recombination through intermediate centers, according to Sah *et al.* (1957) and Shockley and Read (1952) is

$$U = R - G = \frac{\sigma_p \sigma_n v_{th} N_t (pn - n_i^2)}{\sigma_n (n + n_i e^{(E_t - E_i)/kT}) + \sigma_p (p + n_i e^{-(E_t - E_i)/kT})} \tag{1.64}$$

where N_t is the density of empty states at energy E_t, v_{th} the thermal velocity of the charge carriers and σ_n and σ_p the capture cross-section for electrons and holes. If the equation is applied to an n-type semiconductor under low-level condition, with $n_n \gg n_i \exp.\, (E_t - E_i)/kT$ the net rate reduces to

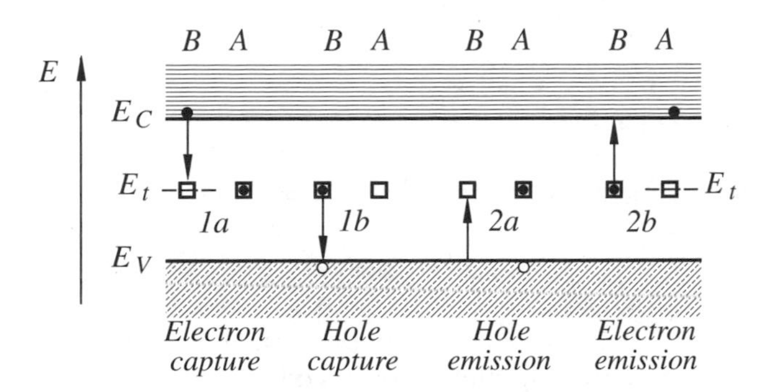

Figure 1.26 Generation and recombination through intermediate energy states (B, before; A, after)

$$U = \frac{\sigma_p \sigma_n v_{th} N_t (n_n p_n - n_i^2)}{\sigma_n n_n}$$

$$U = \sigma_p v_{th} N_t (p_n - p_{no}) \tag{1.65}$$

$$U = \frac{1}{\tau_p}(p_n - p_{no})$$

where τ_p is called the lifetime of holes (Problem 1.5), with the following parameter dependence:

$$\tau_p = \frac{1}{\sigma_p v_{th} N_t} \tag{1.66}$$

Using the continuity equation (Equation 1.53) in conjunction with the one for the net recombination rate (Equation 1.64) yields the differential equation

$$O = D_p \frac{d^2 p'_n}{dx^2} - \frac{p'_n}{\tau_p} \tag{1.67}$$

Applying the same border condition as used before, the following solution

$$p'_n(x) = p'_L \frac{\sinh \frac{w_n - x}{L_p}}{\sinh \frac{w_n}{L_p}} \tag{1.68}$$

results. In this equation

$$L_p = \sqrt{D_p \tau_p} \tag{1.69}$$

is called the diffusion length of holes. This is a general solution of the differential equation with two particularly interesting results, namely for small lateral dimensions – as is considered so far – and for long ones.

Long lateral dimension $w_n \gg L_p$

Under this condition, Equation (1.68) leads to the solution

$$p'_n(x) = p'_L e^{-x/L_p} \tag{1.70}$$

This is shown in Figure 1.27 for the injection ($p'_L > 0$) and extraction ($p'_L < 0$) case.

The spatial behavior during injection is governed by the increased recombination rate $R(n,p) > G$. Contrary to this is the extraction case, where the behavior is dominantly controlled by the thermal generation of carriers $R(n,p) < G$. In either case the total

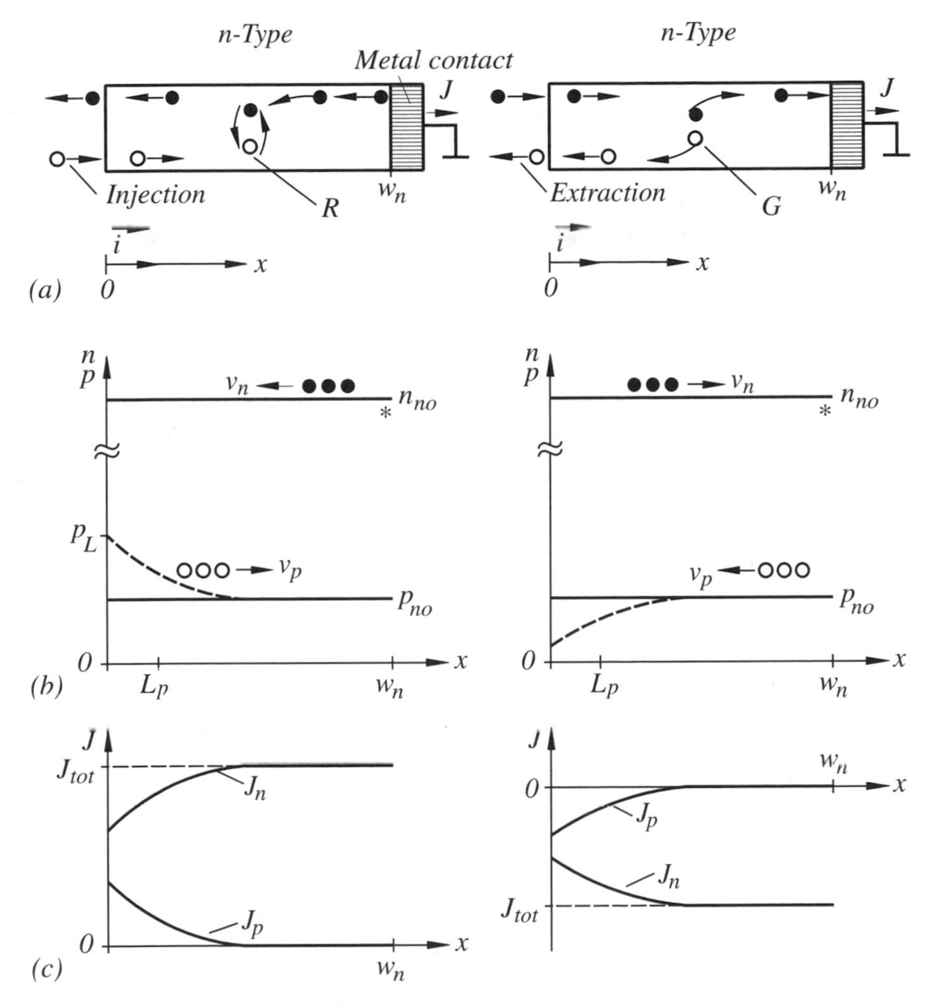

Figure 1.27 (a) One-sided injection and extraction of minority carriers at long dimensions, (b) carrier density distributions and (c) current density distributions (* negligible change in majority carrier density)

current density $J_{tot} = J_n(x) + J_p(x)$ must be constant throughout the sample from a continuous point of view.

Short lateral dimension $w_n \ll L_p$

In this case, Equation (1.68) yields a linear carrier density distribution of

$$p_n'(x) = p_L'\left(1 - \frac{x}{w_n}\right) \tag{1.71}$$

which is identical to Equation (1.57). This is not surprising, since at short dimensions the generation and recombination of carriers can be neglected.

Summary of the most important results

The three most important semiconductor equations – the carrier density equation, current density equation, and continuity equation – are presented. Furthermore it is shown that the carrier density product is equal to the square of intrinsic density at equilibrium. This leads to a description of the intrinsic carrier density as a function of the temperature and the band-gap.

A theoretical experiment is used to demonstrate the effect of injection and extraction of minority carriers into a semiconductor sample. It is proved that the minority carrier current can be considered to be a diffusion current only, as long as low-level injection is guaranteed. The derived equations of the minority carrier distribution and current distribution can be applied directly to forward and reverse biased pn-junctions and therefore bipolar transistors. Substantial mathematical simplifications result in the following chapters.

Problems

1.1 The Fermi level is 0.15 eV below the conduction band edge of a Ge semiconductor sample. Calculate:

(a) The probability that the valence band edge and the conduction band edge are occupied by electrons, and compare the values found with those of an intrinsic Ge semiconductor at room temperature ($kT = 26\,\text{meV}$).

(b) The electron density in the conduction band and the hole density in the valence band. Name the equation which links the carrier densities found in part (b).

1.2 The energy of the donor states is 0.05 eV below the conduction band edge. The density of the ionized donor atoms N_D^+ (Figure 1.11) can be calculated by the equation

$$N_D^+ = N_D[1 - F(E_D)]$$

$$= N_D\left[1 - \frac{1}{1 + \frac{1}{2}\exp\frac{E_D - E_F}{kT}}\right]$$

where $[1 - F(E_D)]$ is the probability of missing electrons, and the factor ½ takes into account that single ionized donors are present.

(a) Calculate for room temperature the Fermi levels for the following doping densities: $N_D = 10^{16}\,\text{cm}^{-3}$, $10^{18}\,\text{cm}^{-3}$, and $10^{19}\,\text{cm}^{-3}$. To start the calculation assume 100% ionization.

(b) Use the calculated Fermi levels to verify the assumption of 100% ionization.

1.3 An Si semiconductor has $5 \cdot 10^{16}$ boron atoms/cm^3 and 10^{15} phosphorus atoms/cm^3. Calculate at room temperature:

(a) Electron and hole concentration.

(b) Conductance.

(c) Energy difference between Fermi level and intrinsic level.

The mobility values of Figure 1.17(b) may be used.

1.4 In a homogeneously p-doped semiconductor with very long dimensions, at $x = 0$, electrons are continuously injected at room temperature. The example shown is part of a pn-junction with a highly doped n-region (not shown).

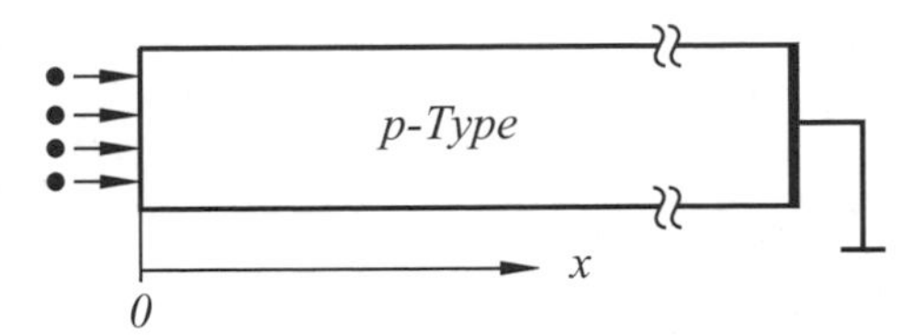

The electron density at $x = 0$ is $10^{10}\,\text{cm}^{-3}$ and the electron and hole mobility are $\mu_n = 1200\,\text{cm}^2/\text{Vs}$ and $\mu_p = 400\,\text{cm}^2/\text{Vs}$, respectively. The doping concentration of the semiconductor is $N_A = 10^{15}\,\text{cm}^{-3}$ and the diffusion length of the electrons has a value of 22 µm.

(a) Draw the local current density of electrons and holes.

(b) Determine the total current density.

(c) Calculate the electrical field at $x \to \infty$.

In order to solve the problem, it would be useful to study the 'Extended view of the experiment' section 1.5 first.

1.5 The goal of this problem is to demonstrate the term lifetime. This can be done by studying the following experiment. In the n-type semiconductor example

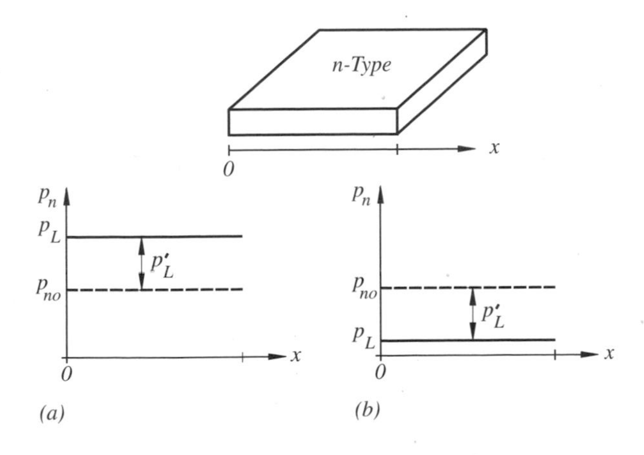

shown, majority and minority carriers are (a) increased or (b) decreased in comparison to their equilibrium values. The increase can be caused, e.g., by exposing the sample to light, which is able to generate electrons and holes by breaking covalent bonds of the crystal lattice. The reduction of charge carriers is not so simple to describe. But it is assumed that it can be done by electrical means.

When the cause of the carrier disturbance is eliminated the question arises how long does it take for the charge carriers to move back to equilibrium?

The solutions to the problems can be found under: *www.unibw-muenchen.de/campus/ET4/index.html*

REFERENCES

A.S. Grove (1967) *Physics and Technology of Semiconductor Devices*. New York: John Wiley & Sons.

E.J. Ryder (1953) Mobility of holes and electrons in high electric fields. *Phys. Rev.*, **90**(5), 760–69.

C.T. Sah, R.N. Noyce and W. Shockley (1957) Carrier generation and recombination in p-n junction. *Proc. IRE*, **45**, 1228–43.

W. Shockley and W.T. Read (1952) Statistics of the recombination of holes and electrons. *Phys. Rev.*, **87**, 835–42.

S.M. Sze (1981) *Physics of Semiconductor Devices* 2nd edn. New York: John Wiley & Sons.

C.D. Thurmond (1975) The standard thermodynamic function of the formation of electrons and holes in Ge, Si, GaAs and GaP. *J. Electrochem. Soc.*, **122**, 1133–41.

FURTHER READING

C. Kittel (1986) *Introduction to Solid State Physics*, 6th edn. New York: John Wiley & Sons.

2
pn-Junction

This junction can be considered to be a basic element of all integrated transistors. Understanding of it is therefore prerequisite for the following chapters, which deal with bipolar and field-effect transistors. The starting point of the derivation is the results of the theoretical experiment of Chapter 1, which are used to determine the current–voltage equation of a pn-junction. Two different physical effects lead to a depletion and a diffusion capacitance which are analyzed. In conjunction with this derivation, the width of the depletion region is determined and the switching and breakdown performance of the junction considered. The chapter ends with a discussion about modeling the pn-junction for computer-aided design (CAD), and with the presentation of easy to use equations, useful for a first estimate of the junction's static and dynamic behavior in a circuit environment.

2.1 INHOMOGENEOUSLY DOPED N-TYPE SEMICONDUCTOR

In order to analyze a pn-junction, it is useful to consider first an inhomogeneously doped, e.g. n-type, semiconductor. This aids in understanding the carrier behavior and the band diagram presentation under equilibrium conditions. Figure 2.1 shows as an example an n-type semiconductor junction with a step doping profile. The difference in doping concentration causes – 100% ionization assumed – an equivalent difference in electron density, which starts a diffusion process, described in Section 1.4.3. This process leads to an increase in the electron density at $x < 0$ and an equal sized electron density reduction at $x > 0$. The latter one is responsible for the domination of positive charge of uncompensated donor atoms in this region. The charge difference at the junction gives rise to an electrical field, causing a drift movement of electrons opposing to the movement of electrons caused by the diffusion process. Both electron movements compensate each other, and equilibrium is reached. In this case a current density of (Equation 1.48)

$$J_n = q\mu_n n_n \mathcal{E} + qD_n \frac{dn_n}{dx} = 0 \tag{2.1}$$

System Integration: From Transistor Design to Large Scale Integrated Circuits. Kurt Hoffmann.
 ISBN: 0-470-85407-3

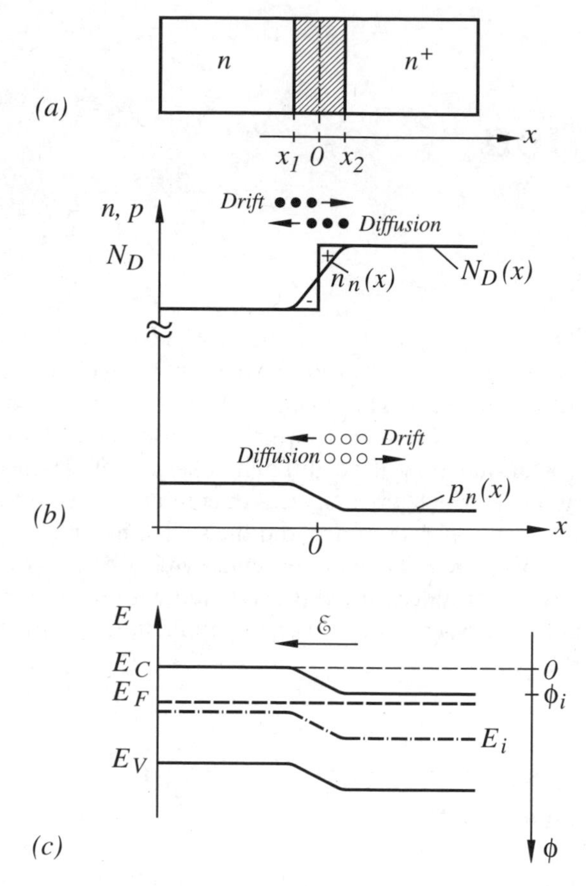

Figure 2.1 (a) Inhomogeneously doped n-type semiconductor, (b) charge carrier distribution (not to scale) and (c) band diagram

results. This equation can be solved for the electrical field

$$\mathcal{E}\mathrm{d}x = -\frac{D_n}{\mu_n}\frac{\mathrm{d}n_n}{n_n} \tag{2.2}$$

which leads to a voltage between the differently doped areas of

$$\phi_i = -\int_{x_1}^{x_2} \mathcal{E}\mathrm{d}x \tag{2.3}$$

This voltage is called the built-in voltage. The integration bounds are the areas where charge neutrality exists. Using the Einstein relationship (Equation 1.47) the following built-in voltage

$$\phi_i = \phi_t \int_{n_n(x_1)}^{n_n(x_2)} \frac{\mathrm{d}n_n}{n_n} = \phi_t \ln \frac{n_n(x_2)}{n_n(x_1)} \tag{2.4}$$

results, where the highly doped area is positive with regard to the lightly doped one. It is useful at this point, to compare the sign of the built-in voltage with the sign assignment discussed in Section 1.3.4.

So far, only the behavior of majority carriers is considered, but of interest is also that of the minority carriers. Different hole densities oppose each other. Because $p_n = n_i^2/N_D$, the highest hole density exists in the lightly doped area, causing a diffusion of holes to the highly doped area. The charge imbalance at the junction gives rise to an electrical field, causing holes to drift and oppose the ones which are diffusing. At equilibrium both carrier movements compensate each other, just as in the case of the majority carriers, but the movement of the minority carriers, by drift and diffusion, is opposite to that of the majority carriers.

Fermi level at equilibrium

Under equilibrium conditions the total current density is zero. The behavior of the Fermi level can therefore be derived by setting, e.g., the majority carrier current density (Equation 2.1) to zero

$$J_n = q\mu_n n_n \mathcal{E}(x) + qD_n \frac{\mathrm{d}n_n}{\mathrm{d}x} = 0$$

The needed carrier gradient $\mathrm{d}n/\mathrm{d}x$ can be found from the electron density (Equation 1.14)

$$n_n = n_i e^{[E_F(x) - E_i(x)]/kT} \tag{2.5}$$

to be

$$\frac{\mathrm{d}n_n}{\mathrm{d}x} = n_n \frac{1}{kT} \left[\frac{\mathrm{d}E_F(x)}{\mathrm{d}x} - \frac{\mathrm{d}E_i(x)}{\mathrm{d}x} \right] \tag{2.6}$$

Furthermore a relationship between the electrical field and the intrinsic energy level (Equation 1.34) is needed

$$\mathcal{E} = \frac{1}{q} \frac{\mathrm{d}E(x)}{\mathrm{d}x} = \frac{1}{q} \frac{\mathrm{d}E_i(x)}{\mathrm{d}x} \tag{2.7}$$

where $\mathrm{d}E_i/\mathrm{d}x$ is used to describe the band bending. With these equations and the Einstein relationship (Equation 1.47) an electron current density of

$$J_n = \mu_n n_n \frac{\mathrm{d}E_F(x)}{\mathrm{d}x} = 0 \tag{2.8}$$

results. Since no current flows at equilibrium the term $\mathrm{d}E_F(x)/\mathrm{d}x$ must be zero. This implies, that *under thermal equilibrium conditions, the Fermi level is constant throughout the semiconductor*. In other words, when a local variation of the Fermi level is observed a current must exist. This case will be considered in some of the following sections.

2.2 PN-JUNCTION AT EQUILIBRIUM

Next a pn-junction with a step doping profile under equilibrium conditions is analyzed (Figure 2.2). The diffusion and drift movements of each carrier type oppose each other at the junction interface, similar to the discussed inhomogeneously doped case, except that the density differences are by far much larger. At equilibrium the net current flow is zero and the Fermi level is constant throughout the semiconductor. A charge and field distribution as shown in Figure 2.2 results. The charge is mainly caused by uncompensated donor and acceptor atoms. This region is called the space charge region or depletion region (DR). The electrical field can be used to determine the built-in voltage (Equation 2.3), as is done for the inhomogeneously doped semiconductor

$$\phi_i = -\int_{x_p}^{x_n} \mathcal{E}\mathrm{d}x$$

Since the majority and minority current densities are zero, either one can be used for the determination of the built-in voltage

$$\begin{aligned} J_n &= q\mu_n n\mathcal{E} + qD_n \mathrm{d}n/\mathrm{d}x = 0 \\ \mathcal{E}\mathrm{d}x &= -\frac{D_n}{\mu_n}\frac{\mathrm{d}n}{n} \\ \phi_i &= \phi_t \int_{n_{po}}^{n_{no}} \frac{\mathrm{d}n}{n} = \phi_t \ln\frac{n_{no}}{n_{po}} \end{aligned} \tag{2.9}$$

With $n_{no} = N_D$ and $n_{po} = n_i^2/N_A$ – 100% ionization assumed – Equation (2.9) leads to a built-in voltage of

$$\boxed{\phi_i = \phi_t \ln\frac{N_A N_D}{n_i^2}} \tag{2.10}$$

Typical values of integrated pn-junctions are about 0.9 V. By increasing the doping concentrations a maximum value of $\phi_i \approx E_G/q$ is possible (Figure 2.2(e)).

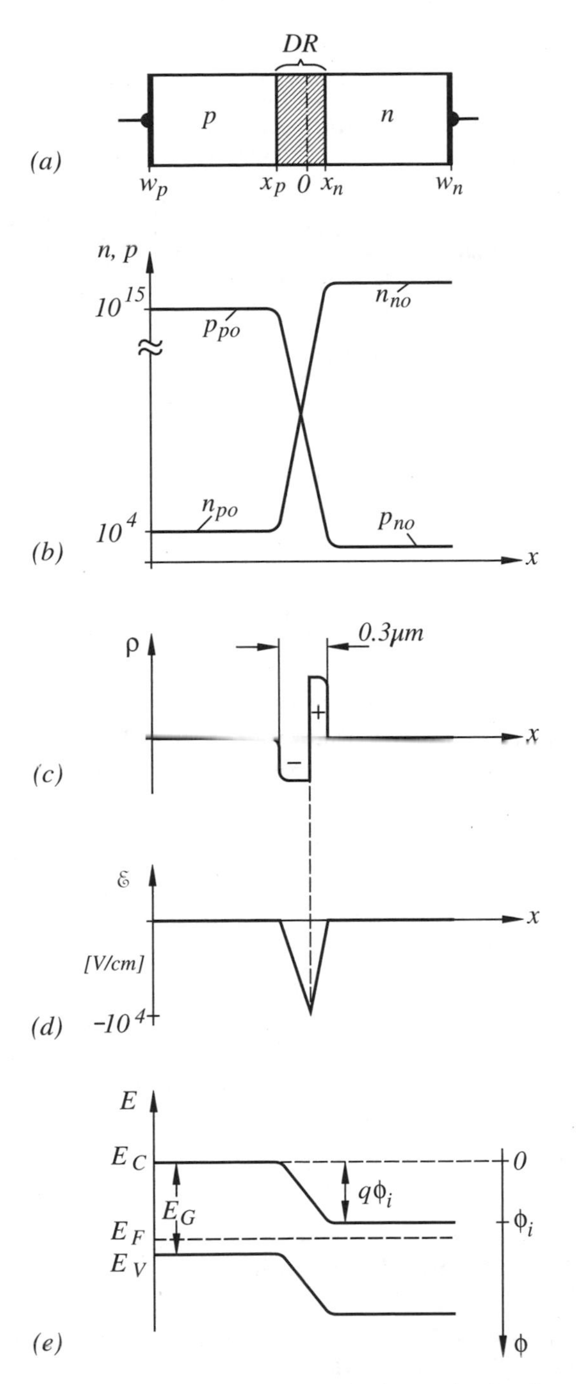

Figure 2.2 (a) pn-Junction with $N_D > N_A$, (b) majority and minority carrier distribution, (c) depletion region charge, (d) electrical field and (e) band diagram

2.3 BIASED PN-JUNCTION

In the following, an external V_{PN} voltage is applied to a pn-junction via metal contacts. If the voltage is different from zero the equilibrium condition will be disturbed. Two biasing cases can be distinguished: one in the forward direction with $V_{PN} > 0$, and one in the reverse direction with $V_{PN} < 0$ (Figure 2.3).

As will be shown, a relatively large current flows in the first case, whereas in the second, the current is almost negligible. Applying a voltage $V_{PN} > 0$ between the p-region and the n-region lowers the energy between these regions from $q\phi_i$ (Figure 2.2(e)) to $q(\phi_i - V_{PN})$, whereas a voltage of $V_{PN} < 0$ (V_{PN} is negative) raises the energy as shown in Figure 2.3(b). At this point it may be useful to compare the effect of an applied voltage to a pn-junction with the example presented in Figure 1.19.

It is obvious that the change in energy between the p-region and the n-region, caused by the applied external voltage source, is responsible for the drift and diffusion currents no longer compensating each other. This gives rise to an injection or an extraction of minority carriers as shown in Figure 2.3(c).

In the forward biased direction with $V_{PN} > 0$, minority carriers are injected into the n-region and p-region. The injected carriers are supplied by the reservoir of the majority carriers, which in turn are supplied by the external voltage source. In the reverse biased direction with $V_{PN} < 0$ minority carriers are extracted from the n-region and p-region, due to the enhanced electrical field. The extracted minority carriers move across the depletion region and as majority carriers to the external voltage source (Figure 2.3(c)).

The behavior of the charge carriers in the n-region is equivalent to that analyzed by the theoretical experiment (Figure 1.24), presented in Section 1.5. In this experiment it is assumed that the generation and recombination of charge carriers are negligible, due to the small volume of the semiconductor sample. This assumption is also applicable to pn-junctions of modern integrated semiconductors. In this case, the minority current of, e.g., holes I_p in the n-region is equal to the majority current of holes I_p in the p-region of the junction. This is obvious, since the injected or extracted minority carriers can only be supplied or extracted by the respective majority carriers. A similar argument is applicable to the electron current I_n. In other words: *the majority carrier current from one side of the pn-junction is the minority carrier current from the other side of the junction*. The total pn-junction current can therefore be determined by simply adding the two minority currents, which can be found directly from the minority carrier distribution.

2.3.1 Density Product under Non-Equilibrium Conditions

In order to derive the minority currents of a pn-junction from the minority density distribution, the minority carrier densities at the edges of the depletion region have to be determined first. The following thought leads, via quasi-Fermi levels, to a density product, which is described as a function of the applied voltage. Using this relationship, the carrier densities at the edges of the depletion region are found.

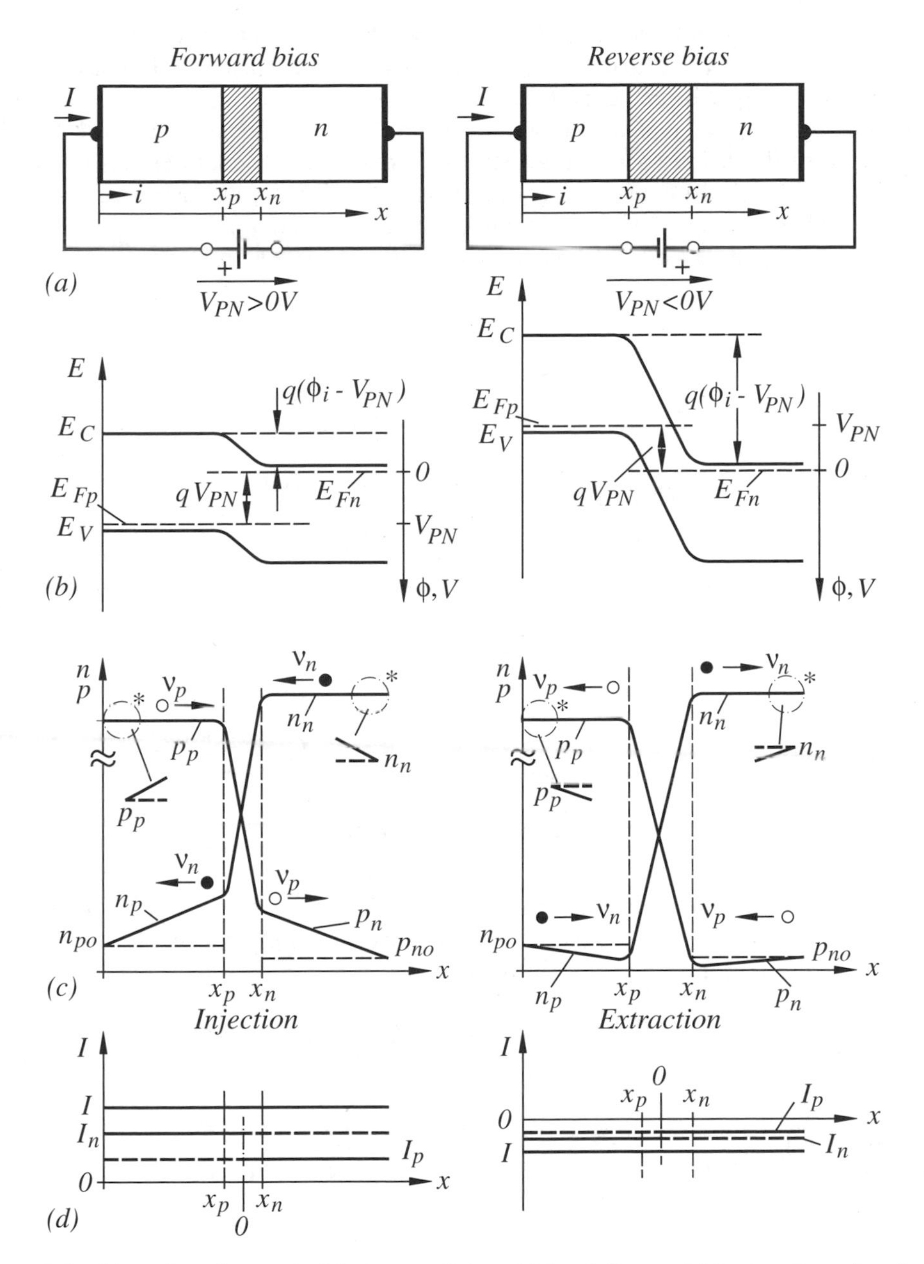

Figure 2.3 (a) pn-Junction under biasing condition, (b) band diagram, (c) minority and majority carriers and (d) current densities (* negligible change in majority carrier density)

The densities of the charge carriers under equilibrium conditions (Equations 1.14 and 1.15)

$$n_o = n_i e^{(E_F - E_i)/kT} \text{ and } p_o = n_i e^{(E_i - E_F)/kT}$$

are characterized by the Fermi level E_F. A change of the Fermi level, e.g. towards higher energy, causes the electron density to increase and the hole density to decrease. When a deviation from equilibrium conditions exists, e.g. by injecting electrons and holes simultaneously into a semiconductor, the Fermi level has to move toward a higher and also toward a lower energy to satisfy the above density equations. This diverging task leads to the definition of quasi-Fermi levels E_{Fn} and E_{Fp}, which are used to describe situations where carriers deviate from equilibrium. Accordingly

$$n = n_i e^{(E_{Fn}-E_i)/kT} \tag{2.11}$$

and

$$p = n_i e^{(E_i-E_{Fp})/kT} \tag{2.12}$$

These densities yield a density product of

$$pn = n_i^2 e^{(E_{Fn}-E_{Fp})/kT} \tag{2.13}$$

Applying this product to the edges of the depletion layer leads to

$$p_n(x_n)n_n(x_n) = n_i^2 e^{(E_{Fn}-E_{Fp})/kT} \tag{2.14}$$

and

$$p_p(x_p)n_p(x_p) = n_i^2 e^{(E_{Fn}-E_{Fp})/kT} \tag{2.15}$$

Since the difference in the quasi-Fermi levels at the junction (Figure 2.3b) is assumed to be proportional to the applied voltage – voltage drops along the n-region and p-region are neglected – the following relationship

$$V_{PN} = \frac{E_{Fp} - E_{Fn}}{-q} \tag{2.16}$$

between the quasi-Fermi levels and the applied voltage exists (Section 1.3.4). This leads to a density product, as a function of the applied voltage, at the depletion layer edges of

$$\boxed{p_n(x_n)n_n(x_n) = p_p(x_p)n_p(x_p) = n_i^2 e^{V_{PN}/\phi_t}} \tag{2.17}$$

This means, that with $V_{PN} > 0$ a carrier increase occurs ($pn > n_i^2$), whereas with $V_{PN} < 0$ a carrier depletion ($pn < n_i^2$) takes place.

An interesting observation can be made by evaluating the density product equation under low-level and high-level injection (Figure 1.23).

Low-level injection

With $n_n(x_n) = N_D$ and $p_p(x_p) = N_A$ the following carrier densities at the edges of the depletion layer

$$p_n(x_n) = \frac{n_i^2}{N_D} e^{V_{PN}/\phi_t}$$

$$\boxed{p_n(x_n) = p_{no} e^{V_{PN}/\phi_t}} \tag{2.18}$$

and

$$n_p(x_p) = \frac{n_i^2}{N_A} e^{V_{PN}/\phi_t}$$

$$\boxed{n_p(x_p) = n_{po} e^{V_{PN}/\phi_t}} \tag{2.19}$$

result. These two equations, which are used in the following chapters, are of the utmost importance for the derivation of current–voltage relationships.

High-level injection

With $n_n(x_n) = p_p(x_n)$ and $n_p(x_p) = p_p(x_p)$ the resulting carrier densities are

$$\begin{aligned} p_n(x_n) &= \frac{n_i^2}{p_n(x_n)} e^{V_{PN}/\phi_t} \\ p_n(x_n) &= n_i e^{V_{PN}/2\phi_t} \end{aligned} \tag{2.20}$$

and

$$\begin{aligned} n_p(x_p) &= \frac{n_i^2}{n_p(x_p)} e^{V_{PN}/\phi_t} \\ n_p(x_p) &= n_i e^{V_{PN}/2\phi_t} \end{aligned} \tag{2.21}$$

The factor ½ at the exponents has the effect that under high-level injection, the minority carrier densities do not increase as much with the voltage as is the case at a low-level situation. This behavior plays an important role in pn-junctions and bipolar transistors, which will be covered when considering second-order effects.

In the above derivation of the density product under non-equilibrium conditions the idea of quasi-Fermi levels was introduced. A maybe more intuitive way leads to the same result, as shown in the following.

Equation (2.9) describes, after rearranging, the relationship between carrier density and built-in voltage

$$n_{po} = n_{no} e^{-\phi_i/\phi_t} \tag{2.22}$$

under equilibrium conditions. If an external voltage is applied to the pn-junction the energy barrier (Figure 2.3b) is reduced to a value of $q(\phi_I - V_{PN})$, assuming – as in the previous derivation – that the voltage drops across the p-region and n-region are negligible. The changed energy barrier leads to a carrier density, e.g. at the *p*-side edge of the depletion layer, of

$$\begin{aligned} n_p(x_p) &= n_{no} e^{-(\phi_i - V_{PN})/\phi_t} \\ n_p(x_p) &= n_{po} e^{V_{PN}/\phi_t} = \frac{n_i^2}{p_p(x_p)} e^{V_{PN}/\phi_t} \end{aligned} \tag{2.23}$$

and to a density product of

$$p_p(x_p) n_p(x_p) = n_i^2 e^{V_{PN}/\phi_t} \tag{2.24}$$

which is identical to that of Equation (2.17).

2.3.2 Current–Voltage Relationship

In this section the current–voltage relationship of the pn-junction is derived, under the following assumptions: the voltage drops across the n-region and p-region are negligible, low-level injection exists, and generation and recombination can be neglected due to the small semiconductor dimensions.

With these prerequisites, the results of the experiment of Section 1.5 can be applied directly to the pn-junction. To ease the illustration details of Figure 2.3 are repeated in Figure 2.4 for convenience.

In the following derivation it is unimportant whether the injection or the extraction of minority carriers is considered, since both processes are described by the same equation. As explained previously, the majority carrier current from one side of the pn-junction is the minority carrier current from the other side of the junction. The total current can therefore be found by simply adding the two minority currents

$$\begin{aligned} I &= I_p + I_n \\ &= -qAD_p \frac{\mathrm{d}p_n}{\mathrm{d}x} + qAD_n \frac{\mathrm{d}n_p}{\mathrm{d}x} \end{aligned} \tag{2.25}$$

where A is the cross-section of the junction. With the gradients – directly from Figure 2.4b – and with the carrier densities at the edge of the depletion layers (Equations 2.18 and 2.19) the current–voltage relationship of the pn-junction is found.

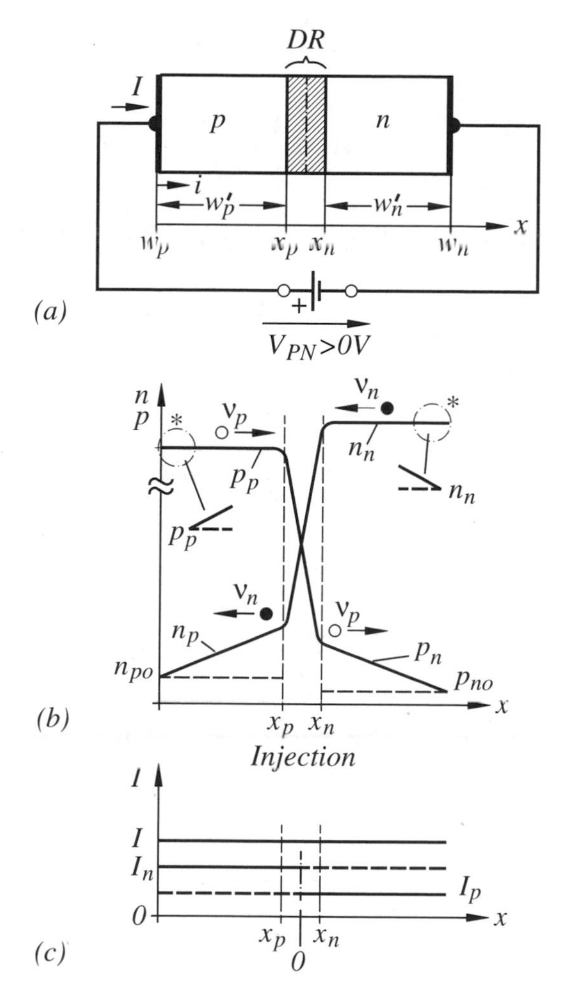

Figure 2.4 (a) pn-Junction under forward biased condition, (b) minority and majority carriers and (c) minority carrier currents and majority carrier currents (dotted lines)

$$I = qAD_p \frac{p_n(x_n) - p_{no}}{w_n'} + qAD_n \frac{n_p(x_p) - n_{po}}{w_p'}$$
$$I = qAD_p \frac{p_{no}}{w_n'} \left(e^{V_{PN}/\phi_t} - 1\right) + qAD_n \frac{n_{po}}{w_p'} \left(e^{V_{PN}/\phi_t} - 1\right) \tag{2.26}$$

This result can be simplified to

$$\boxed{I = I_S\left(e^{V_{PN}/\phi_t} - 1\right)} \tag{2.27}$$

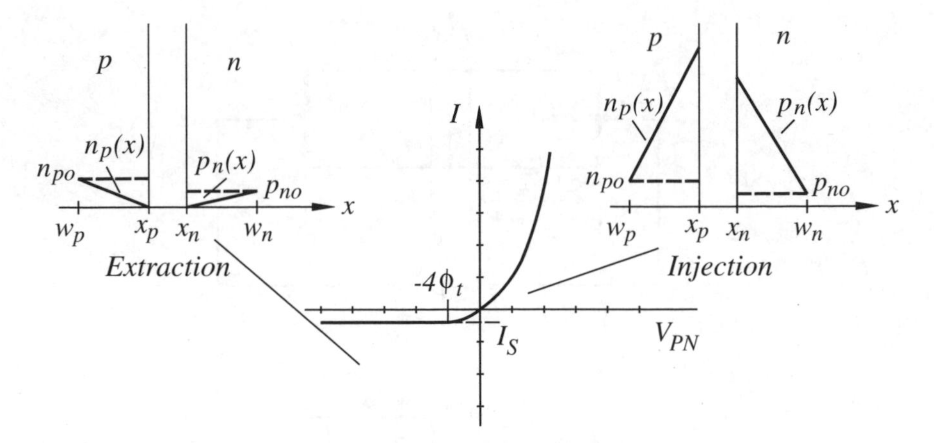

Figure 2.5 Current–voltage relationship of a pn-junction

where

$$I_S = qA\left(\frac{D_p}{w'_n}p_{no} + \frac{D_n}{w'_p}n_{po}\right) \tag{2.28}$$

is named the saturation current, and sometimes the leakage current. The current–voltage relationship is shown in Figure 2.5.

The exponential factor in Equation (2.27) can be neglected compared to the -1 term, when the applied V_{PN} voltage is more negative than $-4\phi \approx -100\,\text{mV}$ at room temperature. In this case the current becomes independent of the voltage. The reason for this behavior is that the density of the minority carriers at the depletion region edges approaches zero. The current $I = -I_S$ remains constant, since the gradients of the minority carriers do not change anymore. Contrary to this situation is the one when V_{PN} is positive. The minority carriers and accordingly the current can be raised until the temperature exceeds its allowed value, causing some kind of melting.

2.3.3 Deviation from the Current–Voltage Relationship

Figure 2.5 shows a saturation current I_S independent of the voltage. In reality this is not the case, since the current increases moderately when a more negative voltage is applied. The reason for this deviation is an increase of the depletion region width DR (Figure 2.4). This increase causes a reduction in the widths w'_p and w'_n of the p-region and n-region, and an increase in the minority carrier gradients, and accordingly an increase in the saturation current. Why the width of the depletion region changes is treated in Section 2.4.1.

The saturation current (Equation 2.28) can be described as a function of the doping concentrations and temperature by using $p_{no} = n_i{}^2/N_D$ and $n_p = n_i{}^2/N_A$

$$I_S = qA\left[\frac{D_p}{w'_n(V_{PN})}\frac{1}{N_D} + \frac{D_n}{w'_p(V_{PN})}\frac{1}{N_A}\right]n_i^2 \tag{2.29}$$

and where the intrinsic density is described by Equation (1.20)

$$n_i = C\left(\frac{T}{[K]}\right)^{3/2} e^{-E_G(T)/2kT} \tag{2.30}$$

The exponential behavior of the intrinsic density with temperature explains the strong dependence of the saturation current on temperature. This may cause undesirable leakage currents in integrated circuits at elevated temperature.

Deviations from the derived current-voltage relationship occur not only in the reverse, but also in the forward biased direction (Figure 2.6). Three regions of deviations can be observed. Region (a) shows a larger current than expected. This is caused by the recombination of carriers (Equation 1.64), which occurs in particular in the depletion region. Since this contribution is small, it can be observed at very small currents only. Region (b) is a transition region, where the junction moves from low-level injection to high-level injection. Under the condition of high-level injection the minority carriers at the edges of the depletion region do not increase as much with the voltage, and thus the current, as is the case under low-level injection (Equations 2.20 and 2.21). Region (c) displays a further reduction in current caused by an increase in voltage drops across the n-region and p-region, which is neglected so far.

The deviations from the current–voltage equation can be expressed with an empirical factor N

$$\boxed{I = I_S\left(e^{V_{PN}/\phi_t N} - 1\right)} \tag{2.31}$$

called the emission coefficient, with values between one and two.

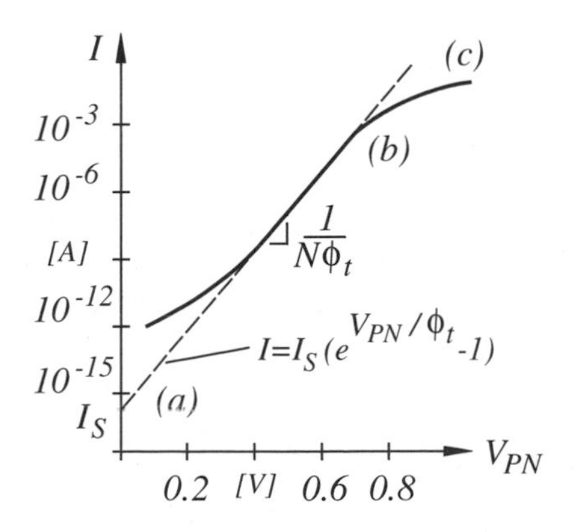

Figure 2.6 Deviations from the current–voltage behavior in forward biased direction (half logarithmic scale)

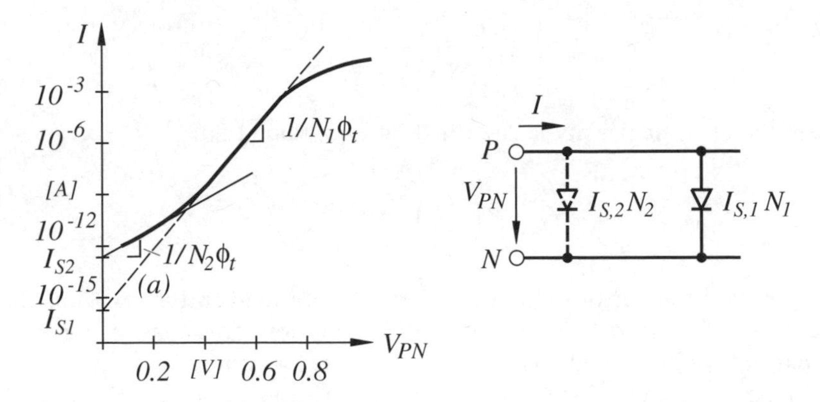

Figure 2.7 Approximation of the current–voltage relationship

Determination of the parameters I_S and N

Under forward biased condition with $V_{PN} > 100\,\text{mV}$ the –1 term in Equation (2.31) can be neglected and the above equation be expressed in logarithmic form

$$\ln\frac{I}{[\text{A}]} \approx \ln\frac{I_S}{[\text{A}]} + \frac{1}{N\phi_t}V_{PN} \tag{2.32}$$

where A is the symbol for Ampère. This is a linear equation on a half-logarithmic scale as shown in Figure 2.6. The saturation current I_S can be found by extrapolating the measured data to the zero V_{PN} value, since in this case $\ln I/[\text{A}] \approx \ln I_S/[\text{A}]$. In reality this value is not necessarily the saturation current. The latter may be increased, e.g. by leakage currents at the semiconductor surface. Nevertheless I_S is an important parameter for the description of the current behavior of a forward biased pn-junction. The emission coefficient can be found from the gradient of the measured data (Figure 2.6) according to Equation (2.32). Typical values in the middle region of the function are $N \approx 1$ and current densities are about $J_S \approx 10^{-18}\,\text{A}/\mu\text{m}^2$.

In computer-aided design programs, so-called compact models are used for the description of semiconductor devices. These models are useful only when a set of appropriately measured parameters can be generated. In the case of the pn-junction this is done by approximating the measured data by the behavior of two parallel connected diodes (Figure 2.7). One describes the low current region with the parameters $I_{S,2}$ and N_2 and the other the middle region with the parameters $I_{S,1}$ and N_1. This procedure is commonly used for junctions at bipolar transistors, which are considered in detail in Chapter 3.

2.3.4 Voltage Reference Point

In the preceding sections is shown how an external voltage, e.g. in Figure 2.3, modifies the band diagram. In this diagram the energy is related to a reference energy, and the reference energy in turn to a reference potential of 0 V. In the following, the question is raised of how this reference point relates to an externally applied voltage source, where the most negative potential is usually used as a 0 V reference point in integrated circuit design. In order to find this out, a pn-junction including metal contacts is considered (Figure 2.8). At the interface between semiconductor and metal, contact voltages exist. These are caused by the different carrier densities between the materials. At the interface the charge carrier balances out to equilibrium conditions, thereby generating a contact voltage. In this example it is assumed that the composition of the materials is such, that a resistive character results. When the pn-junction is short-circuited, equilibrium is reached – in the order of 10^{-12} s – and current ceases to flow. In this case the sum of all voltages is $\phi_{C1} + (-\phi_i) + \phi_{C2} = 0$, or in other words, the built-in voltage must balance the contact voltages

$$\phi_i = \phi_{C1} + \phi_{C2} \tag{2.33}$$

When a V_{PN} voltage is applied to the pn-junction (Figure 2.8(b)), only the voltage V at the depletion region changes to a value of

$$\begin{aligned} V &= \phi_{C1} + \phi_{C2} - V_{PN} \\ &= \phi_i - V_{PN} \end{aligned} \tag{2.34}$$

This is understandable, since the metal–semiconductor interfaces and the n-region and p-region are assumed to have negligible resistances. Equation (2.34) is thus in agreement with the used assumption, that an externally applied voltage has an effect on the depletion region only. This is the reason why the contact voltages do not appear in the derived semiconductor equations. But the reference point of the externally applied voltage (Figure 2.8(b)) differs by the contact voltage ϕ_{C2} from that of the n-region in Figure 2.3.

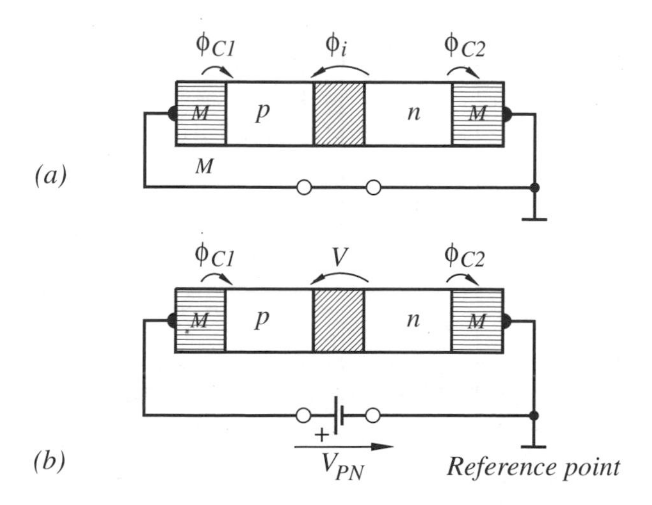

Figure 2.8 (a) Short-circuited pn-junction and (b) pn-junction with applied voltage

2.4 CAPACITANCE CHARACTERISTIC

Two totally different physical effects lead to a depletion and a diffusion capacitance of the pn-junction. The first is based on the fact that the charge of the depletion region changes with the voltage, whereas the second is caused by the injection of charge carriers into the n-region and p-region of the junction.

2.4.1 Depletion Capacitance

In order to determine the depletion capacitance C_j, which is sometimes called junction capacitance, the width of the depletion region has to be determined first. As an example a step junction will be considered (Figure 2.9).

Depletion region width

The depletion region charge can be assumed to consist of ionized donor and acceptor atoms only. The carrier charge can be neglected when one disregards the small

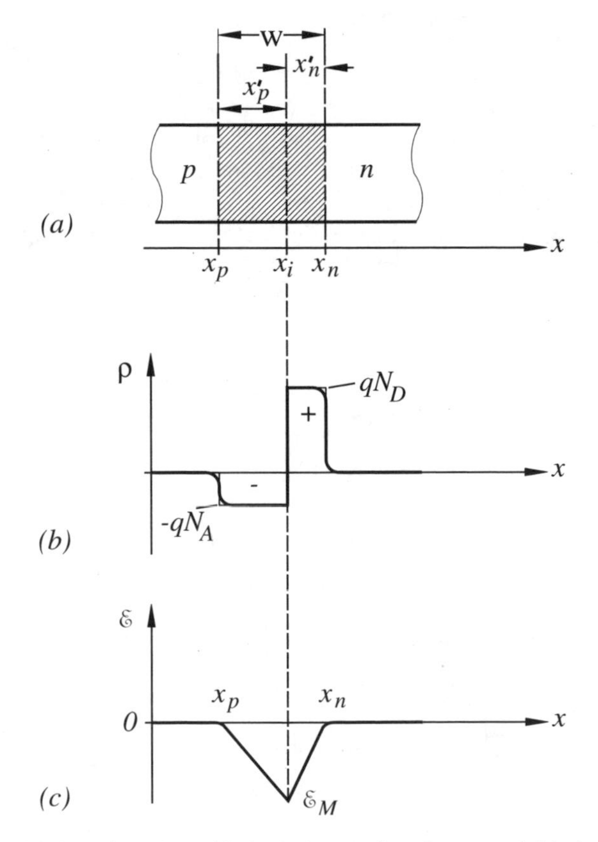

Figure 2.9 (a) Step junction, (b) depletion region charge and (c) electrical field

transition region at the edge of the depletion region (Figure 2.9(b)). This approximation is called depletion approximation and is used to avoid the solution of an implicit equation. According to Poisson's equation, the electrical field can be found by integrating over the charge distribution

$$\mathcal{E}(x) = \frac{1}{\varepsilon_o \varepsilon_{si}} \int_{x_p}^{x_n} \rho(x)\, \mathrm{d}x \tag{2.35}$$

A maximum field at the metallurgical junction x_i of

$$\mathcal{E}(x = x_i) = \mathcal{E}_M = \frac{1}{\varepsilon_o \varepsilon_{si}} \int_{x_p}^{x_i} -qN_A \mathrm{d}x = -\frac{qN_A}{\varepsilon_o \varepsilon_{si}} x'_p \tag{2.36}$$

results. The voltage $\phi_i - V_{PN}$ across the depletion region is linked to the electrical field by

$$\phi_i - V_{PN} = -\int_{x_p}^{x_n} \mathcal{E}(x)\mathrm{d}x \tag{2.37}$$

It is obvious that *the triangular area of the electrical field in the diagram represents the voltage across the depletion region* (Figure 2.9(c)). This is useful to remember, since it is a nice way to visualize the relationship between an applied voltage and an electrical field, which is used for illustration in some of the following sections and chapters. Inspecting Figure 2.9(c) yields

$$\phi_i - V_{PN} = -\frac{1}{2}\mathcal{E}_M w \tag{2.38}$$

where the width of the depletion region is

$$w = x'_p + x'_n \tag{2.39}$$

Since charge neutrality exists, the charge on each side of the depletion region has to be the same.

$$N_A x'_p = N_D x'_n \tag{2.40}$$

This leads to the width of the p-depletion region

$$x'_p = \sqrt{\frac{2\varepsilon_o \varepsilon_r (\phi_i - V_{PN})}{qN_A\left(1 + \frac{N_A}{N_D}\right)}} \tag{2.41}$$

and that of the n-depletion region

$$x'_n = \sqrt{\frac{2\varepsilon_o\varepsilon_r(\phi_i - V_{PN})}{qN_D\left(1 + \frac{N_D}{N_A}\right)}} \tag{2.42}$$

Adding both results in the total depletion region width of

$$\boxed{w = \sqrt{\frac{2\varepsilon_o\varepsilon_r}{q}\left(\frac{1}{N_A} + \frac{1}{N_D}\right)(\phi_i - V_{PN})}} \tag{2.43}$$

Considering this equation it is obvious that the width decreases when the V_{PN} voltage is increased. This relationship was pointed out to exist in conjunction with the deviation of the saturation current from its ideal behavior (Equation 2.29).

In semiconductor technology the case is often encountered, where the doping of one side of the junction is much larger than that of the other side. With, e.g., $N_D \gg N_A$ a depletion region width of

$$w \approx \sqrt{\frac{2\varepsilon_o\varepsilon_r}{q}\frac{1}{N_A}(\phi_i - V_{PN})} \tag{2.44}$$

results. This means that in almost all practical cases the lower doped side determines the depletion region width.

The preceding equations demonstrate, that a voltage change causes a variation in depletion width and therefore a change in the depletion charge. This behavior can be described by a small-signal depletion capacitance. When a positive voltage change $\mathrm{d}V_{PN} > 0$ is superimposed on the V_{PN} voltage the depletion width reduces by $\mathrm{d}x'_p$ in the p-region and $\mathrm{d}x'_n$ in the n-region of the junction, since majority carriers neutralize the excess depletion charge in the $\mathrm{d}x'_n$ and $\mathrm{d}x'_p$ regions (Figure 2.10).

The contrary situation is when a negative voltage change $\mathrm{d}V_{PN} < 0$ occurs, causing excess majority carriers to deplete the $\mathrm{d}x'_n$ and $\mathrm{d}x'_p$ regions. This behavior corresponds to that of a small-signal capacitance

$$C_j = \frac{\mathrm{d}Q}{\mathrm{d}V_{PN}} = \frac{\mathrm{d}Q}{\mathrm{d}x'_p}\frac{\mathrm{d}x'_p}{\mathrm{d}V_{PN}} \tag{2.45}$$

The first quotient $\mathrm{d}Q/\mathrm{d}x'_p$ in this equation can be found from the relationship (Figure 2.10(c))

$$\mathrm{d}Q = qAN_A\,\mathrm{d}x'_p \tag{2.46}$$

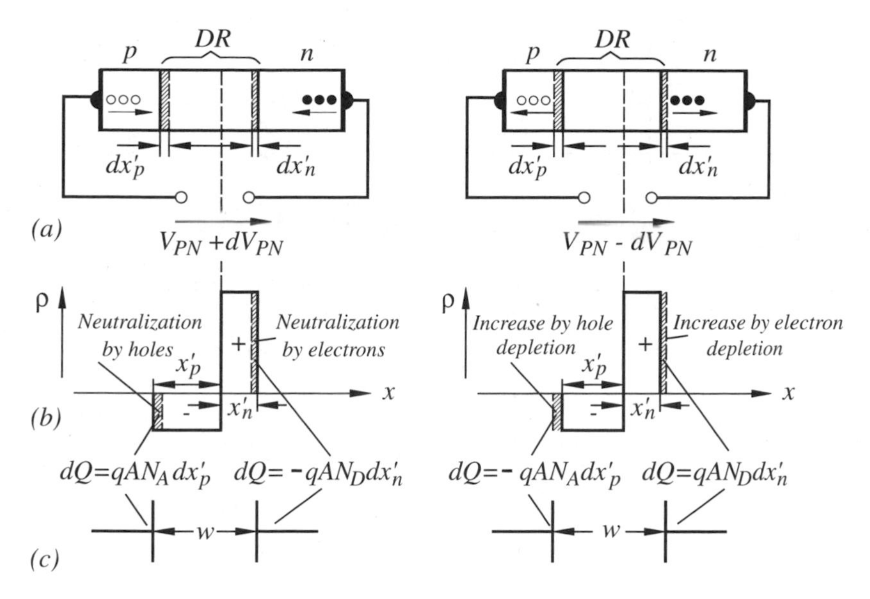

Figure 2.10 (a) Change in depletion region width, (b) change in depletion region charge and (c) plate capacitor equivalent

and the second, $\mathrm{d}x'_p/\mathrm{d}V_{PN}$, from Equation (2.41), yielding a small-signal depletion capacitance of

$$C_j = A\sqrt{\frac{q\varepsilon_o\varepsilon_r}{2\left(\frac{1}{N_A}+\frac{1}{N_D}\right)\phi_i}}\frac{1}{\sqrt{1-\frac{V_{PN}}{\phi_i}}} \tag{2.47}$$

This equation can be presented in the simplified form

$$\boxed{C_j = C_{jo}\left(1-\frac{V_{PN}}{\phi_i}\right)^{-M}} \tag{2.48}$$

where C_{jo} is the small-signal depletion capacitance at $V_{PN} = 0\,\mathrm{V}$. M is called the grading coefficient. These parameters depend strongly on the doping profile. M can assume values between ½ – for the step junction considered – and ⅓ for a linear graded junction (Sze 1981) (Figure 2.11). The dependence of the depletion capacitance on the voltage is shown in Figure 2.12.

In the forward biased direction a discontinuity at $V_{PN} = \phi_i$ exists. The reason for this behavior is the depletion approximation, used to determine the depletion width. This approximation assumes that the charge of the charge carriers can be neglected in comparison to the charge of the ionized atoms. This is definitely not the case at large currents. In reality a decrease of the capacitance can be observed in this case.

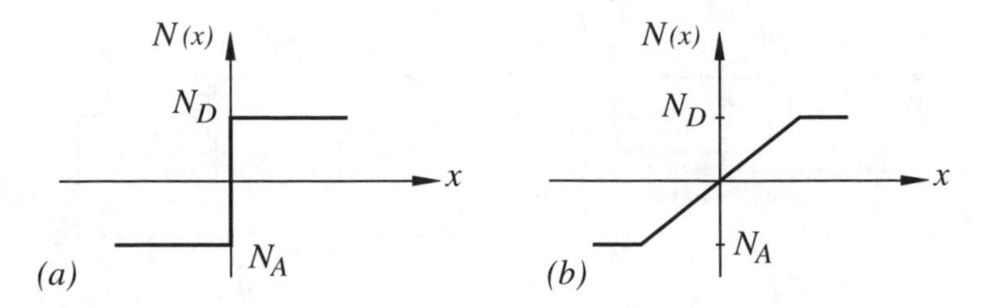

Figure 2.11 Doping profile $N(x)$: (a) step junction and (b) linear graded junction

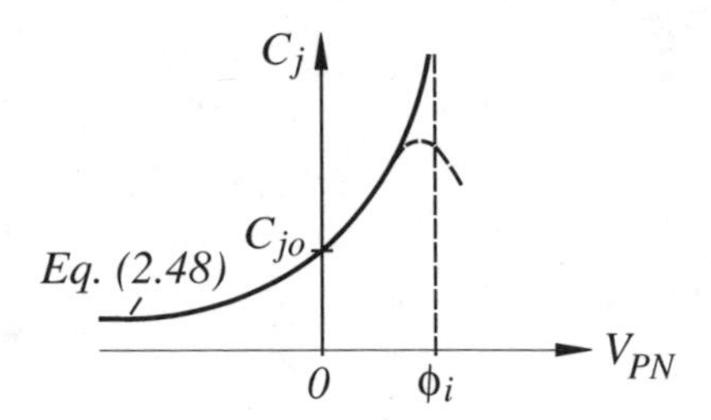

Figure 2.12 Small-signal depletion capacitance C_j as function of V_{PN} voltage

Determination of the parameters C_{jo}, ϕ_i, and M

The value of C_{jo} can be found directly from measured data (Figure 2.12). The measurement should be performed in such a way that the pn-junction remains always in the reverse biased condition. The reason is that in the forward biased condition a diffusion capacitance exists – described in the next section – which is much larger than the depletion capacitance. The parameters ϕ_i and M can be found by adjusting their value – curve fitting – to the measured data. Integrated pn-junctions have a bottom and a side wall contributing to the capacitance value. In order to achieve a high accuracy during circuit simulation, the capacitance parameters of those contributing areas are required. These values allow the designer to cope with all possible layout shapes. The different capacitance values at the bottom and side wall of a pn-junction originate from the diffusion process during manufacturing (Figure 2.13). During this process, dopant atoms, e.g. arsenic once, diffuse due to their concentration gradient from higher concentration regions near the surface toward regions of lower concentration further into the wafer.

Plate capacitor analogy

In Figure 2.10 it is indicated that the depletion capacitance can be viewed as a plate capacitor. If Equation (2.47) is rewritten in the form

$$C_j = \frac{A\varepsilon_o\varepsilon_r}{\sqrt{\frac{2\varepsilon_o\varepsilon_r}{q}\left(\frac{1}{N_A}+\frac{1}{N_D}\right)(\phi_i - V_{PN})}} \tag{2.49}$$

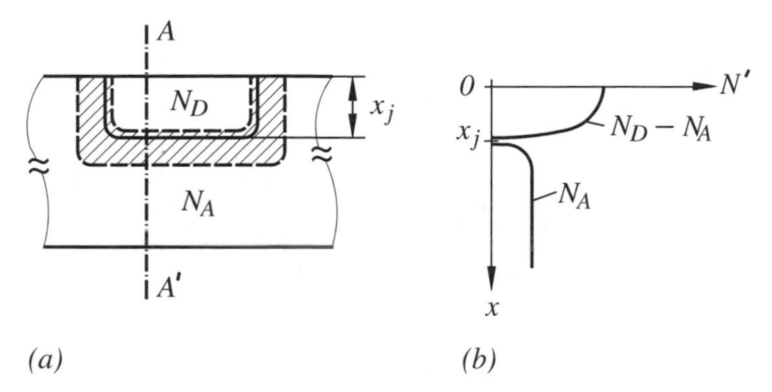

Figure 2.13 (a) Diffused pn-junction and (b) doping concentration $N'(x) = N_D - N_A$ (cross-section A–A')

and the denominator replaced by Equation (2.43), which describes the depletion region width, a junction capacitance of

$$\boxed{C_j = A\frac{\varepsilon_o\varepsilon_r}{w(V_{PN})}} \tag{2.50}$$

results. This is the equation of a parallel plate capacitor with a voltage-dependent plate distance $w(V_{PN})$.

2.4.2 Diffusion Capacitance

As mentioned before, additionally to the depletion capacitance exists another capacitance caused by a completely different mechanism. This mechanism is based on the fact that the n-region and the p-region are able to store minority carrier and majority carrier. How this can be understood is shown in Figure 2.14.

For simplicity, it is assumed that the width of the depletion region remains constant, even when the voltage is changed. Due to the externally applied V_{PN} voltage, minority carriers are injected into the n-region and p-region of the junction. Neutrality demands that the majority carriers assume a distribution which correlates to that of the minority carriers, as described in Section 1.5. This so-called dielectric relaxation plays no important role in integrated circuits, since it takes place within approximately 10^{-12} s. It can therefore be concluded: *the n-region and p-region of the semiconductor behave as if the majority carrier charge and the minority carrier charge are increased or decreased in each region simultaneously*. This fact, that the positive and negative charge can be altered by the applied voltage, corresponds to the behavior of a small-signal capacitance in each region.

In order to derive the capacitance equation, the total positive charge in the n-region and p-region has to be determined first. Differentiating this charge with respect to the voltage yields the small-signal capacitance equation. The positive charge in the n-region

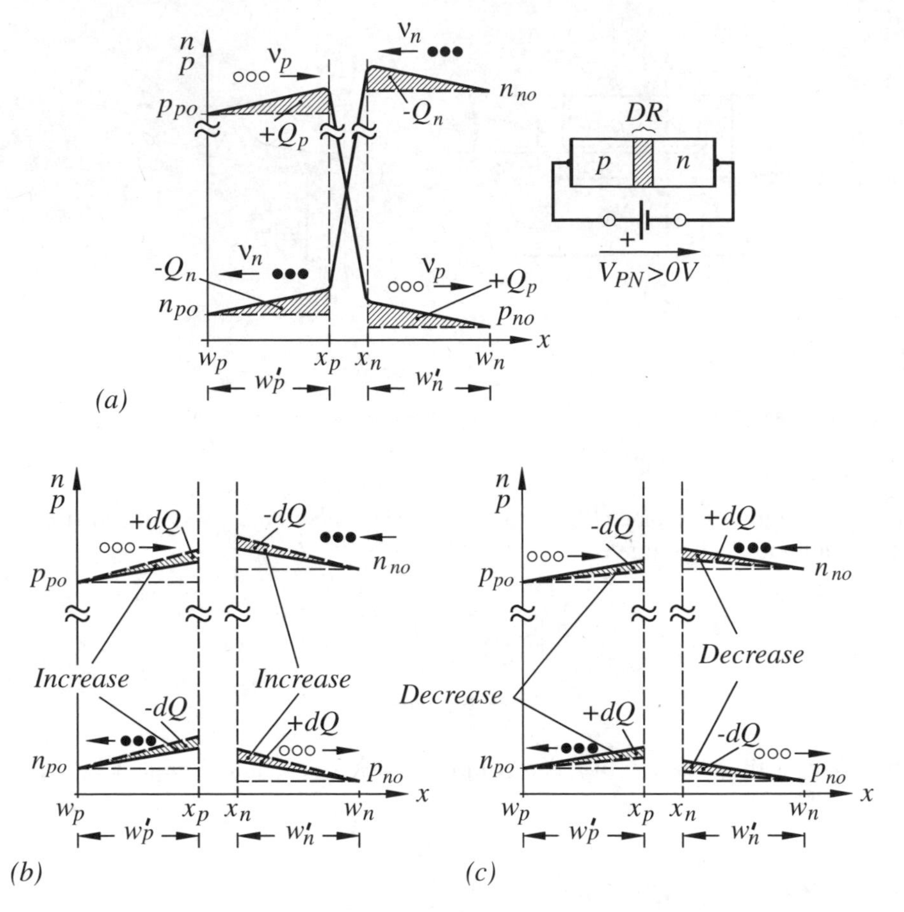

Figure 2.14 (a) Minority and majority carrier distribution, (b) charge change caused by $+dV_{PN}$ and (b) charge change caused by $-dV_{PN}$ (not to scale)

is caused by the injected holes. Its value can be determined directly from Figure 2.14(a) (triangular area) to be

$$
\begin{aligned}
Q_p &= qA\frac{w_n'}{2}[p_n(x_n) - p_{no}] \\
&= qA\frac{w_n'}{2}p_{no}(e^{V_{PN}/\phi_t} - 1)
\end{aligned}
\tag{2.51}
$$

where Equation (2.18) is used to determine the minority carrier density at the edge of the depletion region. Using the hole current part of the saturation current (Equation 2.26)

$$
I_p = qAD_p\frac{p_{no}}{w_n'}(e^{V_{PN}/\phi_t} - 1) \tag{2.52}
$$

the positive charge in the n-region can be expressed as a function of this current

$$Q_p = \frac{(w'_n)^2}{2D_p} I_p$$
$$Q_p = \tau_p I_p \tag{2.53}$$

The constant

$$\boxed{\tau_p = \frac{w'^2_n}{2D_p}} \tag{2.54}$$

has the dimension of time. This can be interpreted in the following way. The electric current is defined (Equation 1.1) by

$$I = A\rho v = A\rho \frac{\mathrm{d}x}{\mathrm{d}t} = \frac{\mathrm{d}Q}{\mathrm{d}t}$$

where $\mathrm{d}x$ is the distance which the charge of density ρ passes in the time $\mathrm{d}t$. This leads to an average time of

$$t = \frac{\int \mathrm{d}Q}{I} = \frac{Q}{I} \tag{2.55}$$

which the charge needs to pass a finite distance. Applying this to the n-region of the junction allows the following interpretation. The holes need an average time of τ_p to pass the distance w'_n. That is the reason why the time τ_p is called the transit time of the holes.

The positive charge in the p-region is caused by the majority carriers. Since the density distribution of the minority carriers – the electrons – can be determined, and the majority carriers assume – within the time of the dialectic relaxation – a distribution which correlates to that of the minority carriers, the excess hole densities can be expressed as

$$Q_p = -Q_n = \tau_n I_n \tag{2.56}$$

where

$$\boxed{\tau_n = \frac{w'^2_p}{2D_n}} \tag{2.57}$$

describes the transit time of the electrons. The total positive charge in the p-region and n-region is therefore

$$Q = \tau_n I_n + \tau_p I_p \tag{2.58}$$

This charge can be expressed with respect to the junction current (Equation 2.25)

$$I = I_n + I_p$$

in the form

$$Q = \tau_T I \tag{2.59}$$

In this case τ_T can be interpreted as the effective transit time of the pn-junction. The transit time plays an important role in the switching and frequency behavior of bipolar transistors, as will be discussed in Chapter 3. Its dependence on currents and the individual transit time of holes and electrons results directly from the previous equation

$$\tau_T = \tau_n \frac{I_n}{I} + \tau_p \frac{I_p}{I} \tag{2.60}$$

A voltage change causes charge variations in the n-region and p-region. This leads to a small-signal diffusion capacitance according to the definition

$$C_d = \frac{\mathrm{d}Q}{\mathrm{d}V_{PN}} \tag{2.61}$$

Using the positive charge, described by Equation (2.59)

$$\begin{aligned} Q &= \tau_T I \\ &= \tau_T I_S \left(e^{V_{PN}/\phi_t} - 1 \right) \end{aligned} \tag{2.62}$$

a diffusion capacitance of

$$\boxed{C_d = \frac{\tau_T}{\phi_t} I_S e^{V_{PN}/\phi_t}} \tag{2.63}$$

results.

The total small-signal pn-junction capacitance, consisting of a diffusion and depletion part, is thus

$$C = C_d + C_j$$

$$\boxed{C = \frac{\tau_T}{\phi_t} I_S e^{V_{PN}/\phi_t} + C_{jo} \left(1 - \frac{V_{PN}}{\phi_i} \right)^{-M}} \tag{2.64}$$

The diffusion capacitance depends exponentially on the V_{PN} voltage. As a result, its capacitance value under reverse biased condition ($V_{PN} < 0$) is small, in comparison to the depletion capacitance. This situation is completely reversed under forward biased conditions ($V_{PN} > 0$), where the diffusion capacitance dominates (Problem 2.6).

So far, generation and recombination are neglected due to the small semiconductor volume assumed. When this is not the case, as with long dimensions – this is discussed in Section 1.5 in the extended view of the experiment – the behavior of the charge carriers as a function of the currents can be expressed by (Problem 2.8)

$$Q = \tau_n I_n(x_p) + \tau_p I_p(x_n) \tag{2.65}$$

This equation is identical to Equation (2.58), but the interpretation differs, since τ_n and τ_p describe the lifetime of the charge carriers. This is important for power semiconductors, where large dimensions are used to achieve high breakdown voltages.

2.5 SWITCHING CHARACTERISTIC

In the preceding sections the depletion and diffusion capacitances of the pn-junction are described and it is pointed out that the diffusion capacity dominates in the forward biased direction, due to the exponential dependence on the voltage. How this influences the switching performance of a pn-junction will be explained next. For this purpose a pn-junction is forward biased by a V_F voltage (Figure 2.15(a)), which causes a current flow of I_F and a respective carrier distribution. The positive charge injected into the pn-junction (Equation 2.59) is

$$Q = \tau_T I_F \tag{2.66}$$

In other words, the diffusion capacitance is charged to a value of V_F. In comparison to this injected charge, that of the depletion capacitance can be neglected. At time $t = 0$ the pn-junction is brought by switch S into the reverse biased direction. Instantly, a perhaps unexpected, large reverse current of

$$I_R \approx V_R/R \tag{2.67}$$

flows, limited by the resistor R only. Hereby it is assumed that $|V_R| \gg V_F$. This behavior is caused by the injected charge during the forward biasing. As long as the charge carriers are present at the edges of the depletion region, the voltage across this region does not change noticeably. At time t_S the carriers at the edges of the depletion region are reduced to their equilibrium condition by the current I_R. The voltage across the pn-junction V_{PN} approaches 0. The depletion capacitance is discharged, and the junction starts to move into the reverse biased direction. From this point on, the current I_R charges the depletion capacitance, which dominates in the reverse direction. Finally the capacitance is charged to a value of V_R and the current I_R reaches the value of the saturation current I_S. The time t_S is called the storage time and is determined next. In this derivation is assumed that the change of charge is caused by the reverse current I_R.

$$\frac{dQ}{dt} = I_R \tag{2.68}$$

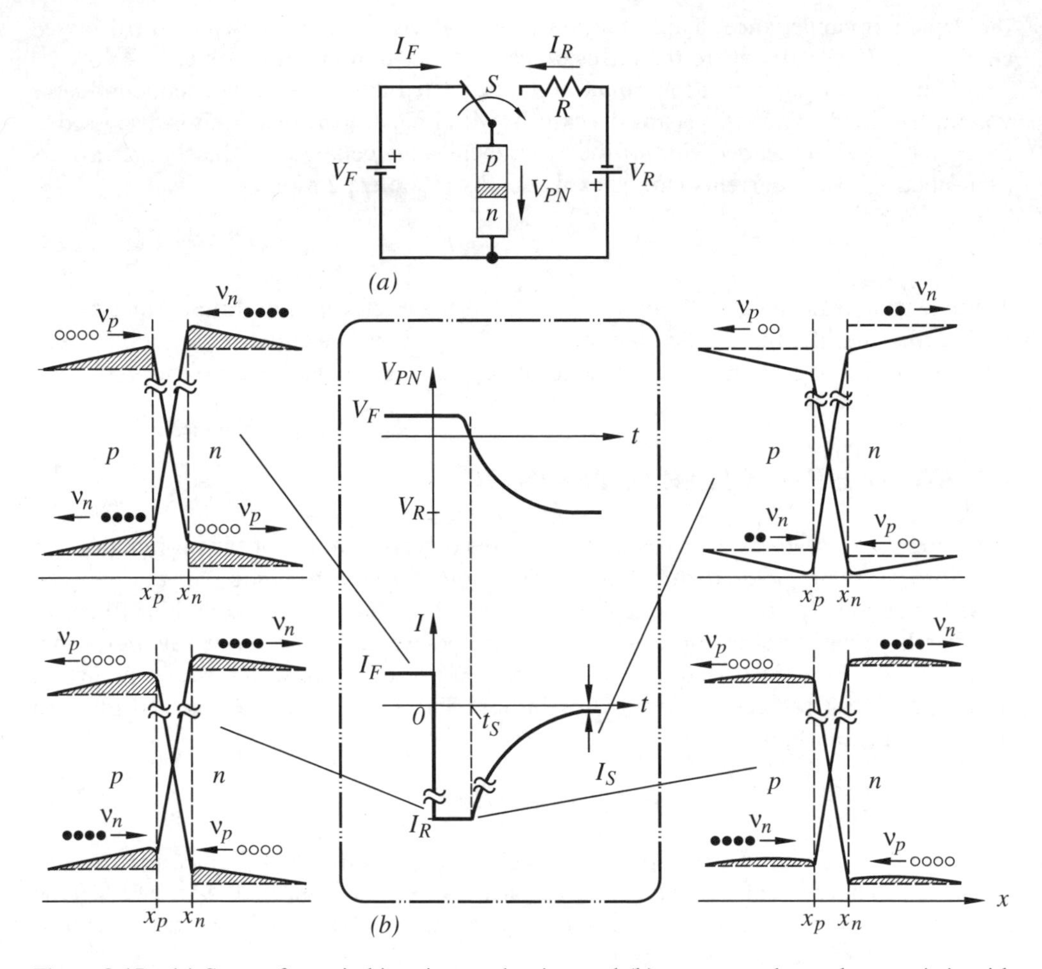

Figure 2.15 (a) Set-up for switching time evaluation and (b) current–voltage characteristic with carrier distribution sketches (not to scale)

Separating the variables and integrating

$$\int_0^{t_s} \mathrm{d}t = \frac{1}{I_R} \int_Q^0 \mathrm{d}Q \tag{2.69}$$

leads together with Equation (2.66) to a storage time of

$$\boxed{t_S = -\frac{Q}{I_R} = -\tau_T \frac{I_F}{I_R}} \tag{2.70}$$

This equation states that the storage time is proportional to the ratio between the current in the forward direction (charge injected) and reverse direction (charge extracted). This fact is important for circuit designers, dealing with the switching performance of bipolar circuits.

A reduction in storage time, particularly when large dimensions are present, can be achieved by introducing energy states into the forbidden gap of the semiconductor (Figure 1.26). This causes an increased recombination rate and, in conjunction, a reduction in the effective transit time (Equation 2.66). A more detailed analysis, including generation and recombination processes, is published in Kraus *et al.* (1992) and Goebel (1994).

2.6 JUNCTION BREAKDOWN

The pn-junction shows in the reverse biased direction a strong increase in current, beginning at the so-called breakdown voltage BV (Figure 2.16). Either an avalanche or a tunnel effect is responsible for the breakdown.

Avalanche breakdown

In a reverse biased pn-junction a saturation current I_S flows. If the electrical field is large enough, free charge carriers in the depletion region are able to gain sufficient energy from the field to break covalent bonds in the lattice of the crystal. Each carrier interacting this way creates two additional carriers, namely an electron and a hole. The two additional carriers then can participate in breaking off further covalent bonds. Since the carriers lose energy when colliding but do not vanish, a carrier multiplication occurs leading to an avalanche breakdown (Figure 2.17(a)).

Zener breakdown

With increasing doping concentration, the width of the depletion region decreases (Equation 2.43) and the probability that avalanche multiplication occurs is reduced. But due to the high doping concentration (Figure 2.17(b)) a tunnel mechanism, called Zener breakdown, is more probable. In this case the field strength is so high that it may exert sufficient force on covalent bonds to make valence electrons tunnel – due to their wave nature – from the valence band of the p-region through the energy barrier of the

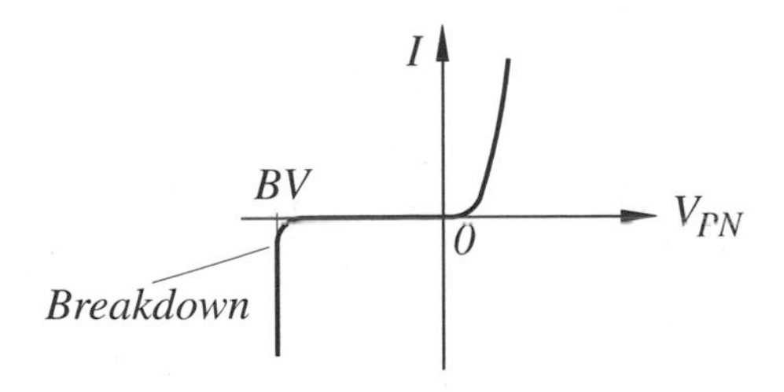

Figure 2.16 Breakdown behavior of a pn-junction

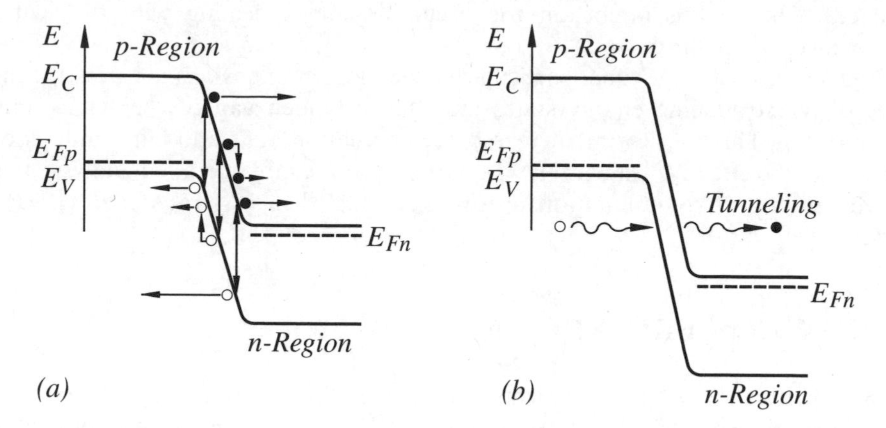

Figure 2.17 Band diagram under reverse biased condition: (a) avalanche effect and (b) tunnel effect

forbidden gap to the conduction band of the n-region. This process creates electrons and holes responsible for the current at breakdown.

The current flow at breakdown is not destructive, unless the high current causes the temperature to exceed its allowed value, causing some kind of melting. The current can be expressed in the form

$$I_{SM} = M \cdot I_S \tag{2.71}$$

where I_S is the saturation current and M a factor expressing the carrier multiplication. An empirical approximation

$$M = \frac{1}{1 - \left(\frac{V_{PN}}{BV}\right)^n} \tag{2.72}$$

is often used for describing the carrier multiplication process. According to this equation, breakdown occurs, when $V_{PN} = BV$ and M approaches infinity.

The starting point for the determination of the breakdown voltage is Figure 2.18. According to Poisson's equation, the electrical field can be found by integrating over the depletion region charge (Equation 2.35)

$$\mathcal{E}(x) = \frac{1}{\varepsilon_o \varepsilon_{Si}} \int_{x_p}^{x_n} \rho(x)\, dx$$

Integrating again – this time over the electrical field (Equations 2.37 and 2.38) – leads to

$$\phi_i - V_{PN} = -\int_{x_p}^{x_n} \mathcal{E}(x)\, dx = -\frac{1}{2}\mathcal{E}_M w$$

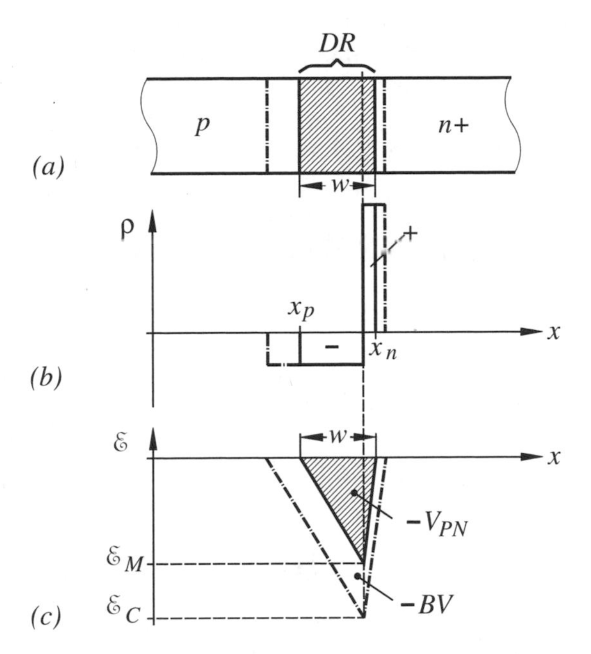

Figure 2.18 (a) Step junction at different voltages ($-V_{PN}$ and $-BV \gg \phi_i$), (b) depletion region charge and (c) electrical field

As explained in Section 2.4.1, *the triangular area of the electrical field in the diagram represents the voltage across the depletion region* (Figure 2.18(c)). If the V_{PN} voltage is increased – in the negative direction – the width of the depletion region extends and the maximum field at the metallurgical junction $\mathcal{E}_M$ increases. When this field approaches the critical field $\mathcal{E}_C$, breakdown occurs at a voltage of

$$BV \approx \frac{1}{2}\mathcal{E}_C w \tag{2.73}$$

The critical field of an ideal step junction (Figure 2.18) depends on the lower doped region of the junction, e.g. of the p-region, and can be determined according to Baliga (1987) by

$$\frac{\mathcal{E}_C}{[V/\text{cm}]} = -4010\left(\frac{N_A}{[\text{cm}^{-3}]}\right)^{1/8} \tag{2.74}$$

Assuming a doping density of $N_A = 10^{14}\,\text{cm}^{-3}$, a critical field of $\mathcal{E}_C \approx -2 \cdot 10^5\,\text{V/cm} = -2 \cdot 10\,\text{V/}\mu\text{m}$ results. This leads, according to Equation 2.73, to a breakdown voltage of $BV \approx -(10\,\text{V/}\mu\text{m}) \cdot w$. This means that *each additional increase in the depletion region width by 1 μm increases the breakdown voltage by 10 V*. The width of the depletion region is according to Equation (2.43)

$$w = \sqrt{\frac{2\varepsilon_0\varepsilon_{Si}}{q}\left(\frac{1}{N_A} + \frac{1}{N_D}\right)(\phi_i - BV)}$$

It is obvious from this equation that the width, and therefore the breakdown voltage, increases when the doping concentration of at least one side of the pn-junction is reduced. Doping densities as low as 10^{13} cm^{-3} are in use when a high breakdown voltage is required.

Substituting of Equation (2.43) into Equation (2.73) gives a breakdown voltage of

$$BV \approx -\frac{1}{2}\frac{\varepsilon_o \varepsilon_{Si}}{q}\left(\frac{1}{N_A}+\frac{1}{N_D}\right)\mathcal{E}_C^2 \tag{2.75}$$

This important relationship has the utmost influence on modern semiconductor technologies. In the course of scaling geometry dimensions, the doping densities must be increased. This causes a reduction in the breakdown voltage and requires therefore a reduction of the applied power-supply voltage.

2.7 MODELING THE PN-JUNCTION

In order to design or analyze the performance of integrated circuits, a circuit simulation has to be performed. This is executed by a simulation program, which needs a model with appropriate parameters, in order to describe the electrical behavior of the semiconductor device. Usually these models are very complex, and are a subject all by themselves. In order to convey the idea behind it, the model of a pn-junction – or a diode as a discrete component – is presented. The considered equations are then modified in order to be useful for static and dynamic hand calculations in a circuit environment.

2.7.1 Diode Model for CAD Applications

This model (Figure 2.19) describes the static and dynamic current–voltage behavior of a diode. It consists of a voltage-controlled current generator, described by Equation (2.31)

$$I' = I_S\left(e^{V'_{PN}/\phi_t N} - 1\right)$$

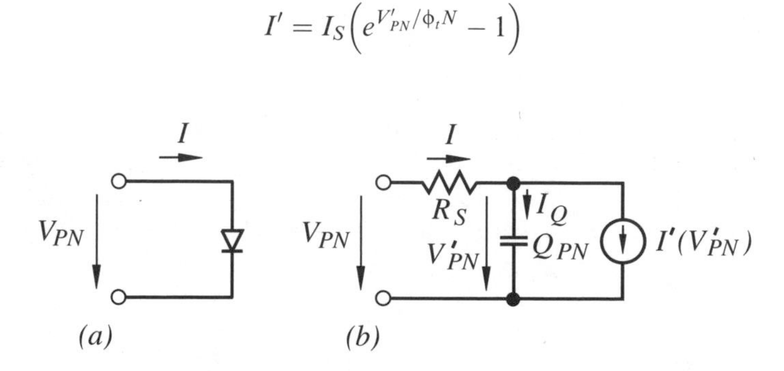

Figure 2.19 (a) Diode and (b) equivalent diode circuit

and a resistor R_S, which takes into account the voltage drops across the n-region and p-region of the junction. This reduces the effective voltages to V'_{PN}. The total charge stored in the junction (Equation 2.64) is modeled by a capacitor with a charge of $Q_{PN} = \int C\mathrm{d}V$, which leads to

$$Q_{PN} = \tau_T I_S\left(e^{V'_{PN}/\phi_t N} - 1\right) + C_{jo}\int_0^{V'_{PN}}\left(1 - \frac{V}{\phi_i}\right)^{-M}\mathrm{d}V \tag{2.76}$$

where the emission factor N is used to adapt the charge equation to that of the current–voltage. The current into the diode is, according to Figure 2.19(b),

$$\begin{aligned} I &= I'(V'_{PN}) + \frac{\mathrm{d}Q_{PN}(V'_{PN})}{\mathrm{d}t} \\ &= I'(V'_{PN}) + C(V'_{PN})\frac{\mathrm{d}V'_{PN}}{\mathrm{d}t} \end{aligned} \tag{2.77}$$

The term $\mathrm{d}Q_{PN}/\mathrm{d}t$ takes care of the fact that the charge in the diode can increase or decrease. In order to demonstrate the application of the model in a circuit simulator, a simplified example is presented.

Example

At time $t = 0$ a voltage step is applied to the diode by closing switch S (Figure 2.20). The diode is charged via the resistor to a final value. The initial condition at the diode is $V_{PN}(t = 0) = 0\,\mathrm{V}$. Of interest is the transient behavior of the voltage across the diode. Since the diode is reverse biased during charging, the depletion capacitance is effective only, causing a current flow of

$$I_Q = C_{jo}\left(1 - \frac{V_{PN}}{\phi_i}\right)^{-M}\frac{\mathrm{d}V_{PN}}{\mathrm{d}t} \tag{2.78}$$

In order to find the charging current as a function of time, the equation is discretized

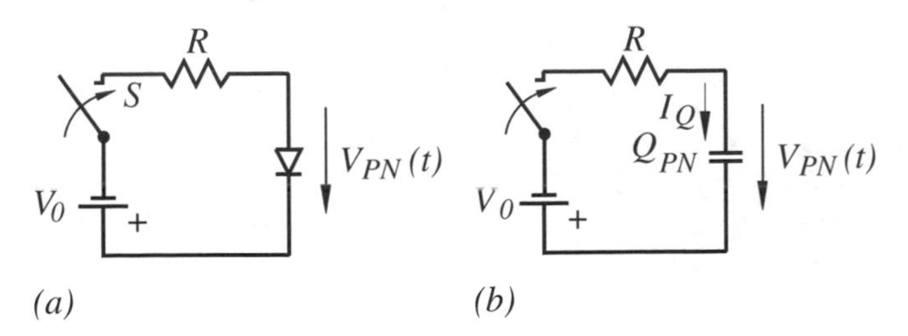

Figure 2.20 (a) Circuit set-up and (b) equivalent circuit of a reverse biased diode

$$I_Q^{n+1} = C_{jo}\left(1 - \frac{V_{PN}^n}{\phi_i}\right)^{-M} \frac{V_{PN}^{n+1} - V_{PN}^n}{\Delta t} \tag{2.79}$$

The current at time $t = n + 1$ is calculated by the difference between the voltages at time $t = n + 1$ and $t = n$. The voltage V_{PN}^n is known and therefore the capacitance value also. The capacitance is considered to be constant in the time interval Δt. In other words, the capacitance value is piecewise linear approximated. Δt can be reduced to suit the required accuracy. By using

$$V_{PN}^{n+1} = V_0 - RI_Q^{n+1} \tag{2.80}$$

the discretized charging behavior of the diode can be found

$$V_{PN}^{n+1} = \frac{V_0 + R\dfrac{C_{jo}}{\Delta t}\left(1 - \dfrac{V_{PN}^n}{\phi_i}\right)^{-M} V_{PN}^n}{1 + R\dfrac{C_{jo}}{\Delta t}\left(1 - \dfrac{V_{PN}^n}{\phi_i}\right)^{-M}} \tag{2.81}$$

With the values $V_{PN}(t = 0) = 0\,\text{V}$, $\phi_i = 0.7\,\text{V}$, $\Delta t = 1.0 \cdot 10^{-9}\,\text{s}$, $M = 0.5$, $C_{jo} = 1\,\text{pf}$, $V_0 = -5\,\text{V}$, and $R = 5\,\text{k}\Omega$ the charging characteristic shown in Figure 2.21 results.

The example presented is particular simple, since it requires a discretization and no iteration. Further discussions are found in Calahan (1972).

An example of diode parameters which are required for the circuit simulation program SPICE 2G (Nagel 1975) is shown in Table 2.1.

2.7.2 Diode Model for Static Calculations

The static behavior of the diode is described by the current–voltage relationship Equation (2.31)

$$I' = I_S\left(e^{V'_{PN}/\phi_t} - 1\right)$$

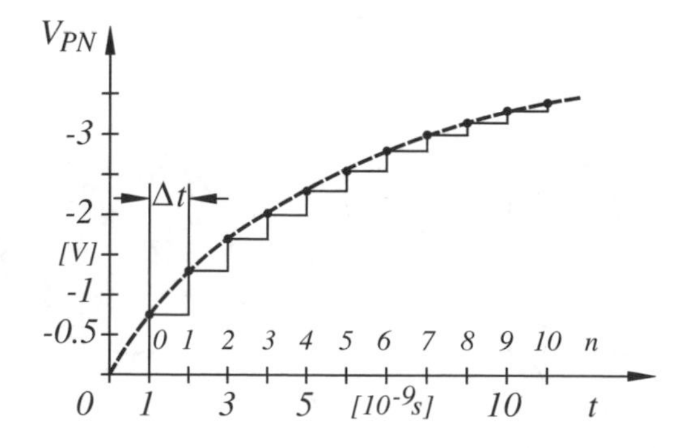

Figure 2.21 Charge behavior of a reverse biased diode

Table 2.1 Parameters of an n^+ p-Si-junction at room temperature

Name	SPICE	Parameter	Example	Unit
I_S	IS	Saturation current	10^{-16}	A
R_S	RSH	Sheet resistance	80	Ω/□
N	N	Emission coefficient	1	
τ_T	TT	Effective transit time	18	μs
C'_{jo}	CJO	Depletion capacitance at $V_{PN} = 0\,\text{V}$ (area)	0.85	fF/μm²
M	M	Grading coefficient (area)	0.32	
C^*_{jo}	CJSW	Depletion capacitance at $V_{PN} = 0\,\text{V}$ (side wall)	0.12	fF/μm
M	M	Grading coefficient (side wall)	0.18	
ϕ_i	PB	Built-in voltage	0.75	V

When one uses this equation in circuit calculations a transcendental equation may result, which can be solved iteratively only. This is no problem for a CAD system, but makes hand estimation awkward, as demonstrated in the following example.

Example

A diode is connected via a $5\,\text{k}\Omega$ resistor to a voltage of $V_0 = 5\,\text{V}$ (Figure 2.22). How large is the current I?

In the forward direction with $V_{PN} > 100\,\text{mV}$ the current–voltage equation reduces to

$$I \approx I_S e^{V_{PN}/\phi_t}$$

With

$$I = \frac{V_0 - V_{PN}}{R} \tag{2.82}$$

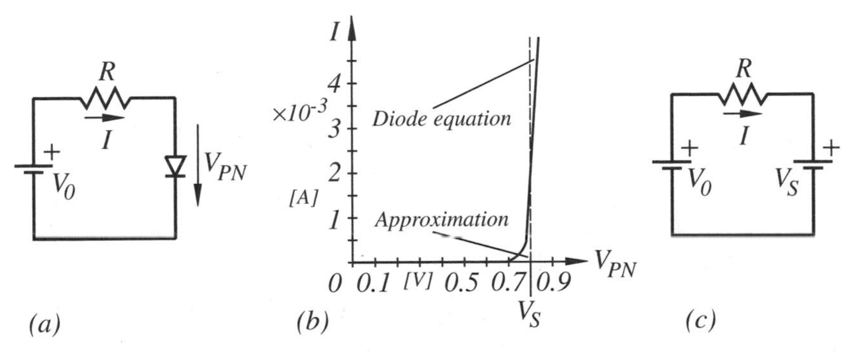

Figure 2.22 (a) Circuit set-up, (b) approximated I–V equation and (c) simplified equivalent circuit

a current of

$$I = \frac{V_0 - \phi_t \ln \frac{I}{I_S}}{R} \tag{2.83}$$

results. This is a transcendental equation, which can only be solved iteratively. In order to avoid this, the diode behavior is approximated in the forward direction by a voltage source with value V_S. The simplification leads to a current of

$$I = \frac{V_0 - V_S}{R} = \frac{5\,\mathrm{V} - 0.8\,\mathrm{V}}{5\,\mathrm{k\Omega}} = 0.84\,\mathrm{mA}$$

In the example it is relatively unimportant whether a value of 0.75 V or 0.85 V is used for the voltage source V_S since V_0 is, at 5 V, relatively large compared to those values. If this is not the case, an iterative solution is unavoidable.

2.7.3 Diode Model for Small-Signal Calculations

Analog circuits usually operate with very small signal levels compared to the bias voltages. Under these circumstances incremental or small-signal models are used for calculations. Figure 2.23 illustrates the derivation of such a model. For this purpose an incremental input signal voltage ΔV_{PN} is applied in series to a bias voltage V_{PN} to the diode. At a fixed bias point A (Figure 2.23(b)), the I–V characteristic can be considered to be linear. This leads to a conductance of

$$\begin{aligned} g_o &= \frac{\partial I}{\partial V'_{PN}} \approx \frac{\Delta I}{\Delta V'_{PN}} \\ &= \frac{I_S}{\phi_t N} e^{V'_{PN}/\phi_t N} \\ &\approx I/\phi_t N \end{aligned} \tag{2.84}$$

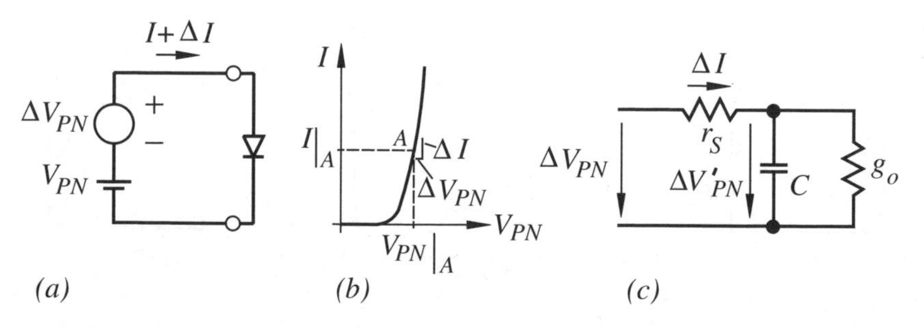

Figure 2.23 (a) Small-signal excitation of a forward biased diode, (b) current–voltage characteristic and (c) small-signal equivalent circuit

which is proportional to the applied current I. The incremental or small-signal capacitance can be found directly from Equation 2.64 for the given biased condition. The resistor r_S takes into account the voltage drops caused by the small-signal voltage ΔV_{PN} on the n-region and p-region of the diode.

The small-signal equivalent circuit is also valid when instead of an incremental voltage change a sinusoidal small-signal voltage is applied.

Summary of the most important results

The injection and extraction of a forward and reverse biased pn-junction with short dimensions is analyzed and the current–voltage relationship derived. A prerequisite for the analyses is the knowledge of the density product under non-equilibrium conditions. It turns out that the density product at the edges of the depletion region is exponentially dependent on the applied V_{PN} voltage. With a voltage of $V_{PN} < -100\,\text{mV}$ the minority carrier density at the edges of the depletion region cannot be reduced any further, causing the current to saturate at a very small value. This is contrary to the situation, when $V_{PN} > 0\,\text{V}$. In this case the minority carriers at the edges of the depletion region can be increased, and therefore the current, until the temperature exceeds its allowed value.

Two totally different physical effects lead to a depletion and diffusion capacitance. The first is based on the effect that the charge of the depletion region changes with the applied voltage, whereas the second is caused by the injection of charge carriers into the n-region and p-region.

The pn-junction breakdown is caused either by an avalanche or a tunnel effect. By increasing the depletion width, the breakdown voltage can be raised. A rule of thumb is that with each additional increase in the depletion width by 1 μm the breakdown voltage increases by about 10 V.

Problems

2.1 A pn-step junction has the doping concentrations $N_A = 10^{15}\,\text{cm}^{-3}$ and $N_D = 2 \cdot 10^{17}\,\text{cm}^{-3}$. Calculate:

(a) The built-in voltage at room temperature.

(b) The width of the depletion region.

(c) The maximum electrical field for an applied voltage of $V_{PN} = 0\,\text{V}$ and $-10\,\text{V}$.

2.2 In a pn-junction drift and diffusion currents compensate each other under equilibrium conditions. Estimate one current component by using the following data: $N_A = 10^{18}\,\text{cm}^{-3}$, $N_D = 5 \cdot 10^{15}\,\text{cm}^{-3}$, depletion region width $6 \cdot 10^{-6}\,\text{cm}$, hole mobility $500\,\text{cm}^2/\text{Vs}$.

2.3 In Section 2.1 an inhomogeneously doped n-type semiconductor is described. What does the current–voltage characteristic look like qualitatively?

2.4 Calculate the depletion capacitance at the bottom and side wall for the shown pn-junction at an applied voltage of $V_{PN} = 0\,\text{V}$. The doping profile can be approximated by a step function.

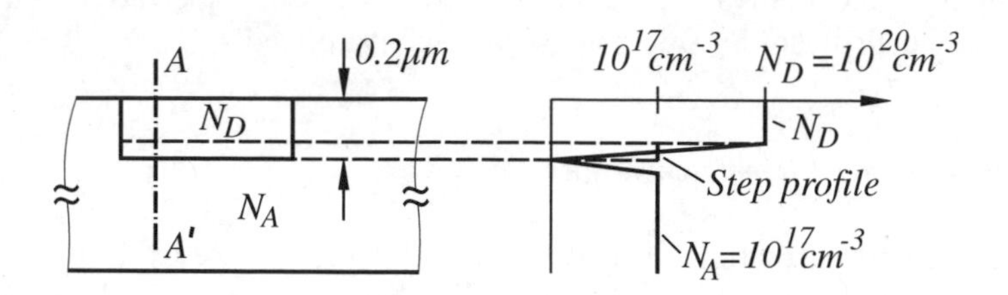

2.5 A pn-junction is influenced by a built-in voltage. Does a current occur when the junction is short-circuited?

2.6 The goal of this problem is to develop a feeling about the parameters of a base emitter junction of a bipolar transistor. The data are: N_D (emitter) $= 5 \cdot 10^{19}\,\text{cm}^{-3}$, $w'_n = 0.2\,\mu\text{m}$, N_A (base) $= 5 \cdot 10^{17}\,\text{cm}^{-3}$, $w'_p = 0.2\,\mu\text{m}$, $D_p = 12\,\text{cm}^2/\text{s}$, $D_n = 21\,\text{cm}^2/\text{s}$, emitter area $A_E = 1\,\mu\text{m}^2$.
Calculate:

(a) The width of the depletion region at $V_{PN} = 0.80\,\text{V}$.

(b) The saturation current.

(c) The current under forward biased conditions at $V_{PN} = 0.80\,\text{V}$.

(d) The depletion and diffusion capacitances at V_{PN} voltages of $0.80\,\text{V}$ and $0\,\text{V}$.

2.7 In an integrated circuit a relatively large capacitor is needed for noise suppression. For this purpose an n-well/p-substrate junction is used. The data are: $N_D = 5 \cdot 10^{17}\,\text{cm}^{-3}$, $N_A = 10^{17}\,\text{cm}^{-3}$, $D_p = 12\,\text{cm}^2/\text{s}$, $D_n = 21\,\text{cm}^2/\text{s}$, $V_{PN} = 0.8\,\text{V}$, $A = 0.1\,\text{mm}^2$.
Calculate at room temperature the current, and the diffusion and junction capacitances by $V_{PN} = 0.8\,\text{V}$.

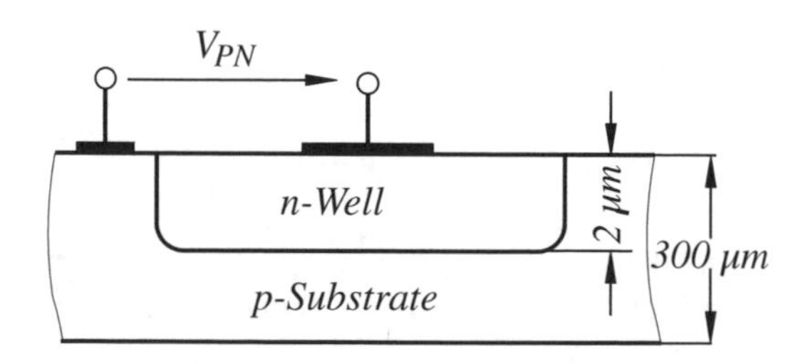

2.8 Derive the equation for the diffusion capacitance, when large dimensions are present. For large dimensions it can be assumed that $w'_n \gg L_p$ and $w'_p \gg L_n$.

The generation and recombination in the depletion region can be neglected. In order to solve the problem study of the 'Extended view of the experiment' section of Section 1.5 is recommended.

The solutions to the problems can be found under: *www.unibw-muenchen.de/campus/ET4/index.html*

REFERENCES

B.J. Baliga (1987) *Modern Power Devices*. New York: John Wiley & Sons.

D.A. Calahan (1972) *Computer-Aided Network Design*. New York: McGraw-Hill.

H. Goebel (1994) A unified method for modeling semiconductor for power devices. *IEEE Trans. Power Electron.*, **9**(5), 497–505.

R. Kraus, K. Hoffman and H.J. Mattausch (1992) A precise model for the transient characteristic of power diodes. In *Proceedings of the 23rd Annual IEEE Power Electronics Specialists Conference (PESC 92)*, vol. 2, pp. 863–9.

L.W. Nagel (1975) SPICE 2: A computer program to simulate semiconductor circuits, memorandum no. ERL-M520, 9. Electronics Research Laboratory, University of California, Berkeley, CA.

S.M. Sze (1981) *Physics of Semiconductor Devices*. New York: John Wiley & Sons.

3

Bipolar Transistor

After a brief presentation of two typical bipolar manufacturing processes the transistor action is presented. A simplified npn structure is used for the derivation of the transport equation. This equation is valid for both a uniform and a non-uniform base doping. The key terms *current gain* and *transport current* are introduced. Important second-order effects, e.g. high current injection, base-width modulation (Early effect), and current crowding, are considered. The chapter ends with a brief discussion of the modeling aspects of the transistor and a presentation of easy to use equations useful for first-hand circuit performance estimation.

3.1 BIPOLAR TECHNOLOGIES

A planar process consists of the selective introduction of dopant atoms into small areas of silicon from the surface of the wafer. The technology is called planar, because all fabrication steps are carried out from the surface of the wafer. The outstanding advantage of planar technology is that millions of separate small doping regions can be fabricated simultaneously on a wafer. This leads to a large number of related semiconductor devices, which are interconnected on one silicon chip to form an integrated circuit.

As an introduction some basic fabrication steps, such as film deposition, lithography, etching, and doping, are considered (Figure 3.1).

A silicon dioxide (SiO_2) layer is usually formed on the silicon wafer surface by a chemical reaction of silicon atoms with oxygen, which is allowed to flow over the semiconductor surface, while the wafer is heated to a very high temperature (Figure 3.1(a)). Once the protective SiO_2 layer has been formed, it must be selectively removed from those areas into which dopant atoms are to be introduced. The selective removal can be accomplished by the deposition of a light-sensitive photoresist. Then the photoresist is exposed via a mask to ultraviolet light, which causes the molecules in the photoresist to become chemically modified in the exposed areas (Figure 3.1(b)). The photoresist is then developed and selectively removed from the exposed areas. The unprotected regions are chemically etched to the bare silicon surface (Figure 3.1(c)). These regions are the locations into which the dopant impurities, e.g. arsenic, can diffuse. This happens in two steps. First, the atoms are placed on or near the surface, and, second, a drive-in diffusion is performed, which moves the dopant atoms further

System Integration: From Transistor Design to Large Scale Integrated Circuits. Kurt Hoffmann.
 ISBN: 0-470-85407-3

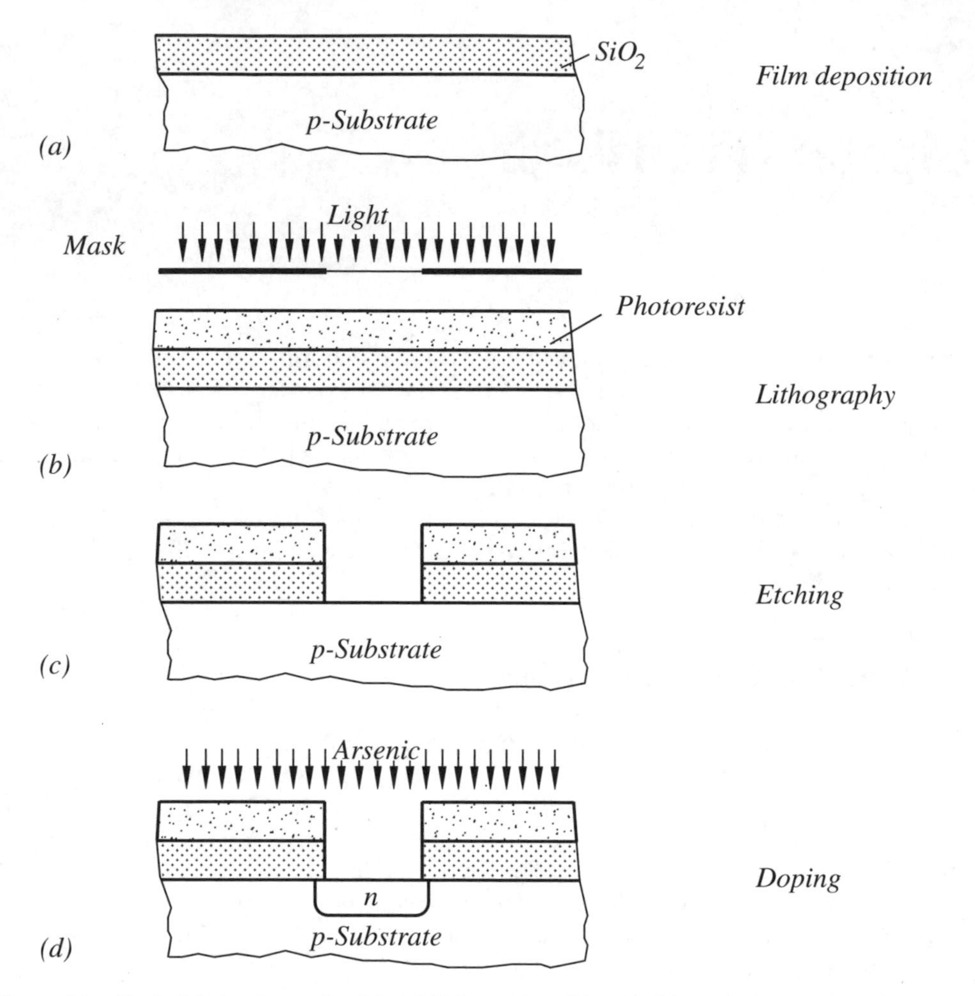

Figure 3.1 Basic fabrication steps: (a) oxide formation, (b) exposition of photoresist, (c) selective oxide and photoresist removal and (d) diffusion of dopant atoms into exposed silicon region

into the wafer when a sufficiently high temperature, e.g. 1050 °C, is applied. The dopant atoms diffuse due to their concentration gradient, from regions of high concentration near the surface toward regions of lower concentration, further into the wafer.

As an alternative technique ion implantation can be used to introduce dopant atoms into selected regions. The ionized dopant atoms are accelerated in an implanter by an electrical field to a high energy. A beam of these ions strikes the semiconductor surface and penetrates the wafer. Unfortunately the impact of the ions causes considerable damage to the semiconductor surface. As a result a large number of vacancies are formed, which turn the exposed crystalline regions into disordered layers. For this reason, it is necessary to anneal the semiconductor to reestablish the crystalline structure. This is done by placing the wafer into a high temperature environment. The advantage of ion implantation, compared to the diffusion technique, is that this process is better controllable and can be performed at lower temperature.

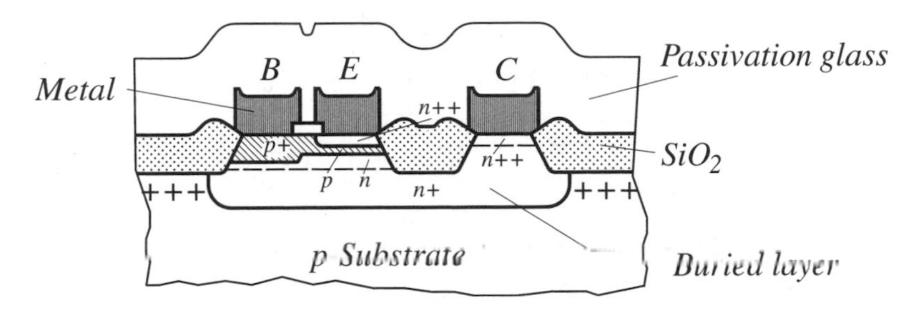

Figure 3.2 Cross-section through an npn transistor

A more complex structure, namely a cross-section through an npn transistor, is shown in Figure 3.2. Its fabrication steps will be covered after giving an overview of the transistor.

The transistor is vertically arranged and consists of an n^{++}-emitter (E) above a p-base (B) and an n-collector (C), connected to the surface via a buried layer. A single or double + sign above the n-type or p-type doping symbol indicates a high or very high doping concentration. Isolation between neighboring transistors is achieved by tying the common p-substrate to the most negative voltage of the circuit. The collector behaves therefore like a back-biased np-junction with respect to the substrate.

In Figure 3.3 the major process steps of a bipolar manufacturing process are shown in some detail.

Buried layer and epitaxy

The starting material is a lightly doped p-wafer with a doping concentration of about $N_A = 10^{15}\,\text{cm}^{-3}$. A silicon dioxide ($SiO_2$) layer is formed on the wafer by a chemical reaction of silicon atoms of the wafer with oxygen, which is allowed to flow over the surface while the wafer is heated to a high temperature in the range 900–1100 °C. Afterwards the wafer is covered with a photoresist and exposed to light via a mask which contains the information of the buried collectors. After developing the photoresist and some further processing steps, the SiO_2 layer is selectively removed by etching (Figure 3.3(a)). The remaining SiO_2 layer acts as a mask during the following implantation step with antimony. After removal of the SiO_2 layer an epitaxy layer with, e.g., $N_D{=}10^{16}\,\text{cm}^{-3}$ is grown (Figure 3.3(b)). Epitaxy is a technique where a crystal layer is grown duplicating the lattice structure of the original crystal. The reason for using this growth technique is the flexibility of impurity control in the epitaxial layer. In the case of the bipolar transistor, epitaxial growth is used to form a lightly doped film on the heavily doped buried layer.

Oxide definition and channel stopper

The LOCal Oxidation of Silicon (LOCOS) utilizes the property that oxygen diffuses through silicon nitride (Si_3N_4) extremely slowly. Therefore no oxide grows where the wafer is covered by this layer. The active regions of the transistor are defined by a

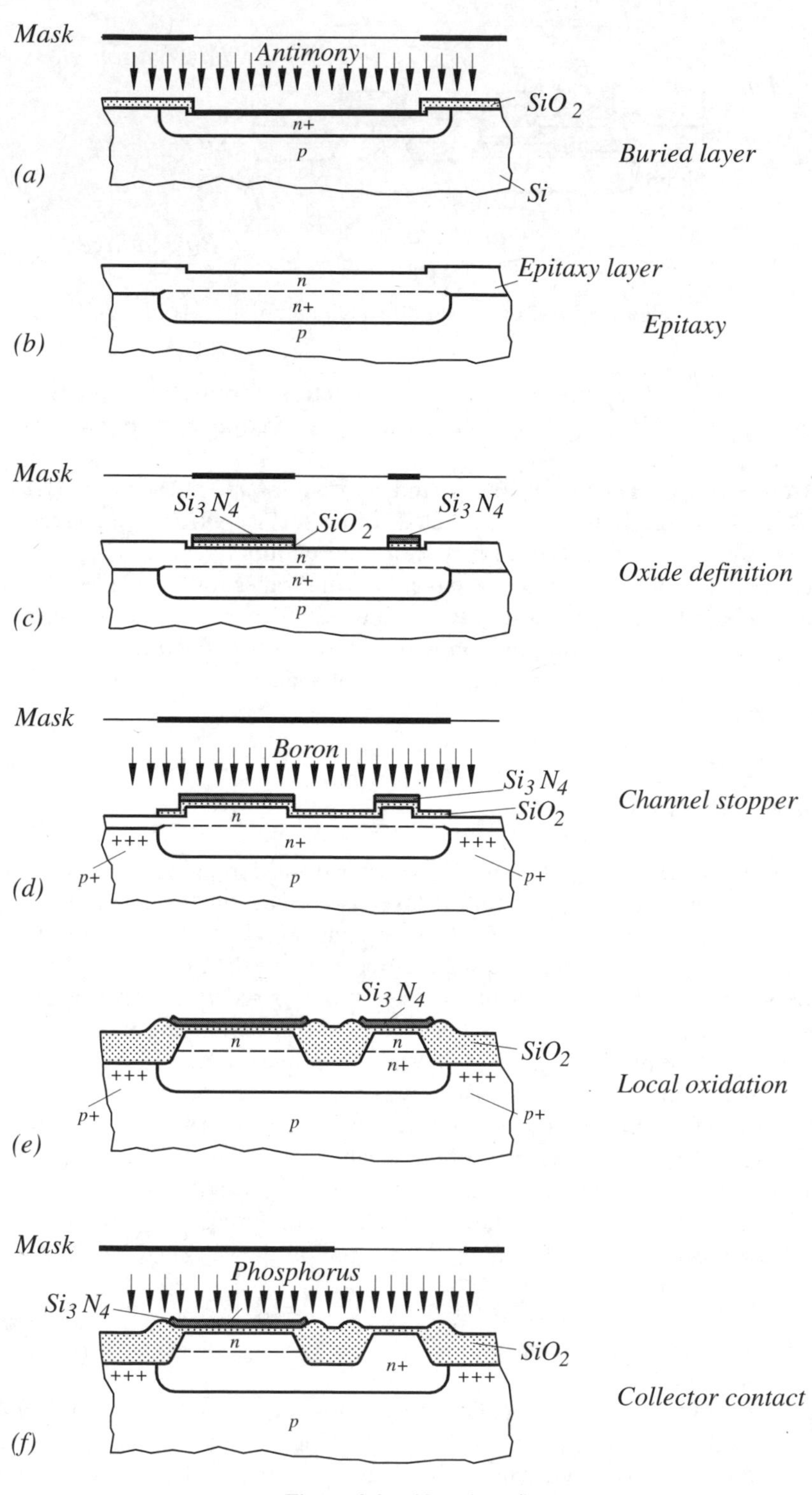

Figure 3.3 (Continued)

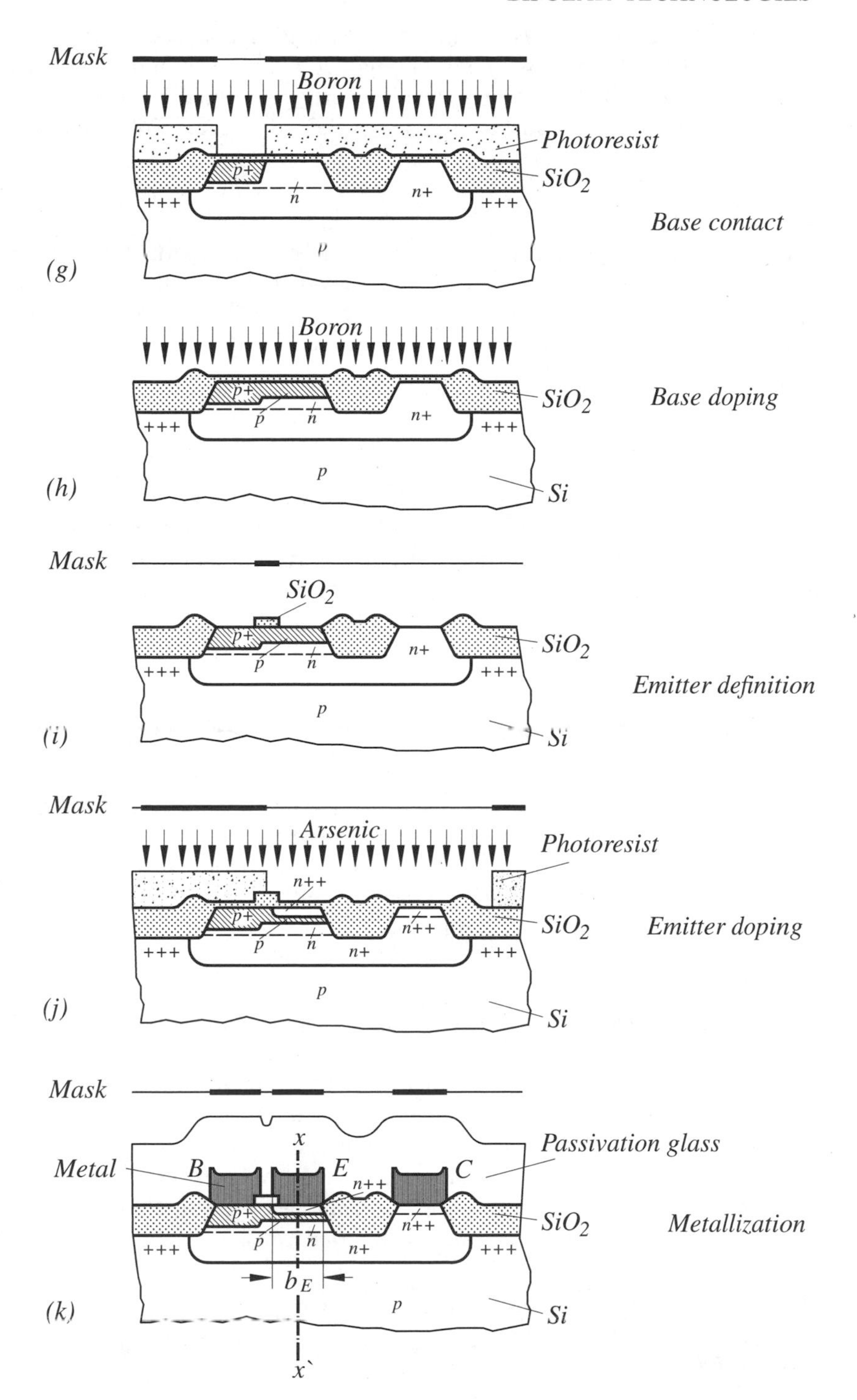

Figure 3.3 Basic fabrication steps of a bipolar process (cross-section of an npn transistor)

lithographic step (Figure 3.3(c)). The composite oxide nitride layers are then plasma-etched as a stack. The wafer is subsequently implanted with boron in order to generate channel stoppers (Figure 3.3(d)). The p^+-channel stoppers avoid parasitic n-channel transistors at the semiconductor surface. This is explained in detail in Section 4.3.3.

Local oxidation

In the following oxidation process, oxide growth is prevented where the wafer is covered by silicon nitride (Figure 3.3(e)). The thick oxide is used for the isolation between neighboring transistors and the separation of emitter and collector.

Collector contact

With an additional lithographic step the oxide is removed from the collector contact areas (Figure 3.3(f)). These windows are used in a following implantation step to fabricate low resistive n^+-collector contacts. A phosphorus implantation is applied since the dopant atoms migrate very far into the semiconductor.

Base contact and doping

Again, a lithographic step is used, but this time to open the base contact areas. By implanting a relatively high dose of boron in this area low resistive p^+-base contacts result (Figure 3.3(g)). In the following process step, the complete base areas are opened and implanted with a relatively low dose of boron (Figure 3.3(h)). This yields the actual p-base of each transistor, which is connected via a p^+-region to the surface of the wafer. The simultaneous doping of the collector contact areas has no effect, since the n^+-contact doping concentration is much larger than the p-base doping.

Emitter definition and doping

The emitter windows and the collector contacts are opened in the following lithographic step (Figure 3.3(i)). Subsequently an arsenic implantation in conjunction with an additional lithographic step is used to generate n^{++}-emitters and low resistive n^{++}-collector contacts (Figure 3.3(j)). Arsenic has the advantage that it diffuses relatively poorly into the semiconductor. This is desirable in order to fabricate very shallow emitters. The emitter diffusion pinches the base beneath the emitter, causing a relatively large sheet resistance of the base as a result.

Contacts and metallization

After oxide removal in the contact windows, metallic films are deposited over the wafer. This is done, e.g., by sputtering a titanium layer as well as an aluminum layer over the wafer. The titanium acts as a diffusion barrier between the aluminum and the silicon and avoids a reaction between the two layers. In the following lithographic step the metal wiring is defined and finally etched (Figure 3.3(k)). Then a passivation glass, e.g. silicon nitride (Si_3N_4), is added as scratch protection. In a final step (not shown) pads are opened for connections to the outside. The silicon wafer is then diced into separate pieces, each containing an integrated circuit. These chips are glued to a package, and wires are connected from the package leads to the pads on the chip. Finally, the package is sealed e.g. with plastic (some examples of plastic packages are shown in Figure 5.49).

A typical impurity concentration profile of this process is shown in Figure 3.4. Its impact on the transistor performance will be discussed in subsequent sections.

The presented fabrication process uses only a small number of lithographic steps and is therefore very cost effective. Due to the relatively large capacitances and resistances, particularly in the base area, this process is not suitable for high frequency applications. These applications can be covered by employing a double-polysilicon process to form a self-alignment between emitter and base. A substantial reduction in transistor area and therefore in capacitance and resistance values results (Figure 3.5).

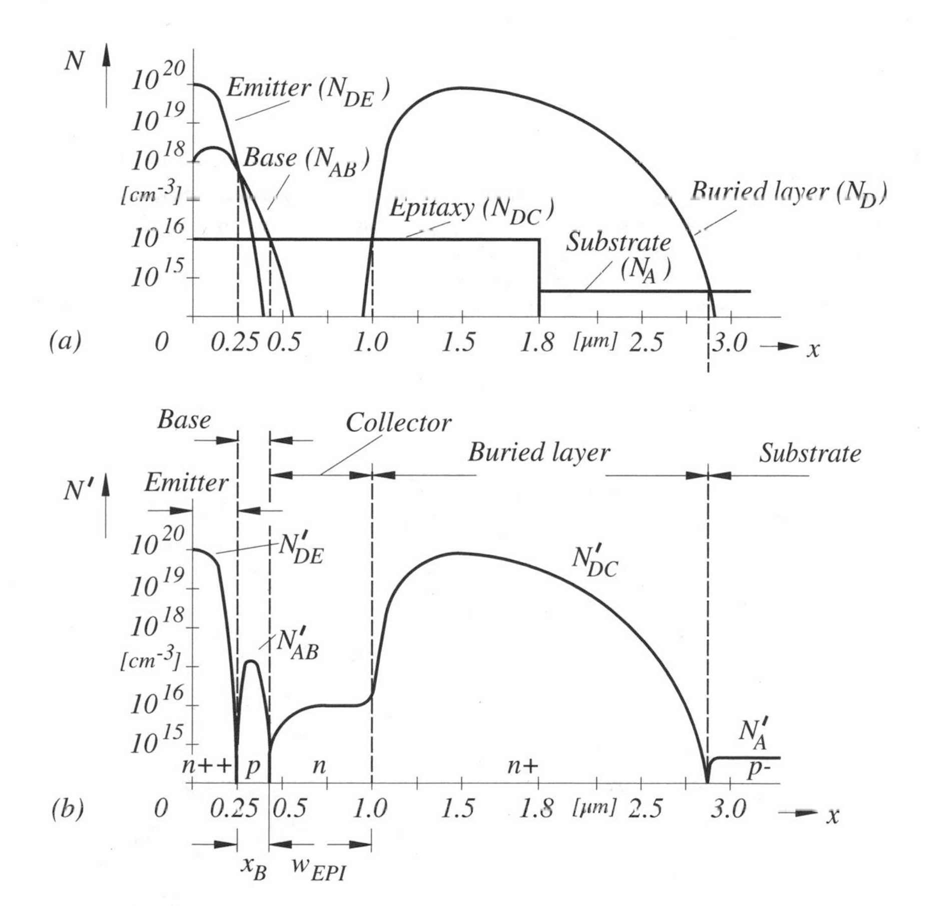

Figure 3.4 (a) Impurity concentration of a pnp transistor (cross-section x–x' Figure 3.3(k)) and (b) net impurity concentration $N'(x) = |N_D - N_A|$

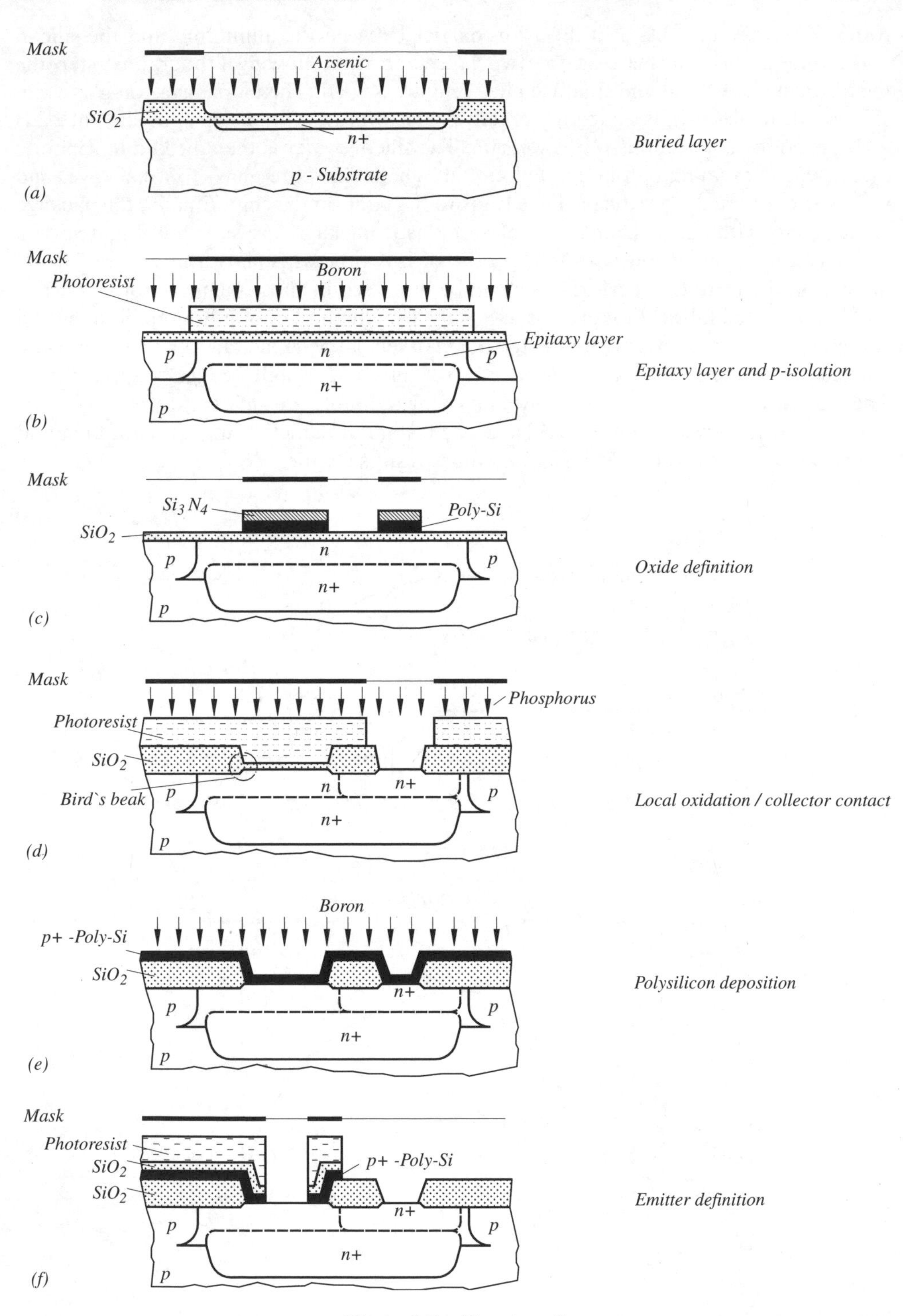

Figure 3.5 (Continued)

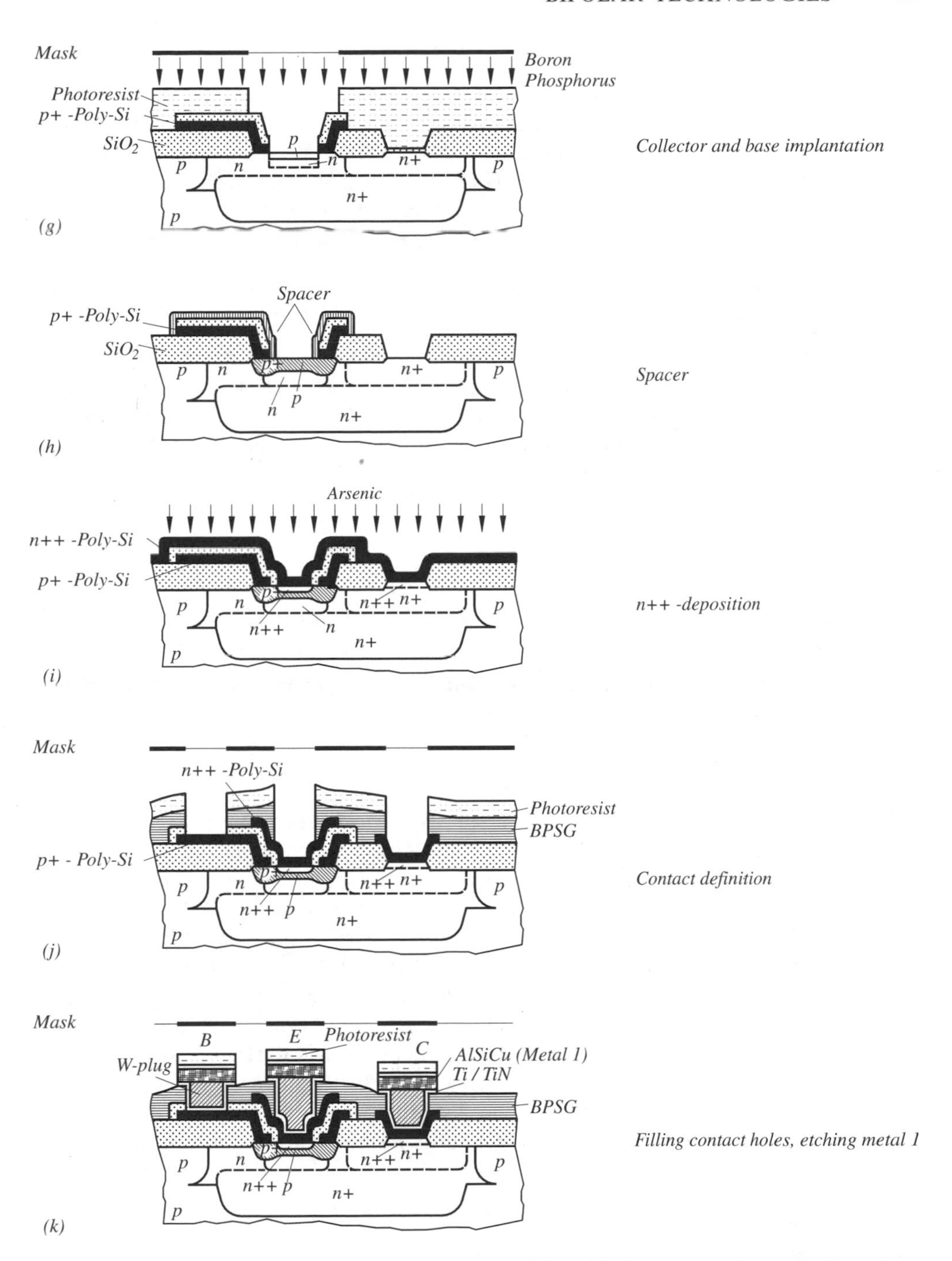

Figure 3.5 Basic fabrication steps of a double-polysilicon bipolar process (cross-section of an npn transistor)

Buried layer and epitaxy

The starting material is a p-doped silicon wafer with a resistivity of about 8 Ωcm. After the oxide (SiO_2) is grown and selectively removed from the surface, arsenic dopant atoms are implanted and n^+-regions are fabricated (Figure 3.5(a)). Then an n-epitaxy layer of approximately 1 μm thickness is grown (Figure 3.5(b)). A subsequent selective boron implantation is used to electrically isolate the n^+-collector contacts from the substrate.

Local oxidation

After stripping the oxide and photoresist, the wafer is covered with SiO_2, polysilicon (Poly-Si), and Si_3N_4 layers, and patterned (Figure 3.5(c)). With the additional polysilicon layer, the so-called bird's beak (Figure 3.5(d)) can be reduced during the following oxidation step. The thick oxide is used for the isolation of neighboring transistors and the separation of emitter and collector.

Collector contact

With an additional lithographic step the oxide is removed for the formation of collector contacts (Figure 3.5(d)). A phosphorus implantation is applied since the dopant atoms migrate far into the semiconductor. This yields low resistive n^+-collector contacts.

Emitter definition and base doping

The wafer is then covered with a polysilicon layer. In this layer there are many small regions, each having a well-organized atom structure, which differ from neighboring regions. This is contrary to a crystalline solid, where the atoms are all arranged in a three-dimensional orderly array. The polysilicon layer is then implanted with boron atoms. This layer serves as an interconnect to the p-base of the transistor, as is shown subsequently. Applying additional processing steps, the emitter areas are opened (Figure 3.5(f)). The collector areas are then covered in a subsequent lithographic step. This is followed by an implantation with phosphorus and boron (BF_2^+). The phosphorus atoms are responsible for the doping of the inner n-collector regions and the boron atoms for the inner p-base zones (Figure 3.5(g)). The key feature of the next process step is that at an elevated temperature the boron of the polysilicon layer diffuses laterally as well as vertically into the silicon. This generates a self-aligned p^+-contact to the inner p-base of each transistor (Figure 3.5(h)).

Emitter implantation

Before the emitter implantation, side wall spacer oxide is grown (Figure 3.5(h)). This self-aligned structure reduces the emitter areas to dimensions smaller than possible with conventional lithographic means. The wafer is then covered with a polysilicon layer, which is subsequently implanted with arsenic atoms to form a highly doped n^{++}-film. In an added temperature step the arsenic atoms diffuse from the polysilicon layer into the silicon. This process yields an n^{++}-emitter region self-aligned to the p-base of each transistor with a very shallow emitter of approximately 0.04 μm. Simultaneously low resistive collector contacts are created (Figure 3.5(i)).

Contacts and metallization

After structuring the n^{++}-polysilicon layer in emitter and collector regions (not shown) an SiO_2 layer is grown (not shown). This layer acts as a diffusion barrier between the silicon and the deposited boron phosphoros silicate glass (BPSG) film. The glass is used for isolation and smoothing of the edges at the semiconductor surface. A lithographic step is then applied for the definition of contacts. In a subsequent etching process the contact holes are open (Figure 3.5(j)) and covered with a Ti/TiN film. The approximately 20 nm thin Ti layer forms with the silicon a low resistive $TiSi_2$ contact, whereas the approximately 100 nm TiN layer acts as a barrier between the Si and the subsequently deposited tungsten. The trenches are filled with tungsten (W-plug) followed by metallization with AlSiCu and a subsequent patterning step (Figure 3.5(k)). In order to increase the packing density multiple independent metal layers – as described in Section 4.1 for a MOS process – may be used. The following processing steps are similar to the ones described in the first manufacturing process.

3.2 TRANSISTOR OPERATION

One way to demonstrate the transistor operation is to consider first two independent pn-junctions and then the interaction between the two. The EB-junction of Figure 3.6, with $V_{BE} > 0$, is forward biased and the BC-junction, with $V_{BC} < 0$, is reverse biased. In the forward biased BE-junction, electrons are injected into the p-region and holes into the n-region. A relatively large current results. Contrary to this case is that of the BC-junction: electrons are extracted from the p-region and holes from the n-region. A very small saturation current with a density below 10^{-18} A/μm^2 results. The minority carrier distribution of Figure 3.6(b) indicates the situation. The subscripts are used according to the semiconductor region. It is obvious that there exists no difference in comparison to the forward and reverse biased pn-junction described in Chapter 2. This changes when the two p-regions are merged (Figure 3.7).

If the base width x_B is sufficiently small, an npn transistor action results. The electrons of the emitter are injected into the base since the energy barrier is reduced by qV_{BE} (see Figure 2.3 for comparison). Contrary to the previous case, the electrons do not move to the base contact but instead toward the nearby BC-junction, which extracts the electrons and sweeps them through the depletion region towards the collector. From here they move as majority carriers to the collector contact. The electron density decreases linearly across the base. At $x = x_B$ the electrons are extracted to a value close to zero. The number of injected holes from the base into the emitter remains unchanged.

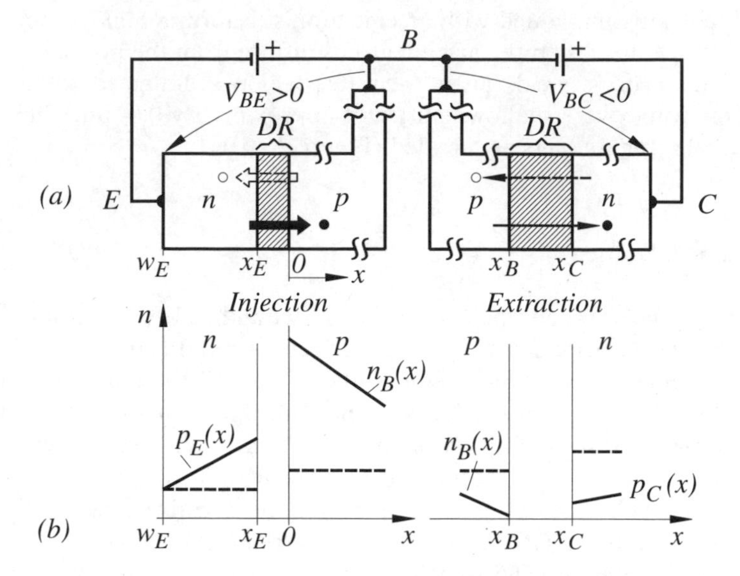

Figure 3.6 (a) Independent pn-junction in forward and reverse biased direction and (b) minority carrier distribution

According to the npn structure the collector current I_C results from the injected electrons into the base and the base current I_B from the injected holes into the emitter. The ratio of the two yields the current gain

$$\boxed{B_F = \frac{I_C}{I_B}} \tag{3.1}$$

under normal forward conditions. In order to avoid confusion the symbol B is used to indicate a static current gain, whereas the symbol β is used for small-signal current gains.

3.2.1 Current–Voltage Relationship

This relationship is a key feature of each transistor and is derived under the following assumptions: low-level injection exists, generation and recombination rates are negligible because of the small dimensions of modern transistors, and the saturation current of the BC-junction is disregarded. Due to the preceding assumption, the minority carrier distribution in the collector is of no importance and can be omitted in the derivation.

The electron density decreases linearly across the base. At x_B the density approaches zero. In reality the density can never be zero since a current is flowing. To ease the derivation it is assumed that the density has a value of n_{Bo} at x_B. The assumption, whether the density is close to zero or has a value of n_{Bo}, has barely any effect on

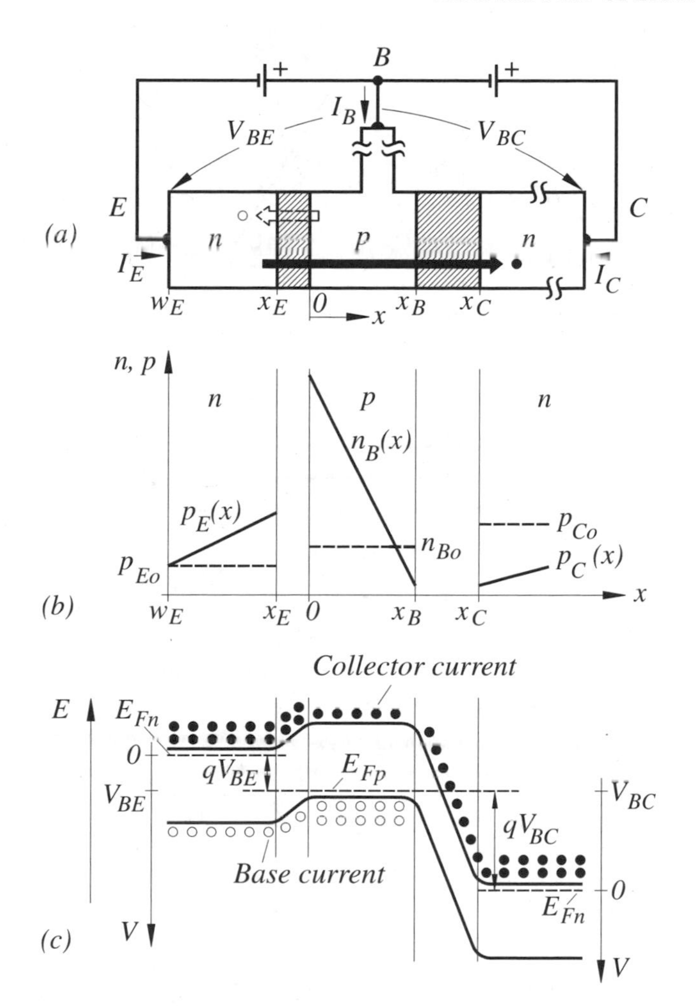

Figure 3.7 (a) npn Transistor, (b) minority carrier distribution and (c) band diagram

the gradient of the charge carriers in the base, and therefore on the current, as long as $V_{BE} > 4\phi_t$. With these assumptions the derivation of the current–voltage equation is identical to that of the pn-junction (Figure 2.4), except that the p-region and the n-region are exchanged on the x-axis (Figure 3.8).

Adding the minority carrier current in the emitter I_p to the one in the base I_n results in an emitter current (Equation 2.25) of

$$\begin{aligned} I_E &= I_p + I_n \\ &= -qAD_{pE}\frac{\mathrm{d}p_E}{\mathrm{d}x} + qAD_{nB}\frac{\mathrm{d}n_B}{\mathrm{d}x} \end{aligned} \tag{3.2}$$

where A describes the emitter area. With the gradients – directly from Figure 3.8 – and the density equations (Equations 2.18 and 2.19) an emitter current of

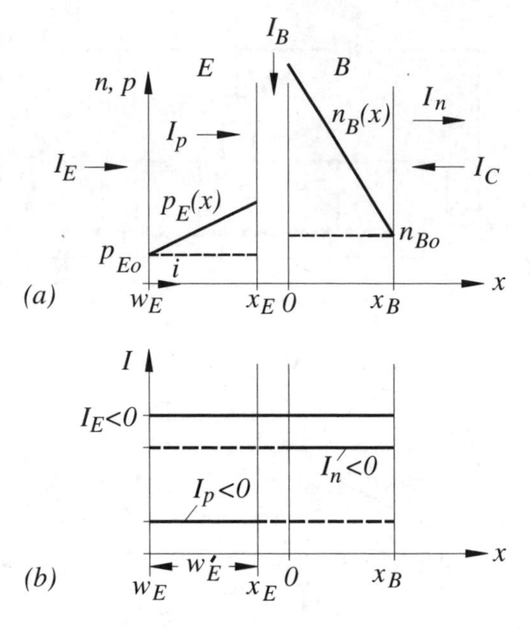

Figure 3.8 Base-emitter region: (a) minority carrier distribution and (b) minority currents I_p, I_n, and emitter current I_E (dashed line majority carrier currents)

$$I_E = -qAD_{pE}\frac{p_E(x_E) - p_{Eo}}{w'_E} - qAD_{nB}\frac{n_B(o) - n_{Bo}}{x_B}$$

$$I_E = -\frac{qAD_{pE}\,p_{Eo}}{w'_E}\left(e^{V_{BE}/\phi_t} - 1\right) - \frac{qAD_{nB}\,n_{Bo}}{x_B}\left(e^{V_{BE}/\phi_t} - 1\right) \tag{3.3}$$

can be derived.

According to the current directions of Figure 3.7 or Figure 3.8, the following collector and base current

$$I_C = -I_n = \frac{qAD_{nB}\,n_{Bo}}{x_B}\left(e^{V_{BE}/\phi_t} - 1\right) \tag{3.4}$$

$$I_B = -I_p = \frac{qAD_{pE}\,p_{Eo}}{w'_E}\left(e^{V_{BE}/\phi_t} - 1\right) \tag{3.5}$$

result.

These equations lead to the following current gain of the transistor (Equation 3.1)

$$B_F = \frac{D_{nB}}{D_{pE}}\frac{n_{Bo}}{p_{Eo}}\frac{w'_E}{x_B}$$

$$B_F = \frac{D_{nB}}{D_{pE}}\frac{N'_{DE}}{N'_{AB}}\frac{n_{iB}^2}{n_{iE}^2}\frac{w'_E}{x_B}$$

$$B_F = \frac{D_{nB}}{D_{pE}} \frac{N'_{DE}}{N'_{AB}} \frac{w'_E}{x_B} \tag{3.6}$$

where $n_{Bo} = n_{iB}^2/N'_{AB}$ and $p_{Eo} = n_{iE}^2/N'_{DE}$ are replaced by the respective doping densities. Identical intrinsic densities are assumed. The effect of different intrinsic densities on the current gain is discussed in conjunction with the temperature behavior of the transistor. Considering Equation (3.6) the question of how to achieve a large current gain arises. The geometry ratio w'_E/x_B can be increased by reducing the base width x_B or by increasing the emitter width w'_E. Increasing the emitter width is not a good idea, since this would lead to a large diffusion capacitance (Section 2.4.2) of the emitter region. High speed technologies therefore use minimum dimensions which lead to a geometry ratio close to one. The parameters left for increasing the current gain are the doping concentrations, or in other words the doping ratio N'_{DE}/N'_{AB}. Due to the vertical structure of the transistor the average emitter doping N'_{DE} is much larger than the average base doping N'_{AB} (Figure 3.4). Thus the doping ratio is mainly responsible for achieving a high current gain. Figure 3.9 shows a typical output characteristic of an npn transistor with a current gain in the order of 120.

Transport current

The collector and base currents (Equations 3.4 and 3.5) can be rewritten in the form

$$I_C = I_{SS}\left(e^{V_{BE}/\phi_t} - 1\right) \tag{3.7}$$

and

$$I_B = \frac{I_{SS}}{B_F}\left(e^{V_{BE}/\phi_t} - 1\right) \tag{3.8}$$

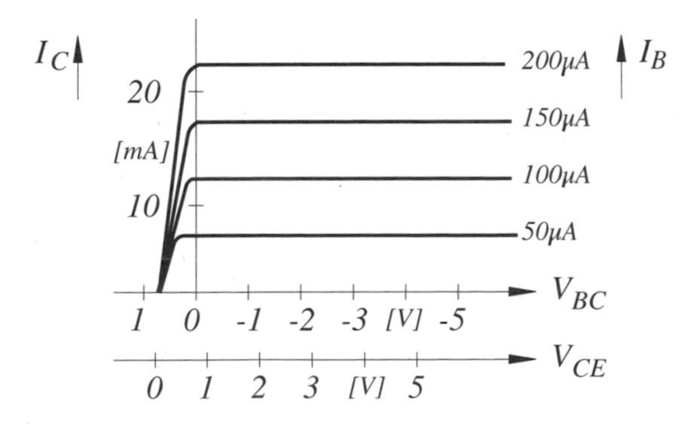

Figure 3.9 Output characteristic of an npn transistor with I_B as parameter

where

$$\boxed{I_{SS} = \frac{qAD_{nB}\,n_{Bo}}{x_B}} \tag{3.9}$$

is called transport current.

Note: this current is not a saturation or leakage current as in the case of a pn-junction but a current parameter depending on the base characteristic only. Typical density values are about $10^{-18}\,\mathrm{A}/\mu\mathrm{m}^2$.

Equation (3.9) can be expressed – under low-level injection conditions with $N'_{AB} = p_B$ – in terms of the majority carrier charge in the base

$$\begin{aligned} I_{SS} &= \frac{qAD_{nB}\,n_{iB}^2}{x_B p_B} \\ &= \frac{q^2A^2D_{nB}\,n_{iB}^2}{Q_B} \end{aligned} \tag{3.10}$$

and where the charge is described by

$$Q_B = qAx_B p_B \tag{3.11}$$

This majority carrier charge plays an important role in the characterization and the modeling of the bipolar transistor, as is indicated in subsequent sections.

Equation (3.10) remains valid even when an inhomogeneous base doping concentration is present

$$Q_B = qA\int_o^{x_B} N'_{AB}(x)\,\mathrm{d}x \tag{3.12}$$

which is the case for all fabrication processes. The proof of this statement is given at the end of this section. The number of doping atoms per area in the base

$$G_B = \int_o^{x_B} N'_{AB}(x)\,\mathrm{d}x \tag{3.13}$$

is called the Gummel number. This important number influences the current gain (Equation 3.6) directly. The smaller the number, the larger the current gain will be. Typical values are around 10^{12} doping atoms/cm^2.

The presented analysis leads to a static large-signal equivalent circuit of the npn transistor, as shown in Figure 3.10, useful for first-order calculations.

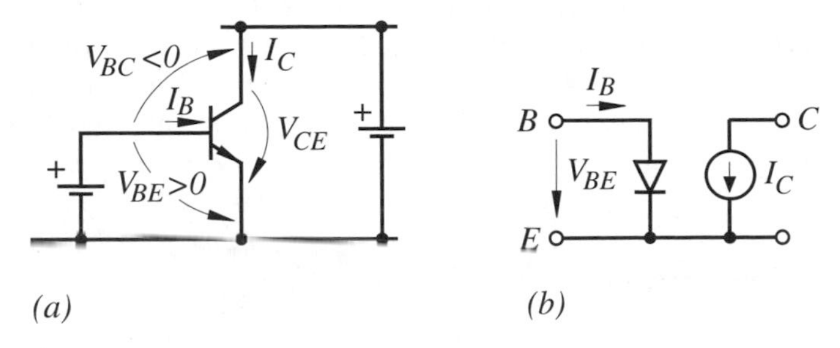

Figure 3.10 (a) npn Transistor under forward biased condition and (b) large-signal equivalent circuit

The circuit consists of a base-emitter diode and a voltage-controlled current generator between collector and emitter, described by Equation (3.7)

$$I_C = I_{SS}\left(e^{V_{BE}/\phi_t} - 1\right)$$

The base current through the diode (Equation 3.8) is described by

$$I_B = \frac{I_{SS}}{B_F}\left(e^{V_{BE}/\phi_t} - 1\right)$$

Determination of I_{SS} and B_F

For the presented large-signal equivalent circuit, the parameters I_{SS} and B_F are needed. Using the half-logarithmic current–voltage plot of Figure 3.11 the transport current can

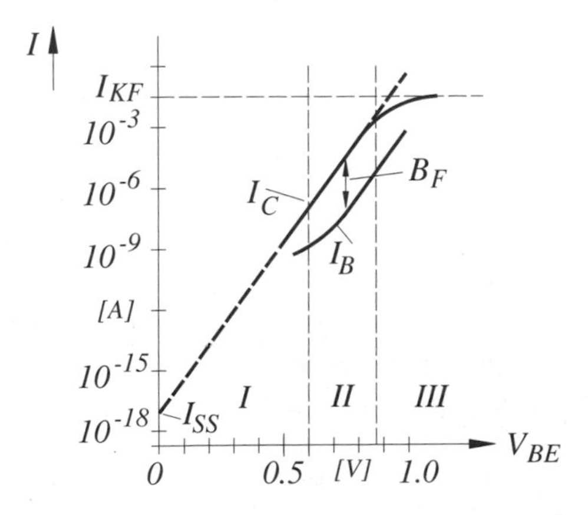

Figure 3.11 Half-logarithmic plot of I_C and I_B as a function of V_{BE}

be found by extrapolating the measured data of the collector current to the V_{BE} zero value. This is the same procedure as the one used to determine the saturation current of the pn-junction (Section 2.3.3). The current gain results from the quotient of the currents I_C/I_B. The reasons for the reduced current gain in the regions *I* and *III* are considered in Section 3.3.1.

Extended view: transport current at inhomogeneous doping

When reading this book this section may be omitted, without losing continuity of the contents.

In the previous section it is claimed that the transport current equation (Equation 3.10) is also valid in the case when inhomogeneous doping is present. The proof of this statement is given next.

Equations (1.48) and (1.49) describe the electron and hole current densities.

$$J_n = q\mu_{nB}\, n_B(x)\mathcal{E}(x) + qD_{nB}\frac{\mathrm{d}n_B(x)}{\mathrm{d}x}$$

$$J_p = q\mu_{pB}\, p_B(x)\mathcal{E}(x) - qD_{pB}\frac{\mathrm{d}p_B(x)}{\mathrm{d}x}$$

For simplicity it is assumed that the mobility and diffusion coefficients have an average value and are constant and thus independent of the position.

To start the derivation it is further assumed that the base current – due to a large current gain – can be assumed to be zero

$$J_p = q\mu_{pB}\, p_B(x)\mathcal{E}(x) - qD_{pB}\frac{\mathrm{d}p_B(x)}{\mathrm{d}x} = 0 \tag{3.14}$$

This leads to an electrical field of

$$\mathcal{E}(x) = \frac{D_{pB}}{\mu_{pB}}\frac{1}{p_B(x)}\frac{\mathrm{d}p_B(x)}{\mathrm{d}x} \tag{3.15}$$

and an electron current of

$$J_n = q\mu_{nB}\, n_B(x)\left(\frac{D_{pB}}{\mu_{pB}}\frac{1}{p_B(x)}\frac{\mathrm{d}p_B(x)}{\mathrm{d}x}\right) + qD_{nB}\frac{\mathrm{d}n_B(x)}{\mathrm{d}x} \tag{3.16}$$

Multiplying both sides of the equation with p_B and using the Einstein relationship (Equation 1.46) the following equation results

$$p_B(x)J_n = qD_{nB}\, n_B(x)\frac{\mathrm{d}p_B(x)}{\mathrm{d}x} + qD_{nB}\, p_B(x)\frac{\mathrm{d}n_B(x)}{\mathrm{d}x} \tag{3.17}$$

Due to the multiplication with p_B the product rule of differential calculus can be used, yielding

$$p_B(x)J_n = qD_{nB}\frac{\mathrm{d}}{\mathrm{d}x}(n_B(x)p_B(x)) \tag{3.18}$$

Since the current in the base is constant the integration across the base leads to

$$J_n = \frac{qD_{nB}\int\limits_0^{x_B}\frac{\mathrm{d}}{\mathrm{d}x}(n_B(x)p_B(x))\,\mathrm{d}x}{\int\limits_0^{x_B}p_B(x)\mathrm{d}x} \tag{3.19}$$

$$J_n = \frac{qD_{nB}[n_B(x_B)p_B(x_B) - n_B(0)p_B(0)]}{\int\limits_0^{x_B}p_B(x)\mathrm{d}x} \tag{3.20}$$

In the previous derivation it is assumed that the electron density at the BC-junction is reduced to its equilibrium value. Accordingly, the density product has a value at x_B of

$$n_B(x_B)p_B(x_B) = n_{Bo}\,p_{Bo} = n_i^2 \tag{3.21}$$

whereas the density product at $x = 0$ is given by Equation (2.17)

$$n_B(0)p_B(0) = n_i^2 e^{V_{BE}/\phi_t} \tag{3.22}$$

This leads to a collector current of

$$I_C = -AJ_n = \frac{qAD_{nB}\,n_i^2(e^{V_{BE}/\phi_t} - 1)}{\int\limits_o^{x_B}N'_{AB}(x)\mathrm{d}x} \tag{3.23}$$

Since at low-level injection $N'_{AB}(x) = p_B(x)$, a transport current of

$$I_{SS} = \frac{qAD_{nB}\,n_i^2}{\int\limits_o^{x_B}N'_{AB}(x)\mathrm{d}x} \tag{3.24}$$

results. This derivation thus proves that the transport current equation is also valid when an inhomogeneous base doping is present.

3.2.2 Transistor under Reverse Biased Condition

In Figure 3.10 the transistor is biased in the forward or normal direction. In the reverse direction with $V_{BC} > 0$ and $B_{BE} < 0$ the functions of emitter and collector are exchanged (Figure 3.12). The resulting current–voltage relationships

$$I_E = I_{SS}\left(e^{V_{BC}/\phi_t} - 1\right) \tag{3.25}$$

$$I_B = \frac{I_{SS}}{B_R}\left(e^{V_{BC}/\phi_t} - 1\right) \tag{3.26}$$

are found in analogy to the previous derivation (indexes E and C exchanged).

The transport current I_{SS} remains unchanged since its characteristic (Equation 3.9) depends on base parameters only. Contrary is the situation for the current gain B_R in the reverse biased direction. In analogy to Equation (3.6)

$$B_R = \frac{I_E}{I_B} = \frac{D_{nB}}{D_{pC}}\frac{N'_{DC}}{N'_{AB}}\frac{w_{EPI}}{x_B} \tag{3.27}$$

results, where w_{EPI} is the effective width of the epitaxy layer (Figure 3.4). Typically the reverse current gain is very small and approximately between one and ten. The primary reason for the gain reduction is the changed doping ratio in the gain equation from the optimized version in the forward direction N'_{DE}/N'_{AB} to the one in the reverse direction N'_{DC}/N'_{AB}.

The currents in the forward direction depend on the V_{BE} voltage whereas those in the reverse direction depend on the V_{BC} voltage. Since the voltages are independent of each other, the two large-signal equivalent circuits can be superimposed into one equivalent circuit (Figure 3.13). According to this equivalent circuit the currents are given by

$$\begin{aligned} I_{CT} &= I_C - I_E = I_{SS}\left(e^{V_{BE}/\phi_t} - 1\right) - I_{SS}\left(e^{V_{BC}/\phi_t} - 1\right) \\ I_{CT} &= I_C - I_E = I_{SS}\left(e^{V_{BE}/\phi_t} - e^{V_{BC}/\phi_t}\right) \end{aligned} \tag{3.28}$$

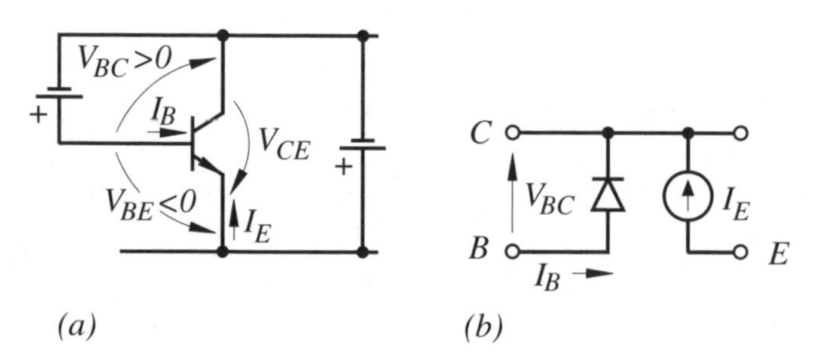

Figure 3.12 (a) npn Transistor under reverse biased condition and (b) large-signal equivalent circuit

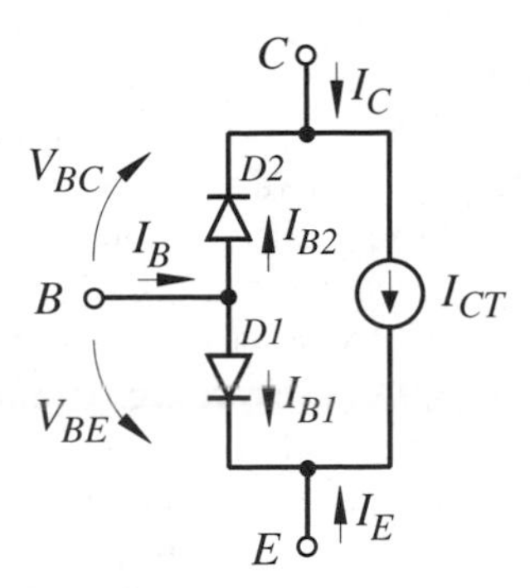

Figure 3.13 Large-signal equivalent circuit for forward and reverse transistor biasing

and

$$I_B = I_{B1} + I_{B2} = \frac{I_{SS}}{B_F}\left(e^{V_{BE}/\phi_t} - 1\right) + \frac{I_{SS}}{B_R}\left(e^{V_{BC}/\phi_t} - 1\right) \tag{3.29}$$

This model is the classical bipolar transport model used for computer-aided design (Getreu 1978). It is analyzed in more detail in Section 3.5.

3.2.3 Voltage Saturation

So far, the transistor is considered to operate in the forward or the reverse direction. In either case one junction is injecting charge whereas the other is extracting the injected charge. In this section the case will be considered where both junctions inject charge simultaneously into the base. This is the case when a bipolar transistor is used, e.g., as an inverter (Figure 3.14(a)).

With an input signal of $V_I = 0\,\text{V}$ the collector current is zero and the output voltage has a value of $V_Q = V_{CC}$. If the input signal changes to a value of, e.g., 0.82 V, a collector current flows, reducing the output voltage to, e.g., 0.3 V. In this case the base

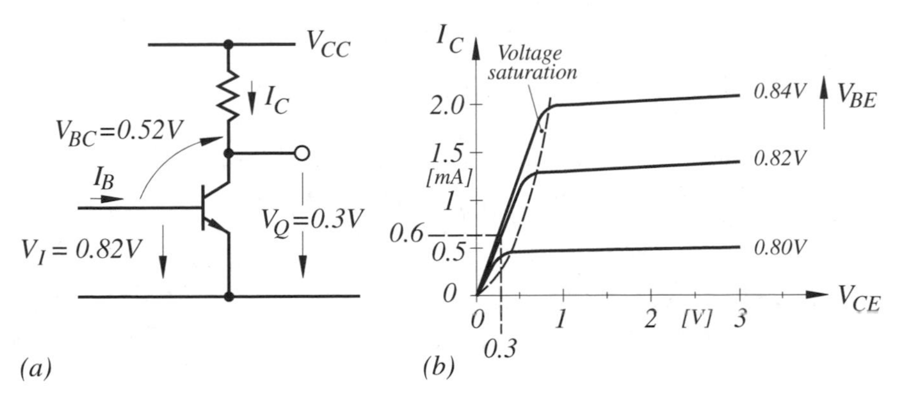

Figure 3.14 (a) Bipolar inverter and (b) output characteristic of an npn transistor with V_{BE} as variable

collector voltage has a positive value of $V_{BC} = 0.52\,\text{V}$, causing the BC-junction to inject charge into the base additionally to the BE-junction.

This situation can be observed at the output characteristic of the npn transistor (Figure 3.14(b)). The transistor is in the forward direction. A reduction of the V_{CE} voltage causes a barely noticeable change in the collector current. This is even the case when the voltages V_{CE} and V_{BE} have the same value, causing V_{BC} to be zero. In this situation the electric field of the BC-junction is still very large and similar to that of the pn-junction shown in Figure 2.2 with $V_{PN} = 0\,\text{V}$. The collector current begins to reduce when $V_{CE} < V_{BE}$ and V_{BC} becomes positive. Both junctions inject charge into the base (Figure 3.15). Charge carriers $n'_B(0)$ and $n'_B(x_B)$ are injected from the emitter and collector side of the base, respectively, causing a total charge distribution $n^*_B(x)$ with a reduced density gradient (Figure 3.15(a)). Since the current I_{CT} (Figure 3.13) is proportional to this density gradient, a current reduction results (Figure 3.14(b)). Under this condition, named voltage saturation, the collector current is smaller than in the normal forward direction

$$I_C < I_B B_F \tag{3.30}$$

This leads to an excess base current

$$I_{BS} = I_B - \frac{I_C}{B_F} \tag{3.31}$$

flowing into the base. This excess current causes an excess base charge Q_{BS} (Figure 3.15(b)), influencing the switching properties of the transistor negatively (Section 2.4.2).

Determination of the saturation voltages

When an inverter drives another one or a similar circuit, the output low-voltage, that is the saturation voltage V_{CEsat}, has to be very low in order to guarantee that the following stage interprets the output information as low signal. The parameters determining this voltage are therefore important to know, when designing such an inverter or a similar circuit.

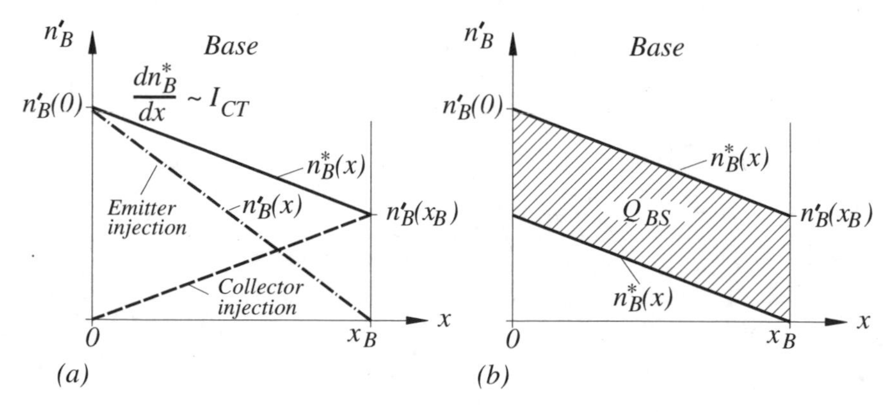

Figure 3.15 (a) Emitter and collector charge carrier injection into the base and (b) excess charge for case (a)

The bipolar transistor is in saturation when the base current has a value $I_B > I_C/B_F$. The resulting voltages across the transistor terminals can be found by applying Kirchhoff's law to the large-signal equivalent circuit of Figure 3.16 and using Equations (3.28) and (3.29). With the assumption that V_{BCsat} and $V_{BEsat} > 100\,\text{mV}$ and $B_F \gg B_R + 1$, the following voltages result

$$V_{BCsat} = \phi_t \ln \frac{B_R(B_F I_B - I_C)}{I_{SS} B_F} \tag{3.32}$$

$$V_{BEsat} = \phi_t \ln \frac{I_C + I_B(1 + B_R)}{I_{SS}} \tag{3.33}$$

$$V_{CEsat} = V_{BEsat} - V_{BCsat}$$

$$\boxed{V_{CEsat} = \phi_t \ln \frac{B_F[I_C + I_B(1 + B_R)]}{B_R(B_F I_B - I_C)}} \tag{3.34}$$

Example

The data of an npn transistor are $B_F = 150$ and $B_R = 10$. The ratio of I_C/I_B is 20. With these data a collector–emitter saturation voltage of

$$V_{CEsat} = \phi_t \ln \frac{B_F[I_C/I_B + (1 + B_R)]}{B_R(B_F - I_C/I_B)}$$

$$= 26\,\text{mV} \ln \frac{150(20 + 11)}{10(150 - 20)} = 33\,\text{mV}$$

results.

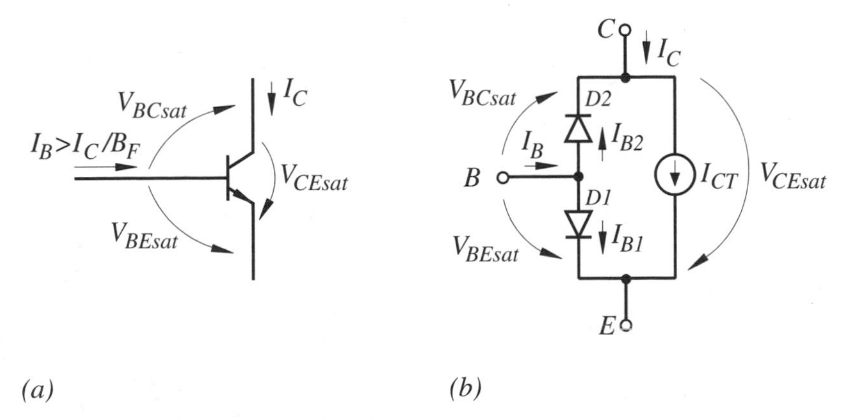

Figure 3.16 (a) Transistor under voltage saturation condition and (b) large-signal equivalent circuit for case (a)

3.2.4 Temperature Behavior

In this section the dependence of the current gain and the collector current on the temperature will be considered. This knowledge is important, e.g. for designing band-gap circuits (Section 10.3) or considering the temperature behavior of a circuit in general.

Temperature behavior of the current gain

A typical temperature behavior of the current gain is shown in Figure 3.17. The increase of the current gain with temperature is caused by the different change in intrinsic density in the emitter n_{iE}, compared to that of the base n_{iB} (Equation 3.6). This effect can be explained in the following way. From a doping density of about $N > 10^{19}\,\mathrm{cm}^{-3}$, quantum mechanical interactions between the doping atoms and the silicon atoms occur. According to the Pauli exclusion principle (Section 1.1), this interaction causes additional energy levels to split into different allowed energy states. The result is a reduction of the band-gap by ΔE_G. Since the emitter of the considered fabrication processes is highly doped, the intrinsic emitter density changes to

$$n_{iE} = C\left(\frac{T}{[\mathrm{K}]}\right)^{3/2} e^{-(E_G(T)-\Delta E_G)/2kT} \tag{3.35}$$

whereas the density in the base (Equation 1.20)

$$n_{iB} = C\left(\frac{T}{[\mathrm{K}]}\right)^{3/2} e^{-E_G(T)/2kT} \tag{3.36}$$

remains almost unchanged. The ratio of these densities (Equation 3.6) yields the following temperature dependence of the current gain

$$B_F(T) = B_F(T_R)e^{-\frac{\Delta E_G}{k}\left(\frac{1}{T}-\frac{1}{T_R}\right)} \tag{3.37}$$

where $B_F\,(T_R)$ is the current gain at a reference temperature T_R.

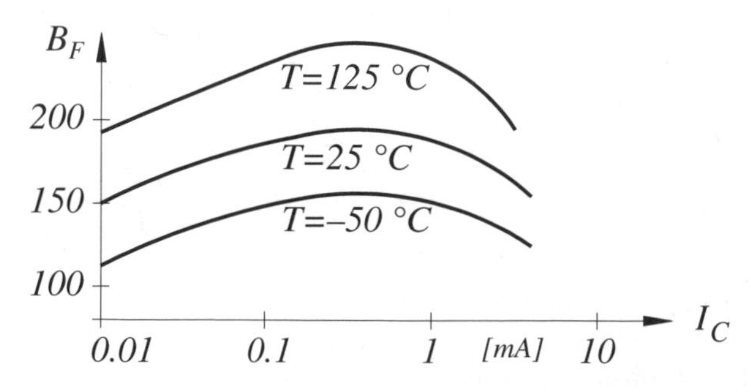

Figure 3.17 Current gain as a function of the collector current and temperature

This equation can be used to determine the band reduction ΔE_G. To do this successfully, the temperature behavior of the different diffusion constants has to be determined first (Rein *et al.* 1978).

Temperature behavior of the collector current

For many circuit design tasks knowledge of the temperature dependence of the collector current is even more important than that of the current gain. Additionally to the information already presented, some further parameters are needed. The electron mobility (Figure 1.17) can be described empirically by the equation

$$\mu_n(T) = \mu_n(300\,\mathrm{K})\left(\frac{T}{300\,\mathrm{K}}\right)^{-a_n} \tag{3.38}$$

where T is the absolute temperature and a_n a parameter with values between 1 and 1.5. This leads to the following temperature dependence of the diffusion constant

$$D_{nB} = \phi_t \mu_n(300\,\mathrm{K})\left(\frac{T}{300\,\mathrm{K}}\right)^{-a_n} \tag{3.39}$$

With the additional Equations (3.4), (3.38), (1.16), and (1.20) a collector current of

$$\begin{aligned} I_C &= \frac{AqD_{nB}n_{Bo}}{x_B}\left(e^{V_{BE}/\phi_t} - 1\right) \\ &= E\left(\frac{T}{300\,\mathrm{K}}\right)^{(4-a_n)} e^{-\frac{E_G(T)}{kT}}\left(e^{V_{BE}/\phi_t} - 1\right) \end{aligned} \tag{3.40}$$

results. In this equation E is a temperature-independent parameter in Ampère and $\phi_t = kT/q$. Furthermore, it has to be considered how the band-gap changes with temperature due to the variation of the lattice constant. The results are summarized in Tsividis (1980). For approximate calculations the linear relationship (Barber 1976)

$$E_G(T)/q = V_G(T) = V_{Go} + \varepsilon T \tag{3.41}$$

may be used, where V_{Go} is the voltage corresponding to the extrapolated value of the band-gap E_{Go}/q to $T \to 0$. The value for ε is approximately $-2.8 \cdot 10^{-4}$ V/K.

3.2.5 Breakdown Behavior

The maximum possible transistor voltages are limited by the breakdown behavior of the transistor junctions. Two breakdown mechanisms can be observed, namely avalanche breakdown and punch-through. These mechanisms are in general not destructive unless the high current causes the temperature to exceed its allowed value. How these effects influence the output characteristic of the transistor is shown in Figure 3.18.

Common-base configuration

This voltage breakdown is denoted by BV_{CBO}, that is, the collector-to-base voltage breakdown with the emitter open-circuited (index O). This breakdown is caused, just as in the case of the pn-junction (Section 2.6), by an avalanche effect. This is noticeable in Figure 3.18(a) from a voltage of about 10 V where the current gain

$$A_F = I_C / -I_E \tag{3.42}$$

is a larger one. This leads to a collector current of

$$I_C = -A_F I_E M \tag{3.43}$$

where

$$M = \frac{1}{1 - \left[\dfrac{V_{CB}}{BV_{CBO}}\right]^n} \tag{3.44}$$

is a factor expressing the carrier multiplication, just as in the case of the np-junction (Equation 2.72). When the V_{CB} voltage approaches the breakdown voltage the factor M becomes infinity.

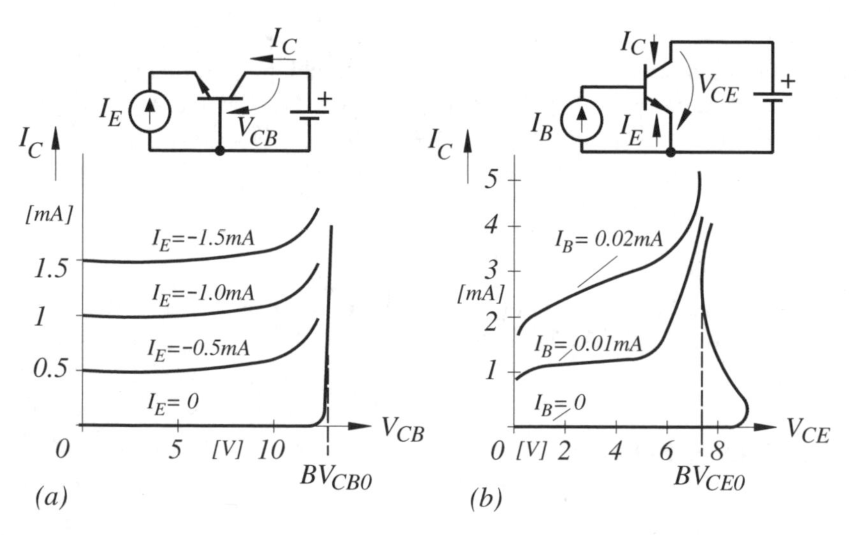

Figure 3.18 Breakdown behavior: (a) common-base configuration and (b) common-emitter configuration

BE-junction breakdown

The BE-junction – collector open-circuited – of the transistor is also subject to avalanche breakdown. Since the base is the more lightly doped region of the BE-junction, the base doping mainly determines the breakdown voltage BV_{BEO} (Equation 2.75). In the case of the BC-junction, this is the more lightly doped side of the collector. Since $N'_{AB} \gg N'_{DC}$ the breakdown voltage $BV_{BEO} \ll BV_{CBO}$.

Common-emitter configuration

The breakdown voltage BV_{CEO} (Figure 3.18(b)), that is the collector-to-emitter voltage breakdown with the base open-circuited, is always lower than the BV_{CBO} voltage considered before. This is because the electron–hole pairs, which are produced by the avalanche mechanism in the BC-junction, are separated by the field, whereby electrons are swept to the collector and holes into the base, where they act as base current supporting an external base current, when available. The result is an increased collector current generated by the gain of the transistor (Figure 3.19(a)). With $I_E + I_B + I_C = 0$ (Figure 3.18(b)) and Equations (3.42) and (3.43), the following collector current

$$\begin{aligned} I_C &= -I_B - I_E \\ &= I_B \frac{A_F M}{1 - A_F M} \end{aligned} \tag{3.45}$$

can be derived. This equation shows that I_C approaches infinity already, when the product $A_F M$ becomes unity, which leads to the following relationship between the breakdown voltages

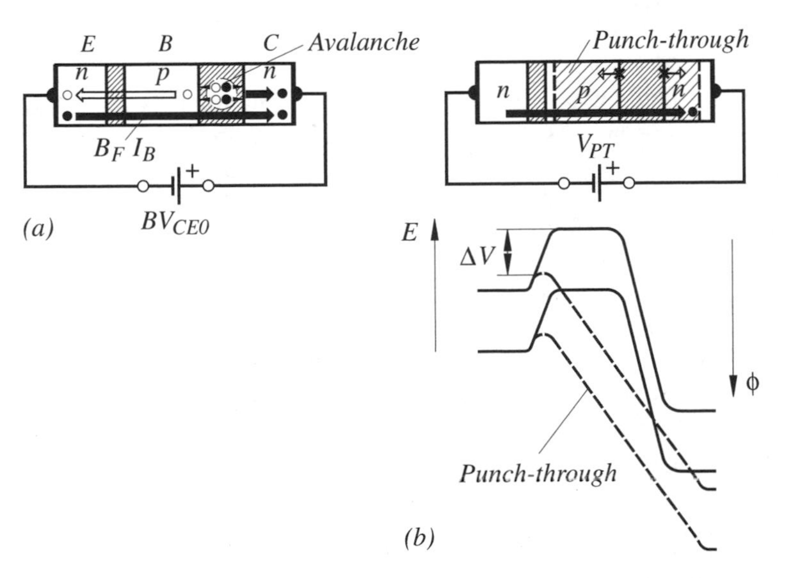

Figure 3.19 Collector-to-emitter breakdown with open base: (a) avalanche effect and (b) punch-through effect

$$\boxed{BV_{CEO} \approx BV_{CBO}(B_F)^{-\frac{1}{n}}} \tag{3.46}$$

This equation can be used to explain the S shape of the breakdown characteristics at $I_B = 0$ (Figure 3.18(b)), when the dependence of B_F on the collector current is considered. As discussed in the following section, B_F is small at low currents and thus the breakdown voltage is high. When the current increases, B_F increases also, causing a decrease in the breakdown voltage and the negative slope shown in Figure 3.18(b).

The BV_{CEO} voltage is limited either by the described avalanche effect or a punch-through effect. A transistor is in this condition when the depletion region of the BC-junction reaches that of the BE-junction before avalanche can take place. When this happens the emitter and collector regions are joined as a continuous depletion region (Figure 3.19(b)). This has the effect that the energy barrier of the BE-junction is reduced by ΔV. As a result a large amount of electrons is injected from the emitter through the base to the collector, causing breakdown. The punch-through voltage can be estimated by Equation (2.41) when a stepped doping profile is assumed.

$$V_{PT} = \frac{x_B^2 q N'_{AB}(1 + N'_{AB}/N'_{DC})}{2\varepsilon_o \varepsilon_r} \tag{3.47}$$

From this equation it is obvious that the base width and its tolerance during manufacturing have a profound effect on the punch-through voltage due to the square-law behavior.

3.3 SECOND-ORDER EFFECTS

Up to now an idealized npn transistor structure is considered. The following second-order effects e.g. high-level injection, base-width modulation, and current crowding, show that the ideal behavior can be substantially impaired.

3.3.1 High Current Effects

In the previous analysis it is assumed that the current gain is constant and independent of the collector current. As Figure 3.11 and Figure 3.20 reveal, this is not the case for real transistors. The observed change in current gain by the V_{BC} voltage is caused by the base-width modulation, which is treated in Section 3.2.2.

Region I

At very low collector current the recombination of electron–holes in the EB depletion region is no longer negligible (Section 1.5). This recombination causes a degradation of the effective base current and therefore a reduced current gain.

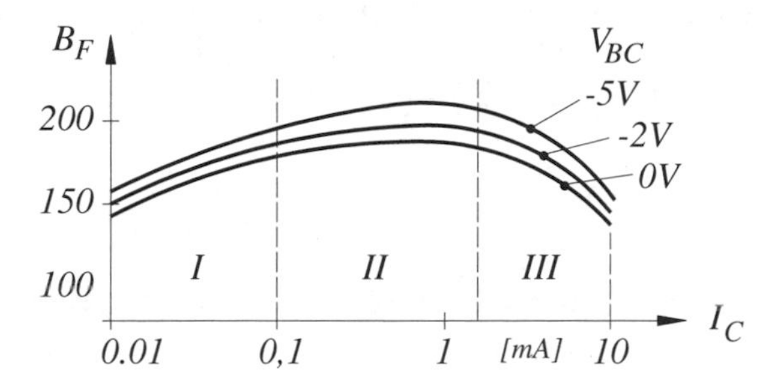

Figure 3.20 Current gain as a function of the collector current with V_{BC} voltage as parameter

Region II

This region is described by the derived equations.

Region III

From a particular collector current the current gain reduces again. This is caused by high-level injection effects at the BE-junction and BC-junction of the transistor.

1. High injections at the BE-junction

Under high-level injection conditions the minority density at the base side of the EB-junction is $n_B(0) \geq N'_{AB}$. This leads to a voltage-dependent injection, similar to that given by Equation (2.21), of

$$n_B(0) \approx n_i e^{V_{BE}/2\phi_t} \tag{3.48}$$

This injection level is reduced by a factor of two in the exponent and therefore explains the reduced collector current

$$I_C \approx \frac{qAD_{nB}n_i}{x_B} e^{V_{BE}/2\phi_t} \tag{3.49}$$

under high-level injection conditions.

2. High-level injection at the BC-junction

Under normal operation conditions the BC-junction extracts the minority carriers, in this case the electrons, from the base and sweeps them through the depletion region to the collector. The electron density decreases linearly across the base. At $x_B = 0$ the density approaches zero. In reality the density can never be zero since a current is flowing. In the case that the electrons approach the saturation velocity v_{sat} (Figure 1.16) the collector current can increase only when the electron density increases. As a result the base region stretches out into the collector, reducing the current gain. This is the well-known Kirk effect (Kirk 1962), which is considered in more detail in the following.

Equation (1.38) leads to the number of electrons passing the BC-junction at saturation speed

$$n = -\frac{I_n}{qAv_{sat}} = \frac{I_C}{qAv_{sat}} \tag{3.50}$$

The charge of the BC-depletion region

$$qN(x) = -qN'_{AB}(x) + qN'_{DC}(x) \tag{3.51}$$

is influenced by the collector current. According to Poisson's equation this leads to

$$\begin{aligned} \frac{\mathrm{d}\mathcal{E}}{\mathrm{d}x} &= \frac{1}{\varepsilon_0\varepsilon_{Si}}[qN(x) - qn] \\ &= \frac{1}{\varepsilon_0\varepsilon_{Si}}\left[qN(x) - \frac{I_C}{Av_{sat}}\right] \end{aligned} \tag{3.52}$$

and to

$$\mathcal{E}(x) = \frac{1}{\varepsilon_o\varepsilon_{Si}}\int_{x_B}^{x_C}\left[qN(x) - \frac{I_C}{Av_{sat}}\right]\mathrm{d}x \tag{3.53}$$

and furthermore to (Equation 2.37)

$$\phi_{iC} - V_{BC} = -\int_{x_B}^{x_C}\mathcal{E}(x)\,\mathrm{d}x \tag{3.54}$$

In this equation ϕ_{iC} is the built-in voltage of the BC-junction. This relationship states that the integral over the field (the area) must be constant as long as the applied V_{BC} voltage does not change. This set of equations is solvable numerically only. The sketch of Figure 3.21 helps clarify the situation. In this figure the collector consists of the actual n-doped collector region w_{EPI} and the highly n^{++}-doped buried collector (Figure 3.4). The metallurgical junction between base and collector is designated by x_i. Assuming a relatively small collector current, the field distribution (1) results (Figure 3.21(c)). This case is comparable to the one of the pn-junction shown in Figure 2.2. An increase in collector current causes the field to move toward the buried collector edge (2). The integral over the field remains constant since the applied voltage has not changed. At a current of $I_C = qAN'_{DC}v_{sat}$ the net charge in the w_{EPI} region is zero. A uniform field distribution exists (3). A further increase in the collector current reduces the field at x_i to almost zero (4). The base region now stretches far out into the collector region, reducing the current gain (Poon *et al.* 1969).

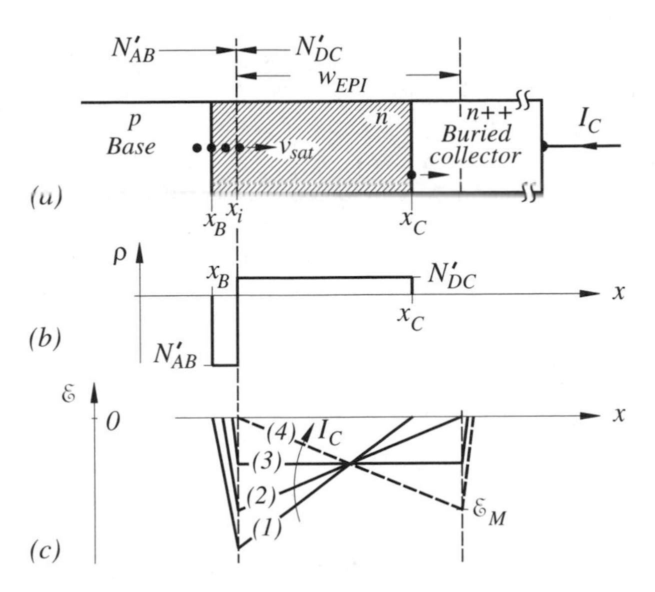

Figure 3.21 (a) High-level injections at the BC-junction, (b) charge distribution at low collector current and (c) field distribution as a function of collector currents at constant V_{BC} voltage

The high-injection current I_{CG} where this situation exists is derived in the following. The field distribution (4) in Figure 3.21(c) is present. Using Equations (3.53) and (3.54) leads to

$$\mathcal{E}_M = \frac{1}{\varepsilon_o \varepsilon_{Si}} \int_{x_i}^{x_i + w_{EPI}} \left(qN'_{DC} - \frac{I_{CG}}{Av_{sat}} \right) \mathrm{d}x$$
$$= \frac{1}{\varepsilon_o \varepsilon_{Si}} \left(qN'_{DC} - \frac{I_{CG}}{Av_{sat}} \right) w_{EPI} \tag{3.55}$$

and

$$\phi_{iC} - V_{BC} = - \int_{x_i}^{x_i + w_{EPI}} \mathcal{E}(x)\, \mathrm{d}x$$
$$= - \frac{w_{EPI}}{2} \mathcal{E}_M \tag{3.56}$$

where $\mathcal{E}_M$ is the maximum field at the edge of the buried collector and ϕ_{iC} is assumed to be constant. Substitution of Equation (3.56) into Equation (3.55) gives a high-injection current of

$$I_{CG} = Av_{sat} \left[qN'_{DC} + (\phi_{iC} - V_{BC}) \frac{2\varepsilon_o \varepsilon_{Si}}{(w_{EPI})^2} \right] \tag{3.57}$$

Example

An npn transistor has the following data: $N'_{DC} = 10^{15}\,\text{cm}^{-3}$, $w_{EPI} = 0.6\,\mu\text{m}$, $V_{BC} = -5\,\text{V}$, $A_E = 24 \cdot 10^{-8}\,\text{cm}^2$, $\phi_{iC} \approx 0.7\,\text{V}$, and $v_{sat} = 10^7\,\text{cm/s}$. With these parameters Equation (3.57) yields a high-injection current of $I_{CG} = 6.5\,\text{mA}$.

This current can be moved to higher values, e.g. by increasing the phosphorus doping of the actual collector as is done in the described manufacturing process of Figure 3.5(g).

It is obvious from the previous discussion that it is not advisable to operate a transistor under the condition of high-level injection due to the strong reduction of current gain and – as is shown in Section 3.5.4 – a substantial increase in transit time. The high-level injection region with $I_C \geq I_{KF}$ is indicated in Figure 3.11. The current I_{KF} is called the forward knee current.

3.3.2 Base-Width Modulation

So far it is assumed that the transistor behaves under normal operation conditions like an ideal current generator. Ideal means the collector current remains absolute constant even when the voltages V_{BC} and V_{CE} change (Figure 3.9). In reality this is not the case, as is indicated in Figure 3.22. The collector current depends on the V_{BC} voltage. This is even more pronounced at higher collector currents and is caused by the changing minority gradient in the base (Figure 3.23). The width w of the BC-depletion region depends on the V_{BC} voltage (Equation 2.43). The width increases when the voltage is increased in the negative direction. This leads to a reduction in base width x_B and accordingly to an increase in the minority carrier gradient. Since the collector current (Equation 3.7) or, to be precise, the transport current (Equation 3.9) is inversely proportional to x_B

$$I_C = \frac{AqD_{nB}\,n_{Bo}}{x_B(V_{BC})}\left(e^{V_{BE}/\phi_t} - 1\right)$$

$$I_C = I_{SS}(V_{BC})\left(e^{V_{BE}/\phi_t} - 1\right)$$

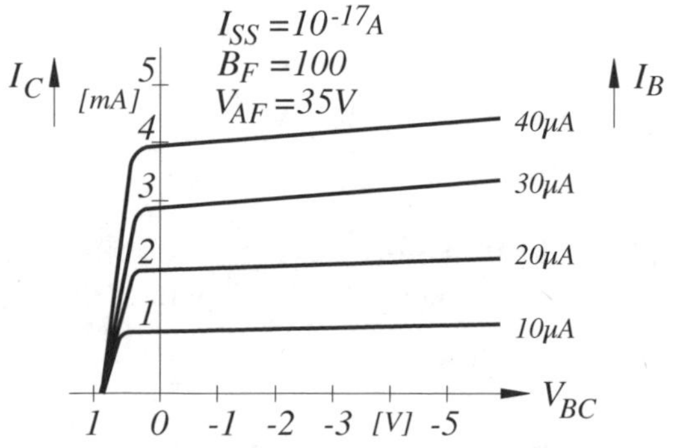

Figure 3.22 Output characteristic of an npn transistor with I_B as parameter

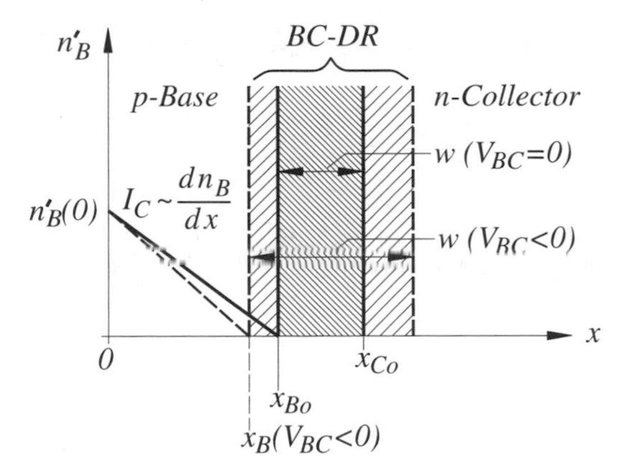

Figure 3.23 Influence of the V_{BC} voltage on the gradient of the minority carriers in the base

and the increase of the collector current can be explained. This effect is called base-width modulation or the Early effect and plays a very important role in the design of analog amplifiers (Section 8.3).

Early voltage

The change in base width x_B (V_{BC}) can be derived by using Equation (2.43). But this would be the case for a homogeneously doped base only. In order to come up with a general solution and an easy to determine parameter, namely the Early voltage, a linear approach is pursued. The base width dependence on the V_{BC} voltage can be expressed by a truncated Taylor series around $V_{BC} = 0$

$$x_B = x_{Bo} + \left.\frac{\mathrm{d}x_B}{\mathrm{d}V_{BC}}\right|_{V_{BC}=0} \cdot V_{BC} \tag{3.58}$$

where x_{Bo} is the base width at $V_{BC} = 0$. Normalizing the equation yields

$$\frac{x_B}{x_{Bo}} = 1 + \frac{1}{x_{Bo}} \left.\frac{\mathrm{d}x_B}{\mathrm{d}V_{BC}}\right|_{V_{BC}=0} \cdot V_{BC}$$

$$\frac{x_B}{x_{Bo}} = 1 + \frac{1}{V_{AF}} V_{BC} \tag{3.59}$$

In this equation

$$V_{AF} = x_{Bo} \left.\frac{\mathrm{d}V_{BC}}{\mathrm{d}x_B}\right|_{V_{BC}=0} \tag{3.60}$$

is the Early voltage (Early 1952) and $1/V_{AF}$ is the gradient in the linear Equation (3.59). Since the transport current is inversely proportional to the base width, its dependence on the Early voltage can be expressed in the following form

$$I_{SS} = I_{SSo}\frac{x_{Bo}}{x_B} = I_{SSo}\frac{1}{1 + V_{BC}/V_{AF}} \approx I_{SSo}\left(1 - \frac{V_{BC}}{V_{AF}}\right) \tag{3.61}$$

where I_{SSo} is the transport current at $V_{BC} = 0\,\text{V}$. In the approximation is assumed that $|V_{BC}/V_{AF}| \ll 1$. Knowing the transport current, the collector current dependence on the V_{BC} voltage can be described by

$$I_C = I_{SSo}\left(1 - \frac{V_{BC}}{V_{AF}}\right)\left(e^{V_{BE}/\phi_t} - 1\right) \tag{3.62}$$

The Early voltage can be interpreted by analyzing the gradient of the $I_C(V_{BC})$ behavior. Equation (3.62) leads directly to

$$\frac{\mathrm{d}I_C}{\mathrm{d}V_{BC}} = -\frac{I_{SSo}}{V_{AF}}\left(e^{V_{BE}/\phi_t} - 1\right)$$

$$\boxed{\frac{\mathrm{d}I_C}{\mathrm{d}V_{BC}} = -\frac{I_{Co}}{V_{AF}}} \tag{3.63}$$

where I_{Co} is the collector current at $V_{BC} = 0\,\text{V}$. This equation reveals that the gradient is proportional to the ratio of I_{Co}/V_{AF} (Figure 3.24). The Early voltage can therefore be found at the intersection of the extrapolated $I_C(V_{BC})$ function. Typical values range between 30 V and 80 V, depending on the base width.

The current gain (Equation 3.6) is – just like the transport current – inversely proportional to the base width and therefore dependent on the V_{BC} voltage as shown in Figure 3.20. This dependence can be considered by rewriting the current gain equation in the following form

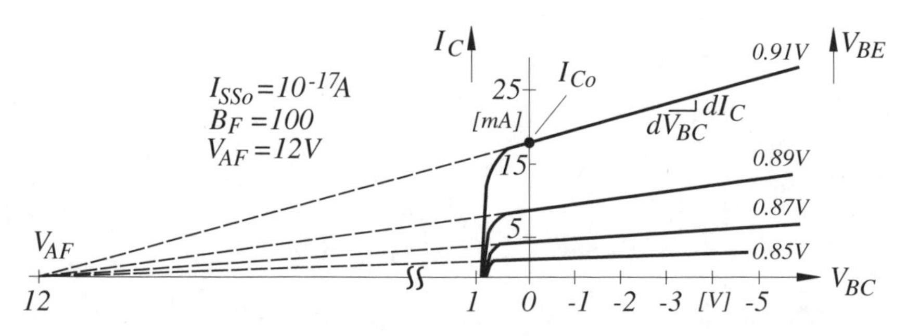

Figure 3.24 Output characteristic of an npn transistor with V_{AF} determination (base-width modulation emphasized)

$$B_F = B_F|_{V_{BC}=0} \cdot \frac{x_{Bo}}{x_B}$$

$$B_F \approx B_F|_{V_{BC}=0} \cdot \left(1 - \frac{V_{BC}}{V_{AF}}\right) \tag{3.64}$$

This means a large negative V_{BC} voltage causes an increase in current gain due to the reduction in base width.

Early voltage as a function of the majority carrier density

In order to optimize the transistor design or to develop an advanced computer-aided model the Early voltage is frequently expressed as a function of the majority carrier charge. Using Equation (3.60) the Early voltage can be rewritten in the form

$$\begin{aligned} V_{AF} &= x_{Bo} \frac{\mathrm{d}V_{BC}}{\mathrm{d}x_B}\bigg|_{V_{BC}=0} \\ &= x_{Bo} \frac{\mathrm{d}V_{BC}}{\mathrm{d}Q_C} \frac{\mathrm{d}Q_C}{\mathrm{d}x_B}\bigg|_{V_{BC}=0} \\ &= x_{Bo} \frac{1}{C_{jco}} \frac{\mathrm{d}Q_C}{\mathrm{d}x_B}\bigg|_{V_{BC}=0} \end{aligned} \tag{3.65}$$

The expansion of the equation by the majority carrier density change $\mathrm{d}Q_C$ in the base (Figure 3.25) makes it possible to replace the respective quotient, in analogy to Equation (2.45), by the depletion capacitance of the BC-junction

$$C_{jco} = \frac{\mathrm{d}Q_C}{\mathrm{d}V_{BC}} \tag{3.66}$$

at $V_{BC} = 0\,\mathrm{V}$.

Since the change of the majority carrier charge is

$$\mathrm{d}Q_C = qAN'_{AB}\,\mathrm{d}x_B \tag{3.67}$$

(compare Figures 3.25 and 2.10) the required equation for the Early voltage can be found by substituting Equation (3.67) in Equation (3.65)

$$\boxed{V_{AF} = \frac{Q_{Bo}}{C_{jco}}} \tag{3.68}$$

where

$$Q_{Bo} = qAx_{Bo}N'_{AB} \tag{3.69}$$

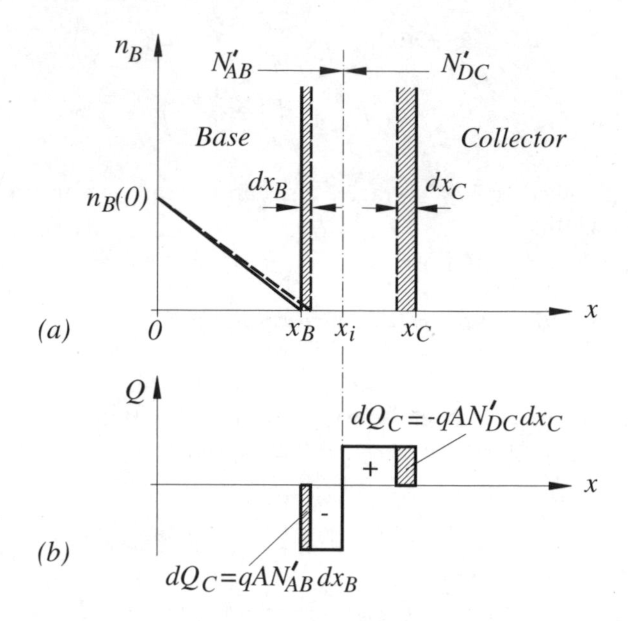

Figure 3.25 Influence of a dV_{BC} change on the BC-junction: (a) base width modulation and (b) charge variation

is the majority carrier charge in the base at $V_{BC} = 0$. This equation demonstrates that by measuring the depletion capacitance C_{jco} and determining the Early voltage, the majority carrier charge in the base can be found.

Extended view: increasing the early voltage

The intrinsic small-signal voltage gain of a bipolar transistor is given by (Section 10.4.2)

$$\nu_o/\nu_i = -V_{AF}/\phi_t \tag{3.70}$$

Considering this equation, it is obvious that a large voltage gain requires a large Early voltage. This can be achieved by increasing the base width (Equation 3.60), which reduces the percent-wise change of the base width. The disadvantage is a reduced current gain (Equation 3.6) and an increased transit time impairing the transistor's speed performance (Section 3.5.4). The influence of the doping density on the Early voltage results directly from Equations (3.68) and (3.69). When an abrupt homogeneously doped BC-junction is considered for simplicity, a depletion capacitance (Equation 2.47) of

$$C_{jco} = A\sqrt{\frac{q\varepsilon_o\varepsilon_{Si}}{2\left(\frac{1}{N'_{AB}} + \frac{1}{N'_{DC}}\right)\phi_{ic}}} \tag{3.71}$$

and an Early voltage (Equations 3.68 and 3.69) of

$$V_{AF} = x_{Bo}\sqrt{\frac{qN'_{AB}}{\varepsilon_o \varepsilon_{Si}}\left(1 + \frac{N'_{AB}}{N'_{DC}}\right)2\phi_{ic}} \tag{3.72}$$

result. This equation states that an increase in the base doping N'_{AB} and a reduction of the collector doping N'_{DC} improve the Early voltage. In other words the change in width at the base side dx_B becomes smaller compared to that at the collector side dx_C (Figure 3.25). Changing the doping concentrations in the way described has the disadvantage that an increase of N'_{AB} reduces the current gain – as mentioned before – and a reduction of N'_{DC} shifts the Kirk effect (Equation 3.57) to lower collector currents. It is obvious from the previous arguments that a series of trade-offs have to be made between the different requirements. These trade-offs are easier to achieve when a heterojunction bipolar transistor, e.g., with an epitaxial grown SiGe base is used (Kasper 1996). The major impact of the SiGe base on the band diagram of the transistor is illustrated in Figure 3.26.

The Ge content causes a reduction of the band-gap in the base and the grading is responsible for an electrical field which supports the movement of electrons through the base. The effect is a reduced transit time and therefore an improved high frequency performance of the transistor. An additional advantage is an improved current gain. This is demonstrated by Equation (3.6), where a homogeneous Ge content and doping is assumed

$$B_F = \frac{D_{nB}}{D_{pE}}\frac{N'_{DE}}{N'_{AB}}\frac{n_{iB}^2}{n_{iE}^2}\frac{w'_E}{x_B}$$

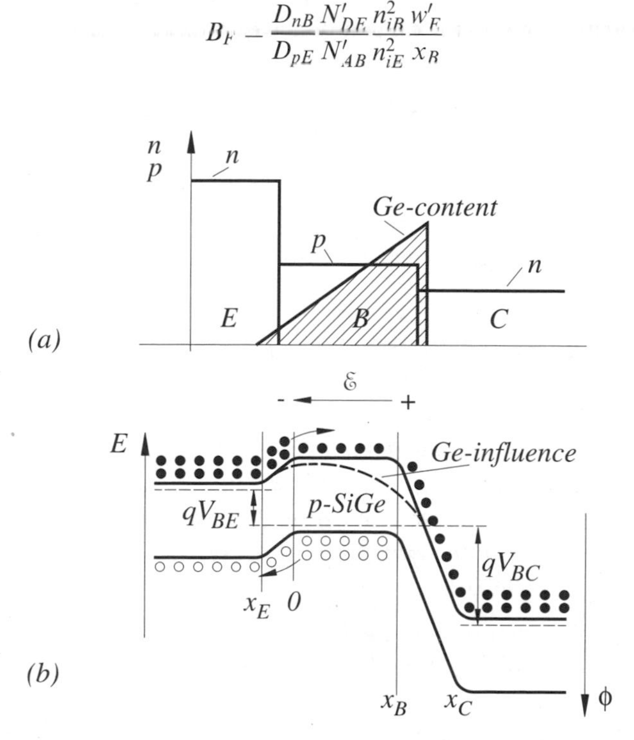

Figure 3.26 Heterojunction bipolar transistor: (a) doping profile with Ge content and (b) band diagram

Since the effective intrinsic density n_{iB} of the SiGe base is much larger than the intrinsic density n_{iE} of the Si emitter (Figure 1.6) the increased current gain can be explained. When an inhomogeneously doped base and a Ge grading is present and, for simplicity, it is assumed that the emitter doping is constant, a current gain of

$$B_F = \frac{N'_{DE} \, w'_E}{D_{pE} \, n_{iE}^2} \left[\int_o^{x_B} \frac{N'_{AB}(x)}{D_{nB}(x) n_{iB}^2(x)} \, \mathrm{d}x \right]^{-1} \tag{3.73}$$

results.

Not only the current gain improves but also the Early voltage, as is shown in the following (Prinz and Sturm 1991). According to Equation (3.63) the Early voltage can be expressed by

$$V_{AF} = -I_{Co} \frac{\mathrm{d}V_{CB}}{\mathrm{d}I_C}$$

and the collector current (Equation 3.7) by

$$I_{Co} = I_{SSo} e^{V_{BE}/\phi_t}$$

where $V_{BE} > 100\,\mathrm{mV}$ and $V_{BC} = 0\,\mathrm{V}$. Considering an inhomogeneous doping situation the transport current (Equation 3.24) can be rewritten in the form

$$I_{SSo} = \frac{qA}{\int_o^{x_{Bo}} \frac{N'_{AB}(x)}{D_{nB}(x) n_{iB}^2(x)} \, \mathrm{d}x} \tag{3.74}$$

Using these equations and Equations (3.67) and (3.68) the following Early voltage

$$V_{AF} = \frac{q n_{iB}^2(x_{Bo}) D_{nB}(x_{Bo})}{C_{jco}/A} \int_o^{x_B} \frac{N'_{AB}(x)}{D_{nB}(x) n_{iB}^2(x)} \, \mathrm{d}x \tag{3.75}$$

results. According to this equation the increase in Early voltage is caused by the fact that the intrinsic density $n_{iB}(x_{Bo})$ in front of the integral is relatively high whereas the average intrinsic density in the base is much smaller. In other words the Ge gradient of the n_i is responsible for the improvement in Early voltage.

A figure of merit frequently used is the product of current gain and Early voltage

$$B_F V_{AF} = \frac{q N'_{DE} w'_E}{C_{jco}/A} \frac{D_{nB}(x_{Bo})}{D_{pE}} \frac{n_{iB}^2(x_{Bo})}{n_{iE}^2} \tag{3.76}$$

where values of up to 100 000 V have been reported (Prinz *et al.* 1991). Furthermore, transit frequencies in the range $f_t > 75\,\mathrm{GHz}$ using a manufacturing process similar to the one of Figure 3.5 but including a Ge content in the base have been observed (Klein and Klepser, 1999).

3.3.3 Current Crowding

So far it is assumed that the current density across the emitter area is constant. Unfortunately this is not the case, as is shown in this section. To illustrate the effect, an idealized emitter base structure is used where the base current I_B is supplied from the left side contact (Figure 3.27). The base current produces a voltage drop along the base resistance. As a result the V_{BE} voltage decreases along the same path. This causes the electron injection from the emitter to fall off starting from the base contact. To make the point, one has to remember that the collector current depends exponentially on the V_{BE} voltage. Therefore an average reduction of this voltage by only $\phi_t = 26\,\text{mV}$ causes the current density to be reduced to one third of its original value. This fall off in electron injection leads to a current crowding at the emitter perimeter at the base contact. This current crowding may cause an unexpected localized heating and a high-level injection at collector currents, which might be acceptable when currents are distributed uniformly. In order to get an idea of the current crowding the effective base resistance, called base spreading resistance, is estimated.

The voltage drop across an incremental base resistance can be calculated by

$$\begin{aligned} \mathrm{d}V &= I_B(y)\,\mathrm{d}R_B \\ &= I_B(y)\frac{\rho_B}{x_B l_E}\,\mathrm{d}y \end{aligned} \tag{3.77}$$

where l_E is the length of the emitter (into the paper) and ρ_B the average resistivity of the base region. The current distribution in the base depends on both the base and emitter currents. To derive the base current a two-dimensional problem has to be solved. This can be done most effectively by using a device simulator. But in order to proceed with the estimation of the spreading resistor, it is assumed that a linear behavior

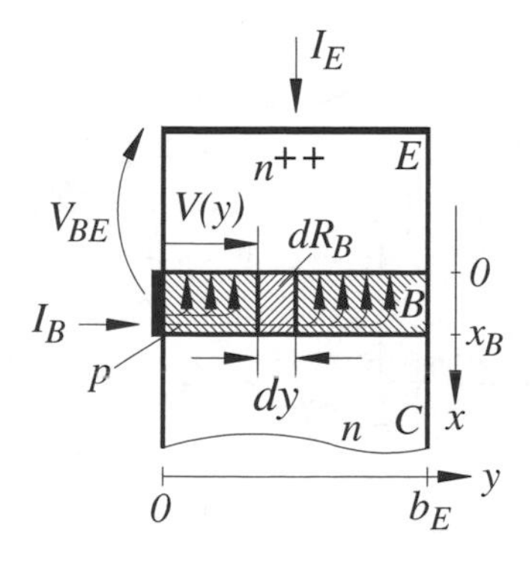

Figure 3.27 Idealized emitter base structure used for the demonstration of current crowding

$$I_B(y) = I_B\left(1 - \frac{y}{b_E}\right) \tag{3.78}$$

can be used to describe the base current dependence, where $I_B(y = 0) = I_B$ and $I_B(y = b_E) = 0$. Substituting of Equation (3.78) into Equation (3.77) and integrating with respect to y gives the voltage distribution across the base as a function of y

$$\begin{aligned} V(y) &= I_B \frac{\rho_B}{x_B L_E} \int_o^y \left(1 - \frac{y}{b_E}\right) \mathrm{d}y \\ &= I_B \frac{\rho_B}{x_B L_E}\left(y - \frac{y^2}{2b_E}\right) \end{aligned} \tag{3.79}$$

In order to calculate the spreading resistance it is necessary first to obtain the average voltage drop across the base

$$\begin{aligned} \overline{V} &= \frac{1}{b_E}\int_o^{b_E} V(y)\,\mathrm{d}y \\ &= I_B \frac{\rho_B}{3x_B}\frac{b_E}{l_E} \end{aligned} \tag{3.80}$$

One speaks of the onset of current crowding when this voltage has a value of about $\phi_t = 26\,\mathrm{mV}$. The possibilities to reduce this voltage by changing ρ_B or x_B are limited by the requirement of a large current gain. This leaves the transistor geometry ratio b_E/l_E for optimization. In practical transistors b_E is usually chosen to be as small as possible and the length l_E is adjusted in order to minimize the current crowding effect. This leads to stripe geometries with large periphery to area ratio (Figure 3.28).

The base spreading resistance for the structure with one base contact only (Figure 3.27) can be estimated to be

$$\boxed{R_{BI} = \frac{\overline{V}}{I_B} = \frac{\rho_B}{3x_B}\frac{b_E}{l_E}} \tag{3.81}$$

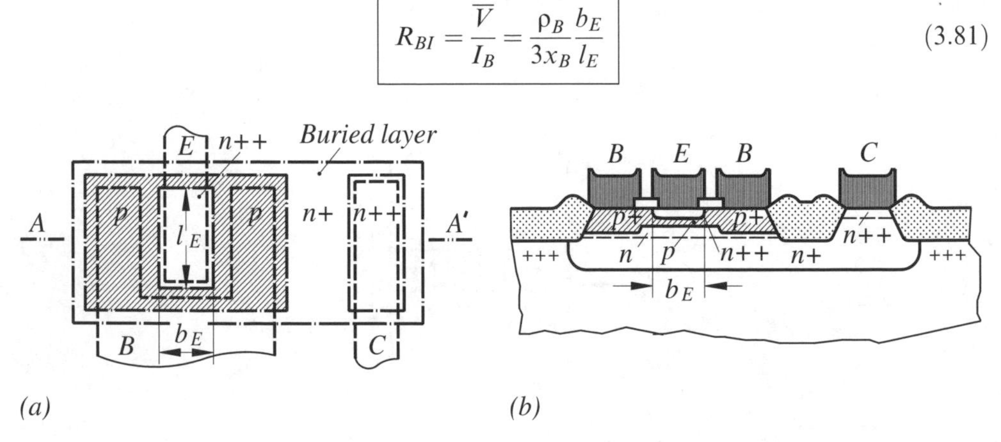

Figure 3.28 (a) Top view of an npn transistor with a stripe geometry and (b) cross-section A–A′

This value reduces to

$$\boxed{R_{BI} = \frac{\rho_B}{12x_B}\frac{b_E}{l_E}} \qquad (3.82)$$

when a structure with two base contacts (Figure 3.28) is used. When higher currents are required the number of stripes can be increased as needed. A summary of resistance formulas for a number of geometrical shapes is included in Baum and Warner (1965). In the double-polysilicon manufacturing process of Figure 3.5 the situation is more favorable, since the base is surrounded by a very low resistive polysilicon base contact.

3.4 ALTERNATIVE TRANSISTOR STRUCTURES

Isolation between neighboring transistors is achieved by tying the common p-substrate to the most negative voltage of the circuit. The positive collector therefore behaves with respect to the substrate like a back-biased pn-junction. This isolation technique together with the vertical structure of the transistor requires a relatively large silicon area. This area can be reduced when more transistors use the same buried layer or, in other words, share the same isolation island toward the substrate. This results in transistors with a common collector (Figure 3.29(a)) or in a transistor with multiple emitters (Figure 3.29(b)). In the past the multiple emitter structures were used for transistor–transistor logic (TTL) (Elmansry 1983). When the functions of emitter and collector are exchanged a multiple collector structure, frequently used for integrated injection logic (I^2L) (Elmansry 1983), results.

pnp-Transistor

The manufacturing processes described in Section 3.1 are optimized for npn transistors. Sometimes an integrated circuit can be improved when additional pnp transistors are available. By keeping the manufacturing process unaltered, this transistor can be implemented by a mask change only. Emitter and collector are generated with the same process step, namely the base implantation of the npn transistor (Figure 3.30). In this lateral pnp transistor, the transistor action is parallel to the semiconductor surface. The distance between collector and emitter determines the base width x_B. Due to tolerances in the lithography process and out-diffusion of the doping material, the distance between the emitter and collector x_B is much larger than in a conventional npn transistor. This results in an inferior transistor with a reduced current gain (Equation 3.6) and an increased transit time (Equation 2.54). Another disadvantage is that high-level injection effects degenerate the transistor action at relatively small collector currents. The low-doped base region (epitaxy n-layer) is responsible for this, as is shown in the following.

The collector current of the pnp transistor is in analogy to Equations (3.7) and (3.9)

$$I_C = \frac{AqD_{pB}}{x_B}p'_B(0) \qquad (3.83)$$

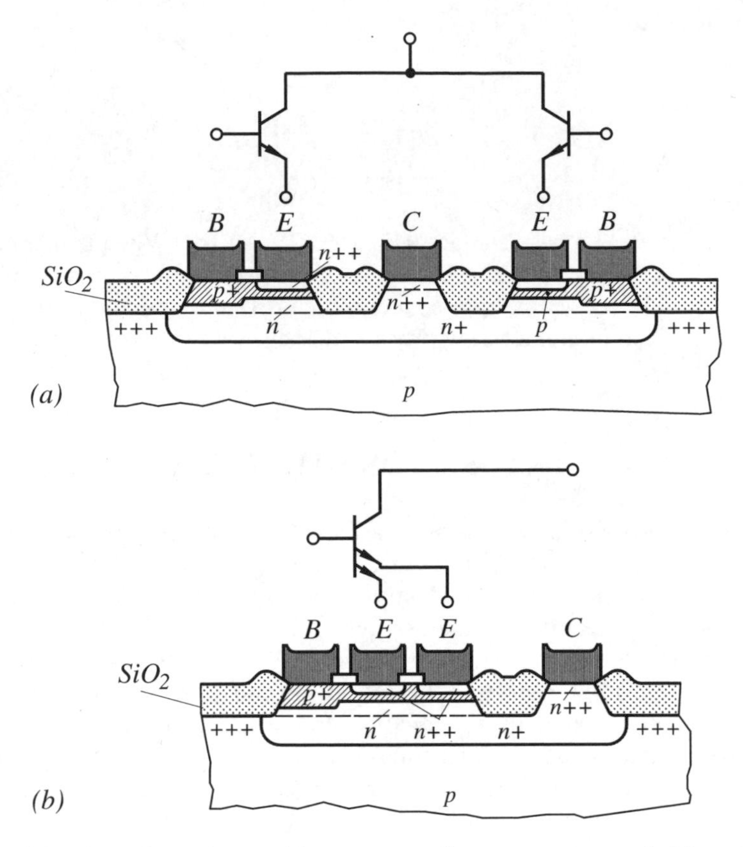

Figure 3.29 Merging of transistors: (a) common collector structure and (b) multiple emitter structure

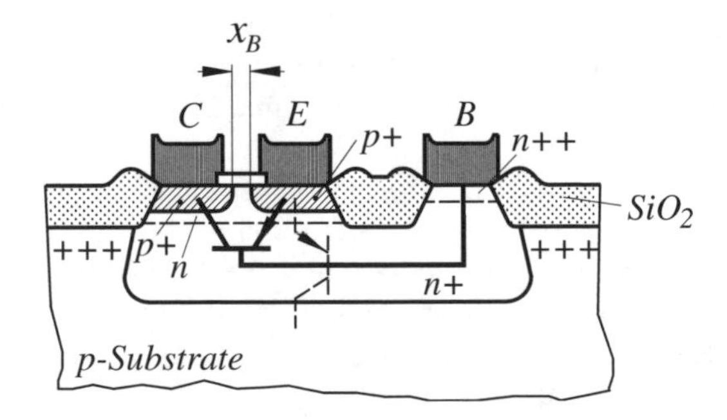

Figure 3.30 Lateral pnp transistor

This leads to a minority carrier density at the emitter side of the base ($x = 0$) of

$$p'_B(0) = \frac{I_C x_B}{qAD_{pB}} \tag{3.84}$$

As long as this concentration is noticeable below the majority carrier density (Figure 1.23), low-level injection exists. Since the majority carrier density N_D is low in the pnp transistor (w_{EPI} region in Figure 3.4) the onset of high-level injection takes place when

$$p'_B(0) \geq N_D \tag{3.85}$$

This leads to a reduced collector current – the following example will demonstrate this – where the transistor can be employed usefully.

Example

The data of a lateral pnp transistor are: $N_D = 10^{16}\,\text{cm}^{-3}$, $A_E = 28\,\mu\text{m}^2$, $x_B = 1\,\mu\text{m}$, $D_{pB} = 9\,\text{cm}^2/\text{s}$.

This results in an onset of high-level injection and a reduction of current gain at a collector current of

$$I_C = \frac{AqD_{pB}}{x_B} N_D = 40.2\,\mu\text{A}$$

An additional effect, which influences the pnp transistor adversely, is an unavoidable vertical pnp transistor toward the substrate (Figure 3.30). This transistor, in effect, reduces the available base current and bypasses the emitter current of the lateral to the vertical transistor. What helps is the inhomogeneous doping between the n$^+$-buried layer and the n base. This situation is comparable to the analyzed inhomogeneously doped semiconductor with a built-in electrical field (Figure 2.1). This field acts as a barrier for the injected holes into the substrate, reducing the current gain of this transistor.

The injection efficiency of the lateral pnp transistor can be increased when the emitter is completely surrounded by the collector (Figure 3.31). This effectively shifts the onset of high-level injection to larger collector currents.

pn-Diode

The transistor has two pn-junctions which can be used as diodes in a circuit. The major differences between the two are the breakdown voltages (Section 3.2.5) and the

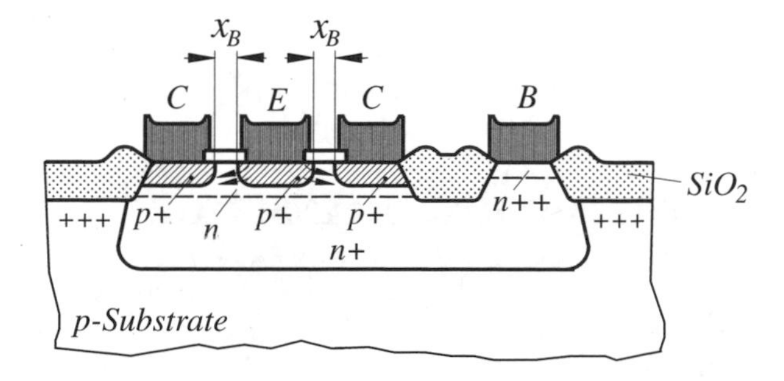

Figure 3.31 Lateral pnp transistor with enclosed emitter

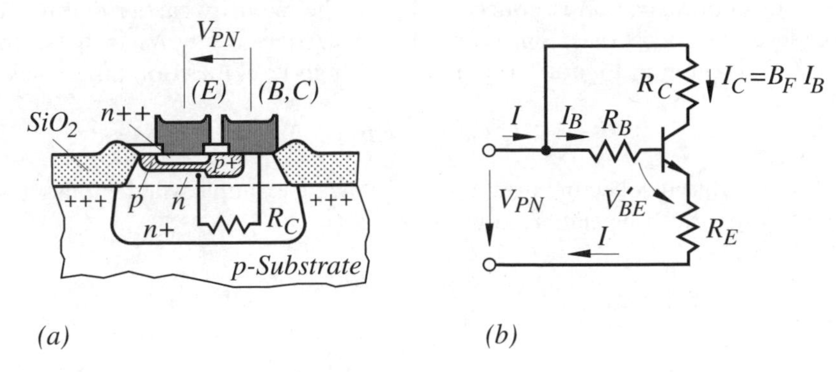

Figure 3.32 (a) pn-Diode with short-circuited BC-junction (R_B and R_E not shown) and (b) equivalent circuit

series resistances. A very common realization uses a transistor with short-circuited BC-junction (Figure 3.32). This configuration has the advantage that the series resistance of the diode can be reduced remarkably. The equivalent circuit leads to the following current–voltage relationship

$$
\begin{aligned}
V_{PN} &= V'_{BE} + I_B R_B + I R_E \\
&= V'_{BE} + I\left(\frac{R_B}{1 + B_F} + R_E\right)
\end{aligned}
\tag{3.86}
$$

where the current description

$$I = I_B + B_F I_B = (1 + B_F) I_B \tag{3.87}$$

is used. The expression in parenthesis of Equation (3.86) can be interpreted as series resistance of the diode

$$\boxed{R_S = \frac{R_B}{1 + B_F} + R_E} \tag{3.88}$$

where the current I produces a voltage drop of $I \cdot R_S$. The reason for the reduction of the series resistance is that most of the current $I_C = B_F I_B$ bypasses the BE-junction. This reduces in effect the relatively large base resistance by $(1 + B_F)$.

3.5 MODELING THE BIPOLAR TRANSISTOR

In this section a procedure similar to the one used at the pn-junction will be employed. First, a brief introduction to the modeling aspects of the bipolar transistor for CAD application is given. Second, the presented equation will then be simplified in order to be used in first-hand circuit calculations.

3.5.1 Transistor Model for CAD Applications

The starting point is the classical transport model of Figure 3.13. The series resistances of base, emitter, and collector are added (Figure 3.33). Due to these resistors one has to distinguish between external and internal transistor voltages. This leads to internal currents of

$$I'_B = I'_{B1} + I'_{B2} = \frac{I_{SSo}}{B_F}\left(e^{V'_{BE}/\phi_t} - 1\right) + \frac{I_{SSo}}{B_R}\left(e^{V'_{BC}/\phi_t} - 1\right) \tag{3.89}$$

and

$$I_{CT} = I_C - I_E = I_{SSo}\left(e^{V'_{BE}/\phi_t} - e^{V'_{BC}/\phi_t}\right) \tag{3.90}$$

The charge elements of the BE-junction and BC-junction are composed of a depletion and a junction part, comparable to those of the pn-junction (Equation 2.76)

$$Q_{BE} = \tau_F I_{SSo}\left(e^{V'_{BE}/\phi_t} - 1\right) + C_{jeo}\int_o^{V'_{BE}}\left(1 - \frac{V}{\phi_{iE}}\right)^{-ME} \mathrm{d}V \tag{3.91}$$

and

$$Q_{BC} = \tau_R I_{SSo}\left(e^{V'_{BC}/\phi_t} - 1\right) + C_{jco}\int_o^{V'_{BC}}\left(1 - \frac{V}{\phi_{iC}}\right)^{-MC} \mathrm{d}V \tag{3.92}$$

where τ_F and τ_R are the effective transit times under forward and reverse operation conditions. Furthermore, a charge element is added

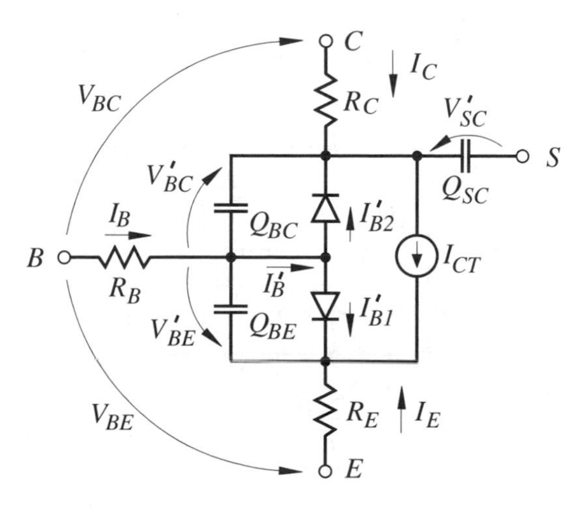

Figure 3.33 Compact model of a bipolar transistor

$$Q_{SC} = C_{jso} \int_{o}^{V'_{SC}} \left(1 - \frac{V}{\phi_{iS}}\right)^{-MS} \mathrm{d}V \tag{3.93}$$

describing the depletion effect of the reverse biased collector to substrate junction. An expansion of the presented transport model leads to the Gummel–Poon model (Gummel and Poon 1970) of Figure 3.34. The two added diodes include the low current region (Section 2.3.3, Figure 2.7) with the parameters I_{SE} and I_{SC} in the forward and reverse direction of the bipolar transistor. This leads to a base current of

$$\begin{aligned} I'_B &= I'_{B1} + I'_{B2} + I'_{B3} + I'_{B4} \\ &= \underbrace{\frac{I_{SSo}}{B_F}\left(e^{V'_{BE}/\phi_t} - 1\right)}_{D1} + \underbrace{\frac{I_{SSo}}{B_R}\left(e^{V'_{BC}/\phi_t} - 1\right)}_{D2} \\ &+ \underbrace{I_{SE}\left(e^{V'_{BE}/N_E\phi_t} - 1\right)}_{D3} + \underbrace{I_{SC}\left(e^{V'_{BC}/N_C\phi_t} - 1\right)}_{D4}. \end{aligned} \tag{3.94}$$

The usefulness of this procedure is demonstrated in Figure 3.35. Under forward biased conditions the currents I'_{B1} and I'_{B3} (Figure 3.34) are effective only. In region I current I'_{B3} passing through diode $D3$ dominates and in region II current I'_{B1} passing through diode $D1$ dominates. The parameters of the diodes can be found in analogy to the procedure outline in Section 2.3.3. Similar, the parameters of $D2$ and $D4$ can be found when the transistor is reverse biased. A summary of typical parameter values, using a manufacturing process similar to the one illustrated in Figure 3.3, is shown in Table 3.1.

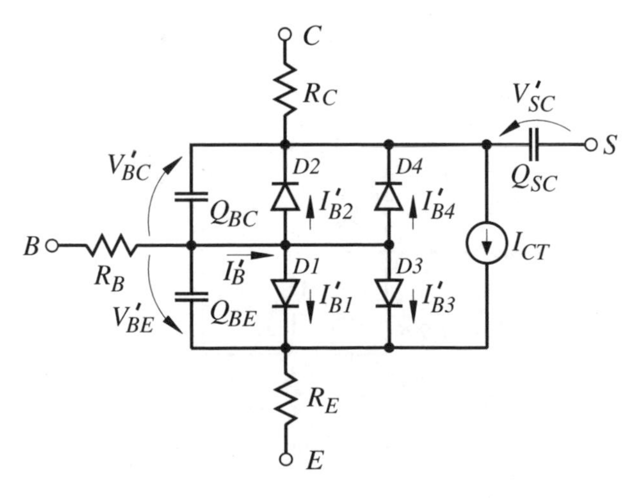

Figure 3.34 Gummel–Poon model

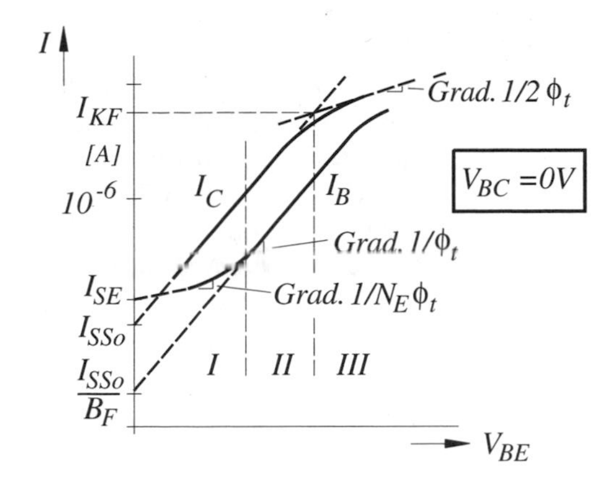

Figure 3.35 Half-logarithmic plot of I_C and I_B as a function of V_{BE}

Table 3.1 Typical npn transistor parameter of a transport model (emitter dimension $b_E = 2.0\,\mu m$, $l_E = 8\,\mu m$)

Name	SPICE	Parameter	Example	Unit
I_{SSo}	IS	Transport current $V_{BC} = 0\,V$	$3.5 \cdot 10^{-17}$	A
I_{SE}	ISE	BE current parameter	10^{-16}	A
I_{SC}	ISC	BC current parameter	$5 \cdot 10^{-16}$	A
N_E	NE	BE emission coefficient	1.5	
N_C	NC	BC emission coefficient	2.0	
V_{AF}	VAF	Early voltage (forward)	35	V
V_{AR}	VAR	Early voltage (reverse)	16	V
I_{KF}	IKF	Knee current (forward)	$4 \cdot 10^{-3}$	A
I_{KR}	IKR	Knee current (reverse)	$0.5 \cdot 10^{-3}$	A
B_F	BF	Maximum static current gain (forward)	175	
B_R	BR	Maximum static current gain (reverse)	15	
C_{jeo}	CJE	BE depletion capacitance at 0 V	65	fF
ϕ_{iE}	VJE	BE built-in voltage	0.85	V
ME	MJE	BE grading coefficient	0.40	
τ_F	TF	Effective transit time (forward)	30	ps
C_{jco}	CJC	BC depletion capacitance at 0 V	120	fF
ϕ_{iC}	VJC	BC built-in voltage	0.45	V
MC	MJC	BC grading coefficient	0.25	
τ_R	TR	Effective transit time (reverse)	250	ps
C_{jso}	CJS	CS depletion capacity at 0 V	240	fF
ϕ_{iS}	VJS	CS built-in voltage	0.60	V
MS	MJS	CS grading coefficient	0.40	
R_B	RBM	Base resistance at large I_C	850	Ω
R_E	RE	Emitter resistance	2	Ω
R_C	RC	Collector resistance	150	Ω

Model Frame

Due to its vertical construction many external components influence the actual internal transistor. These external elements are part of a model frame needed for an adequate description of the terminal behavior of the transistor. In the simplest case this is achieved by adding the resistors as shown in Figures 3.33 and 3.34. Depending on the required accuracy and frequency range, more elaborate model frames may be needed. An example is illustrated in Figure 3.36. By considering the model frame one has to keep in mind that the lumped elements of the frame represent only, to a particular degree, the distributed elements of an actual transistor.

Base-collector diodes

The current and charge behavior of the external base-collector junction is modeled by the two diodes D_{BC1} and D_{BC2}.

Collector-substrate diodes

Similar to the previous case, the diodes D_{CS1} and D_{CS2} describe the collector-substrate junction.

Collector resistors

These resistors include the contributions by the epitaxy layer R_{C1}, the n^+-region of the buried layer R_{C2}, and the contact resistance R_{C3}. The effect the total collector resistance has on the output characteristic of the transistor is shown in Figure 3.37.

As is obvious from this figure, the collector resistance influences the voltage saturation region of the transistor substantially.

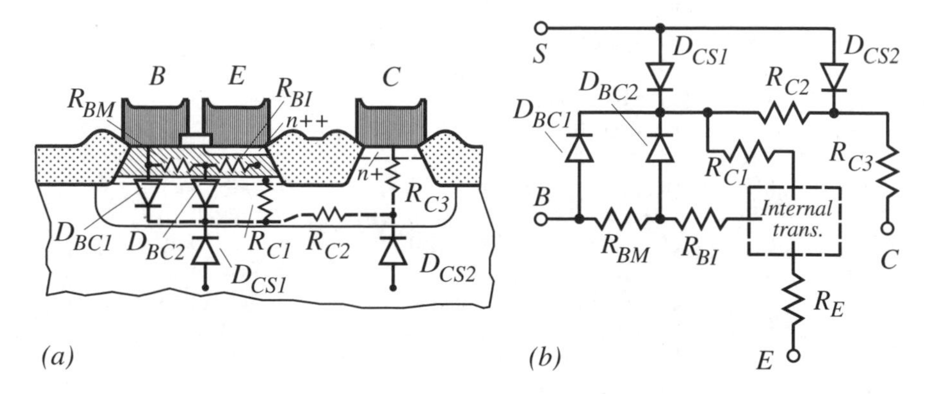

Figure 3.36 (a) Cross-section of an npn transistor (base region exaggerated) and (b) model frame

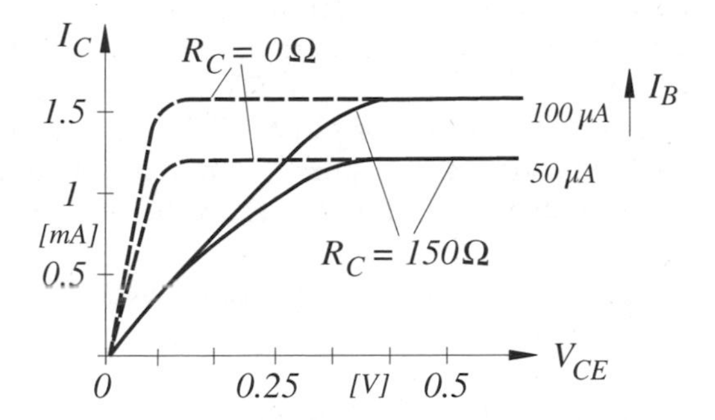

Figure 3.37 Influence of the collector resistance on the output characteristic of a npn transistor

Emitter resistor

The emitter is the region of the transistor with the highest doping concentration. Therefore the emitter resistance R_E is usually very low in the region between 1 and 3 Ω.

Base resistors

The base resistance can be divided into an external resistance R_{BM} and an internal resistance R_{BI}. The first one has a substantially smaller value due to its higher doping and larger out-diffusion of dopant atoms than the second, the internal base spreading resistance (Section 3.3.3). This spreading resistance has a typical collector current dependence as illustrated in Figure 3.38. This characteristic is frequently modeled by the equation

$$R_{BI} = RBM + \frac{RB + RBM}{q_b} \tag{3.95}$$

where the normalized charge (Equation 3.12)

$$q_b = \frac{Q_B}{Q_{Bo}} = \frac{qA\int_{o}^{x_B} p_B(x)\,\mathrm{d}x}{qA\int_{o}^{x_{Bo}} N'_{AB}(x)\,\mathrm{d}x} \tag{3.96}$$

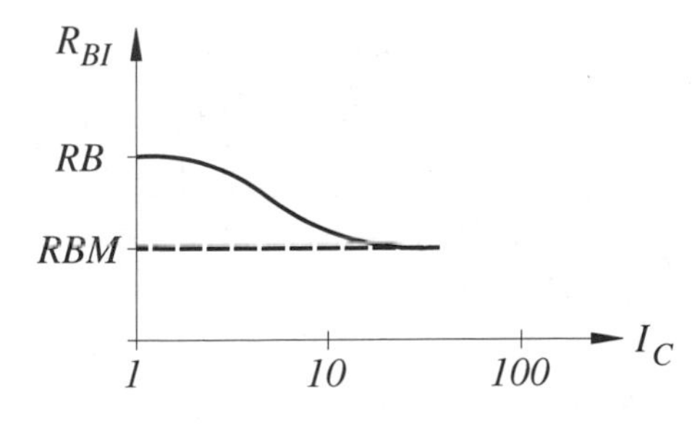

Figure 3.38 Base spreading resistance as a function of the collector current

includes the conductivity modulation effect. An increase in injected minority carriers is compensated simultaneously by majority carriers in the base. This behavior is comparable to that of the majority carriers explained in Section 1.5. The increase of majority carriers causes a reduction of the base resistance and consequently the conductivity modulation. Equation (3.96) takes, furthermore, the base width modulation effect into account (Section 3.3.2), when the integration limits are adjusted accordingly.

3.5.2 Transistor Model for Static Calculations

The presented Gummel–Poon model is used in computer-aided applications and is not suitable for static and dynamic hand calculations due to its complexity. But easy to use equations are required and are very useful in order to perform first-hand circuit calculations before or during the computer-aided design phase.

The starting point is the large-signal model of Figure 3.10, repeated in Figure 3.39 for convenience. The circuit consists of the BE-diode and a voltage-controlled current generator (Equation 3.7)

$$I_C = I_{SS}\left(e^{V_{BE}/\Phi_t} - 1\right)$$

Through the BE-diode passes a base current (Equation 3.8) of

$$I_B = \frac{I_{SS}}{B_F}\left(e^{V_{BE}/\Phi_t} - 1\right)$$

As pointed out in Section 2.7.2, these equations may lead to transcendental equations when used in a circuit configuration. This can be avoided when the equivalent circuit of Figure 3.39(c) is used, where the diode is replaced by a voltage source with a value around 0.8 V.

3.5.3 Transistor Model for Small-Signal Calculations

The small-signal excitation of a bipolar transistor and its effect on the output and input characteristics is the starting point for the derivation of the small-signal model (Figure 3.40).

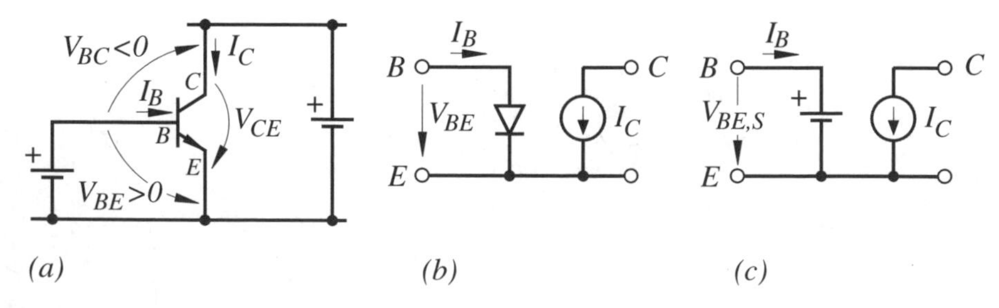

Figure 3.39 (a) npn Transistor under forward biased condition, (b) equivalent circuit and (c) simplified equivalent circuit with an input voltage source

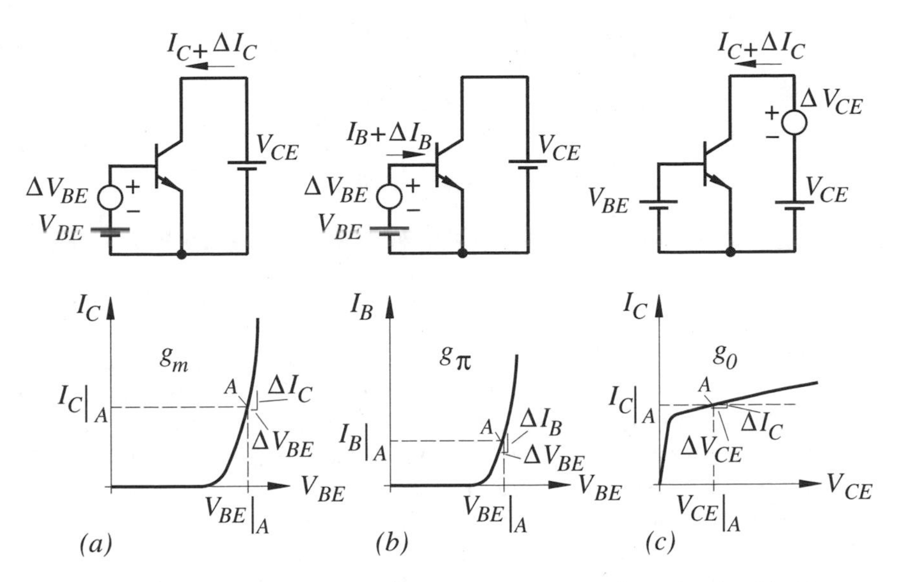

Figure 3.40 Small-signal excitation of a bipolar transistor: (a) I_C (V_{BE}) with V_{CE} constant, (b) I_B (V_{BE}) with V_{CE} constant, and (c) I_C (V_{CE}) with V_{BE} constant

Its effect can be described by the following small-signal conductance parameters, at a particular operation point (A).

Transconductance

$$g_m = \frac{\partial I_C}{\partial V_{BE}} \tag{3.97}$$

Input conductance

$$g_\pi = \frac{\partial I_B}{\partial V_{BE}} \tag{3.98}$$

Output conductance

$$g_o = \frac{\partial I_C}{\partial V_{CE}} \tag{3.99}$$

When all small-signal voltages – all referenced to the emitter – are changed simultaneously (total derivative) the following collector and base current changes occur

$$\begin{aligned}\Delta I_C &= \frac{\partial I_C}{\partial V_{BE}}\Delta V_{BE} + \frac{\partial I_C}{\partial V_{CE}}\Delta V_{CE} \\ &= g_m \Delta V_{BE} + g_o \Delta V_{CE}\end{aligned} \tag{3.100}$$

$$\begin{aligned}\Delta I_B &= \frac{\partial I_B}{\partial V_{BE}}\Delta V_{BE} \\ &= g_\pi \Delta V_{BE}\end{aligned} \tag{3.101}$$

According to Kirchhoff's law these relationships can be represented by the equivalent circuit shown in Figure 3.41(a).

An extended version of the small-signal model is the so-called hybrid-π model (Figure 3.41(b)). It results when the small-signal resistances of the base and the collector regions are included and the small-signal capacitances of the junctions are added. The small-signal equivalent circuits are also valid when, instead of infinitesimal voltage and current changes, infinitesimal sinusoidal excitations are present. *This is indicated throughout the book by small letters, e.g. i_C and v_{be}. Furthermore small letter symbols or indices are used to indicate small-signal components, e.g. r_b and C_{js}.*

Using the transport equations, the following small-signal conductance parameters can be found.

Transconductance

Differentiating Equation (3.7) yields

$$\boxed{g_m = \frac{\partial I_C}{\partial V'_{BE}} = I_C/\phi_t} \tag{3.102}$$

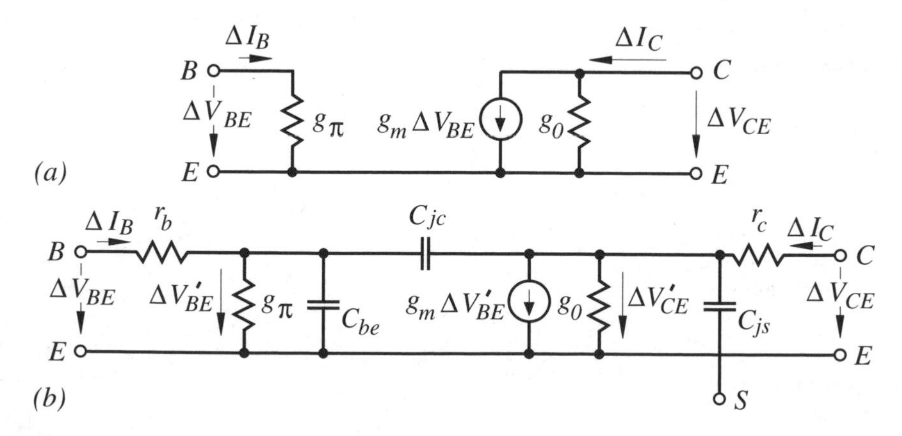

Figure 3.41 Small-signal equivalent circuit: (a) at very low frequency and (b) at medium high frequency (r_E neglected)

The transconductance is proportional to the collector current. This feature is a substantial advantage when the transistor is used for voltage amplification (Section 10.4.2).

Input conductance

Rewriting Equation (3.98) and substituting Equation (3.102) results in an input conductance of

$$\boxed{g_\pi = \frac{\partial I_B}{\partial V'_{BE}} = \frac{\partial I_B}{\partial I_C}\frac{\partial I_C}{\partial V'_{BE}} = \frac{g_m}{\beta_F}} \tag{3.103}$$

where

$$\boxed{\beta_F = \frac{\partial I_C}{\partial I_B}} \tag{3.104}$$

is the small-signal current gain of the transistor. When B_F is independent of I_C around the operation point (A) then $B_F \approx \beta_F$, which is applicable in most practical cases.

Output conductance

An output conductance value different from $g_o = 0$ is caused by the base-width modulation of the transistor (Section 3.3.2). Using the preceding definition and Equation (3.63) leads to

$$g_o = \frac{\partial I_C}{\partial V'_{CE}} = -\frac{\partial I_C}{\partial V'_{BC}}$$

$$\boxed{g_o = \frac{I_{Co}}{V_{AF}} \approx \frac{I_C}{V_{AF}}} \tag{3.105}$$

where the relationship $\partial V'_{CE} = \partial V'_{CB} = -\partial V'_{BC}$ is applied.

Charge behavior

This behavior is included in the small-signal model by the insertion of small-signal junction capacitances. In the forward direction with $V_{BE} > 0$, $V_{BC} < 0$, and $V_{SC} < 0$ Equations (3.91) to (3.93) and Equation (3.102) yield the following relationships

$$C_{bc} = \frac{dQ_{BC}}{dV'_{BC}} = C_{jc} = C_{jco}\left(1 - \frac{V'_{BC}}{\phi_{iC}}\right)^{-MC} \tag{3.106}$$

$$C_{sc} = \frac{dQ_{SC}}{dV'_{SC}} = C_{js} = C_{jso}\left(1 - \frac{V'_{SC}}{\phi_{iS}}\right)^{-MS} \tag{3.107}$$

$$C_{be} = \frac{dQ_{BE}}{dV'_{BE}} = C_{de} + C_{je}$$

$$= \tau_F g_m + C_{jeo}\left(1 - \frac{V'_{BE}}{\phi_{iE}}\right)^{-ME}. \tag{3.108}$$

Since in the forward direction the depletion capacitance C_{je} is difficult to determine (Section 2.4.1), the approximation

$$C_{be} \approx \tau_F g_m + 2C_{jeo} \tag{3.109}$$

is frequently encountered.

3.5.4 Transit Time Determination

The high frequency behavior of the transistor is governed dominantly by the capacitive elements. A figure of merit is the transit frequency f_T. That is the frequency where the magnitude of the small-signal current gain in a short-circuited common emitter configuration falls to unity (Figure 3.42).

A small-signal current i_b is applied to the base, causing a small-signal current i_o at the output of the circuit. Applying Kirchhoff's law to the small-signal equivalent circuit of Figure 3.42(b) and using Equation (3.103) leads to the following transfer function or the small-signal current gain

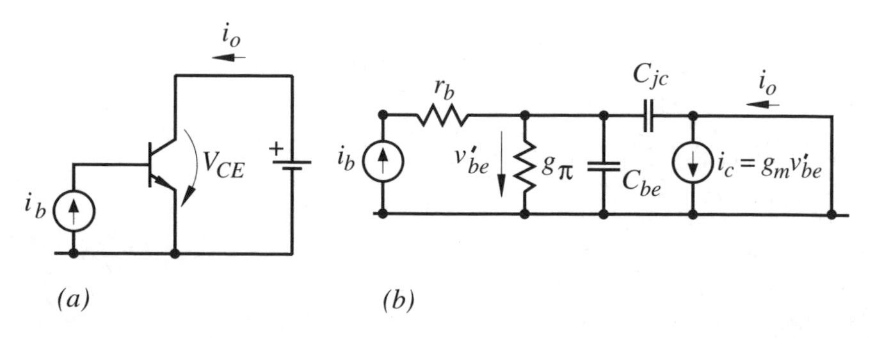

Figure 3.42 (a) Schematic circuit for f_T determination and (b) small-signal equivalent circuit (r_c and r_e neglected)

$$\beta(j\omega) = \frac{i_o}{i_b}(j\omega) = \beta_F \frac{\left(1 - j\frac{\omega}{\omega_z}\right)}{\left(1 + j\frac{\omega}{\omega_p}\right)} \tag{3.110}$$

as a function of the angular frequency $\omega = 2\pi f$ (Problem 3.10). In this equation

$$\beta(j\omega \to 0) = \beta_F \tag{3.111}$$

the small-signal current gain of the circuit is at low frequency. Details for the solutions of transfer functions are included in Chapter 8, Appendix A. The zero-(angular) frequency has a value of

$$\omega_z = \frac{g_m}{C_{jc}} \tag{3.112}$$

and the pole-(angular) frequency (−3 dB bandwidth) a value of

$$\omega_p - \frac{g_m}{\beta_F(C_{be} + C_{jc})} \tag{3.113}$$

Since ω_z is always substantially larger than ω_p the transfer function can be presented in the simplified version

$$\beta(j\omega) = \beta_F \frac{1}{1 + j\frac{\omega}{\omega_p}} \tag{3.114}$$

At high frequencies the imaginary part of the denominator is dominant, allowing the equation to be rewritten in the form

$$\beta(j\omega) = \beta_F \frac{1}{j\frac{\omega}{\omega_p}} \tag{3.115}$$

This leads to the transit (angular) frequency of

$$\omega_T = \omega_p \beta_F = \frac{g_m}{C_{be} + C_{jc}} \tag{3.116}$$

where $|\beta(j\omega)| = 1$. It is probably not unexpected that the transit (angular) frequency is highest when the capacitances are smallest. More insight can be gained when the characteristic frequencies ω_p and ω_T are expressed as a function of the collector current. Using Equations (3.102) and (3.108) these frequencies can be rewritten in the form

$$\omega_p = \frac{1}{\beta_F \left[\tau_F + \frac{\phi_t}{I_C}(C_{je} + C_{jc})\right]} \tag{3.117}$$

and

$$\omega_T = \frac{1}{\tau_F + \frac{\phi_t}{I_C}(C_{je} + C_{jc})} \tag{3.118}$$

An increase in collector current will yield – because of $1/I_C$ – the following maximum values of the characteristic frequencies

$$\omega_{pmax} = \frac{1}{\beta_F \tau_F} \tag{3.119}$$

and

$$\omega_{Tmax} = \frac{1}{\tau_F} \tag{3.120}$$

The transfer function Equation (3.114) is plotted as a Bode plot in Figure 3.43 (Chapter 8, Appendix A), indicating the characteristic frequencies.

In the discussion so far, second-order effects are neglected. How they influence the transit time in reality is shown in Figure 3.44. Equation (3.118) does not predict the observed reduction in transit time when the collector current is increased. The high-level injection is responsible for this effect (Section 3.31). In this case the base region stretches out into the collector and becomes greatly enlarged, which leads to an increased transit

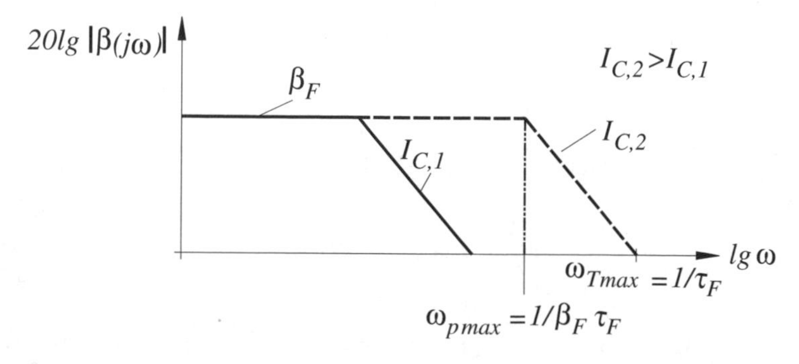

Figure 3.43 Idealized transfer function for the circuit of Figure 3.42

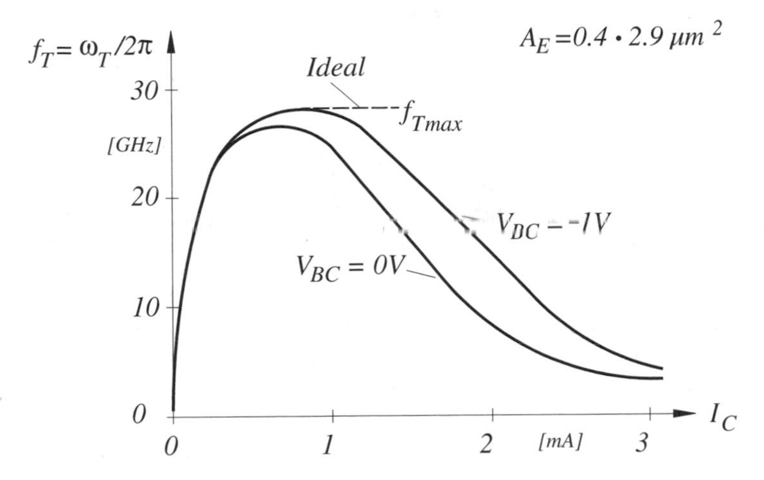

Figure 3.44 Transit frequency as a function of the collector current. Reproduced by permission of Infineon (2000)

time in the base. The observed increase in transit frequency with the V_{BC} voltage is caused by the base-width modulation effect, described in Section 3.3.2.

The maximum transit time has in analogy to Equation (2.57) a value of

$$\boxed{f_{Tmax} = \frac{1}{2\pi}\frac{1}{\tau_F} = \frac{1}{\pi}\frac{D_{nB}}{x_B^2}} \tag{3.121}$$

which is proportional to the square of the base width. This equation demonstrates impressively why modern bipolar processes with very small base width achieve transit

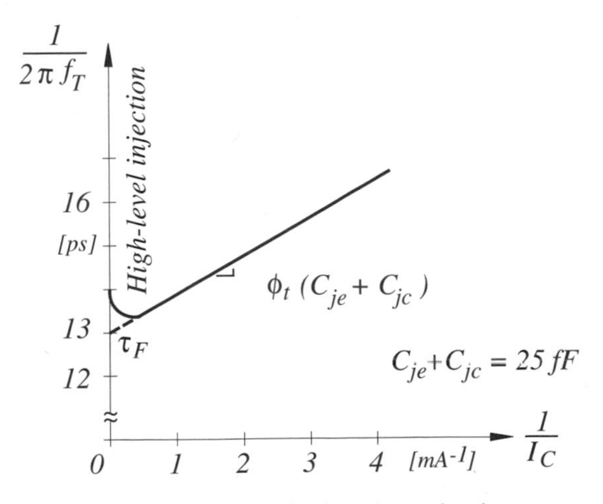

Figure 3.45 Transit time determination

frequencies in the GHz region. But the equation demonstrates also, how a variation in base width during the fabrication process inadvertently can influence the frequency performance.

Sometimes Equation (3.118) is presented in reciprocal form

$$\frac{1}{\omega_T} = \frac{1}{2\pi f_T} = \tau_F + \phi_t(C_{je} + C_{jc})\frac{1}{I_C} \tag{3.122}$$

and linearly plotted as shown in Figure 3.45. Extrapolating the function to $1/I_C$ leads directly to the transit time τ_F. A prerequisite for doing so is that the gradient $\phi_t(C_{je} + C_{jc})$ is almost constant. This is usually the case if the collector current does not vary by more than 1–5 (Problem 3.8).

Summary of the most important results

The transistor action is caused by the fact that with $V_{BE} > 0$ majority carriers of the base are injected into the emitter (base current) and simultaneously majority carriers of the emitter move without losses through the base to the collector, causing a collector current. The ratio of collector to base current describes the static current gain. Just as important as the current gain is the transport current. This is not a saturation current but a current parameter depending on base parameters only.

Second-order effects influence the transistor performance substantially. At high-level injection the base region stretches out into the collector and becomes greatly enlarged, reducing the current gain and transit frequency. It is therefore not advisable to operate the transistor in this region where $I_C > I_{KF}$. An increase in the V_{CE} voltage causes an increase in the BC-depletion region and a reduction of the base width. In conjunction the collector current increases. This base-width modulation effect is characterized by the Early voltage V_{AF}. With I_C/V_{AF} the small-signal output conductance can be determined. The transit frequency – that is, the frequency where the magnitude of the small-signal current gain is 1 – can be considered to be a figure of merit for the frequency behavior of the transistor. Because of its $(1/x_B)^2$ behavior, the transit frequency depends strongly on the base width and its variation during processing.

Problems

3.1 In Figure 3.4 the doping profile of an npn transistor is shown. Estimate – by assuming average uniform doping profiles – the transport current and the collector current when $V_{BE} = 750\,\text{mV}$. How large is the current gain in the forward and reverse biased direction approximately?

Data of the transistor: $A_E = 4\,\mu\text{m}^2$, $D_{pE} \approx 12\,\text{cm}^2/\text{s}$, $D_{nB} \approx 25\,\text{cm}^2/\text{s}$, $D_{pC} \approx 14\,\text{cm}^2/\text{s}$.

Hint: The buried collector contact can be considered to have a very low resistance.

3.2 An npn transistor is implemented in a p-well CMOS process.

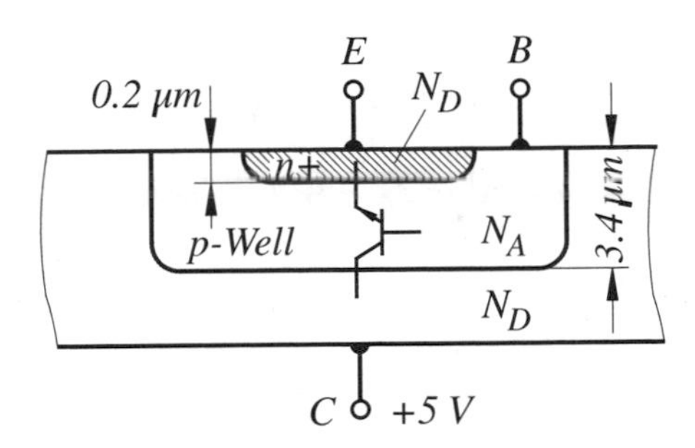

The data of the process are: N_D (source doping) $= 5 \cdot 10^{19}\,\mathrm{cm}^{-3}$, N_A (p-well doping) $= 10^{17}\,\mathrm{cm}^{-3}$, emitter area $A_E = 10 \cdot 10\,\mu\mathrm{m}^2$, $D_{nB} \approx 25\,\mathrm{cm}^2/\mathrm{s}$, $D_{pE} \approx 8\,\mathrm{cm}^2/\mathrm{s}$. Calculate for room temperature:

(a) The transport current.

(b) The collector current at $V_{BE} = 700\,\mathrm{mV}$.

(c) The current gain.

For simplification it can be assumed that the base width corresponds to the metallurgical depth of the p-well at 3.2 μm.

3.3 The circuit shown has the following transistor data: $B_F = 200$, $B_R = 2$, and $I_{SS} = 10^{-15}\,\mathrm{A}$.

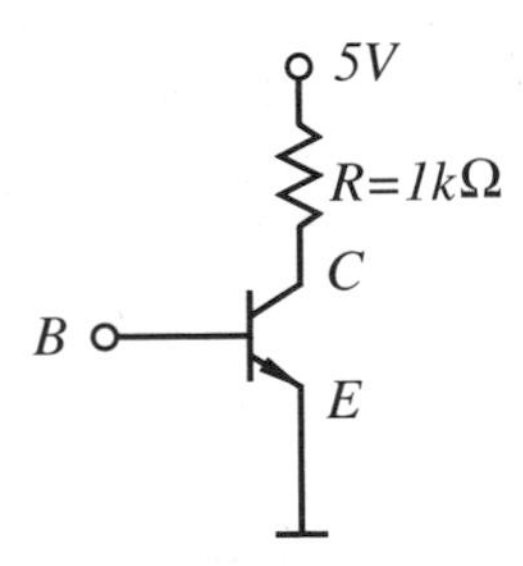

Calculate:

(a) The collector current, base current, and the V_{CE} voltage when the V_{BE} voltage has a value of 700 mV. The transistor operates under current saturation conditions.

(b) The base current needed to drive the transistor into voltage saturation with a saturation voltage V_{CEsat} of 100 mV.

3.4 The transistor in the circuit shown has a current gain of $B_F = 100$.

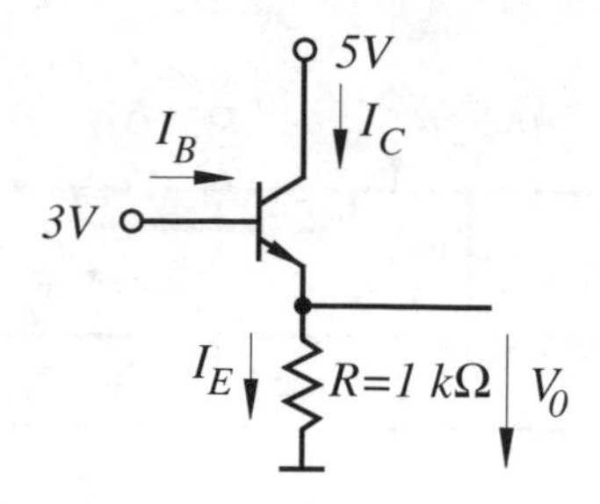

Determine:

(a) The base current.

(b) The collector and emitter current.

(c) The output voltage V_0.

3.5 A transistor has the following technological data: $N'_{AB} = 10^{17}\,\text{cm}^{-3}$, $N'_{DC} = 10^{16}\,\text{cm}^{-3}$, x_B (at $V_{BC} = 0\,\text{V}) = 1\,\mu\text{m}$, $D_{nB} = 18\,\text{cm}^2/\text{s}$, and $A_E = 10\,\mu\text{m}^2$. For simplicity reasons uniform doping densities are assumed. Calculate:

(a) The Early voltage V_{AF}.

(b) The gradient dI_C/dV_{BC} at $V_{BC} = 0\,\text{V}$ and $V_{BE} = 700\,\text{mV}$.

3.6 The criterion for the beginning of current crowding is given when the average voltage drop across the base is larger than ϕ_t. Determine the collector current of an npn transistor at the onset of current crowding at room temperature. The transistor data are: $N'_{AB} = 10^{17}\,\text{cm}^{-3}$, $\mu_p = 400\,\text{cm}^2/\text{Vs}$, $x_B = 1\,\mu\text{m}$, $B_F = 50$, one-sided base contact, $l_E = 30\,\mu\text{m}$, and $b_E = 20\,\mu\text{m}$. For simplicity, uniform doping concentrations are assumed.

3.7 The transistor configuration shown is used as a diode.

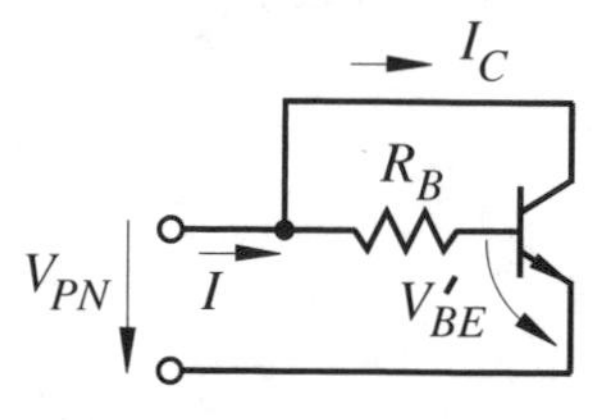

The current through the circuit is $I = 80\,\text{mA}$ and the voltage across the terminals has a value of $V_{PN} = 975\,\text{mV}$. Determine the series resistor R_S (Equation 3.88) and the base resistor R_B. The data of the transistor are: $B_F = 150$, $I_{SS} = 10^{-16}\,\text{A}$, $R_E \approx 0\,\Omega$.

3.8 An npn transistor is connected as shown schematically in Figure 3.42. At 1 GHz the following small-signal current gains are measured: $|\beta(j\omega)| = 40$ at $I_C = 1$ mA and $|\beta(j\omega)| = 48$ mA at $I_C = 3$ mA. The small-signal capacitance C_{jc} has a value of 35 fF. Determine the small-signal capacitance C_{je} and the transit time τ_F. It can be assumed that the transistor remains in low-level injection conditions and that C_{je} and τ_F remain constant. The following relationships have to be considered when solving the problem. In the frequency range $\omega/\omega_p = f/f_p \gg 1$ the small-signal current gain at the frequency f_m is according to Equation (3.115)

$$|\beta(jf_m)| = \beta_F \frac{f_p}{f_m}$$

At the transit frequency this equation yields

$$|\beta(jf_T)| = 1 = \beta_F \frac{f_p}{f_T}$$

The ratio of the small-signal current gains at different frequencies

$$\frac{|\beta(jf_m)|}{|\beta(jf_T)|} = |\beta(jf_m)| = \frac{f_T}{f_m}$$

leads to the result

$$f_T = |\beta(jf_m)| \cdot f_m$$

3.9 (a) Determine the small-signal output voltage v_o of the shown amplifier at low frequency when the small-signal input voltage has a value of $v_i = 50\,\mu$V.

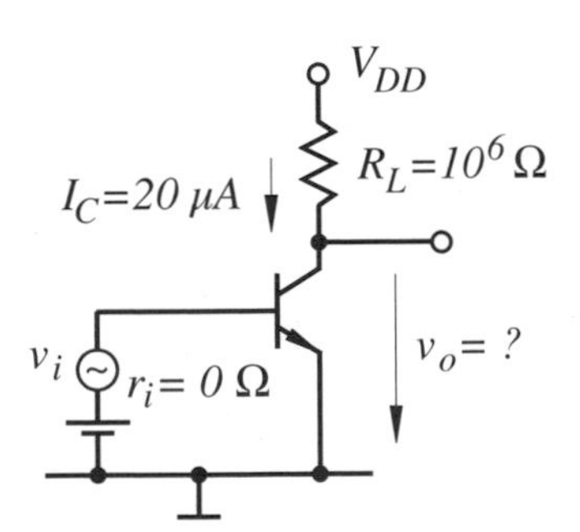

(b) Assuming that the resistor has a value close to infinity – achieved by substituting the resistor with a current generator – determine for this case the small-signal voltage gain. Transistor data are: $V_{AF} = 25$ V and $r_B \approx 0\ \Omega$.

3.10 Use the small-signal equivalent circuit of Figure 3.42(b) to determine the transfer function i_o/i_b of Equation (3.110).

The solutions to the problems can be found under: *www.unibw-muenchen.de/campus/ET4/index.html*

REFERENCES

W.D. Barber (1976) Effective mass and intrinsic concentration in silicon. *Solid-State Electron.*, **10**, 1039–51.

J.R. Baum and R.M. Warner (1965) *Integrated Circuits*. New York: McGraw-Hill.

J.M. Early (1952) Effects of space-charge layer widening in junction transistors. *Proc. IRE*, **40**, 1401–6.

M.I. Elmansry (1983) *Digital Bipolar Integrated Circuits*. New York: John Wiley & Sons.

J.E. Getreu (1978) *Modeling the Bipolar Transistor*. Amsterdam: Elsevier Scientific.

H.K. Gummel and H.C. Poon (1970) An integral charge control model of bipolar transistors. *Bell Syst. Techn. J.*, **49**, 827–52.

Infineon Technology Booklet (2000).

E. Kasper (1996) Silicon germanium hetrodevices. *Appl. Surf. Sci.*, **102**, 189–99.

C.T. Kirk (1962) A theory of transistor cutoff frequency falloff at high current densities. *IRE Trans. Electron. Devices*, **ED-9**, 164–74.

W. Klein and B.U. Klepser (1999) 75 GHz Bipolar-Production Technology for the 21st Century, ESSDERC, pp. 88–94.

H.C. Poon, H.K. Gummel and D.L. Scharfetter (1969) High injection in epitaxial transistors. *IEEE Trans. Electron. Devices*, **ED-16**, 455–7.

E.J. Prinz and J.C. Sturm (1991) Current gain-early voltage products in heterojunction bipolar transistors with nonuniform base bandgaps. *IEEE Electron Device Lett.*, **12**(12), 661–3.

H.M. Rein, Hv. Rohr and P. Wennekers (1978) A contribution to the current gain temperature dependence of bipolar transistors. *Solid-State Electron.*, **21**, 439–42.

Y.P. Tsividis (1980) Accurate analysis of temperature effects in I_C-V_{BE} characteristics with application to bandgap reference source. *IEEE J. Solid-State Circuits*, **SC-15**(6), 1076–84.

FURTHER READING

C.Y. Chang and S.M. Sze (1996) *ULSI Technology*. New York: McGraw-Hill.

G. Massobrio and P. Antognetti (1993) *Semiconductor Device Modeling with SPICE*, 2nd edn. New York: McGraw-Hill.

4

MOS Transistor

The chapter begins with a brief description of a typical CMOS fabrication process. Subsequently a MOS structure is used for the study of the capacity behavior and for the derivation of charge equations. The terms flat-band voltage, body factor, and threshold voltage are introduced. With these prerequisites the current–voltage equations of the MOS transistor are derived. To ease the derivation a distinction between strong and weak inversion is made. Second-order effects, e.g. mobility degradation, short channel effects and hot electron effects, channel length modulation, and breakdown behavior, are analyzed. At the end of this chapter aspects of modeling the transistor for CAD applications are discussed and simple equations, applicable for first hand circuit estimates, are given.

4.1 CMOS TECHNOLOGY

Basically, transistors can be grouped into devices which are current or voltage controlled. As is shown in Chapter 3 bipolar transistors belong to the first group whereas field effect transistors belong to the second one. This group can be divided into junction field effect transistors and metal-oxide-semiconductor (MOS) field effect transistors. The latter are of eminent importance in large scale integrated circuits and are therefore treated exclusively in this chapter. Two types of MOS transistors exist, namely n-channel and p-channel ones. When both types are present in a fabrication technology one speaks of a complementary MOS (CMOS) technology (Figure 4.1).

The gate (G) influences via an insulator (SiO_2) the charge carrier density at the semiconductor surface. Depending on the gate voltage a conducting channel develops between the source (S) and the drain (D) of the transistor. The thick Field OXide (FOX) is used as an insulator between the n-channel and p-channel transistors. The substrate of the two transistors is of p-type and n-type, respectively.

The CMOS technique allows the realization of low power circuits with very good signal to noise ratio on an exceptionally small chip area. These features are some of the reasons for the dominating role this technology plays in very large scale integrated circuits. One popular CMOS process with a self-aligned twin well structure and LOCOS (LOCal Oxidation of Silicon) technique, similar to the one used for the bipolar process, is considered next in some detail.

System Integration: From Transistor Design to Large Scale Integrated Circuits. Kurt Hoffmann.
 ISBN: 0-470-85407-3

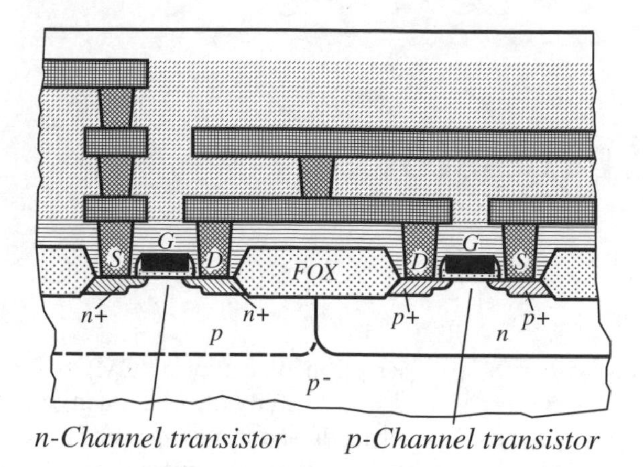

Figure 4.1 Cross-section through an n-channel and p-channel MOS transistor

Well structure

Each CMOS process needs two substrates, namely a p-substrate for the n-channel transistor and an n-substrate for the p-channel transistor. The approach to form the two substrates is called the twin well process. The preferred starting material is a wafer with a p-type substrate. In a first step the wafer is covered with a thin silicon dioxide (SiO_2) and a silicon nitride (Si_3N_4) layer. Afterwards a photoresist is added and exposed by light through the first mask which contains the information of the n-well geometries. After developing the photoresist and some further processing steps the sandwich layer is removed from the exposed area. Subsequently the wafer is implanted with phosphorus atoms, whereby the sandwich layers act as masking (Figure 4.2(a)). During the diffusion of the phosphorus atoms at an elevated temperature a thick oxide layer grows over the n-well regions which are not covered by the sandwich layers. After the localized oxidation of silicon, the masking layers are etched off (Figure 4.2(b)). During the following p-well implantation with boron atoms the thick oxide layer guarantees a self-alignment of the p-well implant and diffusion to the n-well edge (Figure 4.2(b)).

Local oxidation

After stripping the oxide the wafer is covered with SiO_2, polysilicon (Poly-Si), and Si_3N_4 layers, which are subsequently patterned (Figure 4.2(c)). With the additional polysilicon layer the so-called bird's beak (Figure 4.2(d)) is reduced during the following oxidation step. FOX grows where no silicon is covered by the triple stack (Figure 4.2(d)). After removing the masking layers a boron implantation is performed over the wafer (not shown) in order to adjust the threshold voltage of the n-channel transistors. With the additional use of a lithographic step (not shown) the threshold voltage of the p-channel transistors is adjusted by selectively implanting arsenic and boron into the n-wells. Then a thin gate oxide of thickness d_{ox} is grown over the wafer (Figure 4.2(d)).

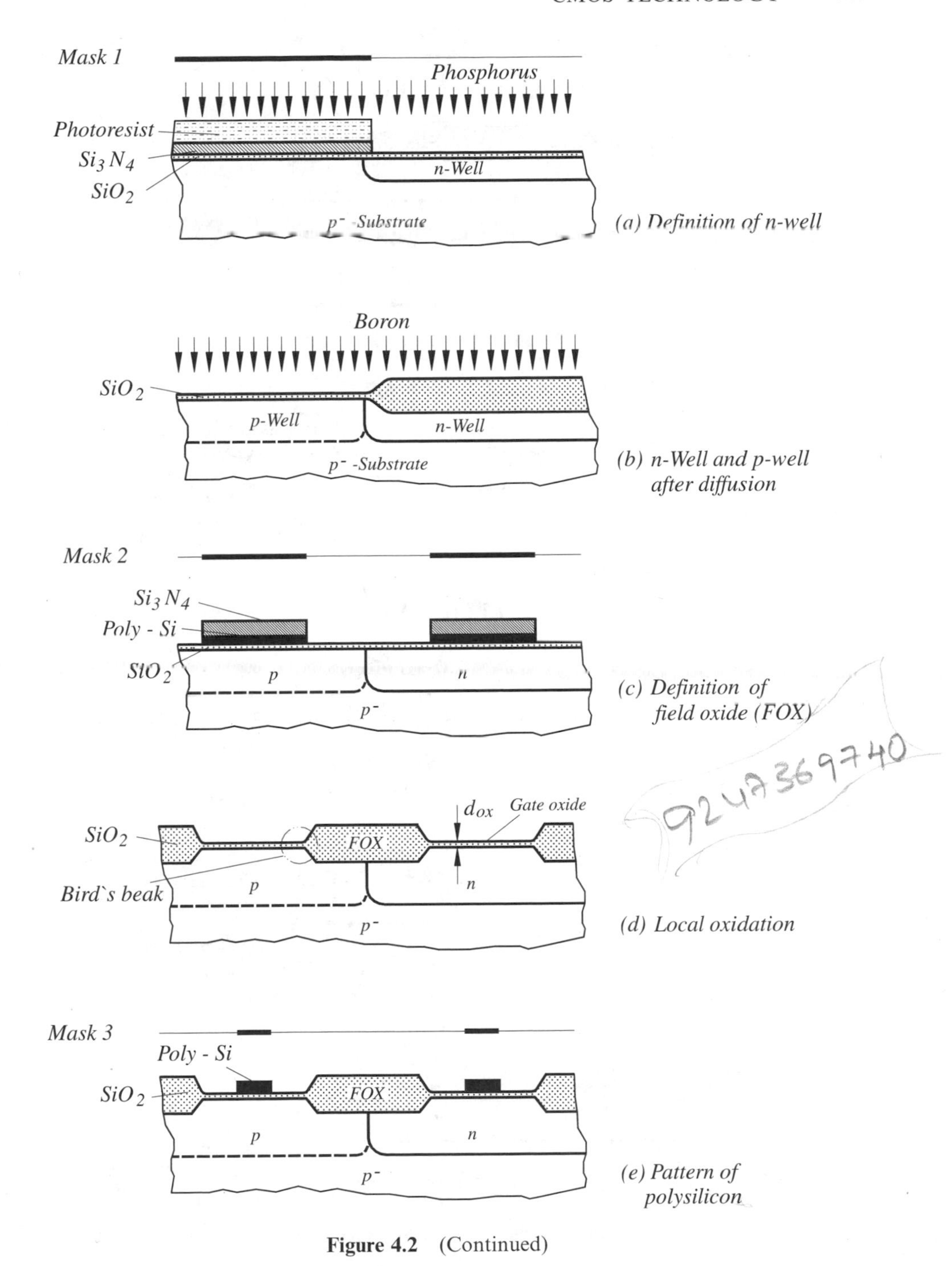

Figure 4.2 (Continued)

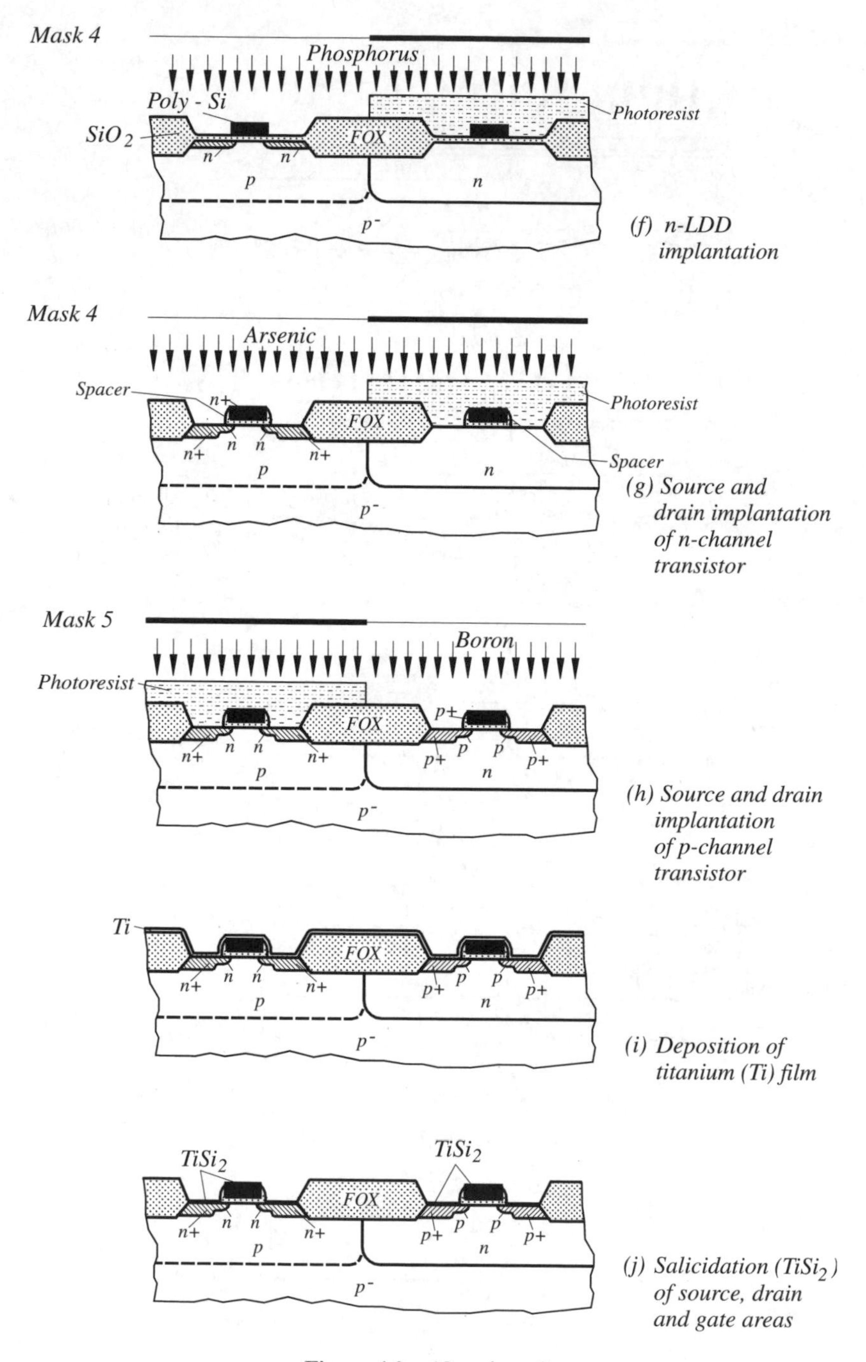

Figure 4.2 (Continued)

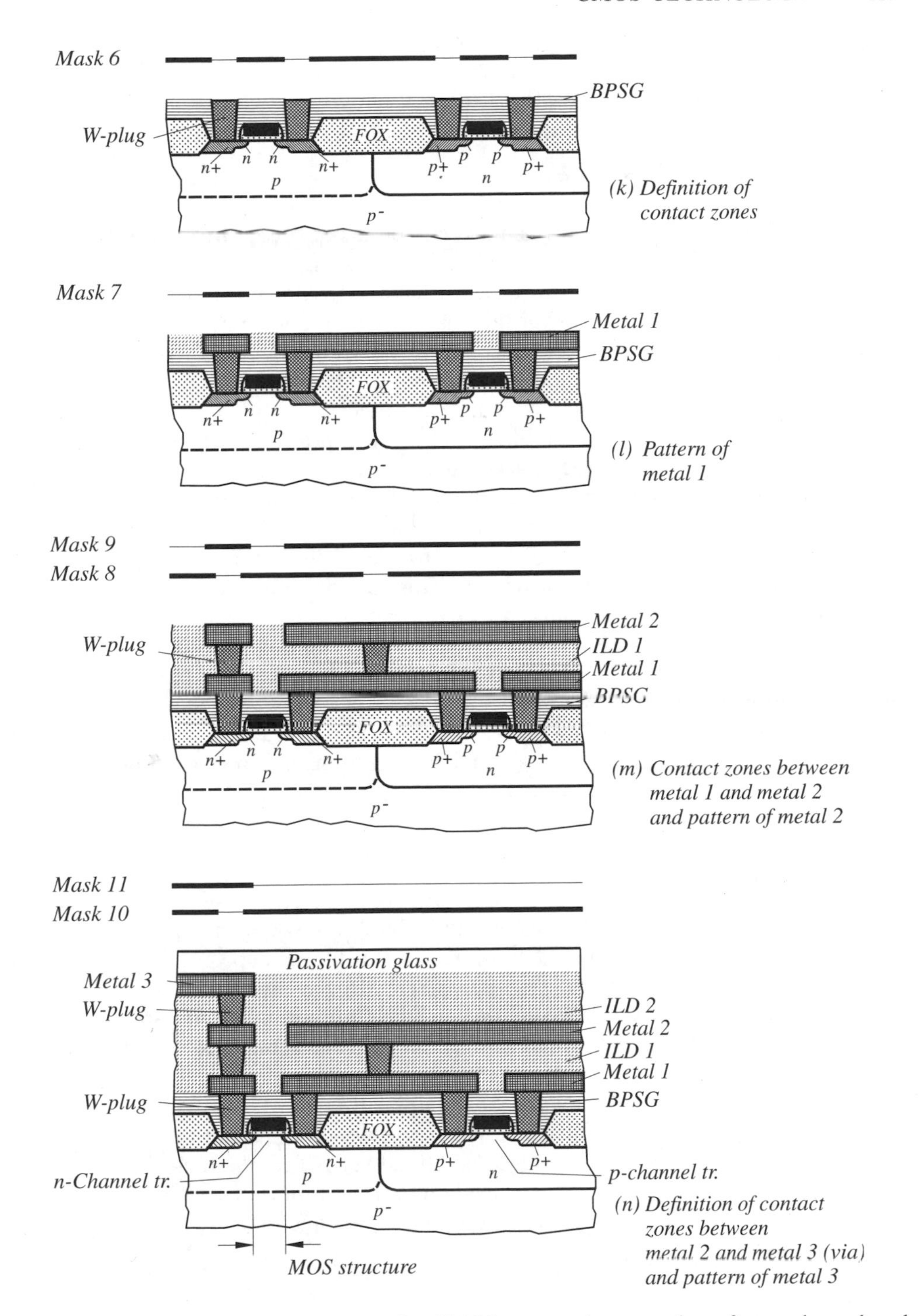

Figure 4.2 Basic fabrication sequence of a CMOS process (cross-section of an n-channel and p-channel MOS transistor)

Pattern of polysilicon

Next polysilicon is deposited over the wafer and patterned (Figure 4.2(e)).

LDD source and drain implantation

In scaled-down devices hot electrons may degrade the device performance. In order to ease the problem lightly doped drain (LDD) (source) structures are used (Section 4.5.4). Since the n-channel and p-channel transistors have different source and drain dopings a lithographic step is required to cover the ones which are not supposed to be implanted (Figure 4.2(f)). First a light phosphorus implantation dose is used to form the LDD regions of the n-channel transistors. A very important feature of this process step is that the gate electrode acts as a mask for the implanted dopant. This guarantees a self-alignment of the source and drain electrodes with respect to the gate. In subsequent processing steps side wall oxide spacers around the edges of the gates are formed. A second heavier implantation dose – this time with arsenic – forms the connecting regions of the source and drain that merge into the lightly doped regions (Figure 4.2(g)). Since the heavier second dose is self-aligned to the spacer and further away from the gate edges its depth can be made larger. This has the advantage that the contact resistance can be decreased substantially.

Next the n-channel transistors are covered with a photoresist mask and a boron implantation is performed. The gate of the p-channel transistors act as a mask – just as in the case of the n-channel transistors – for the self-alignment of the source and drain regions. Due to the spacer and the vertical and horizontal diffusion properties of boron, LDD regions and contact regions are formed in one implantation step (Figure 4.2(h)).

Salicidation

In the Self-Aligned siLICIDATION (SALICIDATION) process all source and drain regions and the polysilicon gates are covered with a low resistive metallic film. To start this process, a thin low resistive titanium (Ti) metal film is deposited over the entire wafer (Figure 4.2(i)). Then the wafer is heated in a nitrogen environment which causes a silicidation reaction, whereby silicon atoms diffuse from source, drain, and polysilicon regions into the titanium. Everywhere else the metal remains unchanged. The unaltered metal is then selectively removed by an etchant. As a consequence only the polysilicon gates and the source and drain regions are completely covered by a self-aligned silicide film (Figure 4.2(j)). Due to this feature low resistive drain, source, and gate regions with only a couple of $\Omega/\Box$ result. An additional advantage is that the highly n-doped and p-doped gates are metallic after the salicidation. Therefore they can be tied together directly in a circuit without producing a polysilicon pn-junction.

Contacts and metallization

After the salicidation an SiO_2 layer is grown. This layer (not shown) acts as a diffusion barrier between the silicon and the deposited boron phosphorus silicate glass (BPSG)

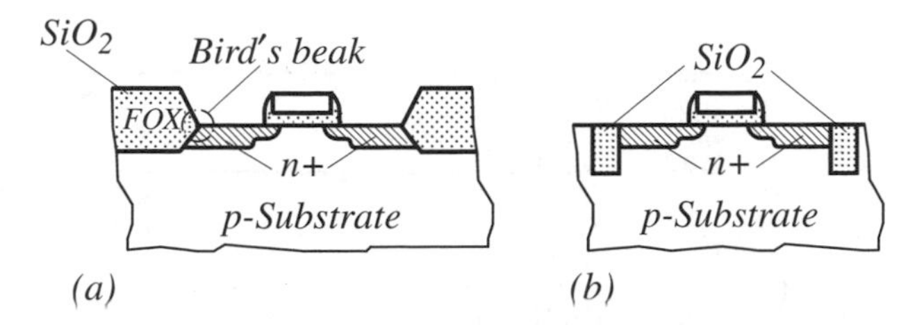

Figure 4.3 Isolation techniques: (a) LOCOS isolation and (b) trench isolation

film. This glass is used for isolation and smoothing of edges at the semiconductor surface. The demand for more levels of metal wiring can only be met when a good planarity of the wafer surface can be achieved. This is done with chemical and mechanical polishing processes. Then a lithographic step is applied for the definition of contacts. In a subsequent etching process the contact holes are open (Figure 4.2(k)) and covered with a Ti/TiN film (not shown). The thin Ti layer forms with the Si a low resistive $TiSi_2$ contact, whereas the TiN layer acts as a barrier between the Si and the subsequent deposited tungsten. This metal is used to fill the trenches (W-plug). Then the wafer is metallized by, e.g. AlSiCu and patterned (Figure 4.2(l)). Next an inter level dielectric (ILD) is deposited, which fills small gaps between metal lines. Subsequently trenches are etched into the ILD using mask 8. Then the trenches are filled with tungsten. Next metal 2 is sputtered and patterned (mask 9). The electrical connection between the two metal layers is called via (hole). Metal layer 3 (Figure 4.2(n)) uses the same process steps as described by metal layer 2 but applies masks 10 and 11 instead. Then a passivation glass is added for scratch protection. In a final step (not shown) pads are opened for connections to the outside. The silicon wafer is then diced into separate pieces, each containing an integrated circuit. These chips are glued to packages. Then wires are connected from the leads of the package to the pads of the chip. Finally the packages are sealed, e.g. with plastic.

As feature sizes in very large scale integrated circuits continue to decrease, a corresponding reduction in the isolation islands between circuit elements is required. This reduction is limited in the presented CMOS process, by the LOCO technique used with its bird's beak, to about 0.25 μm feature size. This can be circumvented by anisotropic etching trenches into the silicon substrate and filling them with suitable dielectric e.g. SiO_2 (Figure 4.3).

4.2 THE MOS STRUCTURE

The MOS structure can be considered to be the basic element of a MOS transistor (Figure 4.2(n)). It consists of a metal electrode – implemented with a highly doped n^+-polysilicon layer – influencing via an thin isolation layer (SiO_2) the condition of the underlying semiconductor surface (Figure 4.4).

A p-type substrate which leads to an n-channel transistor is assumed throughout this chapter. To start with, the MOS structure is considered to be in equilibrium. This implies that the Fermi level is constant throughout the total structure (Section 2.1).

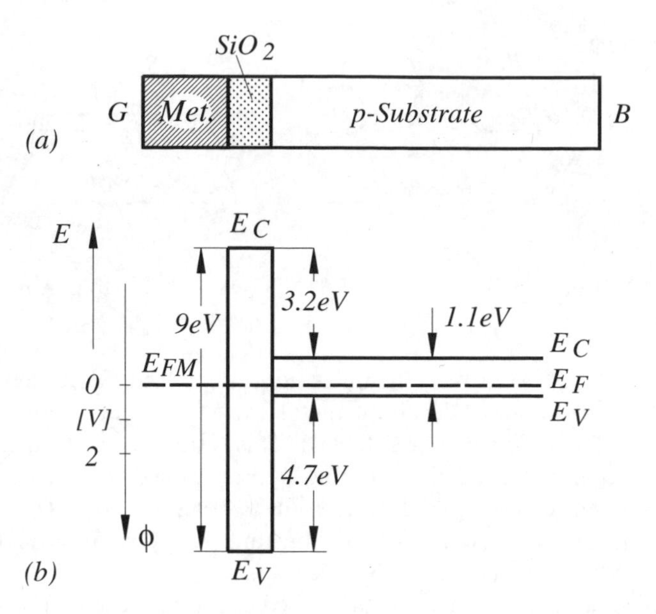

Figure 4.4 (a) MOS structure and (b) band diagram of the MOS structure in equilibrium

Since the isolator is free of charge, the band edges indicate the energy which electrons or holes have to gain in order to overcome the isolation barriers.

4.2.1 Characteristic of the MOS Structure

The MOS structure can be brought into three stationary conditions, namely accumulation, depletion, and inversion, by applying a voltage V_{GB} between the gate G and bulk B terminals. The Fermi level at the backside of the substrate is selected as the reference point (see Section 1.3.4 for the relationship between energy and applied voltages).

Accumulation

If $V_{GB} < 0$, that means the negative pole of the voltage source is applied to the gate, then the electrical field across the insulator is able to attract holes from the p-substrate to the energy barrier of the insulator. Due to the large barrier height the holes are accumulated there and oppose the negative gate charge (Figure 4.5). The band bending corresponds to the voltage drop in each region. At the semiconductor surface the energy difference $E_F - E_V$ is smaller than in the bulk of the substrate. This indicates, according to the density Equation (1.11), an increase in hole density at the surface.

Depletion

The voltage between the terminals of the MOS structure is reversed compared to the previous case. With $V_{GB} > 0$ holes are pushed away from the surface into the substrate.

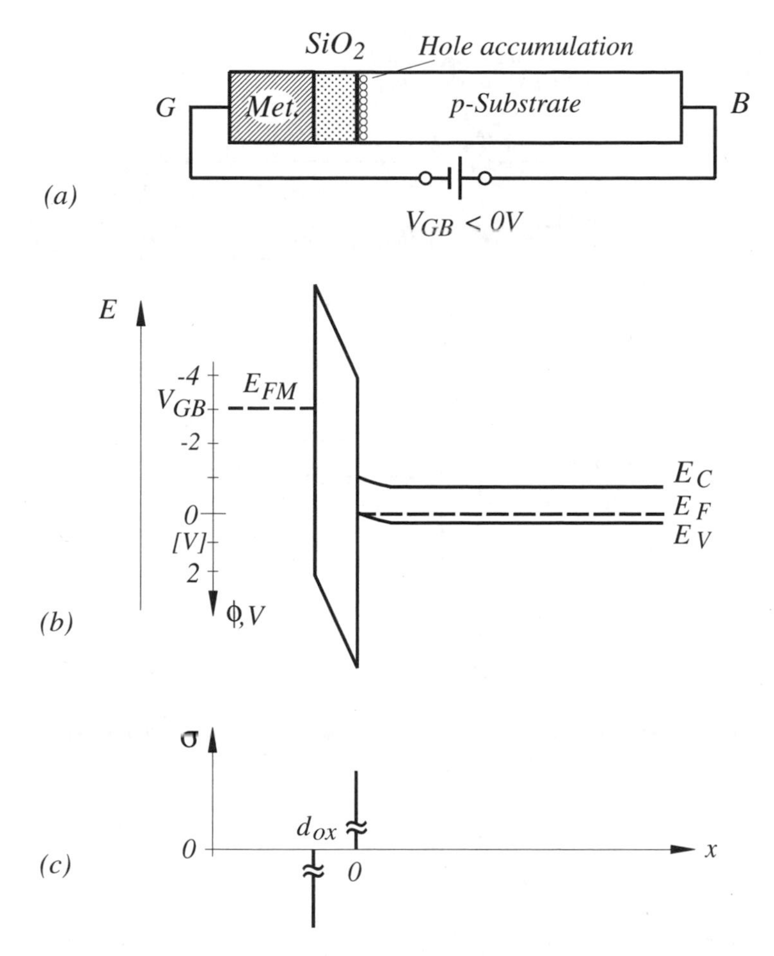

Figure 4.5 (a) MOS structure biased into accumulation, (b) band diagram and (c) charge distribution

This creates at the semiconductor surface a region of width x_d depleted of holes. The negative charge of this depletion region DR consists of uncompensated acceptor atoms opposing the positive gate charge (Figure 4.6).

At the semiconductor side of the MOS structure the Fermi level is constant throughout the substrate. This implies equilibrium condition where generation and recombination of electron–hole pairs compensate each other. The energy difference at the surface $E_F - E_V$ is increased. This is consistent with Equation (1.11), which indicates a hole density reduction.

A depletion effect can also be observed at the polysilicon gate electrode (inset in Figure 4.6(a)) where electrons are pushed away from the gate–oxide interface. Due to the high doping density of the polysilicon gate this effect is much less pronounced than the one in the substrate.

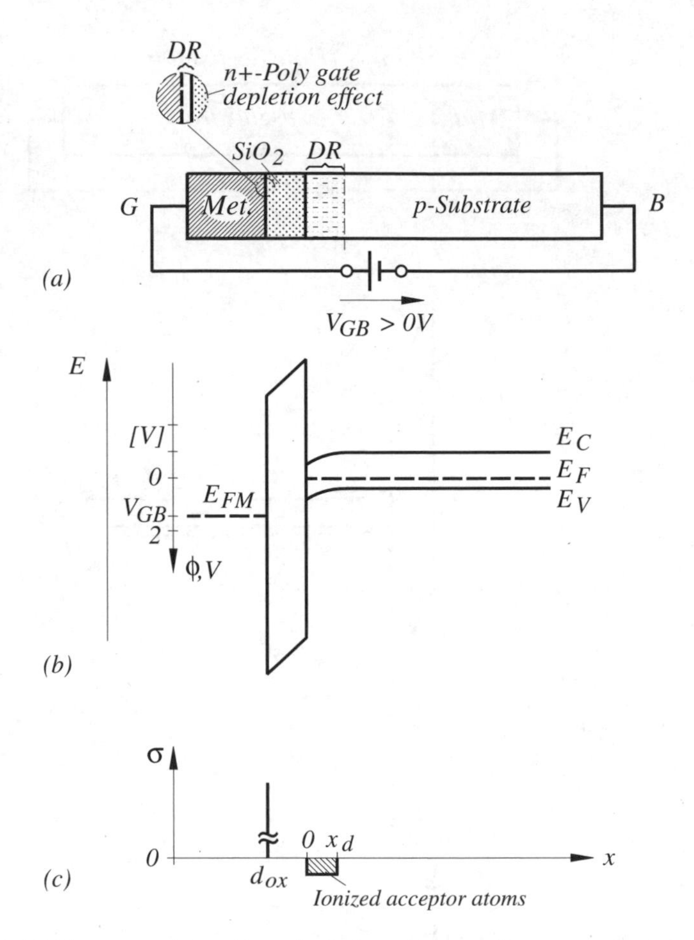

Figure 4.6 (a) MOS structure biased into depletion, (b) band diagram and (c) charge distribution

Deep depletion

An increase of the V_{GB} voltage at the MOS structure causes the depleted acceptor region to extend further into the substrate (Figure 4.7). Accordingly the band bending and the voltage drop across this region increase. The semiconductor is now under non-equilibrium conditions. This is a transient situation which leads the MOS structure into a stationary condition, called inversion.

Inversion

Under non-equilibrium conditions generation and recombination of electron–hole pairs no longer compensate each other. Rather the electron–hole pairs are separated by the large electrical field across the depletion region. Electrons move to the insulator semiconductor interface and holes into the substrate. The electrons build up a very thin inversion layer. This is a layer with a polarity inverted to that of the p-substrate (Figure 4.8).

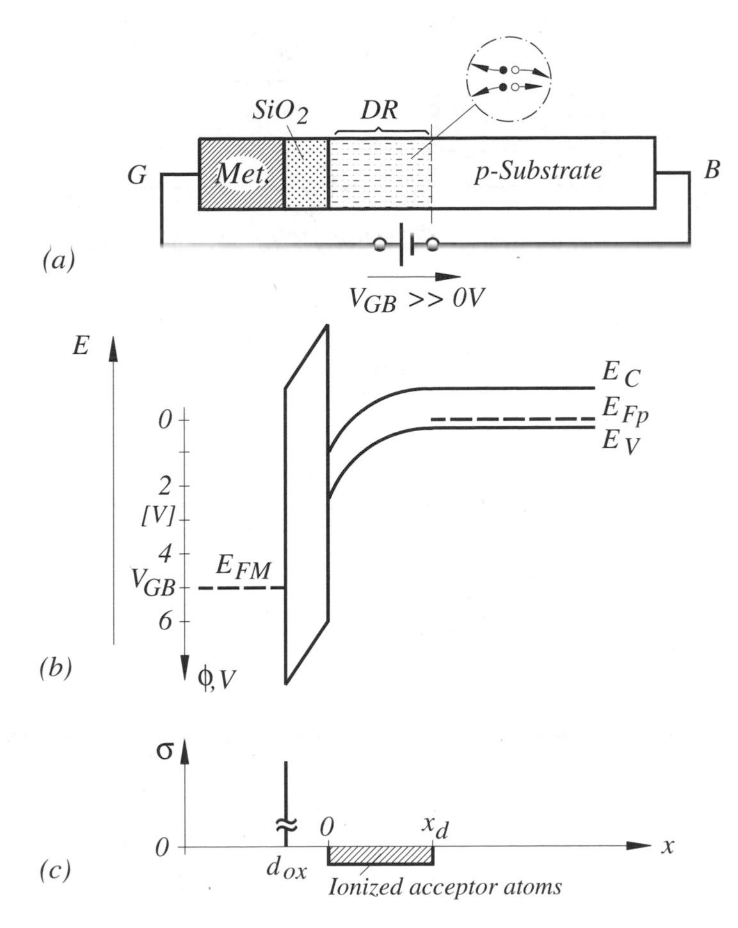

Figure 4.7 (a) MOS structure biased into deep depletion, (b) band diagram and (c) charge distribution

The charge constellation can be interpreted as a field-induced pn-junction. The process leading to this situation usually takes a few seconds until equilibrium is approached at the semiconductor side of the MOS structure. The positive charge at the gate electrode is now compensated by the additional negative charge of the inversion layer, causing the depletion region to reduce (Figure 4.8(b)). A further increase in the V_{GB} voltage has barely an effect on the depletion region width, since the inversion layer density increases. The prerequisite for this happening is that equilibrium is reached at the semiconductor side of the structure. This situation is revisited in conjunction with the determination of the surface potential in Section 4.3.2.

4.2.2 Capacitance Behavior of the MOS Structure

In the foregoing section the biasing conditions which lead the MOS structure into accumulation, depletion, and inversion are considered. When a small-signal voltage

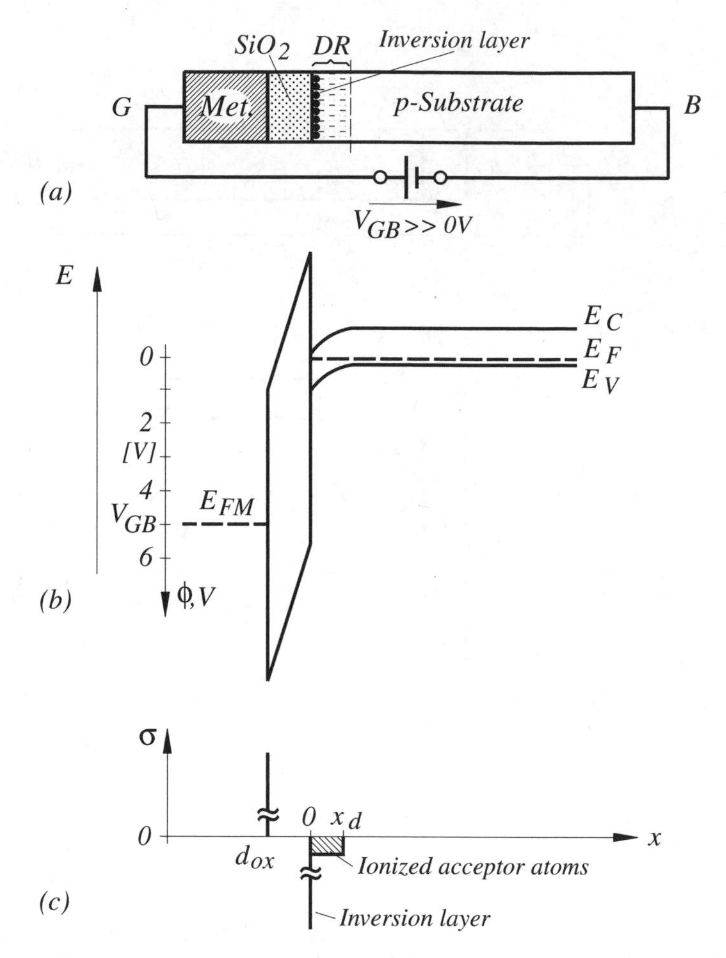

Figure 4.8 (a) MOS structure biased into inversion, (b) band diagram and (c) charge distribution

source is superimposed to the biasing V_{GB} voltage, a small-signal capacitance behavior, as shown in Figure 4.9, can be observed. With $V_{GB} < 0$ the MOS structure is in accumulation. A small-signal voltage, e.g. at a frequency of 1 MHz, causes the charge to change at both sides of the insulator. The structure behaves thus like a parallel plate capacitor with a value of

$$C'_{ox} = \frac{\varepsilon_0 \varepsilon_{ox}}{d_{ox}} \tag{4.1}$$

where C'_{ox} and d_{ox} are the oxide capacitance per area and the oxide thickness, respectively. With $V_{GB} > 0$ a depletion region exists causing the capacitance value to decrease, as the changing charges are further apart. If the voltage is increased further, an inversion layer emerges. This layer limits the depletion region width x_d. A prerequisite is that the semiconductor part of the MOS structure is in equilibrium. Since the width of the

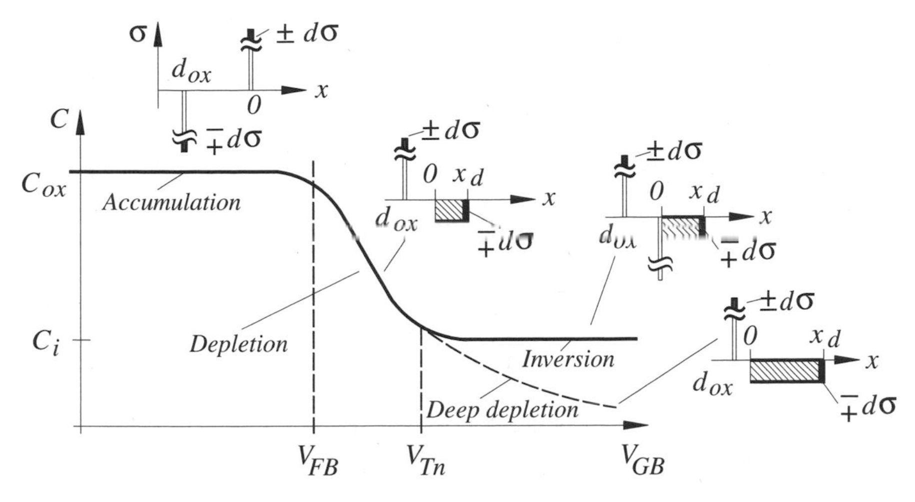

Figure 4.9 Small-signal behavior of the MOS structure as a function of gate bias at a constant medium high frequency

depletion region is almost constant, the capacitance value C_i' of the structure will not change noticeably (Figure 4.9)

$$C_i' = \left(\frac{1}{C_j'} + \frac{1}{C_{ox}'} \right)^{-1} \tag{4.2}$$

In this case the MOS structure can be considered to consist of two in-series connected capacitors, namely one between the gate electrode and the inversion layer and one between the inversion layer and the substrate. The first one is the oxide capacitance C_{ox}' and the second is the depletion capacitance C_j' with a value per area of

$$C_j' = \frac{\varepsilon_0 \varepsilon_{Si}}{x_d} \tag{4.3}$$

This capacitance is comparable to that of a pn-junction described by Equation (2.50).

In the previous analysis it is assumed that the V_{GB} voltage change is so slow that the semiconductor can be considered to be always in equilibrium. If this is not the case, e.g. by increasing the V_{GB} voltage rapidly, it causes the MOS structure to move into deep depletion. This leads to an increase in depletion region width and accordingly to a reduction in capacitance value (Figure 4.9).

The MOS structure is driven by a small-signal voltage source with a frequency of about 1 MHz. This causes, under inversion conditions, charge changes at the gate and at the depletion region edge x_d. It is interesting to observe what happens when the frequency is reduced to such a low value that the generation and recombination process can keep pace with the voltage change (Figure 4.10). In this case the charge change occurs not at the edge of the depletion region but instead at the inversion layer, causing the capacitance to approach the value of the oxide capacitance.

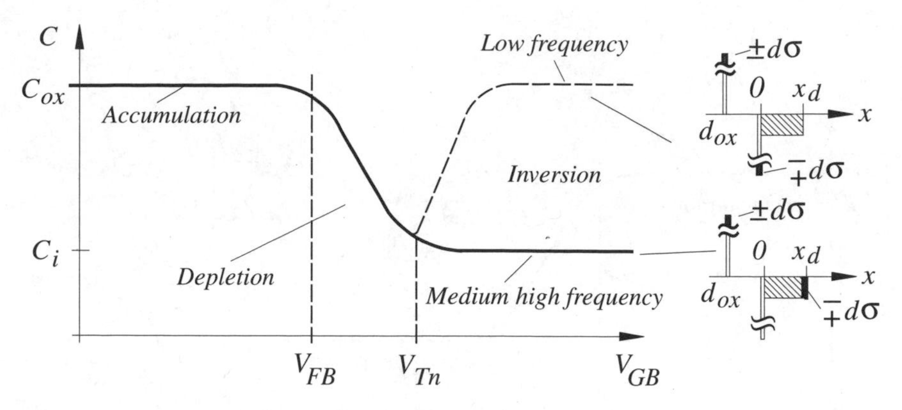

Figure 4.10 Small-signal behavior of the MOS structure as a function of gate bias at two different frequencies

The transition region between accumulation and depletion is characterized by the flat-band voltage V_{FB}, and that between depletion and inversion by the threshold voltage V_{Tn}. The impact of these voltages is discussed in the following sections.

4.2.3 Flat-Band Voltage

The foregoing section is started with a MOS structure considered to be in equilibrium, when the V_{GB} voltage is zero. This implies that the Fermi level is constant throughout the total MOS system (Figure 4.4). This is usually not the case at real MOS structures, since built-in voltages and contact voltages influence the band diagram.

Difference in work functions

The occurrence of these voltages, and how they influence the behavior of the MOS structure, will be discussed next.

Very often the vacuum or free electron energy E_0 is used as a convenient reference energy. This is the energy which an electron would have gained, if it were free of the influence of any given material. The difference between this energy level and the Fermi level is called work function and given the symbol $q\Phi$ in energy units, or sometimes found listed in tables as Φ in volts for different materials. Since in the semiconductor the Fermi level is within the forbidden gap the electron affinity – that is, the energy difference between the vacuum level and the conduction band edge – is used frequently. Figure 4.11 shows the energy levels of separated materials which form the MOS structure.

Additionally a substrate metal contact and different doping concentrations of the substrate are included, in order to indicate the effect an inhomogeneous doping has on the MOS system characteristic. For simplicity, it is assumed that the work function of the bulk contact metal has the same value as that of the gate electrode. What happens when the different materials are brought into contact and the gate electrode is

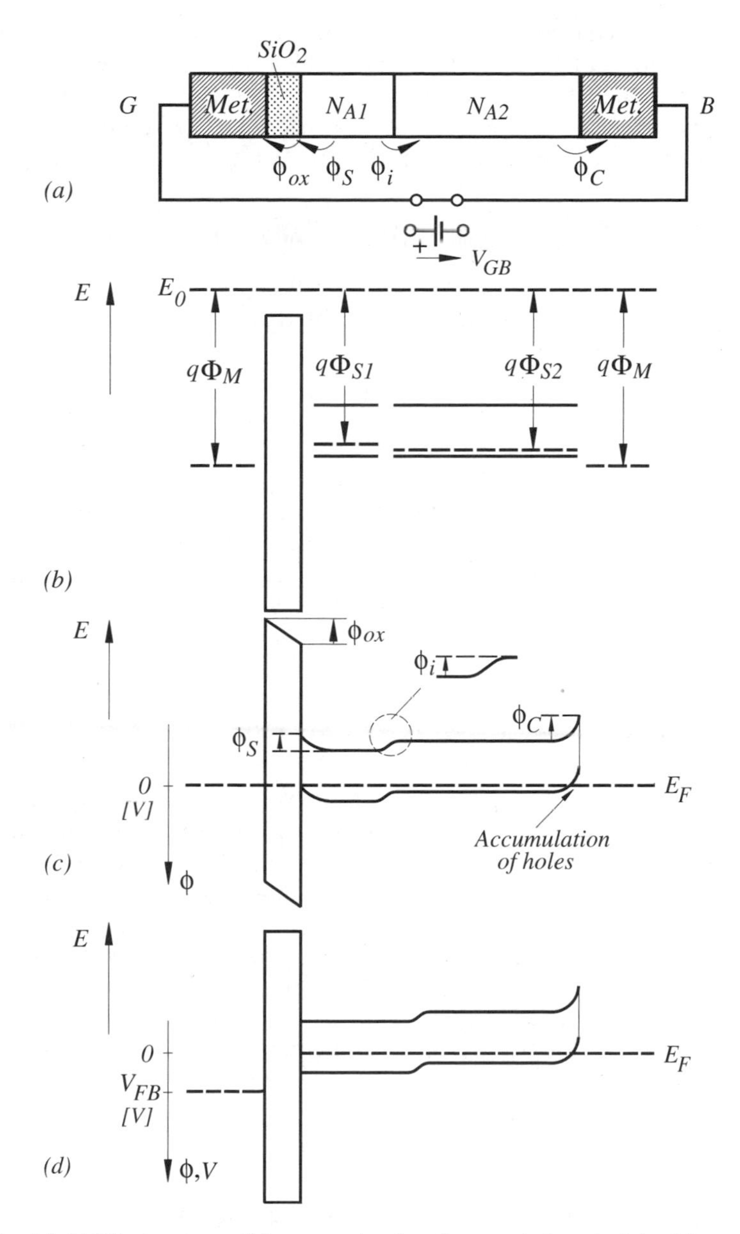

Figure 4.11 (a) MOS structure, (b) energy levels of separated materials, (c) energy band at equilibrium and (d) energy band under flat-band condition

short-circuited with the bulk contact? The disparity in average energies causes the transfer of electrons until equilibrium is reached. To be more precise, the work function of the semiconductor $q\Phi_{S2}$ is smaller than that of the metal contact $q\Phi_M$. This implies that the average energy of the valence electrons in the semiconductor is larger than the average electron energy in the metal contact. Therefore the probability that electrons move by thermal energy from the metal contact to the semiconductor is smaller than the probability that valence electrons move from the semiconductor to the metal contact. This causes a depletion of valence electrons – which amounts to an accumulation of holes – at the semiconductor side of the contact and an electron accumulation at the metal side of the contact, until equilibrium is reached. The result is a low resistive contact with a contact voltage of

$$\phi_C = \Phi_M - \Phi_{S2} \tag{4.4}$$

This voltage has a value of 0.9 V when the work function voltages are $\Phi_M = 5\,\text{V}$ and $\Phi_{S2} = 4.1\,\text{V}$. A similar situation exists at the interface between the different doping densities with a built-in voltage of

$$\phi_i = \Phi_{S2} - \Phi_{S1} = \phi_t \ln \frac{N_{A2}}{N_{A1}} \tag{4.5}$$

The result is probably not very surprising. It is in analogy to the derivation of the built-in voltage of the inhomogeneously doped semiconductor (Equation 2.4) where drift and diffusion currents are used in the derivation of the built-in voltage. In either case a constant Fermi level exists at equilibrium (Problem 4.11).

If the MOS structure is short-circuited, no current flows and the sum of all voltages must be zero

$$\phi_i + \phi_C - \phi_{ox} - \phi_S = 0 \tag{4.6}$$

In this case the contact voltage and the built-in voltage are responsible for voltage drops in regions of high receptivity. This is across the oxide ϕ_{ox} and, depending on the work functions, across the depletion region ϕ_S at the surface of the semiconductor (Figure 4.11(c)). If a V_{GB} voltage is applied the equation above changes to

$$-V_{GB} + \phi_i + \phi_C - \phi_{ox} - \phi_S = 0 \tag{4.7}$$

The V_{GB} voltage can be adjusted in such a way that voltage drops ϕ_{ox} and ϕ_S are compensated by the applied voltage (Figure 4.11(d)). This voltage is called the flat-band voltage and has a value of

$$V_{GB} = V_{FB} = (\phi_i + \phi_C) \tag{4.8}$$

The flat-band voltage therefore characterizes the transition region between accumulation and depletion (Figure 4.10).

So far only the influence of the contact voltages on the flat-band voltage are considered. But there is another one stemming from interface traps. These are defects located at the silicon oxide (Si–SiO_2) interface, causing energy levels within the band-gap. These defects can exchange charge with the silicon by emitting or capturing electrons, similar to the situation illustrated in Figure 1.26. In most practical cases, these centers lead to a positive interface charge σ_{SS} (Figure 4.12) causing mirror charges at the gate and near the surface of the semiconductor. Since in modern manufacturing processes the charge is, at $q \cdot 10^{10}\,\text{cm}^{-2}$, relatively small, the effect on the flat-band

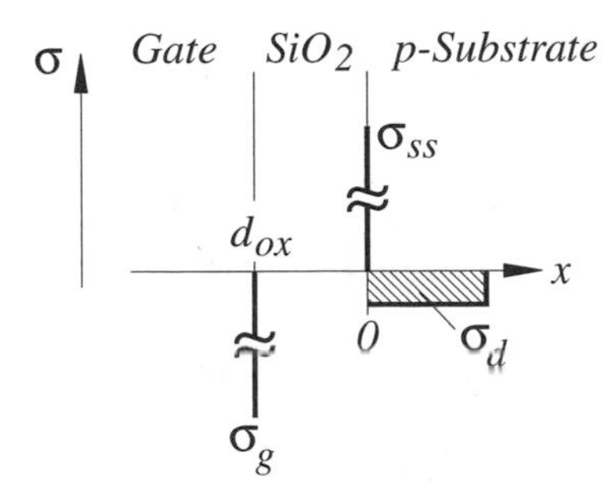

Figure 4.12 Interface charge at the Si–SiO_2 interface

voltage is usually negligible. More critical is the interface charge and charge within the oxide in conjunction with hot electrons. They may lead to instabilities (Leblebici *et al.* 1993; Schwerin *et al.* 1987) and reliability problems (Section 4.5.4).

4.3 EQUATIONS OF THE MOS STRUCTURE

In the preceding section the behavior of the MOS structure is analyzed. Next, the charge of the inversion layer and depletion region are determined as a function of the semiconductor surface voltage. The value of this voltage is then derived at strong inversion in order to avoid the solution of an implicit equation. Knowing this voltage, the threshold voltage of the structure is found. Applying these equations to the MOS transistor leads to the current–voltage relationship of the device.

In the following derivation charges per area, symbolized by σ, are used and will be used throughout this chapter. This stems from the so-called charge-sheet approximation discussed in the following section.

4.3.1 Charge Equations of the MOS Structure

The starting point is the charge distribution of the MOS structure (Figure 4.13). To simplify the analysis, a homogeneously doped substrate N_A is assumed. The field distribution can be found by solving Poisson's equation for the inversion and depletion region. Accordingly it is

$$\int_{\mathcal{E}_{Si}(x)}^{\mathcal{E}_{Si}(x_d)} \mathrm{d}\mathcal{E}_{Si}(x) = -\frac{q}{\varepsilon_0 \varepsilon_{Si}} \int_{x}^{x_d} (n(x) + N_A)\,\mathrm{d}x \tag{4.9}$$

where $p(x) = 0$ is assumed. The equation leads to an implicit equation (see Appendix A at the end of this chapter) which is not very practical in its application. By using the so-called charge-sheet approximation (Brews 1978) under strong inversion conditions – which is discussed in the following section – the implicit solution can be circumvented. In the approximation it is assumed that the thickness of the inversion layer d_i

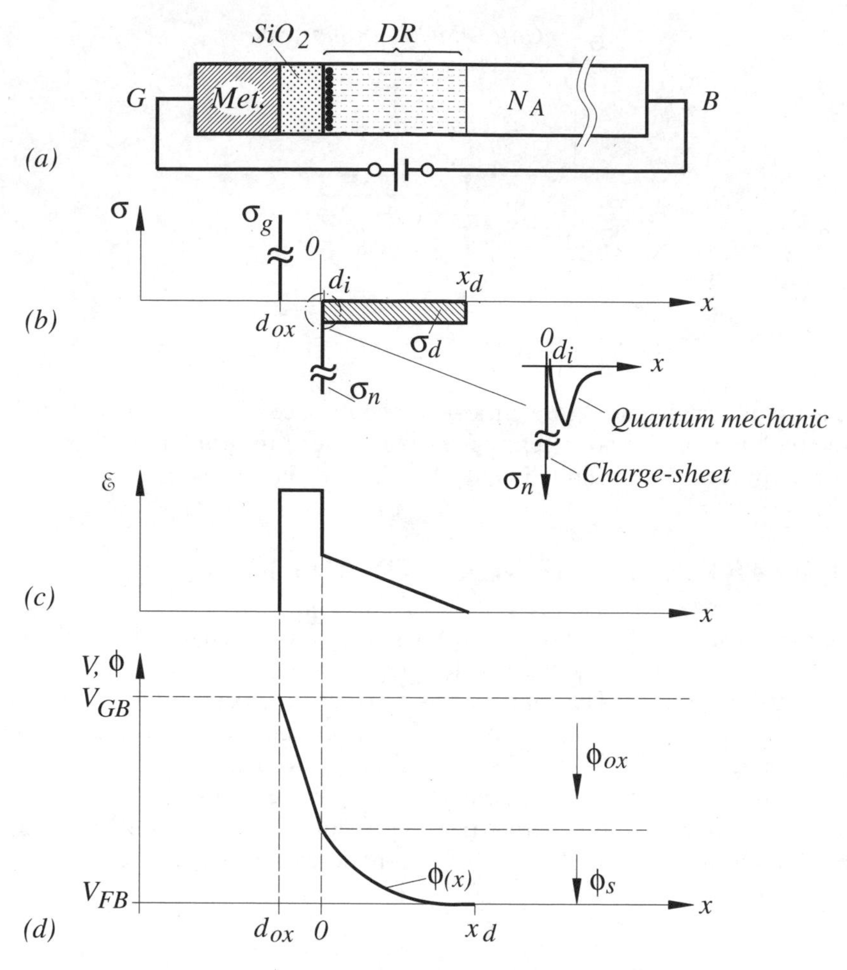

Figure 4.13 (a) MOS structure, (b) charge distribution, (c) field distribution and (d) voltage distribution

approaches zero. Under this assumption, the charge of the inversion layer can be described by a delta function, the so-called charge-sheet. This is a reasonable approximation since the voltage drop across the inversion layer is negligible compared to that across the depletion region. In reality, under quantum mechanical considerations, the center of the charge distribution is not at the Si–SiO_2 interface but slightly displaced into the semiconductor volume (inset in Figure 4.13(b)).

Due to the charge-sheet approximation the field and voltage distributions in the substrate depend on the depletion charge only. Under this condition Equation (4.9) yields a linear field distribution of

$$\mathcal{E}_{Si}(x) = \frac{qN_A}{\varepsilon_0 \varepsilon_{Si}}(x_d - x) \tag{4.10}$$

and a parabolic voltage distribution

$$\phi(x) = -\int_x^{x_d} \mathcal{E}_{Si}(x)\,\mathrm{d}x = \frac{qN_A}{2\varepsilon_0\varepsilon_{Si}}(x_d - x)^2 \tag{4.11}$$

where the integration bounds $\mathcal{E}_{Si}(x = x_d) = 0$ and $\phi(x = x_d) = 0$ are used. This leads to a voltage at the semiconductor surface ($x = 0$), sometimes called surface potential, of

$$\phi(x = 0) = \phi_S = \frac{qN_A}{2\varepsilon_0\varepsilon_{Si}} x_d^2$$

Rearranging this equation results in a relationship between the depletion layer width x_d and the voltage drop ϕ_S across this region

$$x_d = \sqrt{\frac{2\varepsilon_0\varepsilon_{Si}}{qN_A}\phi_S} \tag{4.12}$$

It is interesting to compare this result of the field-induced junction with the one obtained from the metallurgical junction of Equation (2.43). With $N_D > N_A$ the results are identical, except that the voltage across the depletion region is $(\phi_i - V_{PN})$.

Using Equation (4.12) the depletion charge per area can be determined

$$\sigma_d = -qN_A x_d = -\sqrt{qN_A 2\varepsilon_0\varepsilon_{Si}\phi_S} \tag{4.13}$$

Next, the inversion layer charge is derived. For this purpose Gauss' law

$$\oint \vec{D} \cdot \mathrm{d}\vec{A} = Q \tag{4.14}$$

which states that the total electrical displacement $D = \varepsilon\mathcal{E}$ out of a closed surface is equal to the total charge Q enclosed, is applied to the semiconductor as shown in Figure 4.14. The surface encloses the inversion layer charge Q_n and that of the depletion layer Q_d. Since the electrical displacement D_{Si} is zero at $x = x_d$ the equation yields

$$\begin{aligned} Q_n + Q_d &= -D_{ox}\,\mathrm{d}A \\ \sigma_n + \sigma_d &= -D_{ox} \end{aligned} \tag{4.15}$$

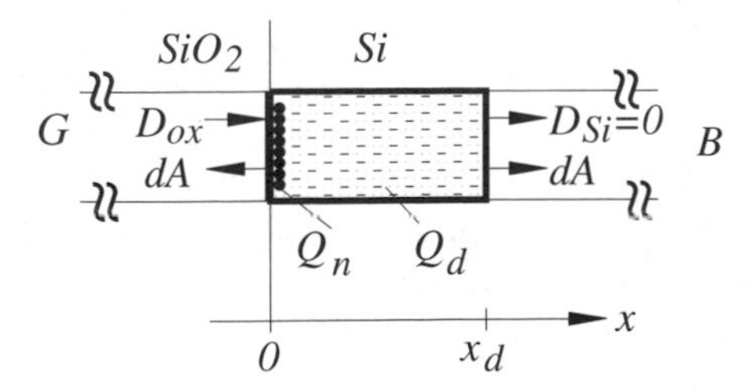

Figure 4.14 Discontinuity of the electrical displacement

where σ_n is the inversion layer charge per area. This equation states the simple fact that the discontinuity of the electrical displacement D_{ox} is caused by the enclosed charges. According to Figure 4.13, the electrical displacement can be expressed as a function of the ϕ_{ox} voltage across the oxide by

$$\begin{aligned} D_{ox} &= \varepsilon_0 \varepsilon_{ox} \mathcal{E}_{ox} \\ &= -\varepsilon_0 \varepsilon_{ox} \frac{d\phi}{dx} \\ &= \varepsilon_0 \varepsilon_{ox} \frac{\phi_{ox}}{d_{ox}} \\ &= C'_{ox} \phi_{ox} \end{aligned} \tag{4.16}$$

where C'_{ox} is the oxide capacitance per area. In order to use this equation in the derivation of the current–voltage relationship of the MOS transistor, the voltage ϕ_{ox} has to be expressed with respect to the terminal voltage of the MOS structure.

Assuming the V_{GB} voltage has a value of V_{FB} flat-band conditions exist (Figure 4.13(d)). The voltage difference $V_{GB} - V_{FB}$ can therefore be viewed as the effective voltage influencing the semiconductor system, causing a voltage drop across the insulator ϕ_{ox} and one across the semiconductor ϕ_S

$$V_{GB} - V_{FB} = \phi_{ox} + \phi_S \tag{4.17}$$

This leads to an electrical displacement (Equation 4.16) of

$$D_{ox} = C'_{ox}(V_{GB} - V_{FB} - \phi_S) \tag{4.18}$$

Substituting of Equations (4.13) and (4.18) into Equation (4.15) results in an inversion layer charge per area of

$$\begin{aligned} \sigma_n &= -C'_{ox}(V_{GB} - V_{FB} - \phi_S) + \sqrt{qN_A 2\varepsilon_0 \varepsilon_{Si} \phi_S} \\ &= -C'_{ox}(V_{GB} - V_{FB} - \phi_S - \gamma\sqrt{\phi_S}) \end{aligned} \tag{4.19}$$

The factor

$$\boxed{\gamma = \frac{1}{C'_{ox}} \sqrt{qN_A 2\varepsilon_0 \varepsilon_{Si}}} \tag{4.20}$$

is called the body-effect parameter. Its significance is explained in conjunction with the threshold voltage derivation.

The charge equations of the MOS structure are derived. In order that these equations are useful for the description of the terminal behavior of a MOS transistor, the surface voltage ϕ_S as a function of the gate voltage is required. Deriving this relationship leads to an implicit equation as shown in Appendix A at the end of this chapter. In order to circumvent such an equation, the surface voltage is determined first under strong inversion conditions and second, in a later section, under weak inversion conditions.

4.3.2 Surface Voltage at Strong Inversion

The starting point is the band diagram of the semiconductor side of the MOS structure (Figure 4.15). It can be constructed by multiplying the voltage distribution function of Equation (4.11) by $-q$ (Section 1.3.4). The bending of the energy bands corresponds to a surface voltage of

$$\phi_S = \phi(0) + \phi_F \tag{4.21}$$

where the individual voltages – in accordance with the nomenclature described in Section 1.3.4 – are given by

$$\phi(0) = \frac{E_i(0) - E_F}{-q} \tag{4.22}$$

and

$$\boxed{\phi_F = \frac{E_F - E_i(x_d)}{-q}} \tag{4.23}$$

ϕ_F is called Fermi potential or, better, Fermi voltage (Equation 1.32). Increasing the gate voltage causes the voltage $\phi(0)$ and the energy difference $(E_F - E_i)$ at the semiconductor surface $(x = 0)$ to increase also. This corresponds to an electron density increase at the silicon surface (Equation 1.14) of

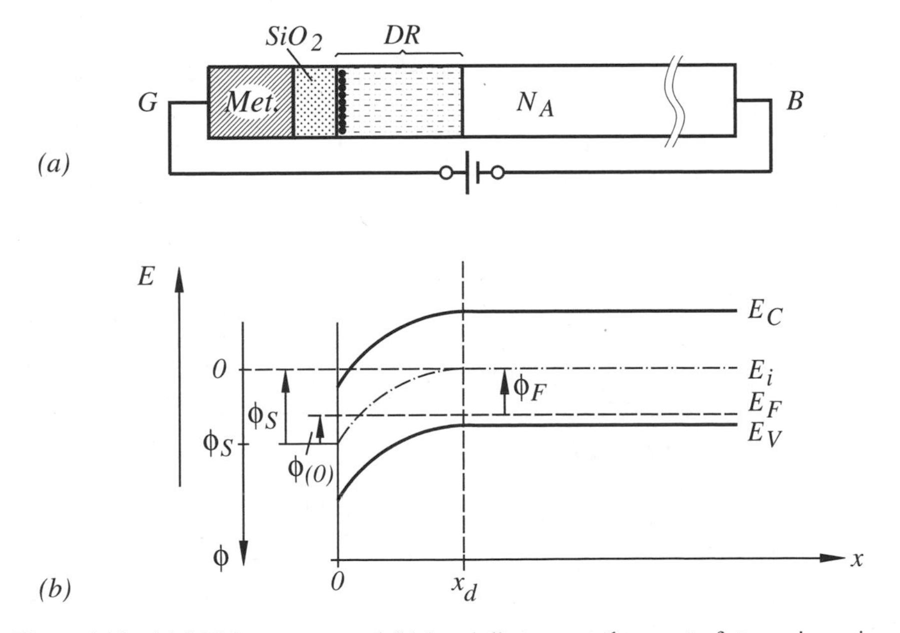

Figure 4.15 (a) MOS structure and (b) band diagram at the onset of strong inversion

$$n(0) = n_i e^{\frac{E_F - E_i(0)}{kT}} = n_i e^{\phi(0)/\phi_t} \tag{4.24}$$

Expressing this relationship as function of the surface voltage results in

$$n(0) = n_i e^{(\phi_S - \phi_F)/\phi_t} \tag{4.25}$$

The required Fermi voltage (Equation 4.23) can be found from the substrate doping density (Equation 1.15)

$$p(x_d) = N_A = n_i e^{\frac{E_i(x_d) - E_F}{kT}} = n_i e^{\phi_F/\phi_t} \tag{4.26}$$

After rearranging the above equation the following Fermi voltage

$$\boxed{\phi_F = \phi_t \ln \frac{N_A}{n_i}} \tag{4.27}$$

results.

The relationship between the electron density at the surface $n(0)$ of the semiconductor and the surface voltage ϕ_S (Equation 4.25) is illustrated in Figure 4.16. This illustration is very useful for the determination of the surface voltage at strong inversion.

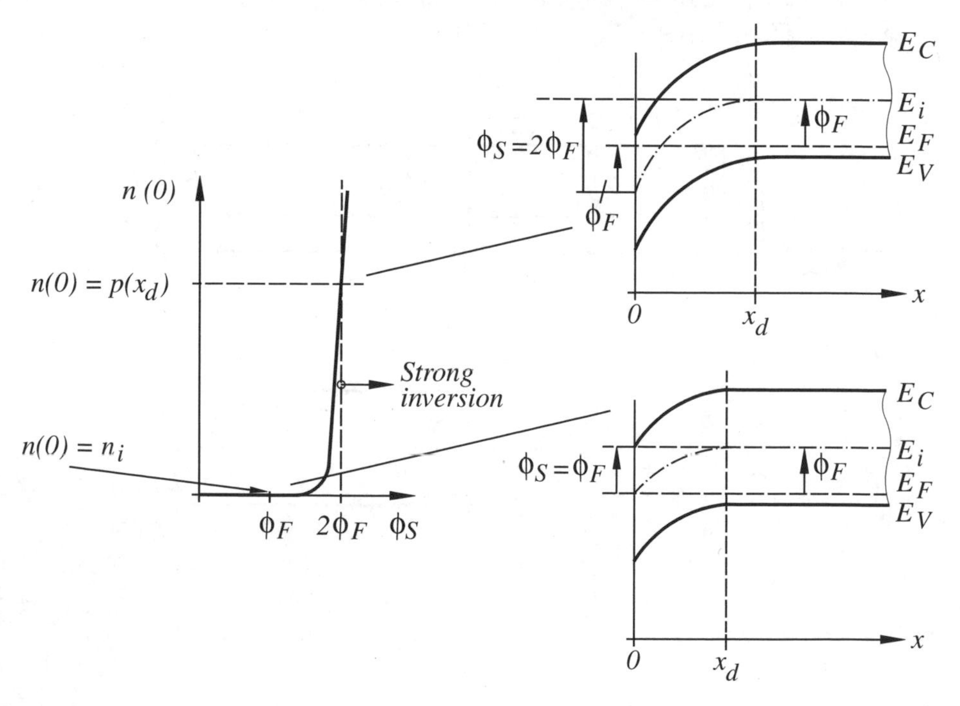

Figure 4.16 Electron density at the semiconductor surface as a function of the surface voltage

If the gate voltage is adjusted in such a way that the surface voltage ϕ_S has a value of ϕ_F, then the electron density has a value according to Equation (4.25) of $n(0) = n_i$. Assuming the surface voltage has a value of $\phi_S = 2\phi_F$ an electron density of

$$n(0) = n_i e^{\phi_F/\phi_t} \tag{4.28}$$

exists. Under this constellation the electron density at the semiconductor surfaces corresponds to the density of holes in the substrate (Equation 4.26). This is the definition for the beginning of strong inversion (SI). Beyond this condition the charge increases but the surface voltage remains basically constant (Figure 4.16). The following approximation

$$\boxed{\phi_S(SI) = 2\phi_F} \tag{4.29}$$

can therefore be used to describe the surface voltage under strong inversion conditions and the charge densities, described by Equations (4.13) and (4.19) accordingly.

4.3.3 Threshold Voltage and Body Effect

The preceding section describes how the inversion layer is built up. Electron–hole pairs are separated by the electrical field across the depletion region. Holes move into the substrate and electrons to the semiconductor surface, building up an inversion layer. This is a relatively slow process, in the order of seconds (Figure 4.17(a)).

If an n^+-region called source (S) is adjacent to the MOS structure (Figure 4.17(b)) – this is the case at the MOS transistor – electrons are able to move out of the source region along the semiconductor surface and build up an inversion layer. Depending on the distance the electrons have to travel this can be very fast (Gondro *et al.* 2001). The V_{SB} voltage reverse biases the metallurgical n^+p-junction

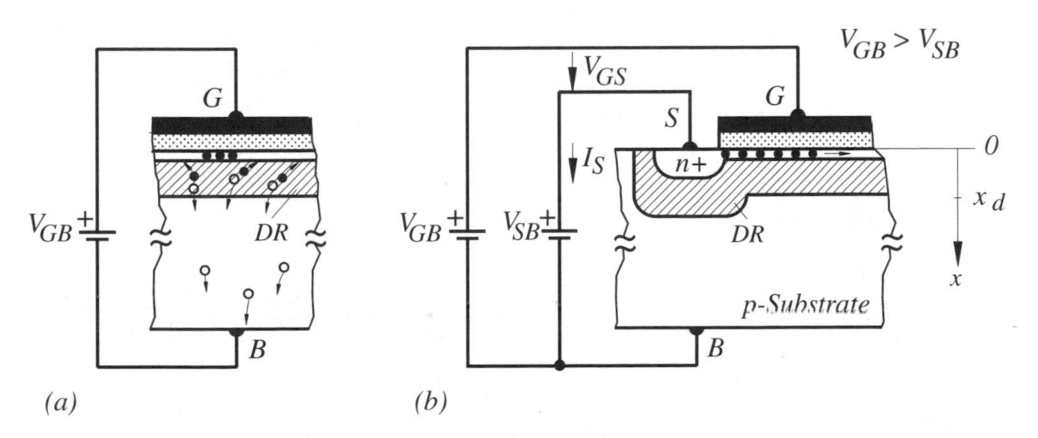

Figure 4.17 (a) MOS structure and (b) MOS structure with n^+-source

and the field-induced n^+p-junction. This leads to a very small saturation or leakage current I_S contributed by both junctions. This is comparable to the situation described by Equation (2.28).

In conjunction with the modified MOS structure the question arises as to which value the charge assumes in the inversion layer. To answer this question the surface voltage has to be determined first.

The V_{SB} voltage reverse biases the field-induced n^+p-junctions. In this case the band diagram of Figure 4.18(a) changes to one, which is characterized by two quasi-Fermi levels E_{Fp} and E_{Fn} of the p-substrate and the inversion layer. These quasi-Fermi levels are separated by the applied external energy qV_{SB} (Figure 4.18(b)).

The concept of quasi-Fermi levels is presented in conjunction with the metallurgical junction (Figure 2.3) in order to describe situations where the semiconductor is under non-equilibrium conditions (Section 2.3.1). Under this condition the surface voltage has a value of

$$\boxed{\phi_S(SI) = 2\phi_F + V_{SB}} \tag{4.30}$$

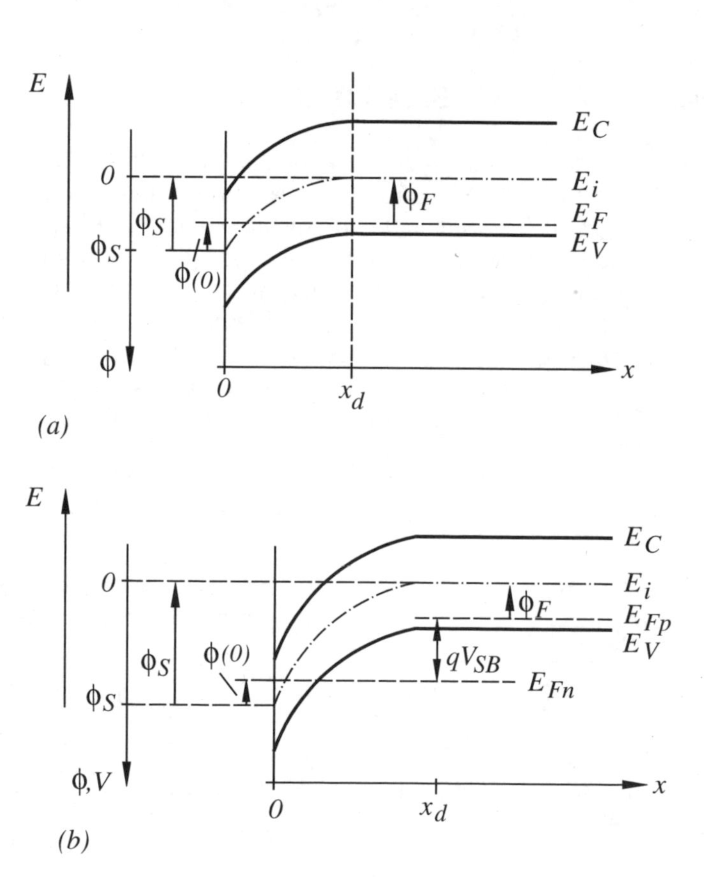

Figure 4.18 Band diagram of the MOS structure: (a) without V_{SB} voltage and (b) with V_{SB} voltage > 0

where strong inversion with $\phi(0) = \phi_F$ is assumed. This results in an inversion layer charge (Equation 4.19) of

$$\sigma_n = -C'_{ox}(V_{GB} - V_{FB} - 2\phi_F - V_{SB} - \gamma\sqrt{2\phi_F + V_{SB}}) \quad (4.31)$$

This equation is an essential prerequisite for the derivation of the current–voltage relationship of the MOS transistor.

Using the transistor in circuit design, it is common practice to reference the gate voltage to the source (Figure 4.17). This modification leads with

$$V_{GB} - V_{SB} = V_{GS} \quad (4.32)$$

to an inversion layer charge of

$$\sigma_n = -C'_{ox}(V_{GS} - V_{FB} - 2\phi_F - \gamma\sqrt{2\phi_F + V_{SB}}) \quad (4.33)$$

The equation is sketched in Figure 4.19.

Threshold voltage

As is obvious from Figure 4.19 the inversion layer charge becomes zero at a particular V_{GS} voltage. The gate voltage where this happens is called the threshold voltage. Its value can be found from Equation (4.33) to be

$$\boxed{V_{Tn} = V_{FB} + 2\phi_F + \gamma\sqrt{2\phi_F + V_{SB}}} \quad (4.34)$$

The threshold voltage is a very useful transistor parameter in circuit design. It tells the designer if a transistor is cut off ($V_{GS} < V_{Tn}$) or tuned on ($V_{GS} > V_{Tn}$).

The preceding derivation is limited to the case of strong inversion. If the charge is reduced beyond this region, weak inversion exists and the derivation is no longer valid.

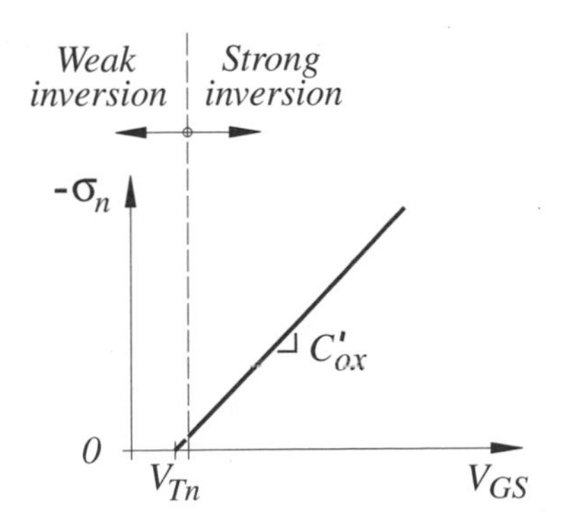

Figure 4.19 Inversion layer charge as a function of the gate source voltage

It is therefore not surprising that in reality the MOS system behaves differently. As it turns out and is described in Section 4.4.3, in weak inversion an exponential current–voltage behavior can be observed. The threshold voltage can therefore be interpreted to be the extrapolated value from the strong inversion description.

If the source bulk voltage V_{SB} is zero, a threshold voltage of

$$\boxed{V_{Ton} = V_{FB} + 2\phi_F + \gamma\sqrt{2\phi_F}} \tag{4.35}$$

results. Substituting Equation (4.35) into Equation (4.34) leads to a modified threshold description

$$\boxed{V_{Tn} = V_{Ton} + \gamma(\sqrt{2\phi_F + V_{SB}} - \sqrt{2\phi_F})} \tag{4.36}$$

which has the advantage that two parameters V_{Ton} and γ are sufficient to characterize the threshold. Typical threshold voltages values are around 0.4 V.

Body effect

The body-effect parameter (Equation 4.20)

$$\boxed{\gamma = \frac{1}{C'_{ox}}\sqrt{qN_A 2\varepsilon_0\varepsilon_{Si}}}$$

is a measure of how the V_{SB} voltage – sometimes called back-bias – influences the threshold voltage (Figure 4.20). The reason for this behavior is illustrated in Figure 4.21. For didactical reasons it is assumed that the gate is charged (not shown) by a positive charge σ_g. Since charge neutrality exists is

$$\sigma_g = |\sigma_n| + |\sigma_d| \tag{4.37}$$

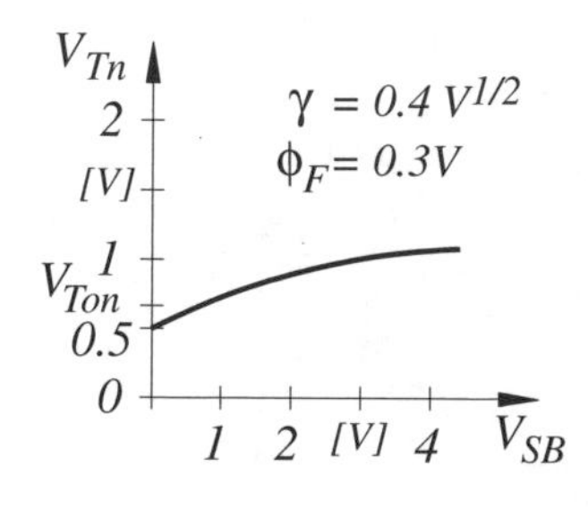

Figure 4.20 Influence of the V_{SB} voltage on the threshold voltage

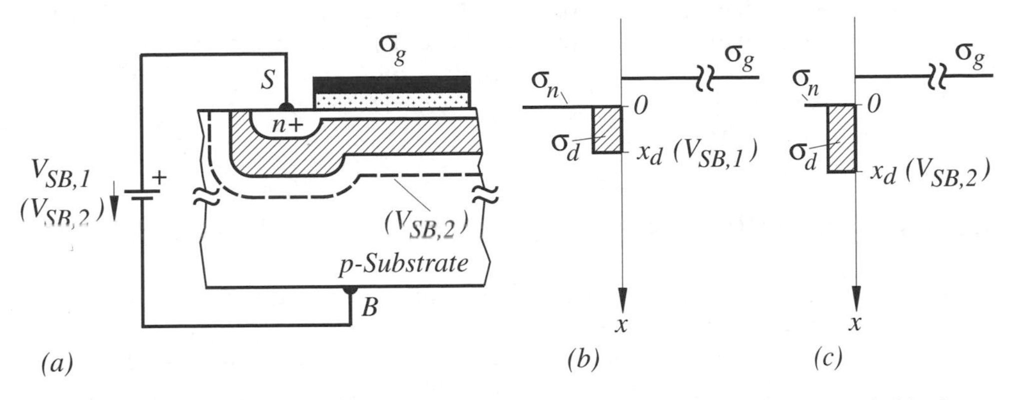

Figure 4.21 (a) Modified MOS structure, (b) charge distribution at $V_{SB,1}$ and (c) charge distribution at $V_{SB,2} > V_{SB,1}$

An increase in the V_{SB} voltage from $V_{SB,1}$ to $V_{SB,2}$ (Figure 4.21(c)) causes the width x_d of the depletion region to extend and in conjunction the depletion region charge σ_d to increase. Since σ_g is constant the charge σ_n must decrease since σ_d increases. This causes the threshold voltage to shift to higher values. If the V_{SB} voltage is increased to such a value that $\sigma_g = |\sigma_d|$, this causes the inversion layer to disappear completely. The effect that the threshold voltage can be varied by a source bulk voltage is usually an undesirable one. For example, noise at the common substrate of an integrated circuit may influence the inversion layer charge and therefore the current of a MOS transistor so much that an unacceptable disturbance in a circuit occurs. Because of this and other reasons it is desirable to have a body-effect factor as small as possible. Typical values are below $0.4\,\mathrm{V}^{1/2}$.

Field threshold

Adjacent transistors or diffusion regions are separated by thick oxide regions called Field OXide (FOX), when, e.g., a manufacturing process, according to Figure 4.2, is used. If a metal layer crosses e.g. two diffusion regions via a thick oxide region, this leads to a field oxide transistor (Figure 4.22). The metal layer acts as gate and the diffusion regions as source (S) and drain (D). A sufficiently large voltage at the metal layer – e.g. the power-supply voltage – is able to create an inversion layer below the thick oxide, causing a current to flow between the two diffusion regions. In order to

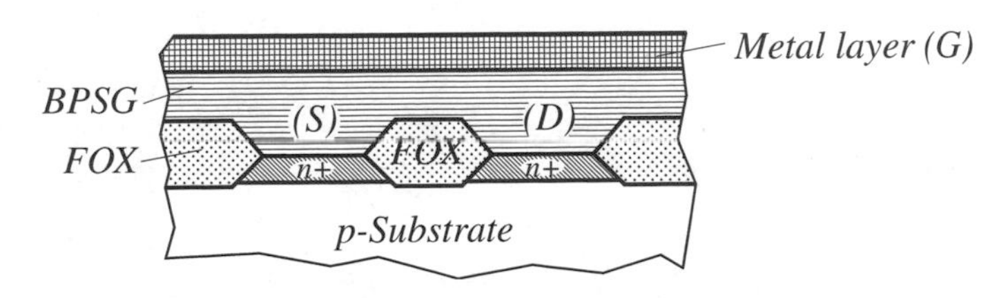

Figure 4.22 Field oxide transistor (FOX)

avoid such a current the threshold voltage of this structure, called the field threshold V_{FT}, has to be larger than the power-supply voltage. This can be achieved, according to the preceding equations, by increasing the insulator thickness – these are the FOX and BPSG layers – or by using a boron implantation, called channel stopper, underneath the field oxide to increase the doping density.

4.4 MOS TRANSISTOR

Adding an additional diffusion region, called a drain, to the modified MOS structure, a MOS transistor results (Figure 4.23). Electrons are able to move from the source to the drain and build up an inversion layer, called a channel, connecting source and drain when the V_{GS} voltage is larger than the threshold voltage V_{Tn}. This leads to a drain current I_{DS} when the voltage $V_{DS} > 0$. An increase in the gate voltage causes an increase in the inversion layer charge and therefore in the drain current. The transistor behaves like a voltage-controlled resistor or a voltage-controlled current generator, depending on the operation condition, which is the topic of the following section.

4.4.1 Current–Voltage Characteristic at Strong Inversion

This characteristic will be derived under the conditions:

(a) The drain voltage V_{DS} is very small. Due to this assumption the field in the vertical direction is much larger than the one in the lateral direction, and the depletion region width can be assumed to be constant from source to drain. Thus it is possible to use the results from the one-dimensional analysis. Otherwise Gauss' law has to be applied to a two-dimensional problem.

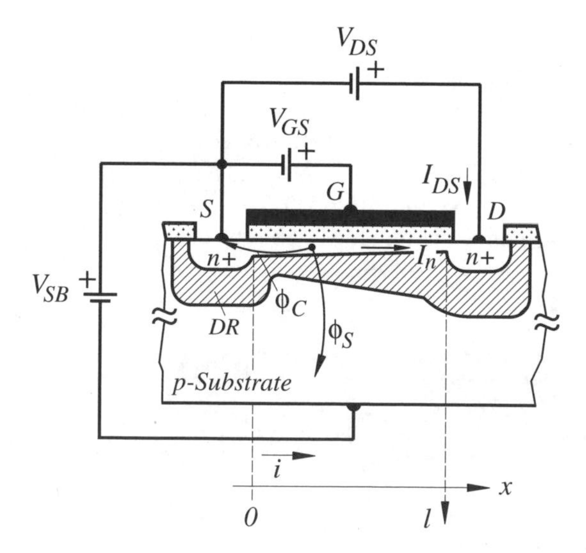

Figure 4.23 n-Channel MOS transistor in the resistive region

(b) The mobility of the electrons in the inversion layer is constant.

(c) Strong inversion conditions exist.

Resistive region

The inversion layer charge has a value according to Equation (4.19) of

$$\sigma_n = -C'_{ox}(V_{GB} - V_{FB} - \phi_S - \gamma\sqrt{\phi_S})$$

Contrary to the MOS structure with a constant surface voltage (Equation 4.30), the voltage of the MOS transistor (Figure 4.23) changes from source to drain

$$\phi_S(SI, x) = 2\phi_F + V_{SB} + \phi_C(x) \tag{4.38}$$

In this equation ϕ_C is the channel voltage with a value of 0 V at the source – reference point – and one of V_{DS} at the drain. This results in variable channel charge from source to drain of

$$\sigma_n(x) = -C'_{ox}\left(V_{GS} - V_{FB} - 2\phi_F - \phi_C(x) - \gamma\sqrt{2\phi_F + V_{SB} + \phi_C(x)}\right) \tag{4.39}$$

where $V_{GS} = V_{GB} - V_{SB}$. The right term in this equation describes the depletion region charge (Equation 4.13) which changes from source to drain

$$\sigma_d = -\sqrt{qN_A 2\varepsilon_0\varepsilon_{Si}(2\phi_F + V_{SB} + \phi_C(x))} \tag{4.40}$$

With the assumption – listed in the introduction of this section – that the depletion region width is constant between source and drain or, in other words, the depletion region charge is independent of the channel voltage ϕ_C, the channel charge equation simplifies to

$$\begin{aligned}\sigma_n(x) &= -C'_{ox}\left(V_{GS} - V_{FB} - 2\phi_F - \phi_C(x) - \gamma\sqrt{2\phi_F + V_{SB}}\right) \\ &= -C'_{ox}(V_{GS} - V_{Tn} - \phi_C(x))\end{aligned} \tag{4.41}$$

after the threshold voltage equation (Equation 4.36) is substituted. This relationship is a prerequisite for the derivation of the following current–voltage description of the MOS transistor.

The starting point is an infinitesimal channel section, shown in Figure 4.24. The application of the current equation (Equation 1.48) to the infinitesimal channel element results in

$$\begin{aligned}I_n(x) &= d_i w J_n(x) \\ &= d_i w\left(q\mu_n n\mathcal{E}(x) + qD_n\frac{dn(x)}{dx}\right) \\ &= \mu_n w\sigma_n(x)\frac{d\phi_C(x)}{dx} - \mu_n w\phi_t\frac{d\sigma_n(x)}{dx}\end{aligned} \tag{4.42}$$

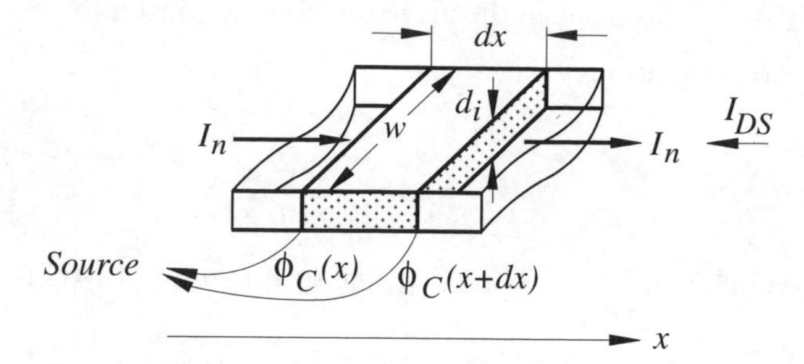

Figure 4.24 Infinitesimal channel section of a MOS transistor

where the Einstein relationship (Equation 1.47) is used and the channel charge description is

$$\sigma_n(x) = -d_i q n(x) \tag{4.43}$$

From a continuity point of view the current $I_n(x)$ is constant throughout the channel. This leads with $I_{DS} = -I_n(x)$ (see Figure 4.23 for current directions) to

$$I_{DS} = -\mu_n w \cdot \sigma_n(x) \frac{\mathrm{d}\phi_C(x)}{\mathrm{d}x} + \mu_n w \phi_t \frac{\mathrm{d}\sigma_n(x)}{\mathrm{d}x} \tag{4.44}$$

This equation states that the drain current consists of two current terms, a *drift current term* (field term $\mathrm{d}\phi_C/\mathrm{d}x$ exists)

$$I_{\mathrm{Drift}}(x) = -\mu_n w \sigma_n(x) \frac{\mathrm{d}\phi_C(x)}{\mathrm{d}x} \tag{4.45}$$

and a *diffusion current term* (charge gradient $\mathrm{d}\sigma_n/\mathrm{d}x$ exists)

$$I_{\mathrm{Diff}}(x) = \mu_n w \phi_t \frac{\mathrm{d}\sigma_n(x)}{\mathrm{d}x} \tag{4.46}$$

As mentioned before, the drain current I_{DS} is constant throughout the channel from a continuity point of view. This cannot be said for the drift and diffusion parts of the current.

Equations (4.45) and (4.46) are coupled differential equations which cannot be integrated independently. But in order to come up with an analytical solution it is useful to compare the size of each current term separately.

Drift term

According to Equation (4.45) is

$$I_{DS} = -\mu_n w \sigma_n(x) \frac{\mathrm{d}\phi_C(x)}{\mathrm{d}x} \tag{4.47}$$

Separating the variables and integrating from source ($x = 0$) to drain ($x = l$) results in

$$\int_{x=0}^{x=l} I_{DS}\mathrm{d}x = -\mu_n w \int_{\phi_C=0}^{\phi_C=V_{DS}} \sigma_n(x)\mathrm{d}\phi_C$$

$$= \mu_n w C'_{ox} \int_{\phi_C=0}^{\phi_C=V_{DS}} (V_{GS} - V_{Tn} - \phi_C(x))\mathrm{d}\phi_C \qquad (4.48)$$

where Equation (4.41) is used for the description of the channel charge. After carrying out the integration a current of

$$\boxed{I_{DS} = \mu_n C'_{ox} \frac{w}{l} [(V_{GS} - V_{Tn})V_{DS} - V_{DS}^2/2]} \qquad (4.49)$$

results.

Diffusion term

Equation (4.46) leads to a current of

$$I_{DS} = \mu_n w \phi_t \frac{\mathrm{d}\sigma_n(x)}{\mathrm{d}x} \qquad (4.50)$$

Separation of the variables and integration results in

$$\int_{x=0}^{x=l} I_{DS}\mathrm{d}x = \mu_n w \phi_t \int_{\sigma_n(\text{Source})}^{\sigma_n(\text{Drain})} \mathrm{d}\sigma_n \qquad (4.51)$$

Substituting Equation (4.41) in Equation (4.51) leads to a drain current of

$$\boxed{I_{DS} = \mu_n C'_{ox} \frac{w}{l} \phi_t V_{DS}} \qquad (4.52)$$

A comparison of the two current terms reveals that the drift current dominates as long as the ($V_{GS} - V_{Tn}$) term is larger than ϕ_t, which is 26 mV at room temperature (Bagheri *et al.* 1985; Tsividis 1987; Turchetti 1983).

With this assumption the drain current can be described by

$$\boxed{I_{DS} = \beta_n[(V_{GS} - V_{Tn})V_{DS} - V_{DS}^2/2]} \qquad (4.53)$$

with

$$\beta_n = k_n \frac{w}{l} \text{ and } k_n = \mu_n C'_{ox}$$

β_n is called the gain factor of the transistor and k_n sometimes the gain factor of the process. Typical values of k_n are around 200 μA/V^2. Due to its simplicity the above current–voltage equation is very useful for first hand calculations and is used throughout the chapters dealing with circuit design aspects.

In order to achieve a high drain current it is possible to increase the oxide capacitance C'_{ox} or to decrease the gate length l. This is obvious since these transistor parameters are multiplication factors in front of the current–voltage expression (Equation 4.53). The oxide capacitance (Equation 4.1) can be increased by decreasing the oxide thickness d_{ox} to a couple of nm or exchanging the insulator material for one with a higher dielectric constant. The influence of the channel length is illustrated in Figure 4.25. In both cases shown, the drain current has the same value since both transistors have the same geometry ratio. But transistor T_2 has only 1/4 of the gate capacitance C_g compared to transistor T_1. Therefore, circuits with transistor T_2 show a faster switching performance since lower capacitance values have to be charged. The preferred choice of a small channel length is obvious. But this is only true for digital circuits. Analog circuits require transistors with a very high output resistance in order to reduce the influence on the channel length modulation (Section 4.5.2). This usually leads to transistors with a relatively large gate length in the order of $l > 1$ μm.

The presented equations describe the transistor behavior in the resistive region, sometimes called the linear region. If the drain current is plotted as a function of the drain voltage the current shows an unexpected decrease (Figure 4.26). This effect should not be too surprising, since the equation is derived under the condition of a small drain voltage V_{DS}, which made it possible to use the results of the one-dimensional analysis.

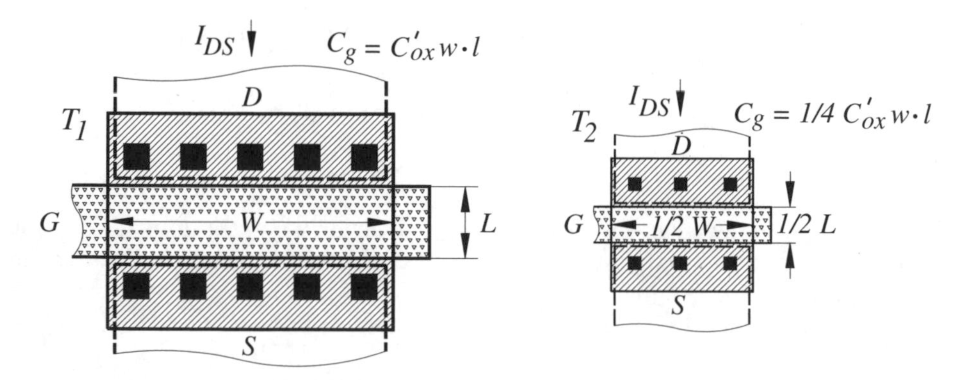

Figure 4.25 Comparison between two MOS transistors with different geometry ratio (W, L are drawn geometries)

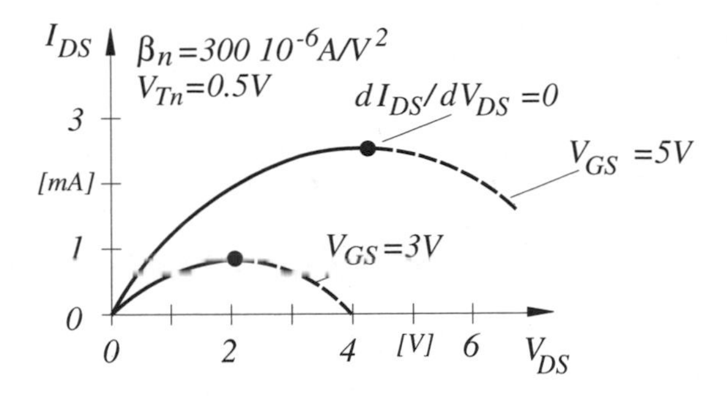

Figure 4.26 Calculated output characteristic of a MOS transistor

With an increasing drain voltage this assumption is violated and leads to an unphysical transistor behavior. The voltage at the onset of the current reduction, where $\mathrm{d}I_{DS}/\mathrm{d}V_{DS} = 0$, is called saturation voltage and can be determined by differentiation equation (Equation 4.53)

$$V_{DS} = V_{DSsat} = V_{GS} - V_{Tn} \tag{4.54}$$

What happens inside the transistor when the drain voltage is increased is considered in the following section.

Current saturation region

For this purpose it is useful to analyze the behavior of the channel voltage ϕ_C in more detail. This voltage can be found by replacing the integration limits l and V_{DS} of Equation (4.48) by the variables x and ϕ_C

$$I_{DS} = \mu_n C'_{ox}\frac{w}{x}\left[(V_{GS} - V_{Tn})\phi_C - \frac{\phi_C^2}{2}\right] \tag{4.55}$$

and solving for the channel voltage

$$\phi_C(x) = (V_{GS} - V_{Tn}) - \sqrt{(V_{GS} - V_{Tn})^2 - 2[(V_{GS} - V_{Tn})V_{DS} - V_{DS}^2/2]x/l} \tag{4.56}$$

In this equation it is assumed that the substituted drain current description (Equation 4.49) is valid for voltages up to the saturation voltage. The result is shown in Figure 4.27 for two different V_{DS} voltages.

As stated previously, the drain current I_{DS} (Equation 4.47) must be constant from source to drain, from a continuity point of view

$$\begin{aligned} I_{DS} &= -\mu_n w \sigma_n(x)\frac{\mathrm{d}\phi_C(x)}{\mathrm{d}x} \\ I_{DS} &= -v_n w \sigma_n(x) \end{aligned} \tag{4.57}$$

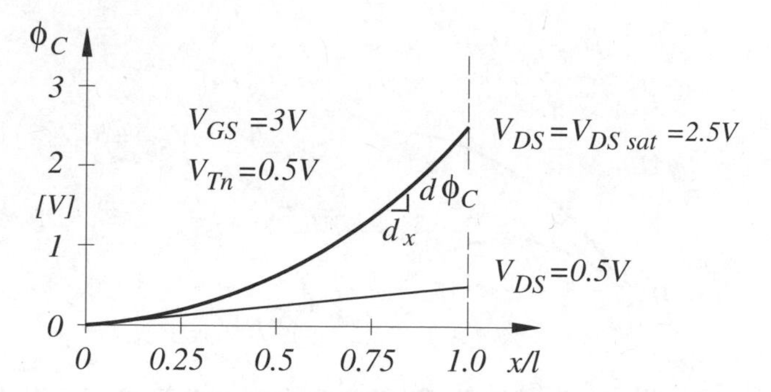

Figure 4.27 Channel voltage between source and drain for different V_{DS} voltages

But Figure 4.27 demonstrates that the electrical field – indicated by the term $d\phi_C/dx$ – varies from source to drain and is highest at the drain end of the channel. Accordingly, the charge density σ_n and the electron speed v_n must adjust. This leads to the fact that the electron speed (Equation 1.36) is highest and the charge density lowest at the drain end of the channel. According to Equation 4.41 the channel charge approaches zero there, when the channel voltage has a value of $\phi_C(l) = V_{DSsat} = V_{GS} - V_{Tn}$. In reality this is not the case; rather, the charge density approaches a small value and the electrons assume an elevated speed. The location in the channel where this happens is called the pinch-off point and is illustrated in the example of Figure 4.28.

At a small drain voltage a continuous channel between source and drain at the semiconductor surface exists. With a V_{DS} voltage of $V_{GS} - V_{Tn}$ – in this example 2.5 V – the channel voltage at the drain $\phi_C(l)$ and the V_{DS} voltage still have the same value, but the channel starts to pinch-off (Figure 4.28(b)). A further increase in the V_{DS} voltage does not change the channel voltage at the pinch-off point $\phi_C(l)$ noticeably (Figure 4.28(c)). Instead the electrons move from the pinch-off point through the depletion zone to the most positive region, the drain (inset, Figure 4.28(c)).

In conclusion, the channel voltage at the pinch-off point can be considered to be constant even when the drain voltage is increased beyond the saturation voltage of $V_{DSsat} = V_{GS} - V_{Tn}$.

With a constant channel voltage $\phi_C(l)$ the current becomes independent of the drain voltage and saturates at a voltage of (Equation 4.54)

$$V_{DS} \geq V_{DSsat} = V_{GS} - V_{Tn}$$

In this case a current (Equation 4.53) of

$$\boxed{I_{DS} = \frac{\beta_n}{2}(V_{GS} - V_{Tn})^2} \tag{4.58}$$

results.

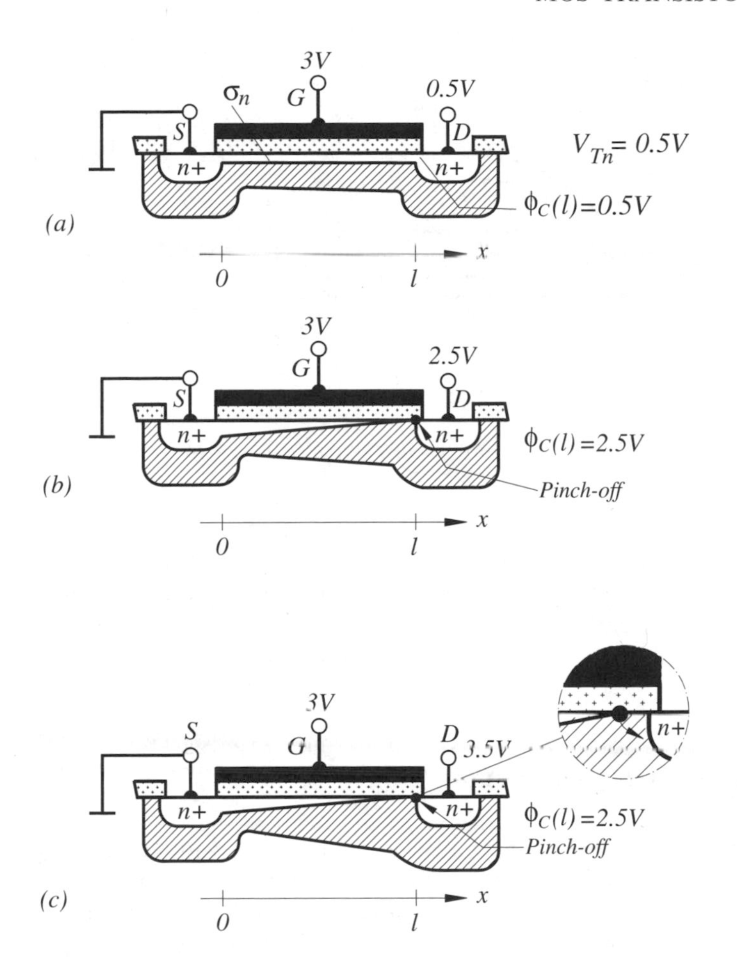

Figure 4.28 Channel charge distribution at different drain voltages: (a) continuous channel $V_{DS} = 0.5\,\text{V}$, (b) onset of channel pinch-off $V_{DS} = V_{DSsat} = 2.5\,\text{V}$ and (c) $V_{DS} > V_{DSsat}$ (threshold voltage $V_{Tn} = 0.5\,\text{V}$)

In conjunction with the occurring pinch-off a channel length modulation effect can be observed, which is treated in Section 4.5.2.

Assuming that the current description in the resistive region (Equation 4.53) is valid up to the saturation voltage (Equation 4.54), then the equation can be merged with that of the saturation description (Equation 4.58) in order to characterize the total output characteristic of the MOS transistor (Figure 4.29). It is obvious that these two merged equations describe the transistor behavior only approximately since the derivation is based on a one-dimensional analysis. Despite this drawback, the equations were sufficient for many years. The reason was that initially MOS transistors were used in digital integrated circuits only. In these applications the transistor is either turned on – this means the transistor is in the resistive region with a small V_{DS} value – or turned off with no current flowing. Inaccuracies occurred mainly during transient analyses.

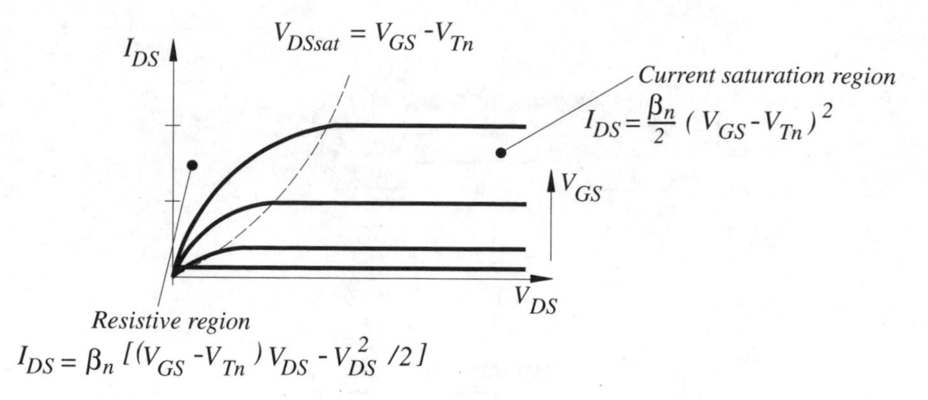

Figure 4.29 Output characteristic of a MOS transistor

With scaled dimensions and the application of MOS transistors to analog circuits, the situation changed significantly. This led to complex descriptions, called compact models, used in computer-aided designs. The value of the simple equations above can therefore be seen in their easy use for first hand circuit performance estimations.

4.4.2 Improved Transistor Equation

In the derived transistor equations it is assumed that the width of the depletion region is constant from source to drain. In reality the width changes and with it the depletion region charge σ_d. In the following this effect is taken into consideration. As a result a more precise equation set is available, which is sometimes used in simple compact models or for first hand calculations.

The charge of the inversion layer is according to Equation (4.39)

$$\sigma_n(x) = -C'_{ox}\left(V_{GS} - V_{FB} - 2\phi_F - \phi_C(x) - \gamma\sqrt{2\phi_F + V_{SB} + \phi_C(x)}\right)$$

If the square root term is linearized by the first two terms of the Taylor series around the source voltage

$$\sqrt{2\phi_F + V_{SB} + 2\phi_C(x)} \approx \sqrt{2\phi_F + V_{SB}} + \frac{\phi_C(x)}{2\sqrt{2\phi_F + V_{SB}}} \tag{4.59}$$

an improved description of the inversion layer charge

$$\begin{aligned}\sigma_n(x) &= -C'_{ox}\left(V_{GS} - V_{FB} - 2\phi_F - \phi_C(x) - \gamma\left(\sqrt{2\phi_F + V_{SB}} + \frac{\phi_C(x)}{2\sqrt{2\phi_F + V_{SB}}}\right)\right)\\ &= -C'_{ox}(V_{GS} - V_{Tn} - (1 + F_B)\phi_C(x))\end{aligned} \tag{4.60}$$

results. In this equation

$$F_B = \frac{\gamma}{2\sqrt{2\phi_F + V_{SB}}} \tag{4.61}$$

is a factor taking the local change in depletion region width into account. Using this equation the drain current in the resistive region

$$I_{DS} = \beta_n \left[(V_{GS} - V_{Tn})V_{DS} - \frac{(1 + F_B)}{2} V_{DS}^2 \right] \tag{4.62}$$

and the one in the saturation region

$$I_{DS} = \frac{\beta_n}{2} \frac{(V_{GS} - V_{Tn})^2}{1 + F_B} \tag{4.63}$$

can be derived similarly to the procedure presented in the preceding section. In this case the saturation voltage has a value of

$$V_{DSsat} = \frac{V_{GS} - V_{Tn}}{1 + F_B} \tag{4.64}$$

The current–voltage equations are identical to the ones already derived, with the exception that the factor F_B takes the variation in depletion layer width into account. These equations predict a smaller current. This current reduction is even more pronounced when a high substrate doping or, in other words, a larger body-effect factor is present.

4.4.3 Current–Voltage Characteristic at Weak Inversion

The threshold voltage tells the designer when a transistor is cut off ($V_{GS} < V_{Tn}$) or tuned on ($V_{GS} > V_{Tn}$). This behavior stems from the assumption that the transistor operates in strong inversion (Section 4.3.2). In reality, a MOS system or transistor does not cut off when biased from strong inversion into weak inversion; rather, in the biasing condition $V_{GS} \leq V_{Tn}$ an exponential current–voltage relationship

$$\boxed{I_{DS} = \beta_n (n - 1)\phi_t^2 e^{(V_{GS} - V_{Tn})/\phi_t n}(1 - e^{-V_{DS}/\phi_t})} \tag{4.65}$$

results, where

$$\boxed{n = 1 + C_j'/C_{ox}'} \tag{4.66}$$

describes a capacitor coupling factor of the transistor. The depletion capacitance of this equation can be determined according to Equation (4.12)

$$C'_j = \frac{\varepsilon_0 \varepsilon_{Si}}{x_d} \tag{4.67}$$

Typical values of n are between 1.5 and 2.5. The influence of the drain voltage on the sub-threshold current is neglible when the value of $V_{DS} > 100\,\text{mV}$ at room temperature. Under this condition the transistor behaves like a voltage-controlled current generator.

For the interested reader the derivation of the current–voltage relationship under weak inversion condition is included in Appendix A at the end of this chapter.

An example is used to give an idea about the size of the currents in weak inversion.

Example

A transistor has the following data: $\beta_n = 500 \cdot 10^{-6}\,\text{A/V}^2$, $V_{Tn} = 0.5\,\text{V}$, and $n = 2$. Calculate the sub-threshold current I_{DS} at room temperature when $V_{GS} = 0.4\,\text{V}$ and 0.1 V. The drain voltage is larger than 100 mV.

$$I_{DS}(V_{GS} = 0.4\,\text{V}) = 500\,\mu\text{A/V}^2 \cdot (26 \cdot 10^{-3}\,\text{V})^2 e^{(0.4\,\text{V} - 0.5\,\text{V})/2 \cdot 26 \cdot 10^{-3}\,V} = 49\,\text{nA}$$

$$I_{DS}(V_{GS} = 0.1\,\text{V}) = 500\,\mu\text{A/V}^2 \cdot (26 \cdot 10^{-3}\,\text{V})^2 e^{(0.1\,\text{V} - 0.5\,\text{V})/2 \cdot 26 \cdot 10^{-3}\,V} = 150\,\text{pA}$$

Sub-threshold swing

A very useful expression or, better, a figure of merit is the sub-threshold swing S of a transistor (Figure 4.30). That is the gate voltage swing needed to change the drain current by one decade

$$S = \frac{\mathrm{d}V_{GS}}{\mathrm{d}\log_{10} I_{DS}} = \ln 10 \frac{\mathrm{d}V_{GS}}{\mathrm{d}\ln I_{DS}} \tag{4.68}$$

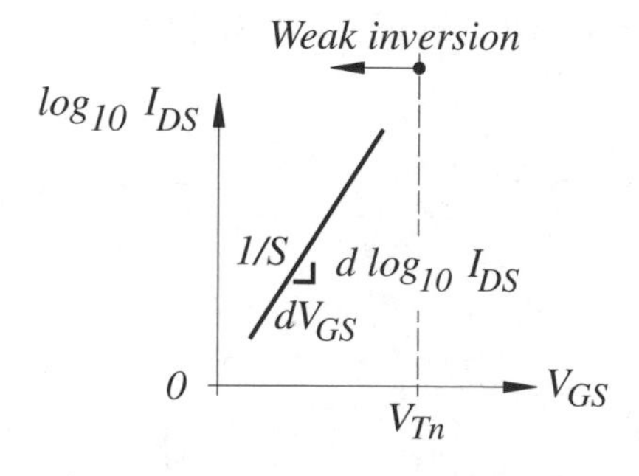

Figure 4.30 Sub-threshold swing of a MOS transistor

This figure of merit can be found from Equation (4.65), which yields with $V_{DS} > 100\,\mathrm{mV}$ the expression

$$\ln I_{DS} = \ln \beta_n (n-1)\phi_t^2 + (V_{GS} - V_{Tn})/\phi_t n \tag{4.69}$$

Differentiating this equation leads to a sub-threshold swing of

$$\boxed{S = \phi_t (1 + C_j'/C_{ox}') \ln 10} \tag{4.70}$$

When one assumes the best case, that is $C_j'/C_{ox}' \ll 1$, a sub-threshold swing at room temperature of

$$S = 26\,\mathrm{mV} \cdot \ln 10 = 60\,\mathrm{mV/decade}$$

results. This means that a reduction of the gate voltage by 60 mV reduces the sub-threshold current by one decade. Typical sub-threshold swing values are between 80 mV and 120 mV per decade.

Example

A transistor has a threshold voltage of $V_{Tn} = 0.5\,\mathrm{V}$ and a sub-threshold swing of $S = 120\,\mathrm{mV/decade}$. At a gate voltage of $V_{GS} = 0.35\,\mathrm{V}$ a current of $1.2\,\mu\mathrm{A}$ can be observed flowing through the transistor. A gate voltage of $V_{GS} = 0\,\mathrm{V}$ reduces the current by how much?

The 350 mV reduction in gate voltage results in a current of about

$$I_{DS}(V_{GS} = 0\,\mathrm{V}) \approx 1.2\,\mathrm{nA}$$

which is a reduction of three decades at room temperature.

This simple example demonstrates that the threshold voltage of a MOS transistor cannot be reduced without considering what effect an increased sub-threshold current has on the operation of the circuit.

4.4.4 Temperature Behavior

The temperature behavior of the MOS transistor is caused by the temperature-dependent parameters, threshold voltage, and gain factor of the process. The last one is, according to Equation (4.53)

$$k_n(T) = \mu_n(T) C_{ox}'$$

dependent on the electron mobility (Figure 1.17), which can be described empirically by

$$\mu_n(T) = \mu_n(300\,\mathrm{K})\left(\frac{T}{300\,\mathrm{K}}\right)^{-a_n} \tag{4.71}$$

The factor a_n typically has a value between 1.5 and 2. With a_n of 1.5 this equation predicts a mobility reduction of about 35%, when the temperature is increased from room temperature to 100 °C.

The temperature influence on the threshold voltage (Equation 4.35)

$$V_{Ton}(T) = V_{FB} + 2\phi_F(T) + \gamma\sqrt{2\phi_F(T)}$$

is mainly caused by the changing Fermi voltage (Equation 4.27)

$$\phi_F = \frac{kT}{q}\ln\frac{N_A}{n_i}$$

and the intrinsic density (Equation 1.20)

$$n_i = C\left(\frac{T}{[\mathrm{K}]}\right)^{3/2} e^{-E_G(T)/2kT}$$

The temperature coefficient of the threshold voltage can be derived, when Equation (4.35) is differentiated with respect to the temperature

$$\begin{aligned} \frac{\mathrm{d}V_{Ton}}{\mathrm{d}T} &= \frac{\mathrm{d}\phi_F}{\mathrm{d}T}\left[2 + \frac{\gamma}{\sqrt{2\phi_F}}\right] \\ \frac{\mathrm{d}V_{Ton}}{\mathrm{d}T} &= \frac{\mathrm{d}\phi_F}{\mathrm{d}T}\left[2 + \frac{1}{C'_{ox}}\sqrt{\frac{qN_A\varepsilon_0\varepsilon_{Si}}{\phi_F}}\right] \end{aligned} \tag{4.72}$$

with

$$\frac{\mathrm{d}\phi_F}{\mathrm{d}T} = \frac{1}{T}\left(\phi_F - \frac{E_G}{2q}\right) - \frac{3\,k}{2\,q}$$

In this equation it is assumed that the band-gap E_G is independent of the temperature. Since $E_G/2q > \phi_F$, the term $\mathrm{d}\phi_F/\mathrm{d}T$ and therefore the temperature coefficient of the threshold voltage $V_{Tn}/\mathrm{d}T$ is negative. Typical values are around $-2\,\mathrm{mV/°C}$.

It is interesting to observe the influence of the temperature-dependent parameters on a transistor, e.g. in saturation (Equation 4.58)

$$I_{DS}(T) = \frac{\mu_n(T)C'_{ox}}{2}\frac{w}{l}(V_{GS} - V_{Tn}(T))^2$$

The drain current decreases with increasing temperature due to the reduction in electron mobility. But simultaneously the current increases due to an increase of the term $(V_{GS} - V_{Tn})$, caused by a reduced threshold voltage (Figure 4.31).

At large gate voltages the effect of the mobility reduction dominates and causes a decrease in current with temperature, whereas at small gate voltages the opposite behavior can be observed due to the threshold reduction with temperature (Filanovsky 2000).

The analysis of the temperature behavior so far dealt with a transistor being in strong inversion. The following analyzes what happens when the transistor is in weak inversion or, in other words, in the sub-threshold region. The answer can be found directly from Equation (4.65)

$$I_{DS} = \beta_n(n-1)\phi_t^2 e^{(V_{GS}-V_{Tn})/\phi_t n}\left(1 - e^{-V_{DS}/\phi_t}\right)$$

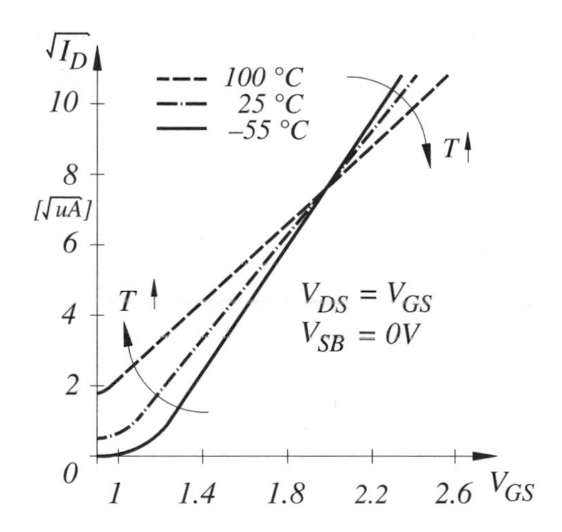

Figure 4.31 Temperature behavior of a MOS transistor in saturation and strong inversion

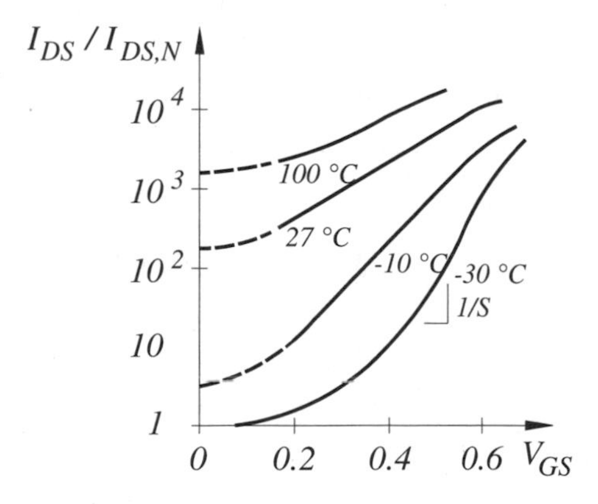

Figure 4.32 Normalized temperature behavior of a MOS transistor in weak inversion (dashed: leakage currents, Section 4.5.5)

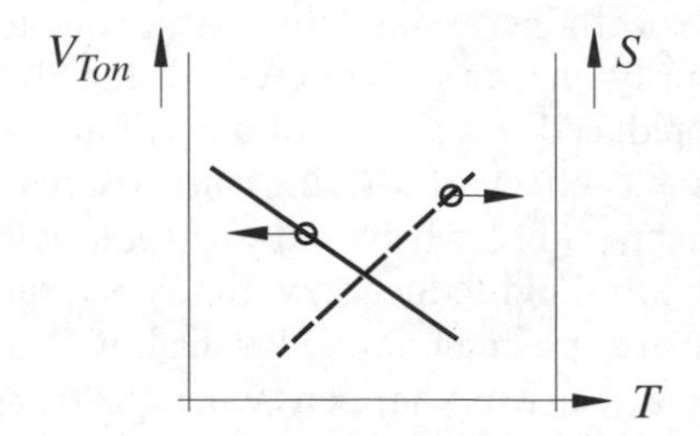

Figure 4.33 Temperature sketch of threshold voltage V_{Ton} and sub-threshold swing S

The transistor has a pronounced positive temperature dependence due to the $\phi_t = kT/q$ term in the exponent (Figure 4.32). This behavior can also be described by the fact that an increase in temperature causes an increase in the sub-threshold swing S (Equation 4.70) and simultaneously a reduction in the threshold voltage (Figure 4.33). A substantial reduction of the number of decades the transistor current can be reduced is the result.

The sub-threshold behavior is of particular importance when a charge is stored at a source or drain region. This is the case, e.g. at dynamic memories, which are covered in Section 7.5.1.

Summary

In order to derive simple analytical equations for the description of the current–voltage behavior of a MOS transistor, it is necessary to divide the transistor into the regions sub-threshold, resistive, and saturation as shown in Figure 4.34. The three equations describing the operation regions of the transistor are of the utmost importance for first

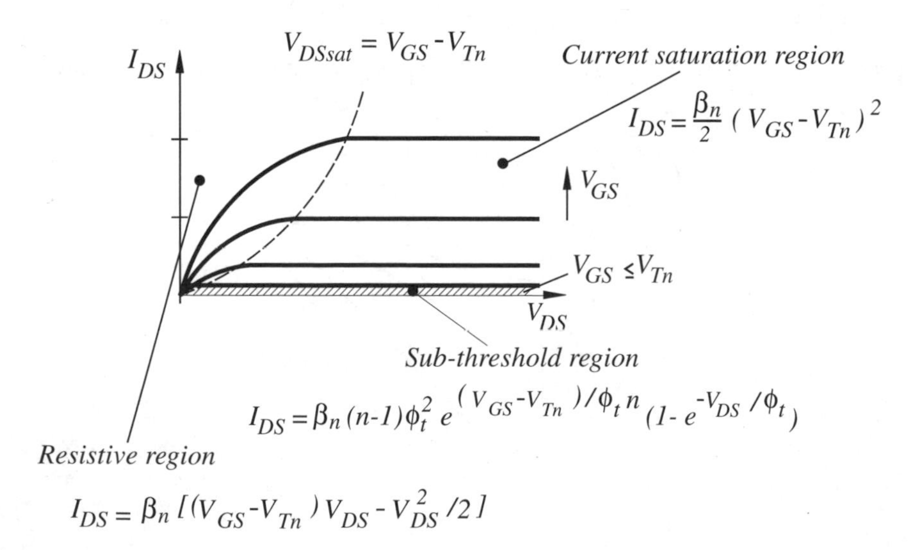

Figure 4.34 Partition of the output characteristic of a MOS transistor into three operation regions

hand calculations, particularly when the transistor is employed in a circuit environment. This is the case, starting in Chapter 5 of this book, where these equations are used exclusively.

4.5 SECOND-ORDER EFFECTS

These effects include mobility degradation, channel length modulation, and the impact a short channel has on the current–voltage characteristic of the MOS transistor. Furthermore, the breakdown behavior and parasitic bipolar effects are considered.

4.5.1 Mobility Degradation

In the previous derivation the mobility of the electrons μ_n is assumed to be constant. In reality this is not the case, since the mobility is under the influence of a vertical and a lateral electrical field of the MOS transistor. This leads to a substantial reduction in current, particularly at small dimensions.

Influence of the vertical electrical field

This field – the gate field – acts perpendicularly on the electron movement from source to drain. This leads to an electron acceleration toward the SiO_2–Si interface. Here the electrons suffer collisions, which cause a reduction in electron mobility. This effect is the more pronounced the larger the gate voltage is. This effect is empirically described by (Schwerin *et al.* 1987)

$$\mu_s = \frac{\mu_n}{1 + \theta(V_{GS} - V_{Tn})} \tag{4.73}$$

where θ is an adjustable parameter and μ_n the electron mobility in the inversion layer at the semiconductor surface under a low vertical field intensity. A typical value for θ is around $0.1\,\mathrm{V}^{-1}$

Influence of the lateral electrical field

Figure 4.27 illustrates how the electrical field changes from source to drain. At the drain end of the channel the electrical field is largest and in conjunction the electron velocity. Between the electron velocity and the electrical field exists a nonlinear relationship (Figure 1.16), which is repeated in Figure 4.35 for convenience. At a field of $\mathcal{E}_M$ the electron velocity saturates. This implies that the electron mobility μ_s is reduced when the drain voltage and therefore the lateral field is increased beyond this point. This effect can be described by (Caughy and Thomas 1967)

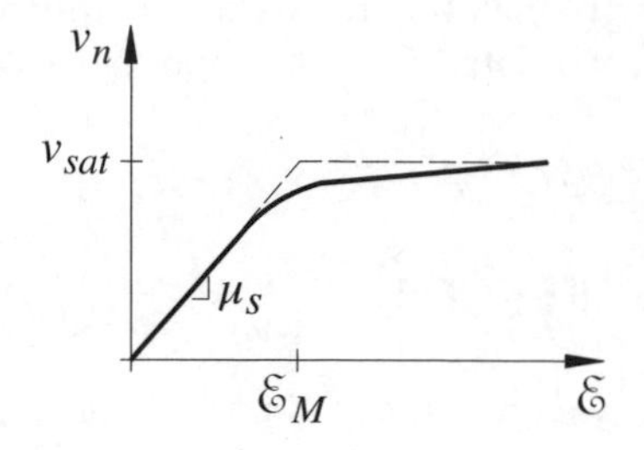

Figure 4.35 Electron velocity as a function of an electrical field

$$v_n = \frac{-\mu_s \mathcal{E}}{1 + \dfrac{\mathcal{E}}{\mathcal{E}_M}} \tag{4.74}$$

If $\mathcal{E} \ll \mathcal{E}_M$ an electron velocity of

$$v_n = -\mu_s \mathcal{E}$$

results. When $\mathcal{E} \gg \mathcal{E}_M$, the electron velocity approaches its saturation velocity of

$$v_n = v_{sat} = -\mu_s \mathcal{E}_M \tag{4.75}$$

Equation (4.74) therefore leads to an effective mobility of

$$\mu_{eff} = -\frac{v_n}{\mathcal{E}} = \frac{\mu_s}{\left(1 + \dfrac{\mathcal{E}}{\mathcal{E}_M}\right)} \tag{4.76}$$

which is used in many compact models. If the horizontal electrical field is approximated by V_{DS}/l, then the relationship

$$\boxed{\mu_{eff} = \frac{\mu_s}{1 + \dfrac{V_{DS}/l}{v_{sat}/\mu_s}}} \tag{4.77}$$

results. It is recommended to use this equation in hand calculations when the channel length of the transistor is below 1.5 μm and to assume, for simplicity, that $\mu_s = \mu_n$.

4.5.2 Channel Length Modulation

Up to now it is assumed that the drain current saturates and remains constant even when $V_{DS} > V_{DSsat}$. Real transistors show a deviation from this theoretical case, as shown in Figure 4.36. In order to analyze this effect it is useful to consider Figure 4.28

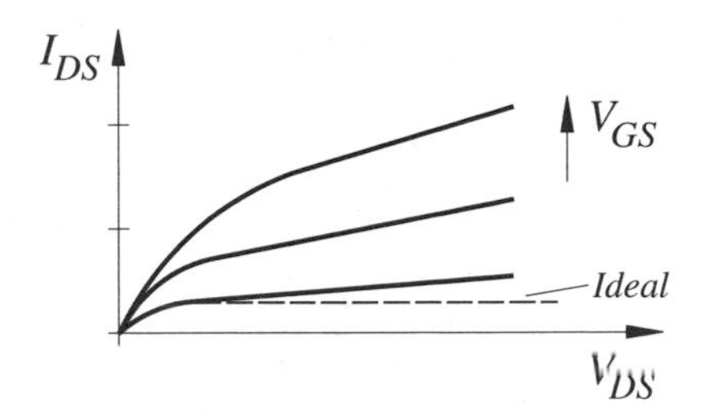

Figure 4.36 Output characteristic of a MOS transistor with emphasized channel length modulation

again. The channel pinches off at a voltage of $V_{DS} = V_{DSsat} = V_{GS} - V_{Tn}$ and the channel voltage at the pinch-off point has a value of $\phi_C = V_{DSsat}$. An increase in the drain voltage causes a slide shift of the pinch-off point towards the source (Figure 4.37). This behavior is similar to that of a pn-junction where the depletion region width is a function of the applied voltage. The shift of the pinch-off point toward the source can be interpreted as a reduction in effective channel length, responsible for an increase in drain current

$$I_{DS} = \frac{\mu_n C'_{ox}}{2} \frac{w}{l(V_{DS})} (V_{GS} - V_{Tn})^2 \tag{4.78}$$

It is obvious that the so-called channel length modulation is the more pronounced the shorter the channel length is.

In order to include this effect in a compact model, a two-dimensional field problem has to be solved. In order to circumvent this, a one-dimensional approach (Baum *et al.* 1970) is often used. The charge in the region $l > x < l'$ consists of ionized acceptor atoms. Applying Poisson's equation, a field distribution in analogy to Equation (2.36)

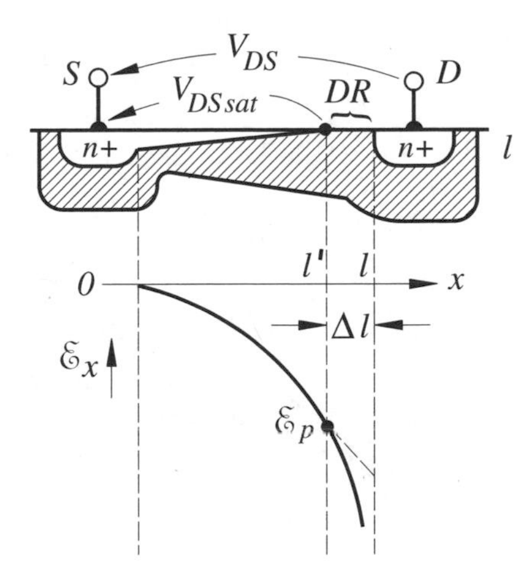

Figure 4.37 Cross-section of a MOS transistor with lateral field at $V_{DS} > V_{DSsat}$

$$\mathcal{E}_x(x) = \mathcal{E}_p - \frac{qN_A}{\varepsilon_0 \varepsilon_{Si}} x \tag{4.79}$$

results, where $\mathcal{E}_p$ is the field at the pinch-off point l'. Integrating

$$V_{DS} = V_{DSsat} - \int_{l'}^{l} \mathcal{E}_x(x)\, \mathrm{d}x \tag{4.80}$$

leads to an effective channel length of

$$l' = l - \sqrt{(\mathcal{E}_p/2\alpha)^2 + \frac{K}{\alpha}(V_{DS} - V_{DSsat})} - |\mathcal{E}_p|/2\alpha \tag{4.81}$$

where $\alpha = qN_A/(2\varepsilon_0\varepsilon_{Si})$ is used as an abbreviation.

The one-dimensional approach is inaccurate and leads to differences between model description and measured data. By adding a factor K this problem is minimized.

A common phenomenological approach (Shichman *et al.* 1968) uses for the voltage range $V_{DS} \geq V_{GS} - V_{Tn}$

$$\boxed{I_{DS} = \frac{\beta_n}{2}(V_{GS} - V_{Tn})^2(1 + \lambda V_{DS})} \tag{4.82}$$

where the factor $(1 + \lambda V_{DS})$ is a correction term. $1/\lambda$ can be found at the intersection of the extrapolated current lines (Figure 4.38). A typical value for λ is $0.05\,\mathrm{V}^{-1}$. The advantage of Equation (4.82) is its ease of use in first hand circuit calculations. For this reason the equation is used exclusively in the analog sections of this book.

4.5.3 Short Channel Effects

The analysis of the MOS transistor so far is based on the assumption of long and wide channel dimensions. At channel sizes in the order of the depletion region width of source

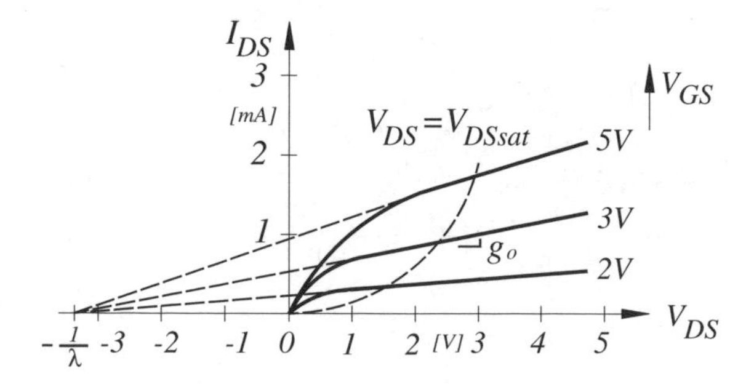

Figure 4.38 Output characteristic of a MOS transistor with channel length modulation

and drain second order effects occur resulting in a substantial change in transistor performance.

Transistor with short channel length

In this case the depletion region of the metallurgical drain junction and source junction reduces the depletion region controlled by the gate voltage, causing a reduction in the threshold voltage. The smaller the channel length l, the more dominating this effect will be (Figure 4.39(a)). This so-called roll-down effect depends to a large extent on the actual size of the depletion region and consequently on the substrate doping density. The disadvantage of this effect is that in the roll-down region the manufacturing tolerance in gate length translates into an additional tolerance in threshold voltage. Therefore, in circuit design, the gate length is usually limited to a minimum value, where the roll-down effect is still minor.

In order to analyze this effect it is assumed that the gate-controlled depletion region can be described by a trapezoidal shape (Yau 1974) (Figure 4.40). The threshold voltage Equations (4.34), (4.13), and (4.20) is given by

$$\begin{aligned} V_{Tn} &= V_{FB} + 2\phi_F + \gamma\sqrt{2\phi_F + V_{SB}} \\ &= V_{FB} + 2\phi_F - \frac{Q_d}{C_{ox}} \end{aligned} \tag{4.83}$$

where

$$Q_d = \sigma_d \cdot w \cdot l \tag{4.84}$$

represents the gate-controlled depletion charge. This charge changes to

$$Q_d^* = \sigma_d \cdot w(l - x) \tag{4.85}$$

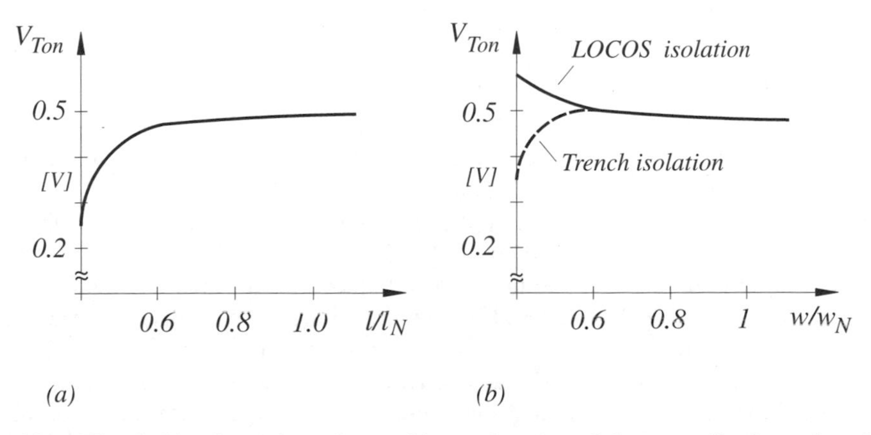

Figure 4.39 Threshold voltage dependence: (a) as a function of the normalized gate length and (b) as a function of the normalized gate width

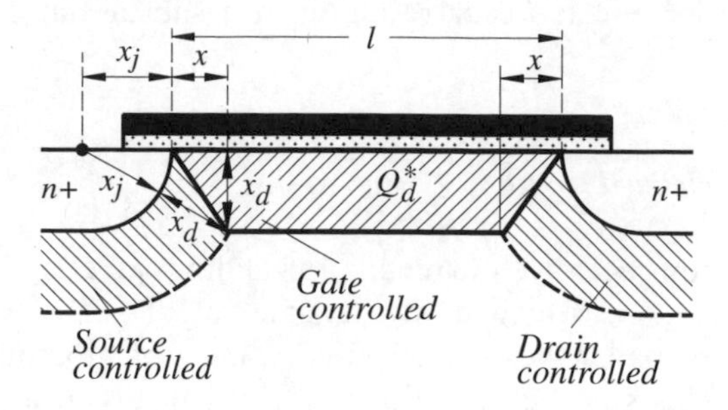

Figure 4.40 Gate-controlled depletion region modeled by a trapezoidal shape at $V_{DS} = 0$

if a trapezoidal shape is assumed. With

$$(x_j + x)^2 + x_d^2 = (x_j + x_d)^2 \tag{4.86}$$

the threshold voltage dependence on the channel length l can be expressed as

$$\begin{aligned} V_{Tn} &= V_{FB} + 2\phi_F - \frac{Q_d^*}{Q_d}\frac{Q_d}{C_{ox}} \\ &= V_{FB} + 2\phi_F + F_S\gamma\sqrt{2\phi_F + V_{SB}} \end{aligned} \tag{4.87}$$

where

$$F_S = 1 - \frac{x_j}{l}\left[\sqrt{1 + 2\frac{x_d}{x_j}} - 1\right] \tag{4.88}$$

is a factor describing the threshold voltage reduction.

A deviation in gate length between neighboring transistors therefore is responsible for a difference in threshold voltage of

$$\begin{aligned} \Delta V_{Tn} &= V_{Tn,1} - V_{Tn,2} \\ \Delta V_{Tn} &= \left(\frac{1}{l_2} - \frac{1}{l_1}\right)x_j\left[\sqrt{1 + 2\frac{x_d}{x_j}} - 1\right]\gamma\sqrt{2\phi_F + V_{SB}} \end{aligned} \tag{4.89}$$

This voltage difference is part of the so-called offset voltage in differential transistor stages, which are covered in more detail in Section 10.4.1.

An applied drain voltage causes the depletion region at the drain side to increase (Figure 4.41), which leads to a further reduction in threshold voltage (Skotnicki *et al.* 1986). This effect is called drain induced barrier lowering (DIBL) and is illustrated in Figure 4.42.

The analyzed short channel effect leads to the following conclusion: during manufacturing a slightly smaller than allowed minimum gate length may cause a reduction in threshold voltage and consequently, as described in Section 4.4.3, a deteriorated sub-threshold behavior. It is therefore recommended to use larger than minimum allowed gate length at transistors, where an increased sub-threshold current may cause an adverse effect on the circuit performance.

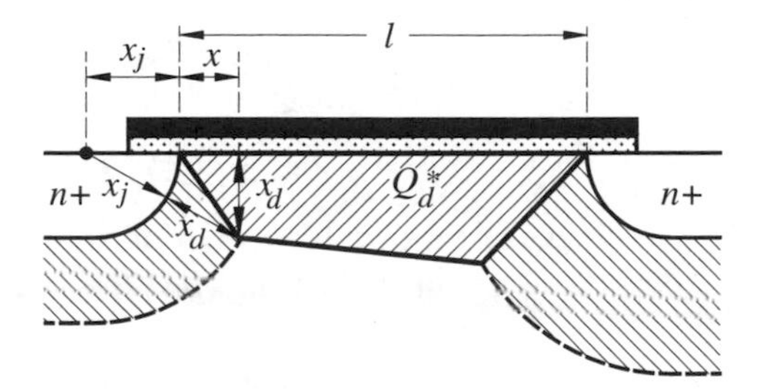

Figure 4.41 Gate-controlled depletion region modeled by a trapezoidal shape at $V_{DS} > 0$ V

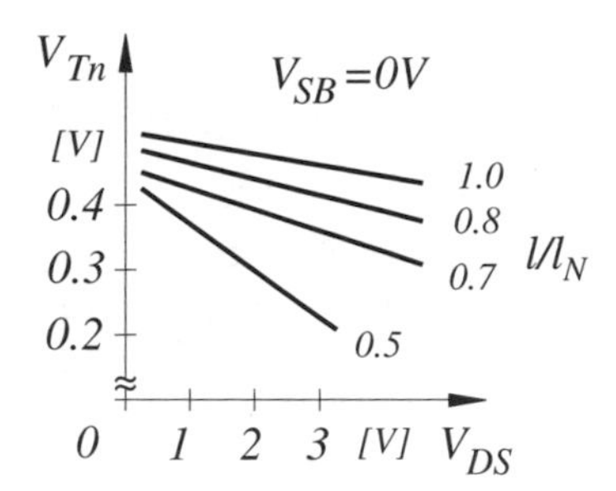

Figure 4.42 Threshold voltage as a function of V_{DS} voltage and normalized gate lengths

Transistor with short channel width

Depending on the isolation technology used (Section 4.1, Figure 4.3) a reduction or increase in threshold voltage can be observed (Figure 4.39(b)) when the gate width is reduced (Figure 4.43). The threshold voltage (Equation 4.83) can be written in the form

$$V_{Tn} = V_{FB} + 2\phi_F - \frac{Q_d}{C_{ox}}$$

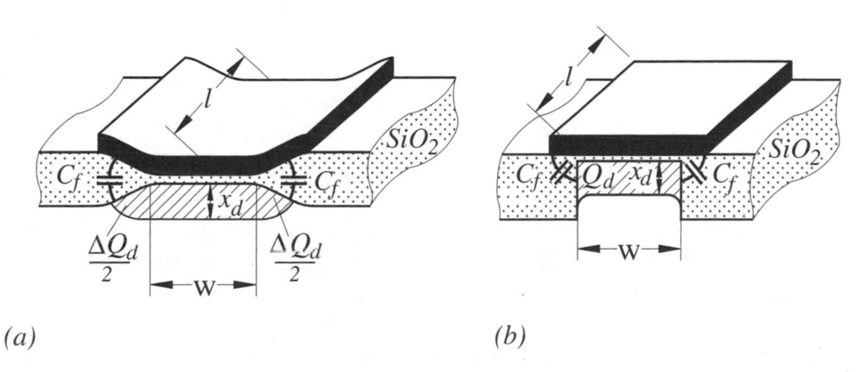

Figure 4.43 Cross-section through the width of a MOS transistor: (a) LOCOS insulation and (b) trench insulation

By modifying this equation

$$V_{Tn} = V_{FB} + 2\phi_F - \frac{Q_d + \Delta Q_d}{C_{ox} + C_f} \tag{4.90}$$

it is possible to analyze the influence of the width reduction. In this equation ΔQ_d is an additional charge of the depletion region, extending into the lateral direction, and C_f a side wall capacitance.

In the case of the LOCOS insulation, the capacitance C_f is negligible compared to that of the oxide C_{ox}. Contrary to this situation, the additional charge ΔQ_d is not negligible when the gate width, and consequently the charge Q_d, is small. This leads to an increase in threshold voltage

$$V_{Tn} = V_{FB} + 2\phi_F - \frac{\sigma_d}{C'_{ox}} - \frac{\Delta Q_d}{C'_{ox} wl} \tag{4.91}$$

when the gate width is reduced (note: all charges have negative signs).

In the case of the trench insulation the situation is reversed. C_f is, due to the overlapping into the trench, not negligible compared to the oxide capacitance. An additional depletion region charge does not exist. With respect to Equation (4.90) this leads to a reduction in threshold voltage, when the gate width is reduced

$$V_{Tn} = V_{FB} + 2\phi_F - \frac{\sigma_d}{C'_{ox} + C_f/(wl)} \tag{4.92}$$

The short channel effects are usually included in advanced compact models used for CAD applications. A drawback is that these additional mathematical descriptions require a substantial number of geometrical and fitting parameters. But the alternative would be to characterize or parameterize all possible transistor sizes a designer plans to use. Not a thrilling task!

4.5.4 Hot Electrons

Figure 4.27 shows how the channel voltage ϕ_C and the electrical field – indicated by the term $d\phi_C/dx$ – change from source to drain. The largest electrical field exists at the drain side of the channel. According to the field distribution the charge density σ_n and the electron speed v_n must adjust along the channel since the drain current (Equation 4.57)

$$I_{DS} = -v_n w \sigma_n(x)$$

is constant. Depending on the channel length and the voltages applied to the transistor, the channel electrons may approach the saturation velocity v_{sat} of about 10^7 cm/s (Figure 1.16) at the drain side of the channel. In this case, the electrons assume a high kinetic energy of $E - E_C = kT_{eff}$ where T_{eff} can be interpreted as effective temperature, which is significantly higher than room temperature. The highly energetic electrons therefore are called hot electrons. A certain percentage of these hot electrons create

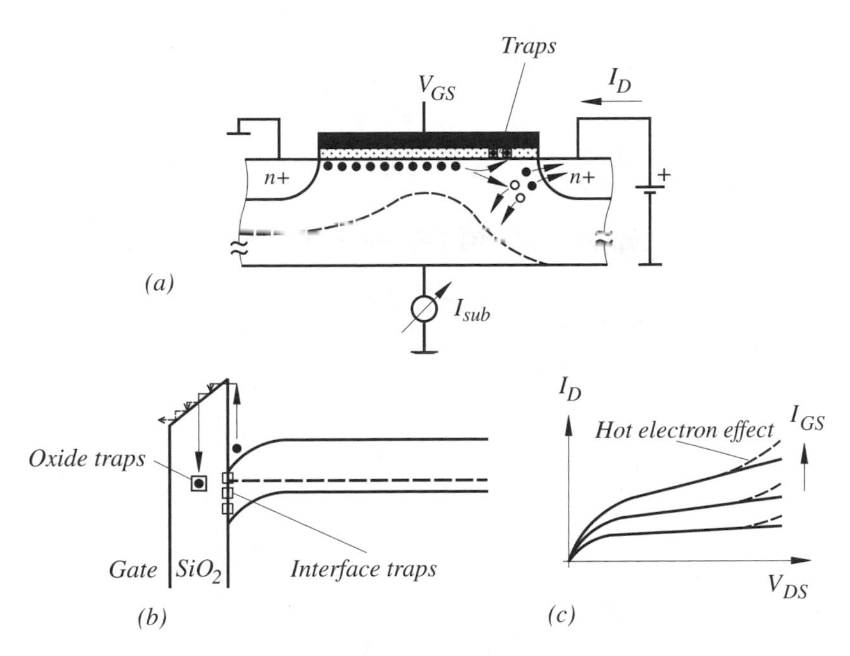

Figure 4.44 (a) MOS transistor with hot electron injection, (b) band diagram including interface and oxide traps and (c) output characteristic including hot electron effect

electron–hole pairs by an avalanche process (Section 2.6) in the pinch-off region of the transistor (Figure 4.44). The holes created by this process are collected at the substrate and the electrons at the drain, causing an increase in drain current (Figure 4.44(c)). But some electrons and holes in this avalanche process are able to gain sufficient kinetic energy to charge interface traps or surmount the Si–SiO_2 energy barrier and charge oxide traps (Figure 4.44(b)). These trapped charges play an important role in the gradual degradation of the oxide and consequently the transistor characteristic. One way to reduce this undesirable degradation effect is obviously the reduction of the power-supply voltage of the circuit and thus the drain voltage. Additionally the doping of the drain region, and in conjunction that of the source region, can be reduced. The so-called lightly-doped drain (source) (LDD) structure (Figure 4.2(h)) causes an increase in the depletion region width in the drain (source) side of the junction and consequently a reduction in the electrical field (Section 2.6). This is an improvement compared to a situation where highly-doped drain (source) (HDD) regions are used and where the depletion region extends mostly into the substrate. A device simulation (*Medici User's Manual* 1997) reveals the reduction in the lateral electrical field in the LDD case (Figure 4.45).

4.5.5 Gate-Induced Drain Leakage

If a transistor is used as a switch in a circuit application, it is important in many cases to have a precise knowledge of the leakage current of the switch or, in other words, of the leakage current at the drain of the transistor (Figure 4.46). Since in this situation

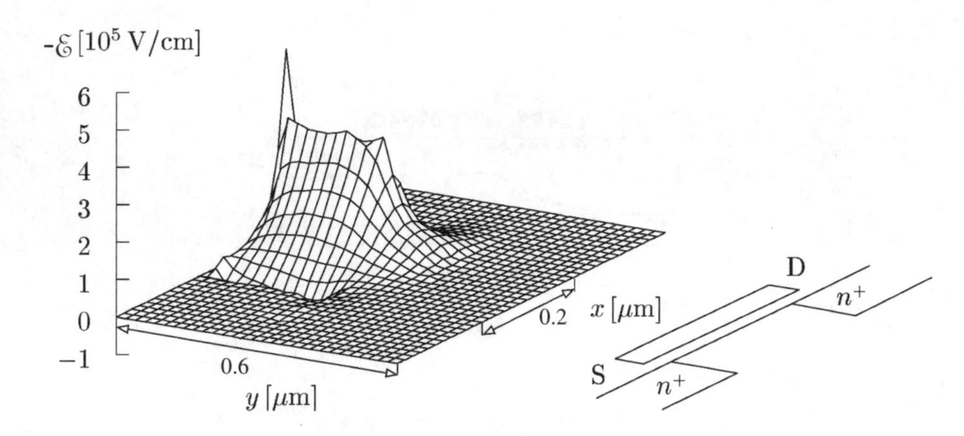

a)

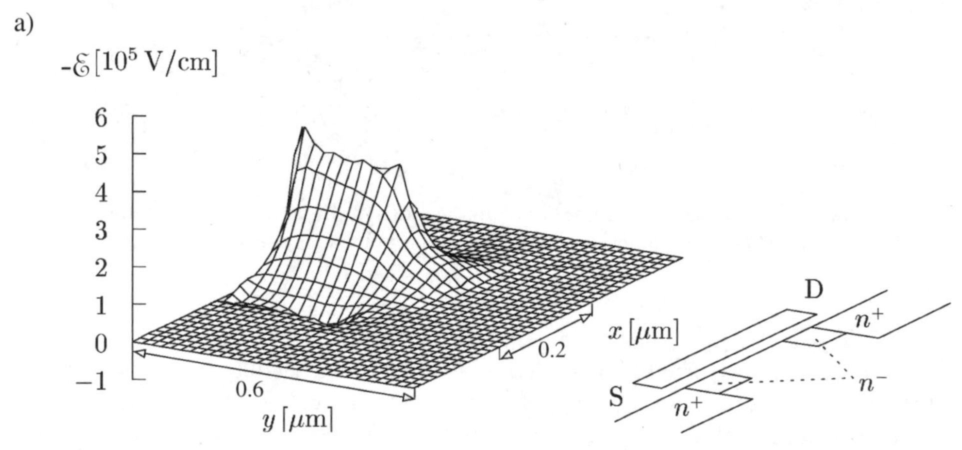

b)

Figure 4.45 Lateral electrical field in a MOS transistor: (a) HDD structure and (b) LDD structure (parameters: $V_{GS} = 1.0\,\text{V}$, $V_{DS} = 2.5\,\text{V}$, $V_{Tn} = 0.4\,\text{V}$, $d_{ox} = 6\,\text{nm}$, $l = 0.25\,\mu\text{m}$)

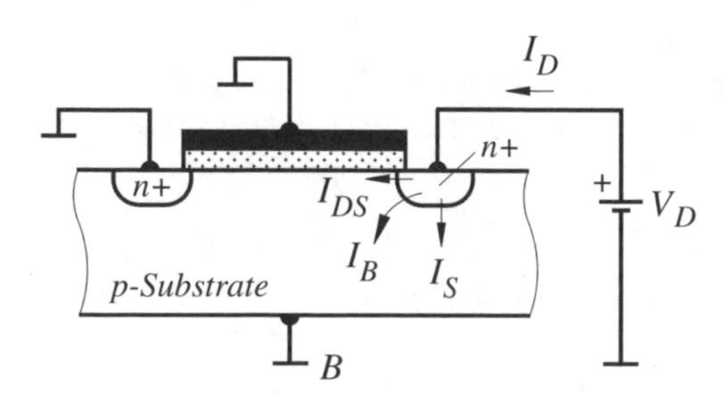

Figure 4.46 Drain leakage current of a MOS transistor: sub-threshold current I_{DS}, saturation current of pn-junction I_S, gate induced drain leakage current I_B

the gate voltage has a value of $0 \leq V_{GS} \leq V_{Tn}$ the transistor is in the sub-threshold region and a drain current I_{DS} is present according to Equation (4.65). Additionally a saturation current I_S of the back-biased drain–substrate junction exists (Equation 2.28). But there is another current, called gate-induced drain leakage (GIDL) I_B, which is considered in the following. For this purpose the cross-section of the drain side of the MOS transistor is shown in Figure 4.47. The indicated voltage conditions are responsible for a depletion region in the drain region where the gate overlaps the n^+-region. This is an analogous situation to the one shown in Figure 4.7. With a sufficiently thin gate oxide and large enough voltage, interband tunneling is generated by the high electrical field in the surface depletion region (Chang *et al.* 1995; Tanaka 1995). In this case valence electrons tunnel into the conduction band and are collected at the drain. The holes left behind in the valence band are collected at the back side of the p-substrate. The drain-induced leakage current (Koyanagi *et al.* 1987) can be expressed by

$$I_B = A\mathcal{E}_S e^{-B/\mathcal{E}_s} \tag{4.93}$$

where the electrical field has the relationship

$$\mathcal{E}_S = \frac{V_{DG} - qE_G}{3d_{ox}} \tag{4.94}$$

In these equations V_{DG} is the voltage between drain and gate and A and B are material parameters.

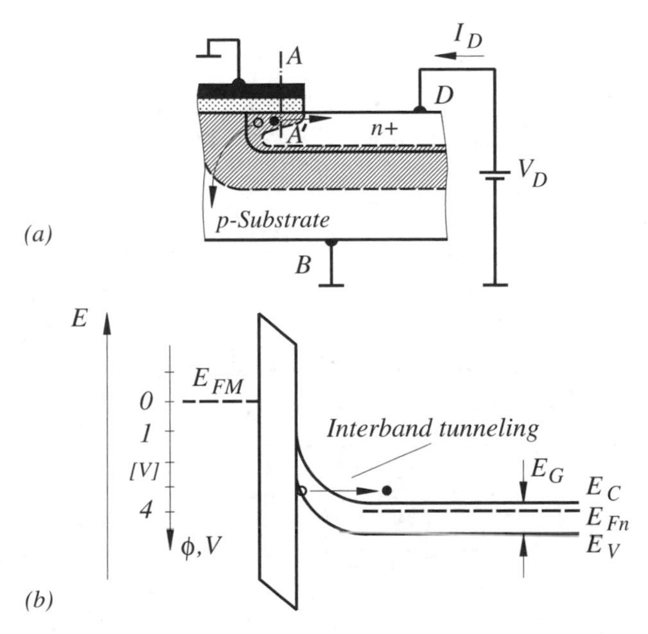

Figure 4.47 (a) Cross-section of the drain side of a MOS transistor and (b) band diagram with interband tunneling (cross-section A–A')

4.5.6 Breakdown Behavior

At the MOS transistor two dominant breakdown mechanisms can be observed, namely avalanche and punch-through (Figure 4.48). In order to avoid the destruction of the transistor, the drain current is limited to 1 μA. At a relatively large gate length, the maximum drain voltage at this current is limited by the avalanche breakdown of the drain–bulk junction and is comparable to the one described in Section 2.6. A different situation exists when a transistor with reduced gate length is considered. In this case it can be observed that the smaller the gate length, the lower the drain voltage will be. This behavior is caused by the so-called punch-through effect where the depletion regions of drain and source touch. In this case electrons are able to move directly from the source through the joined depletion region to the drain.

An increase in substrate doping concentration leads to a reduced depletion layer width at source and drain region. This causes the onset of punch-through to shift toward smaller gate length. But simultaneously the avalanche breakdown is reduced, as sketched in Figure 4.48.

If the drain current is not limited to 1 μA, a so-called snap-back effect can be observed (Figure 4.49). The voltage condition $V_{DB} = BV$ generates an avalanche breakdown at the drain–bulk junction. Electrons move to the drain electrode and holes into the substrate, causing the substrate to become slightly positively biased, which leads to a bipolar transistor action. The base B (substrate) is positively biased, causing holes to be injected into the emitter E (source) and electrons to move through the base to the collector C (drain). In this case the base width corresponds to the gate length of the MOS transistor. An increase in current causes an increase in the current gain of the transistor (Section 3.2.5, Figure 3.18(b)) and in conjunction a reduction in the breakdown voltage. This so-called snap-back effect is characterized by I_{SP} and V_{SP}. A further increase in the V_{DB} voltage leads to thermal destruction of the device (Amerasekera *et al.* 1991).

The snap-back behavior of the MOS transistor is frequently used in conjunction with input protection devices (Section 5.6.3). It has the advantage that in the snap-back conditions the static power dissipation $I_{SP} \cdot V_{SP}$ is relatively low.

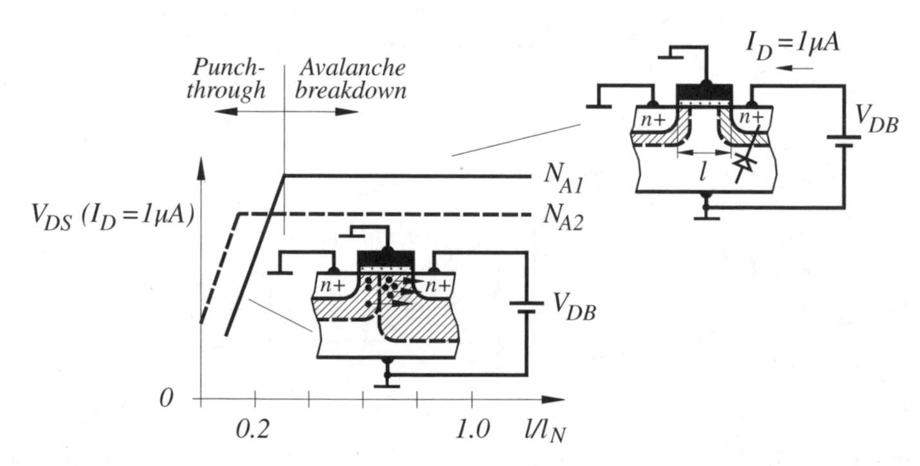

Figure 4.48 Drain voltage at $I_{DS} = 1\mu A$ as a function of a normalized gate length ($N_{A2} > N_{A1}$)

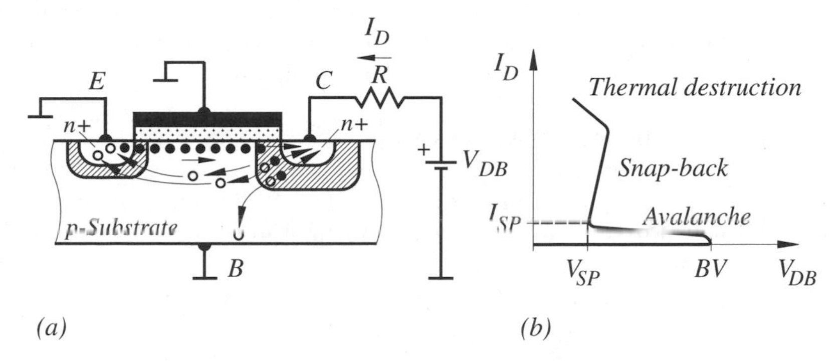

Figure 4.49 (a) MOS transistor under snap-back condition and (b) output behavior

4.5.7 Latch-up Effect

In all integrated CMOS circuits parasitic npn transistors and pnp transistors are present. These transistors can act as silicon-controlled rectifiers (SCR) which, when triggered, are able to cause a large current to flow, damaging the integrated circuit permanently. How this may happen is demonstrated with an example using a cross-section of a CMOS process, described in Section 4.1 (Figure 4.50). The n-well is connected via an n^+-region to V_{CC} and the p-substrate via a p^+-region to ground. The p^+-region at V_{CC}, the n-well, and the p-substrate form a vertical pnp transistor (T_1), whereas the n^+-region, the p-substrate, and the n-well act as an npn transistor (T_2). To each of the base emitter junctions a resistor is parallel connected. These resistors represent the resistance of the p-substrate and the n-well to their respective electrode. These lumped components form an equivalent circuit (Figure 4.50(b)), which acts as SCR.

Assuming that by some means – which will be discussed in the following – the emitter-base junction of the npn transistor T_2 is forward biased. This leads to a collector current I_{C2}. Part of this collector current consists of the base current of the pnp transistor $T1$, which causes a collector current I_{C1} to flow. Part of this current supplies the base current to T_2. The circuit is in a low resistive condition. When the means for getting the circuit

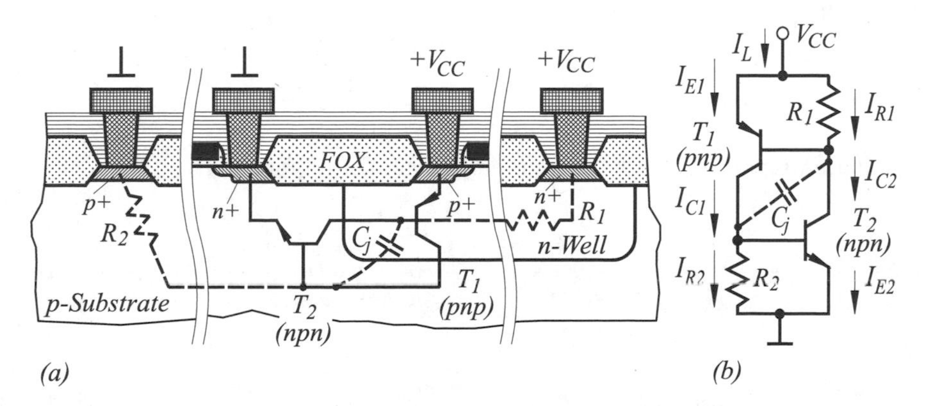

Figure 4.50 (a) Cross-section of a CMOS process and (b) equivalent SCR circuit

into this state are removed, and the circuit remains in the low resistive condition, one speaks of a latch-up situation. The means for getting the circuit into the latch-up state, or in other words, the triggering mechanisms causing this situation are, e.g.:

(a) A fast rise time of the power-supply voltage dV_{CC}/dt may cause a current of $i \approx C_j dV_{CC}/dt$ to flow through the depletion capacitance C_j of the n-well/p-substrate junction. Part of this current is injected into the base of T_2 and may trigger the SCR structure. Usually this does not happen in a system where the power-supply changes very slowly due to the large capacitors in the power-supply unit. But when power is available in a system and the integrated circuit is plugged into a socket, an unintentionally large voltage change may occur.

(b) Large noise amplitudes on the power supply may exceed the breakdown voltage of the SCR structure.

(c) Voltage ringing on the inputs of an integrated circuit, assembled on a printed circuit board. This ringing is usually caused by the inductances of the wiring on the circuit board in conjunction with the capacitance of the circuits. In Section 5.6.3 this topic is revisited when the ESD protection is analyzed.

(d) Injection of charge carriers, e.g. by light or by capacitance coupling. The latter may unintentionally forward bias a junction briefly and trigger the SCR structure.

A very large current may flow, and finally damage the circuit, when after the triggering occurred, the SCR remains in the latch-up state. In order that this happens, the bipolar transistors must have a particular current gain.

In the latch-up state the current into the circuit (Figure 4.50(b)) is

$$
\begin{aligned}
I_L &= I_{C1} + I_{C2} \\
I_L &= A_1 I_{E1} + A_2 I_{E2} \\
I_L &= A_1(I_{E1} + I_{R1}) - A_1 I_{R1} + A_2(I_{E2} + I_{R2}) - A_2 I_{R2}
\end{aligned}
\tag{4.95}
$$

where $A_1 = I_{C1}/I_{E1}$ and $A_2 = I_{C2}/I_{E2}$ are the static current gains of transistor T_1 and T_2. Rearranging the above equation leads to

$$1 = A_1 + A_2 - A_1 \frac{I_{R1}}{I_L} - A_2 \frac{I_{R2}}{I_L} \tag{4.96}$$

This equation is valid when the current gains meet the requirement

$$A_1 + A_2 = 1 + A_1 \frac{I_{R1}}{I_L} + A_2 \frac{I_{R2}}{I_L} \tag{4.97}$$

or

$$B_1 B_2 = 1 + B_1(1 + B_2)\frac{I_{R1}}{I_L} + B_2(1 + B_1)\frac{I_{R2}}{I_L} \tag{4.98}$$

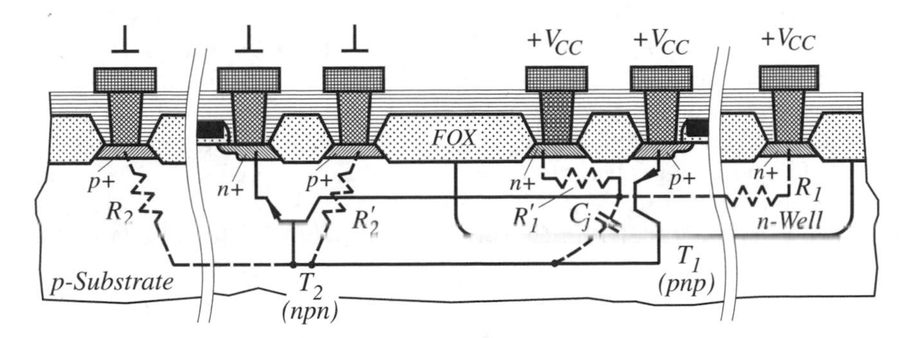

Figure 4.51 Cross-section of a CMOS process with guard rings

where the relationship between the different current gains $B = A/(1 - A)$ is used. The above equations describe the general latch-up conditions of an SCR (Genda 1984). When the resistor values are infinitive and therefore I_{R1} and I_{R2} are zero a current gain of

$$\boxed{A_1 + A_2 = 1} \text{ or } \boxed{B_1 B_2 = 1} \tag{4.99}$$

is sufficient for the SCR to remain in the latch-up condition. Typical current gain values B for the lateral transistor are 0.1 to 10 and for the vertical one about 20 to 50.

A very effective way to reduce the latch-up sensitivity of a CMOS circuit is to reduce the resistor values. This can be done by using many low resistive interconnections to the p-substrate and the n-well. In circuits which are particularly endangered, like inputs of an integrated circuit, guard rings may be used around sensitive circuits (Figure 4.51). The guard ring in the n-well consists of an n^+-ring connected to V_{CC} and in the p-substrate of a p^+-ring connected to ground. They effectively reduce the resistance values since they act as parallel resistors to the base emitter regions.

The most effective way to reduce the latch-up sensitivity by technological means is to use a very low resistive substrate material and connect the back side of the circuit to ground. The effect is that basically all base–emitter junctions of the npn transistors are short-circuited. This is the case at very small transistor dimensions, which require a highly doped substrate in order to reduce the width of the depletion regions. If a highly doped and thus low resistive substrate cannot be used, the alternative is to employ a very low resistive starting material with an appropriately doped epitaxial layer on top of the starting material.

4.6 POWER DEVICES

So far standard MOS transistors are considered, with breakdown voltages mainly determined by the substrate doping (Equation 2.75). But many industrial and automotive applications require higher breakdown voltages and current densities than are possible with a standard CMOS technology. How this can be achieved is the topic of the following section.

To start the discussion about power devices it is appropriate to consider first some basic ideas behind them. For this purpose a sketch of a MOS transistor is shown in Figure 4.52. When the gate length l is sufficiently large that no punch-through occurs, then the maximum drain–bulk voltage is limited by the breakdown voltage of the drain–bulk junction. Solving Poisson's equation leads to the relationship between the charge, electrical field, and voltage distribution of the depletion regions DR. The result is, as stated in Section 2.6, that the *triangular area of the electrical field represents the voltage across the depletion region*. If the V_{DB} voltage is increased, this causes the depletion region and thus the electric field $\mathcal{E}$ to increase also. When the electrical field approaches the critical field $\mathcal{E}_C$ avalanche breakdown occurs at the breakdown voltage BV. This voltage can be increased when the doping density is reduced. In order not to influence the transistor behavior, the substrate doping remains unchanged, but the drain doping is reduced to n^- (Figure 4.52(b)). An additional increase in breakdown voltage is possible when the n^--doping of the drain is further reduced and a low resistive n^+-drain connection is provided (Figure 4.52(c)). According to the larger area under the electrical field, the breakdown voltage is increased. The lightly doped n^--region is called the drift zone or ν-zone. If the doping in this region has a value of about $N_D = 10^{14}\,\text{cm}^{-3}$, then each additional increase of the ν-zone width by 1 μm leads to an increase in breakdown voltage by about 10 V, as described in Section 2.6.

In a real transistor corner regions of the drain–bulk depletion region near the surface occur and may result in an increased electrical field and thus to a reduction in break-

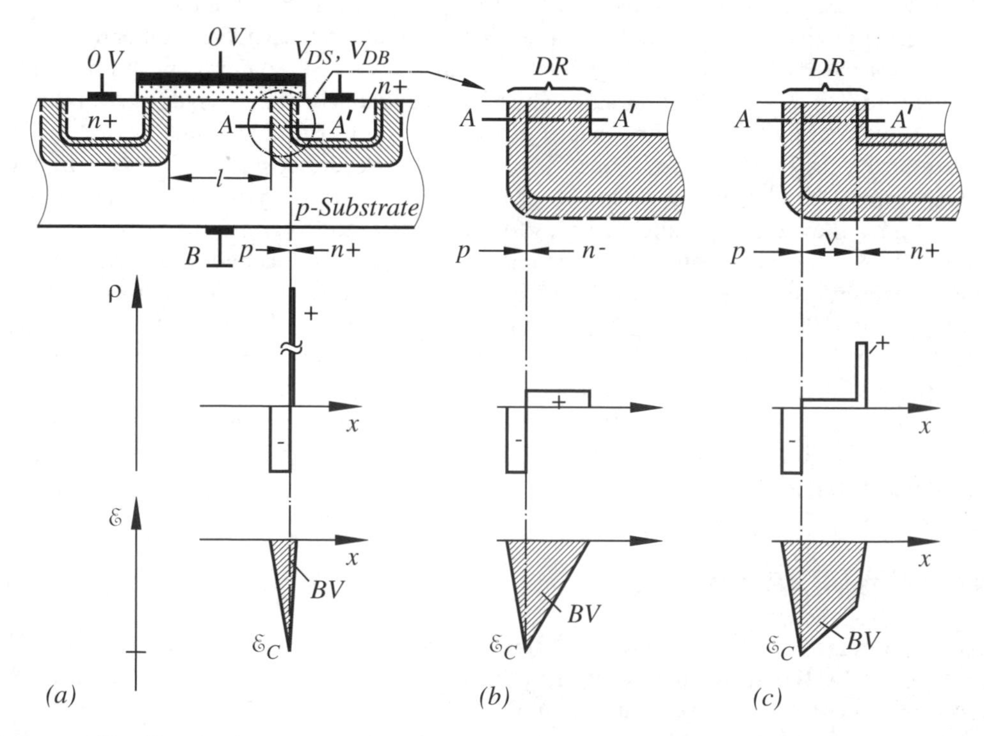

Figure 4.52 Sketch of MOS transistor with charge and field distribution at the drain (cross-section A–A'): (a) pn^+-junction, (b) pn^--junction and (c) $p\nu n^+$-junction

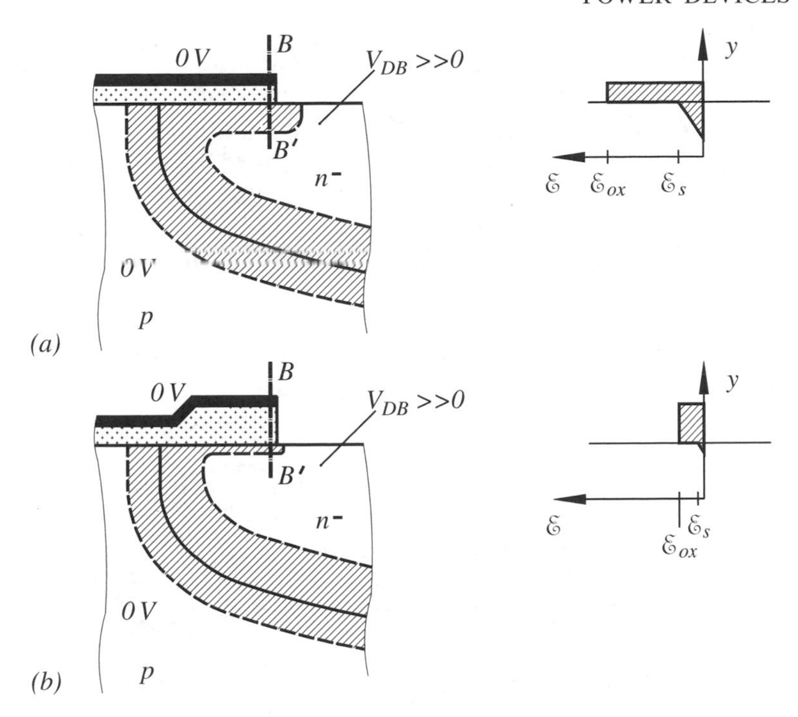

Figure 4.53 Overlap region drain gate with electrical field distribution (cross-section B–B'): (a) overlap with gate oxide and (b) overlap with expanded oxide thickness

down voltage. Furthermore, an increased electrical field between the gate and the drain of the transistor exists, which may lead to an oxide breakdown. This is shown for a gate drain overlap region in Figure 4.53. In case (a) the depletion region extends into the n^--region at the surface of the drain, caused by the positive voltage between drain and gate (Section 4.2.1). This leads to electrical fields at the oxide $\mathcal{E}_{ox}$ and at the semiconductor surface $\mathcal{E}_S$, responsible for enhanced breakdown mechanisms. These fields can be reduced when the gate influence on the semiconductor surface is decreased by an expanded gate oxide (Figure 4.53(b)).

This expanded gate oxide leads to transistor structures as sketched with doping profile in Figure 4.54. This is an asymmetrical transistor, where the drain–bulk and drain–gate voltages support large power-supply voltages. The remaining maximum voltages are unchanged by the drain modifications. Due to the relatively few transistor modifications, this device is an ideal candidate to be incorporated into an existing CMOS or BICMOS process.

Due to the additional ν-region, the area consumption of the transistor is relatively large, particularly when high breakdown voltages are required. This problem can be coped with, when a so-called DMOS transistor (double-diffused MOS) is used (Figure 4.55). At the time of the development of this transistor fine geometries were not available and small channel lengths could not be achieved by lithographic means. For this purpose, the higher diffusion rate of boron N_A (p-body), compared to that of phosphorus N_D (n^+-source), was used for the generation of a small channel length (Tihany *et al.* 1977). When this transistor structure is modified such that the drain is

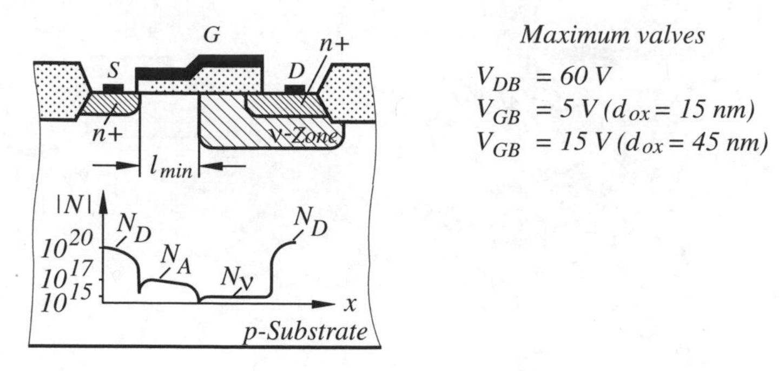

Figure 4.54 MOS transistor with increased drain breakdown voltage and typical maximum values

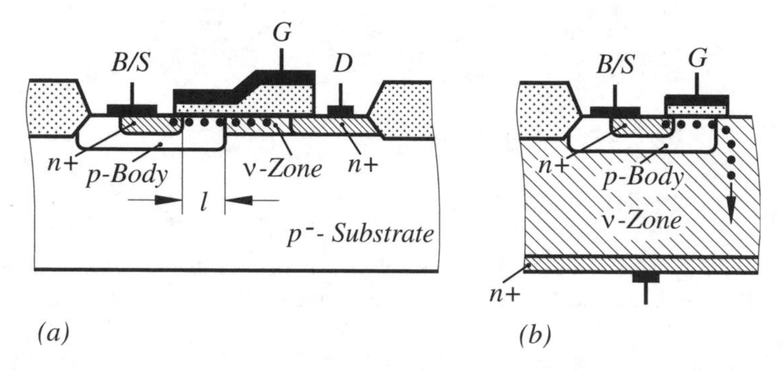

Figure 4.55 (a) Lateral DMOS transistor and (b) vertical DMOS transistor

placed via an n^+-region at the back side of the device, then the lateral DMOS transistor is changed into a vertical DMOS one (Figure 4.55(b)). This has the advantage that the lateral dimensions are independent of the required size of the drift zone. To get a feeling for the dimensions involved, the following example illustrates the area gain.

Example

The breakdown voltage of a DMOS transistor should be larger than 200 V. From Section 2.6 it is known that approximately each additional increase in drift zone width by 1 μm leads to an increase in breakdown voltage by about 10 V. Accordingly, a drift zone or ν-zone of about 20 μm is required.

High voltage transistors are usually used as switches. These require a very low on-resistance in order to cope with the power dissipation on the chip. For this purpose hundreds or even thousands of such transistors may be parallel connected in a cell array (Figure 4.56).

The annular DMOS transistor consists of a gate area and n^+-source regions which are connected to the p-body via a p^+-contact. The drain on the back side is brought up to the surface via a buried layer and an n^+-sinker for ease of interconnection. The

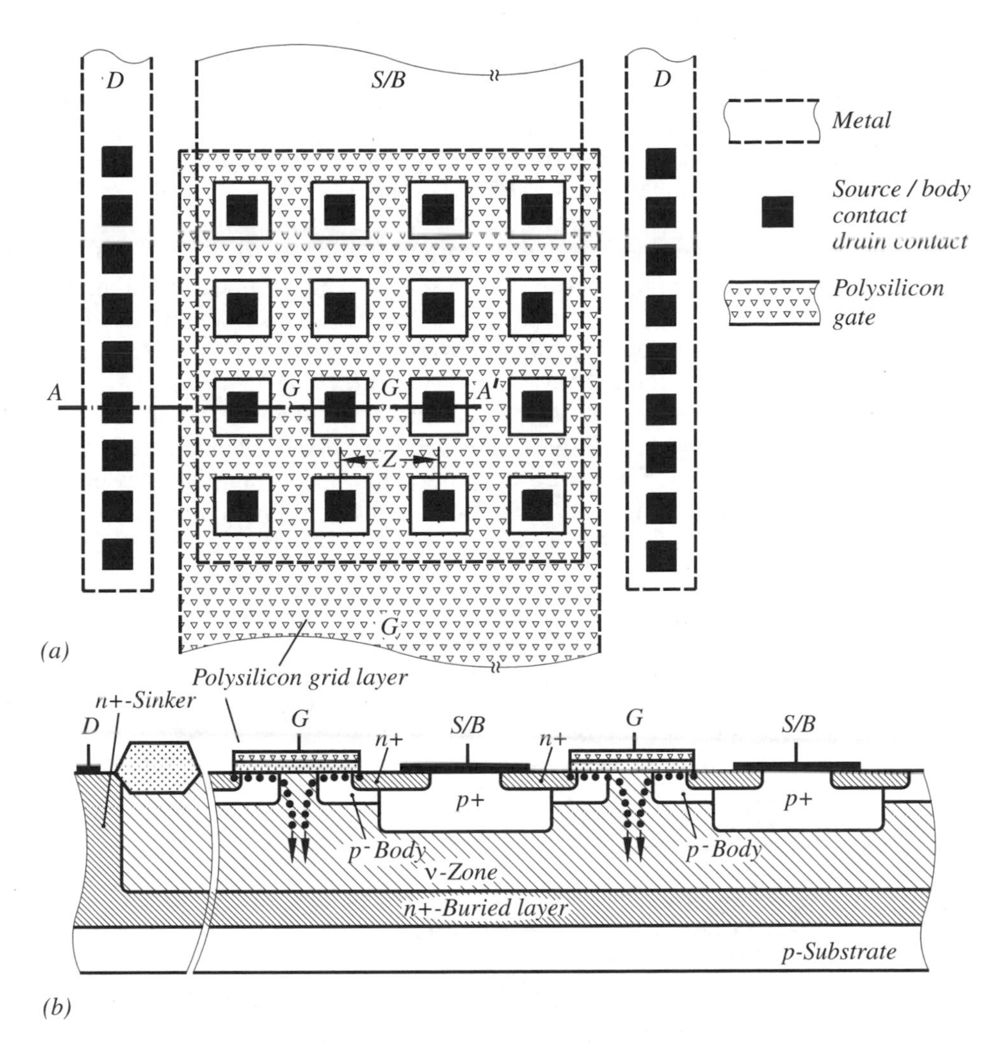

Figure 4.56 Parallel connected DMOS transistor cells with up-drain connection: (a) layout and (b) cross-section A–A'

gate connections consist of a polysilicon grid layer. The source S and bulk B contacts to the individual cells are made by a sheet of metal over the hole array.

A figure of merit of such an array is the specific on-resistance $R_{on} \cdot A$ achievable in an area of 1 mm^2. When, e.g., the on-resistance of one cell at nominal voltage conditions is 1 kΩ and the cell consumes an area of $20 \cdot 20\,\mu\text{m}^2$ this results in an array with 2500 cells per 1 mm^2 and in a specific on-resistance of 0.4 Ωmm^2. This is an important number for the designer since the percentage of the total chip area needed for the power switch or switches can easily be determined. It is obvious from this discussion that it is highly desirable to achieve an on-resistance as small as possible. How this can be achieved by technological means is discussed next.

Self-aligned DMOS cell

The principle of this cell structure is based on a self-aligned contact, discussed in conjunction with Figure 3.5 of Chapter 3. After gate oxidation, polysilicon deposition, and opening the contact area, a boron (p) and arsenic (n^+) implant is carried out and a side wall oxide spacer is formed (Figure 4.57(a)). This is followed by an anisotropic etchback (Figure 4.57(b)), where the spacer masks the silicon groove etch and the following p^+-contact implant. Finally, a metallization layer is deposited (Figure 4.57(c)). This scheme offers the advantage that the size of the polysilicon opening is limited by the required diffusion area of the p-body of the DMOS transistor and not by the lithography. This leads to a reduced area consumption and thus to an improved specific on-resistance, typically around $0.2\,\Omega\text{mm}^2$ (Preussger *et al.* 1991).

Another equally important advantage of the scheme is that the resistance of the p-body is reduced, which leads to an improved second breakdown behavior. How this can be understood is illustrated in Figure 4.58. At a particular high drain field, the onset of avalanche breakdown occurs. This causes electrons to move to the drain and holes into

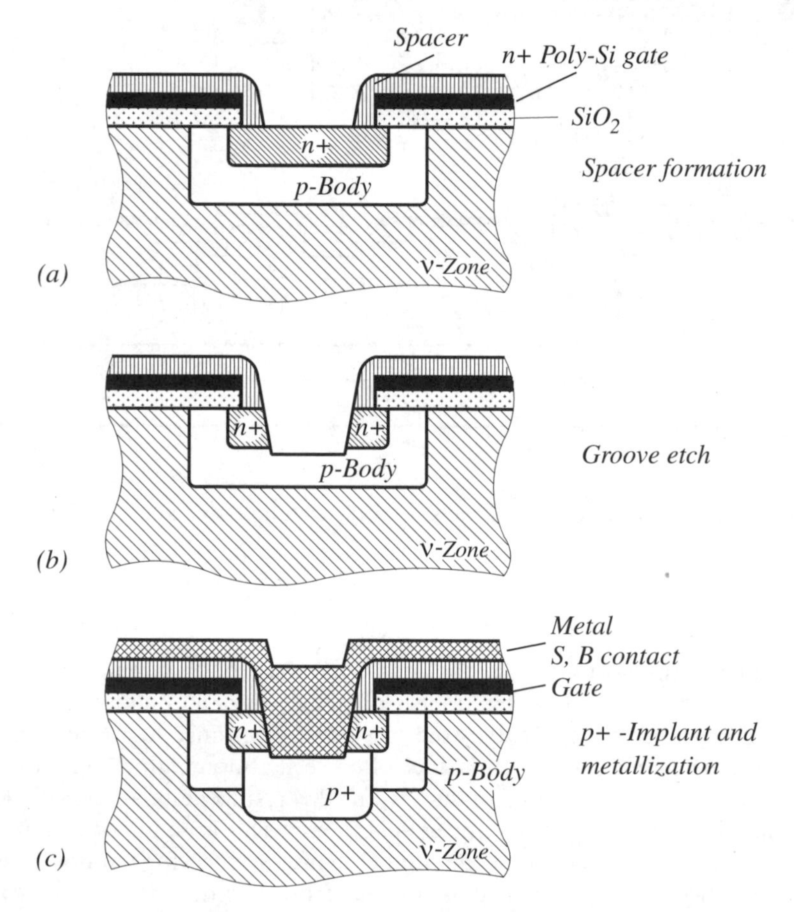

Figure 4.57 Self-aligned DMOS cell: (a) after spacer formation, (b) after groove etch and (c) after p^+-implant and metallization

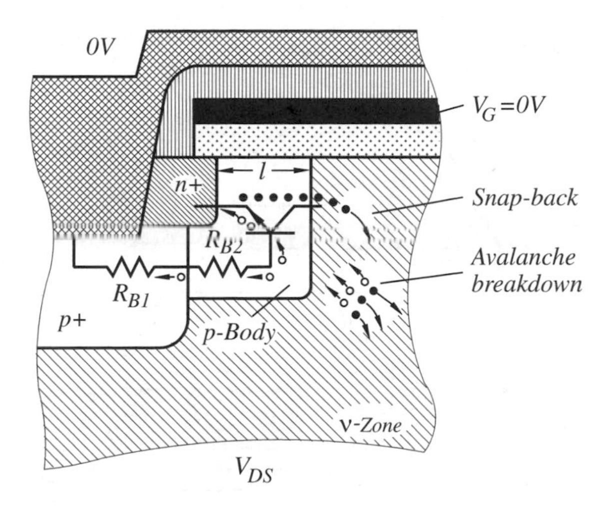

Figure 4.58 Schematic representation of second-order breakdown mechanisms at a DMOS cell

the p-body. The hole current generates at the p-body a positive voltage drop, which forward biases the BE-junction of the parasitic npn transistor, which leads to a snap-back behavior, as described in Section 4.5.6. Additional, the positive voltage drop at the p body decreases the threshold voltage of the DMOS transistor via the body effect (Section 4.3.3), causing a channel current (Hu *et al.* 1982).

The output characteristic of a DMOS cell or array reveals a distinct difference to a standard MOS transistor (Figure 4.59). In the cut-off region (A) with $V_{GS} = 0$, the width of the depletion region *DR* and the voltage across it depend, as expected, on the V_{DS} voltage only. In the saturation region (B) the ν-zone has only a minor effect on the drain current. The reason is that the channel of the structure behaves pretty much like a standard MOS transistor, which pinches-off in saturation (Section 4.4.1). Under this condition the maximum voltage drop occurs across the ν-zone. A different situation exists when the V_{GS} voltage is increased significantly and the device operates in quasi-saturation (C). In this case, the device starts to behave as if the n^+-source is short-circuited to the ν-zone, creating an n^+-ν-n^+ structure. This causes an increase of the injected electron current from the n^+-source into the ν-zone (Kreuzer *et al.* 1996). In the resistive region (D) the current flow is almost homogeneously distributed over the available area between neighboring cells and spreads out into the ν-zone at the end of the depletion region. In this region the electrical field is small and the electron mobility is almost constant. Thus, the ν-zone can be considered to behave like a linear resistor. This is also the case for the channel of the structure and explains therefore the almost linear behavior of the output characteristic in this operation region.

The discussion about the on-resistance reduction so far is based on reducing the source bulk contact area via a self-aligned source bulk contact. Additionally the distance *z* between cells (Figure 4.56(a)) can be reduced. Unfortunately this is only possible up to a value in the order of 12–14 μm, since otherwise the on-resistance of the ν-zone between the p-bodies increases (Figure 4.59(e)).

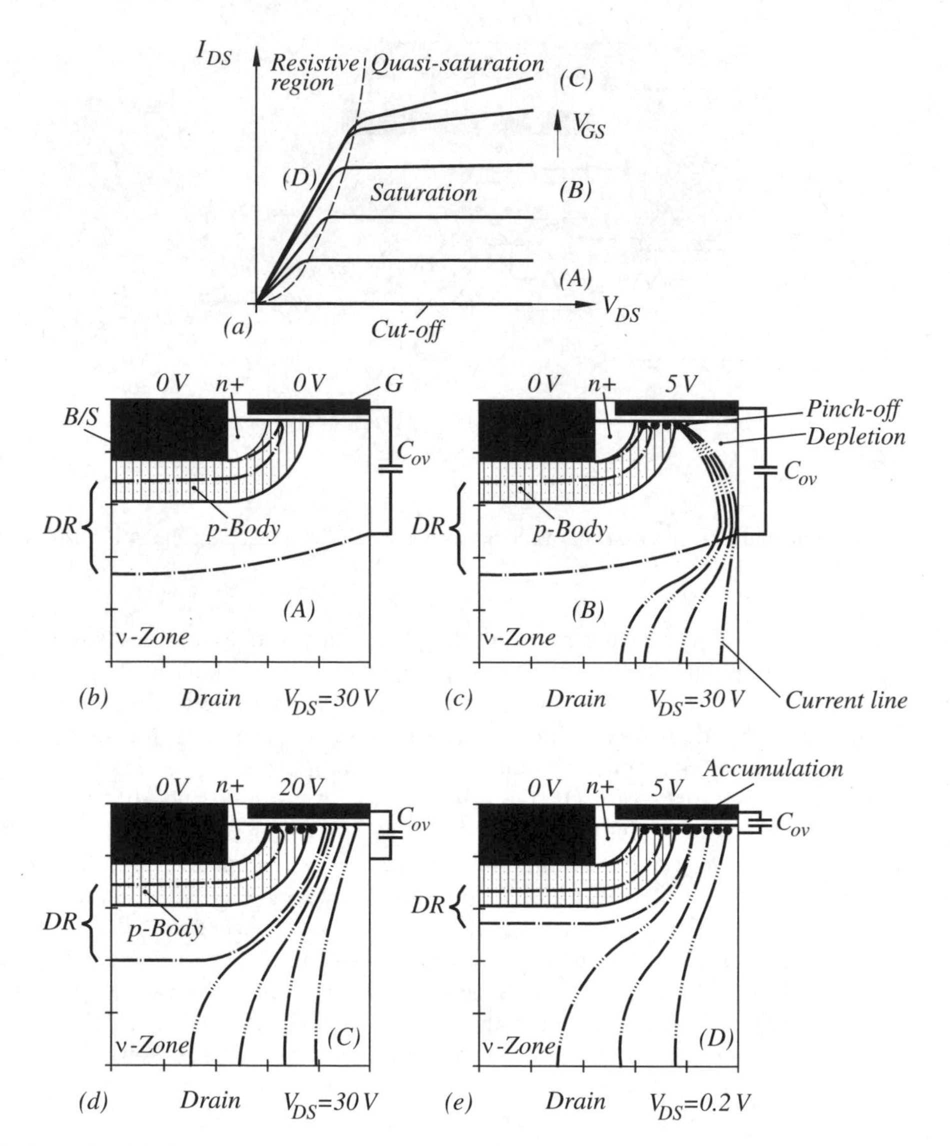

Figure 4.59 (a) Output characteristic of a DMOS transistor and (b–e) device simulations at different operation conditions

From a modeling point of view, a DMOS cell can be represented by a subcircuit, as shown in Figure 4.60 (Kraus *et al.* 1998).

MOS transistor

The channel length of this transistor is determined by the difference between the diffusion rate of boron and arsenic. This leads to a channel length in the order of

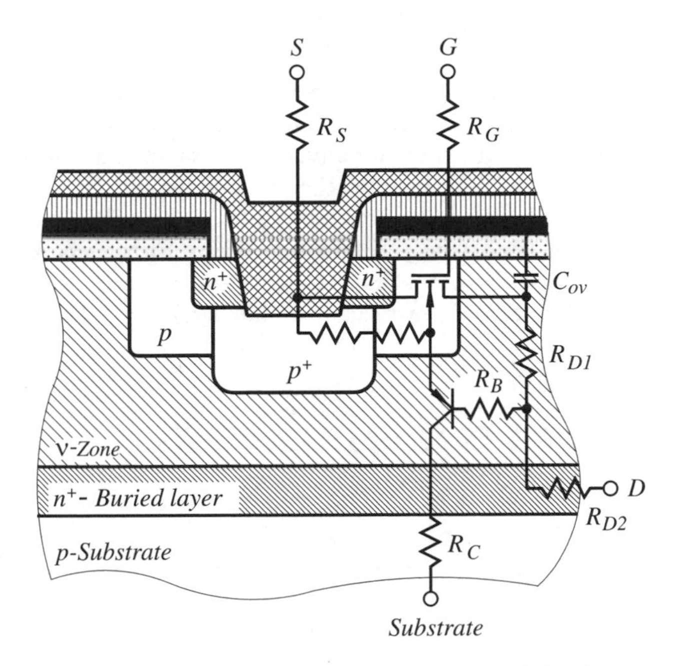

Figure 4.60 DMOS transistor cell with subcircuit

1 μm (Figure 4.57(a)). With this relatively long channel, short channel effects do not have to be considered and a relatively simple CAD model, similar to the one described in the following Section, may be used.

pnp transistor

Besides the already discussed npn transistor, there exists a pnp transistor, which plays a role when DMOS transistors are used in applications with inductive loads. In this case, the device is used additionally as a free-running diode under inverse biased conditions. This requires that the bipolar transistor, including the base resistance R_B and collector resistance R_C, is modeled appropriately. This can be done in a similar way to the procedure described in Section 3.5.

Resistors

The drain resistance R_{D1} includes the nonlinear behavior of the ν-zone, illustrated in Figure 4.59 (Kreuzer *et al.* 1996), and the resistance R_{D2} the contribution of the buried layer. In a large cell array the polysilicon resistance of the gate R_G has to be included as well as the source resistance R_S, in order to guarantee the correct determination of the switching behavior.

Overlap capacitance

If the transistor is cut off, a large depletion zone exists (Figure 4.59(b)) and the gate and drain overlap capacitance C_{ov} is relatively small. If the gate voltage is turned on, the device changes into the resistive region where the inversion layer spreads out into the overlap region and forms an accumulation layer at the semiconductor surface (Figure 4.59(e)). This causes the overlap capacitance to increase to a final value of C'_{ox} (Wunderlich *et al.* 1994). The overlap capacitance plays a dominating rule in the switching performance of the device, since the Miller effect amplifies its adverse effect (Sections 5.4 and 8.3.1).

The behavior of the internal transistor capacitances is discussed in the following section.

4.7 MODELING OF THE MOS TRANSISTOR

After a brief introduction to the modeling aspects of the MOS transistor for CAD applications, some simplified equations will be revisited in order to be used in first hand static and dynamic performance estimations, when the transistor is in a circuit environment.

4.7.1 Transistor Model for CAD Applications

Today's circuit simulation tools use transistor models of different complexity. Common to all of these is the partitioning of the model into a model frame and an internal transistor model. The advantage in doing so is that the model frame can be adjusted to the individual transistor layout or design, whereas the internal transistor remains unchanged.

Model frame

In order to analyze the model frame, it is useful to consider the cross-section of a MOS transistor again (Figure 4.61). As illustrated in this figure, the model frame consists of the following elements:

- 1, 2: Gate source overlap capacitance C^*_{gso} and gate drain overlap capacitance C^*_{dgo} per channel width
- 3, 4: Drain and source depletion capacitance C'_j per area
- 5, 6: Drain and source depletion capacitance C^*_j per side wall length
- 7, 8: Source and drain diodes to include the saturation (leakage) current of the junction
- 9, 10: Source and drain resistors.

As pointed out in Section 2.4.1, pn-junctions, or in this case source and drain regions, have a bottom and side wall part contributing differently to the total depletion capaci-

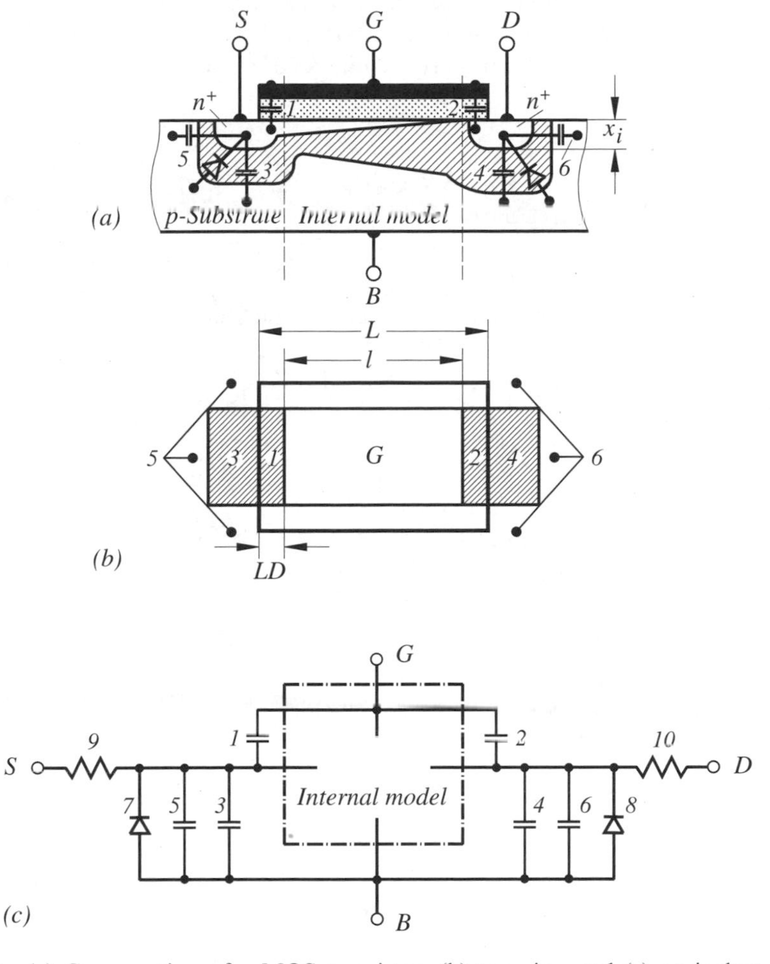

Figure 4.61 (a) Cross-section of a MOS transistor, (b) top view and (c) equivalent circuit of transistor frame

tance. In order to achieve a high accuracy during circuit simulation, the capacitance parameters of these contributing areas have to be known. The parameters allow the designer to cope with all possible transistor frames. They have units per area and per side wall length. This has the advantage that the depletion capacitance can be determined directly from the topology (top view) of the transistor or layout.

Internal transistor model

The accuracy of a circuit simulation cannot be better than that of the transistor model including its parameters. This led to the development of complex transistor models which differ between manufacturers and also between technologies.

In order to develop an understanding of such a model part of the basic SPICE level 3 version (Vladimirescu and Liu 1980) is considered. Some of these equations are presented in the preceding sections and are summarized in the following.

Threshold voltage:

$$V_{Tn} = V_{Ton} + \gamma(\sqrt{2\phi_F + V_{SB}} - \sqrt{2\phi_F})$$

$$V_{Ton} = V_{FB} + 2\phi_F + \gamma\sqrt{2\phi_F}$$

Current–voltage relationship under strong inversion conditions

$$I_{DS} = \beta_n\left[(V_{GS} - V_{Tn}) - \frac{1 + F_B}{2}V_{DSX}\right]V_{DSX}$$

$$V_{DSX} = \begin{cases} V_{DS} & \text{if } V_{GS} - V_{Tn} > V_{DS} \\ V_{DSsat} & \text{if } V_{GS} - V_{Tn} \leq V_{DS} \end{cases}$$

$$\beta_n = \mu_{eff} C'_{ox} \frac{w}{l}$$

$$F_B = \frac{\gamma}{2\sqrt{V_{SB} + 2\phi_F}}$$

$$\mu_{eff} = \frac{\mu_S}{1 + \dfrac{V_{DS}/l}{v_{sat}/\mu_S}}$$

$$\mu_S = \frac{\mu_n}{1 + \theta(V_{GS} - V_{Tn})}$$

$$V_{DSsat} = \frac{v_{sat} l}{\mu_S}\left(\sqrt{1 + 2\frac{\mu_S}{v_{sat} l}\frac{V_{GS} - V_{Tn}}{1 + F_B}} - 1\right)$$

Channel length modulation:

$$\Delta l = l - l' = \sqrt{\left(\frac{\mathcal{E}_p}{2\alpha}\right)^2 + \frac{K}{\alpha}(V_{DS} - V_{DSsat})} - \frac{|\mathcal{E}_p|}{2\alpha}$$

$$\alpha = \frac{qN_A}{2\varepsilon_0\varepsilon_{Si}}$$

Current–voltage relationship under weak inversion conditions **($V_{GS} \leq V_{TN}$):**

$$I_{DS} = \beta_n(n-1)\phi_t^2 e^{(V_{GS} - V_{Tn})/\phi_t n}(1 - e^{-V_{DS}/\phi_t})$$

The most important transistor parameters of a 1.5 μm CMOS technology are summarized in Table 4.1. The mentioned technology is a high voltage version due to the relatively thick gate oxide of 20 nm and a breakdown voltage above 12 V.

Charge model of the internal transistor

The charge behavior of the internal transistor can be described phenomenologically by bias dependent capacitances, as is discussed in conjunction with Figure 4.63. This was the standard approach taken in the early days of model development (Meyer 1971). But due to charge inconsistencies during transistor switching, charge oriented models are preferred today (Ward and Dutton 1978). In the following an introduction is given. To ease the presentation, second-order effects are not included.

Table 4.1 Parameter of a 1.5 μm CMOS process

Name	SPICE	Parameter	n-Channel transistor	p-Channel transistor	Unit
V_{Ton}	VTO	Zero-biased threshold voltage	0.8	−0.8	V
k_n, k_p	KP	Gain factor	$120 \cdot 10^{-6}$	$40 \cdot 10^{-6}$	A/V^2
γ	GAMMA	Body factor	0.3	0.4	$\sqrt{V}$
$2\phi_F$	PHI	Surface potential/voltage	0.78	0.70	V
d_{ox}	TOX	Gate oxide thickness	$20 \cdot 10^{-9}$	$20 \cdot 10^{-9}$	m
$N_{A(D)}$	NSUB	Substrate doping	5.10^{16}	10^{16}	cm^{-3}
x_j	XJ	Metallurgical junction depth	$0.3 \cdot 10^{-6}$	$0.4 \cdot 10^{-6}$	m
μ_o	U0	Nominal surface mobility	695	232	cm^2/Vs
v_m	VMAX	Maximum carrier drift velocity	$1.5 \cdot 10^5$	$0.8 \cdot 10^5$	m/s
δ	DELTA	Channel width factor	0.04	0.09	
θ	THETA	Mobility degradation	0.10	0.19	1/V
	ETA	Static feedback	0.25	0.30	
K	KAPPA	Saturation field factor	1	5	
R_D, R_S	RSH	Drain and source resistance	40	60	Ω/□
C^*_{gso}	CGSO	Gate-source overlap capacitance per channel width of source	$0.34 \cdot 10^{-9}$	$0.34 \cdot 10^{-9}$	F/m
C^*_{gdo}	CGDO	Gate-drain overlap capacitance per channel width drain	$0.34 \cdot 10^{-9}$	$0.34 \cdot 10^{-9}$	F/m
C'_j	CJ	Zero-biased bulk junction bottom capacitance per area	$0.3 \cdot 10^{-3}$	$0.5 \cdot 10^{-4}$	F/m^2
M	MJ	Bulk junction bottom grading coefficient	0.5	0.5	
C^*_j	CJSW	Zero-biased bulk junction capacitance per sidewall	$0.1 \cdot 10^{-9}$	$0.1 \cdot 10^{-9}$	F/m
M	MJSW	Bulk junction sidewall grading coefficient	0.33	0.33	
ϕ_i	PB	Bulk built-in voltage	0.70	0.64	V
J_S	JS	Bulk junction saturation current	10^{-6}	10^{-6}	A/m^2

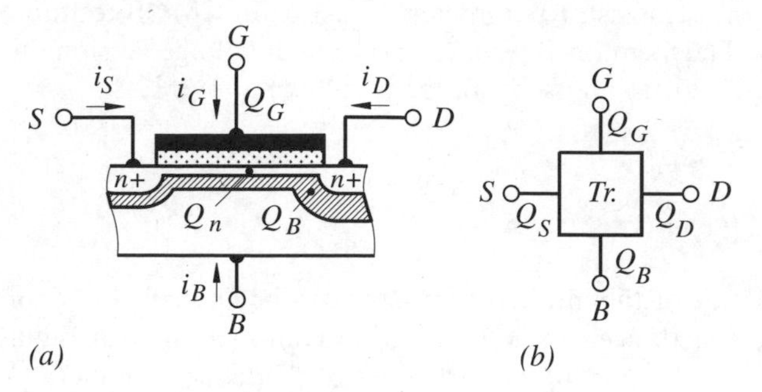

Figure 4.62 (a) MOS transistor charges and (b) lumped terminal charges of the transistor

In the charge model the lumped charges at the internal transistor terminals have to be known (Figure 4.62). The current flowing into or out of the transistor terminals results from the charge changes

$$i_G = \frac{\mathrm{d}Q_G}{\mathrm{d}t} \qquad i_B = \frac{\mathrm{d}Q_B}{\mathrm{d}t} \qquad i_S + i_D = \frac{\mathrm{d}(Q_S + Q_D)}{\mathrm{d}t} \tag{4.100}$$

Hereby it is assumed that the charges can be determined any time from the biasing condition of the transistor. This is a quasi-static description, where drift and diffusion processes are not included. The gate charge Q_G is the mirror charge of the inversion layer charge Q_n and the depletion region charge Q_B. Rearranging Equation (4.60) leads to

$$\begin{aligned} \sigma_g(x) &= -[\sigma_n(x) + \sigma_d(x)] \\ &= C'_{ox}[V_{GS} - V_{FB} - 2\phi_F - \phi_C(x)] \end{aligned} \tag{4.101}$$

where the depletion region charge is given by

$$\sigma_d(\phi_C) = -C'_{ox}\gamma\left(\sqrt{2\phi_F + V_{SB}} + \frac{\phi_C(x)}{2\sqrt{2\phi_F + V_{SB}}}\right) \tag{4.102}$$

The total gate charge can be found by integrating Equation (4.101) over the channel length

$$Q_G = w\int_{o}^{l} \sigma_g(x)\,\mathrm{d}x \tag{4.103}$$

Since the local charge dependence is not known, a variable transformation (Equation 4.47)

$$\mathrm{d}x = \frac{-w\sigma_n\mu_n}{I_{DS}}\,\mathrm{d}\phi_C \tag{4.104}$$

is used as well as Equations (4.101), (4.60), and (4.62) to find the total gate charge

$$\begin{aligned} Q_G &= -\frac{w^2\mu_n}{I_{DS}}\int_0^{V_{DS}} \sigma_g(\phi_C)\sigma_n(\phi_C)\,\mathrm{d}\phi_C \\ &= wlC'_{ox}\left(V_{GS}-V_{FB}-2\phi_F-\frac{V_{DS}}{2}+\frac{1+F_B}{12F_I}V_{DS}^2\right) \end{aligned} \tag{4.105}$$

where

$$F_I = V_{GS} - V_{Tn} - \frac{1+F_B}{2}V_{DS} \tag{4.106}$$

is used for simplification. The charge at the substrate electrode can be found with a similar derivation, which yields

$$\begin{aligned} Q_B &= w\int_0^l \sigma_d(x)\,\mathrm{d}x \\ &= -\frac{w^2\mu_n}{I_{DS}}\int_0^{V_{DS}} \sigma_d(\phi_C)\sigma_n(\phi_C)\,\mathrm{d}\phi_C \\ &= -wlC'_{ox}\left[\gamma\sqrt{2\phi_F+V_{SB}}+\frac{F_B}{2}V_{DS}-\frac{(1+F_B)}{12F_I}V_{DS}^2\right] \end{aligned} \tag{4.107}$$

where Equations (4.60), (4.62), (4.101), and (4.102) are employed. So far the charges are related to the terminals G and B. This is different for the channel charge (Figure 4.62)

$$Q_n = -(Q_G + Q_B) \tag{4.108}$$

which has to be divided between the source and drain terminals. This is done in most transistor models by the relationships

$$\begin{aligned} Q_D &= w\int_0^l \frac{x}{l}Q_n\,\mathrm{d}x \\ Q_S &= w\int_0^l \left(1-\frac{x}{l}\right)Q_n\,\mathrm{d}x \end{aligned} \tag{4.109}$$

Advanced transistor models

Due to the continued reduction or shrinking of structural dimensions in integrated circuits, second-order effects – as discussed in Section 4.5 – are becoming more and more dominating. As a consequence, advanced transistor models must include all these effects and are thus much more complex than the presented example. A relative widely used model is the one from Berkeley, called BSIM (Berkeley Short channel IGFET Model), which is described in Cheng and Hu (1999) and Foty (1997). When this model is used as an exercise in a circuit simulation, it is recommended to use the parameters of Table 4.2 or the following website: *www.unibw-muenchen.de/campus/ET4/index.html*. The parameters listed are from transistors with a channel length of 0.13 μm. The transistors are characterized by the following values:

n-channel MOS transistor:

$$I_{DS} = 600\,\mu\text{A}/\mu\text{m width at } V_{CC} = V_{GS} = 1.3\,\text{V}$$
$$V_{Tn} = 0.3\,\text{V at } 100\,\text{nA} \cdot (\text{w/l})$$

p-channel MOS-transistor:

$$I_{DS} = 260\,\mu\text{A}/\mu\text{m width at } V_{CC} = -V_{GS} = 1.3\,\text{V}$$
$$V_{Tp} = -0.3\,\text{V at } 40\,\text{nA} \cdot (\text{w/l})$$

Table 4.2 Model parameter BSIM3V3 of a 0.13 μm CMOS process

MODEL NMOS LEVEL = 7					
+A0	= 8.50000E − 01	A1	= 0.00000E + 00	A2	= 1.00000E + 00
+AF	= 1.00000E + 00	AGS	= 6.00000E − 01	ALPHA0	= 1.30000E − 08
+AT	= 4.10000E + 04	B0	= 8.70000E − 08	B1	= 5.80000E − 09
+BETA0	= 1.00000E + 01	CAPMOD	= 2.00000E + 00	CDSC	= 4.30000E − 04
+CDSCB	= 0.00000E + 00	CDSCD	= 1.00000E − 03	CF	= 1.00000E − 10
+CGBO	= 1.00000E − 12	CGDL	= 0.00000E + 00	CGDO	= 1.80000E − 10
+CGSL	= 0.00000E + 00	CGSO	= 1.80000E − 10	CIT	= 0.00000E + 00
+CJ	= 1.00000E − 03	CJSW	= 1.00000E − 10		
+CKAPPA	= 6.00000E − 01	CLC	= 1.00000E − 07	CLE	= 6.00000E − 01
+DELTA	= 2.90000E − 02	DLC	= 1.50000E − 08	DROUT	= 9.00000E − 01
+DSUB	= 1.70000E + 00	DVT0	= 7.50000E + 00	DVT0W	= 0.00000E + 00
+DVT1	= 1.90000E + 00	DVT1W	= 5.30000E + 06	DVT2	= 0.00000E + 00
+DVT2W	= −3.2000E − 02	DWB	= 0.00000E + 00	DWC	= 0.00000E + 00
+DWG	= 0.00000E + 00	EF	= 1.00000E + 00	EM	= 4.10000E + 07
+ETA0	= 1.00000E + 00	ETAB	= −5.0000E − 04	JS	= 1.00000E − 07
+JSW	= 1.00000E − 14	K1	= 3.90000E − 01	K2	= −8.7000E − 03
+K3	= 1.00000E − 03	K3B	= 0.00000E + 00	KETA	= 0.00000E + 00
+KF	= 0.00000E + 00	KT1	= −2.5000E − 01	KT1L	= 0.00000E + 00
+KT2	= 0.00000E + 00	LINT	= 2.00000E − 08	LMAX	= 1.00000E − 05
+LMIN	= 1.30000E − 07	MJ	= 5.00000E − 01	MJSW	= 4.00000E − 01
+MJSWG	= 6.00000E − 01	MOBMOD	= 1.00000E + 00	N	= 1.00000E + 00

+NCH = 5.80000E + 17 NFACTOR = 1.90000E + 00 NLX = 1.20000E - 07
+NOIMOD = 1.00000E + 00 NQSMOD = 0.00000E + 00 PB = 8.00000E - 01
+PBSW = 8.00000E - 01 PBSWG = 8.00000E - 01 PCLM = 1.00000E - 01
+PDIBLC1 = 1.80000E - 01 DIBLC2 = 1.00000E - 02 PDIBLCB = 0.00000E + 00
+PRT = 0.00000E + 00 PRWB = 0.00000E + 00 PRWG = 0.00000E + 00
+PSCBE1 = 7.50000E + 08 PSCBE2 = 5.00000E - 05 PVAG = 0.00000E + 00
+RDSW = 1.40000E + 02 TNOM = 2.50000E + 01 TOX = 3.00000E - 09
+U0 = 6.00000E + 02 UA = 6.00000E - 10 UA1 = 2.00000E - 09
+UB = 2.20000E - 18 UB1 = -2.5000E - 18 UC = 2.40000E - 10
+UC1 = 0.00000E + 00 UTE = -1.5000E + 00 VERSION = 3.20000E + 00
+VOFF = -9.3000E - 02 VSAT = 1.15000E + 05 VTH0 = 2.50000E - 01
+W0 = 6.00000E - 07 WINT = 0.00000E + 00 WMAX = 1.00000E - 05
+WMIN = 1.50000E - 07 WR = 1.00000E + 00 XJ = 5.00000E - 08
+XPART = 0.00000E + 00

MODEL PMOS LEVEL = 7
+A0 = 5.10000E - 01 A1 = 0.00000E + 00 A2 = 1.00000E + 00
+AF = 1.00000E + 00 AGS = 3.40000E - 01 ALPHA0 = 5.00000E - 09
+AT = 1.30000E + 03 B0 = 0.00000E + 00 B1 = 0.00000E + 00
+BETA0 = 1.20000E + 01 CAPMOD = 2.00000E + 00 CDSC = 9.00000E - 04
+CDSCB = 4.80000E - 04 CDSCD = 5.20000E - 04 CF = 1.20000E - 10
+CGBO = 1.00000E - 12 CGDL = 0.00000E + 00 CGDO = 1.90000E - 10
+CGSL = 0.00000E + 00 CGSO = 1.90000E - 10 CIT = 0.00000E + 00
+CJ = 1.00000E - 03 CJSW = 1.00000E - 10
+CLC = 1.00000E - 07 CLE = 6.00000E - 01 DELTA = 1.50000E - 02
+DLC = 1.80000E - 08 DROUT = 8.00000E - 07 DSUB = 8.50000E - 01
+DVT0 = 1.40000E + 01 DVT0W = 0.00000E + 00 DVT1 = 1.90000E + 00
+DVT1W = 5.30000E + 06 DVT2 = 0.00000E + 00 DVT2W = -3.2000E - 02
+DWB = 0.00000E + 00 DWC = 0.00000E + 00 DWG = 0.00000E + 00
+EF = 1.00000E + 00 ETA0 = 4.00000E - 01 ETAB = 0.00000E + 00
+JS = 1.00000E - 08 JSW = 1.00000E - 14 K1 = 3.20000E - 01
+K2 = 0.00000E + 00 K3 = 1.00000E - 03 K3B = 0.00000E + 00
+KETA = 0.00000E + 00 KF = 0.00000E + 00 KT1 = -2.5000E - 01
+KT1L = 0.00000E + 00 KT2 = 0.00000E + 00 LINT = 1.50000E - 08
+LMAX = 1.00000E - 05 LMIN = 1.30000E - 07 MJ = 5.00000E - 01
+MJSW = 4.00000E - 01 MJSWG = 4.00000E - 01 MOBMOD = 1.00000E + 00
+N = 1.00000E + 00 NCH = 9.00000E + 17 NFACTOR = 1.13000E + 00
+NLX = 1.60000E - 07 NOIMOD = 1.00000E + 00 NQSMOD = 0.00000E + 00
+PB = 8.00000E - 01 PBSW = 8.00000E - 01 PBSWG = 8.00000E - 01
+PCLM = 1.40000E + 00 PDIBLC1 = 1.30000E - 03 PDIBLC2 = 1.00000E - 05
+PDIBLCB = -1.0000E - 03 PRT = 0.00000E + 00 PRWB = 0.00000E + 00
+PRWG = 0.00000E + 00 PSCBE1 = 7.20000E + 08 PSCBE2 = 8.00000E - 10
+PVAG = 0.00000E + 00 RDSW = 2.20000E + 02 RSH = 0.00000E + 00
+TNOM = 2.50000E + 01 TOX = 3.00000E - 09 U0 = 1.30000E + 02
+UA = 4.90000E - 10 UA1 = 5.80000E - 12 UB = 1.30000E - 18
+UB1 = -1.0000E - 18 UC = -1.40000E - 12 UC1 = 0.00000E + 00
+UTE = -1.2000E + 00 VERSION = 3.20000E + 00 VOFF = -1.4000E - 01
+VSAT = 8.00000E + 04 VTH0 = -2.80000E - 01 W0 = 3.40000E - 07
+WINT = 0.00000E + 00 WMAX = 1.00000E - 05 WMIN = 1.50000E - 07
+WR = 1.00000E + 00 XJ = 5.00000E - 08 XPART = 0.00000E + 00

4.7.2 Transistor Model for Static and Dynamic Calculations

As stated in Chapter 3, easy to use equations are needed for first hand circuit calculations. They do not necessarily have to be precise, but should be useful to estimate static and dynamic circuit performances and lead to a more detailed knowledge of the circuit function. This is particularly useful for analog circuits, where e.g. the parameter influence on the gain, the pole locations, and zero locations of an amplifier have to be known (Chapters 6 and 7). But there is no question about it: only with an appropriate circuit simulation tool a reliable prediction of the circuit performance is possible. A summary of the most important easy to use equations, describing the static behavior of the MOS transistor, is shown in Table 4.3.

In order to estimate the transient behavior of a circuit, as is done in Chapter 5, the charge behavior of the transistor has to be known. Using the presented charge model in a circuit environment is not very practical. Therefore, capacitance presentations are used instead. The starting point is the cross-section of the MOS transistor (Figure 4.63).

Table 4.3 Summary of n-channel and p-channel transistor equations (Figure 4.34)

n-Channel transistor	p-Channel transistor
Strong inversion	
Resistive region	
$I_{DS} = \beta_n\left[(V_{GS} - V_{Tn})V_{DS} - \frac{V_{DS}^2}{2}\right]$	$I_{DS} = -\beta_p\left[(V_{GS} - V_{Tp})V_{DS} - \frac{V_{DS}^2}{2}\right]$
if $V_{GS} - V_{Tn} > V_{DS}$	if $\lvert V_{GS} - V_{Tp}\rvert > \lvert V_{DS}\rvert$
Saturation region	
$I_{DS} = \frac{\beta_n}{2}(V_{GS} - V_{Tn})^2$	$I_{DS} = -\frac{\beta_p}{2}(V_{GS} - V_{Tp})^2$
if $V_{GS} - V_{Tn} \leq V_{DS}$	if $\lvert V_{GS} - V_{Tp}\rvert \leq \lvert V_{DS}\rvert$
Threshold voltage	
$V_{Tn} = V_{Ton} + \gamma(\sqrt{2\phi_F + V_{SB}} - \sqrt{2\phi_F})$	$V_{Tp} = V_{Top} - \gamma(\sqrt{-2\phi_F - V_{SB}} - \sqrt{-2\phi_F})$
$V_{Ton} = V_{FB} + 2\phi_F + \gamma\sqrt{2\phi_F}$	$V_{Top} = V_{FB} + 2\phi_F - \gamma\sqrt{-2\phi_F}$
$\gamma = \frac{\sqrt{qN_A 2\varepsilon_0\varepsilon_{Si}}}{C'_{ox}}$	$\gamma = \frac{\sqrt{qN_D 2\varepsilon_0\varepsilon_{Si}}}{C'_{ox}}$
$\phi_F = \phi_t \ln N_A/n_i$	$\phi_F = -\phi_t \ln N_D/n_i$
$\beta_n = k_n \frac{w}{l}$; $k_n = \mu_{eff} C'_{ox}$	$\beta_p = k_p \frac{w}{l}$; $k_p = \mu_{eff} C'_{ox}$
Mobility ($l < 1.5\,\mu$m)	
$\mu_{eff} = \frac{\mu_n}{1 + \frac{V_{DS}/l}{v_{sat}/\mu_n}}$	$\mu_{eff} = \frac{\mu_p}{1 - \frac{V_{DS}/l}{v_{sat}/\mu_p}}$
Weak inversion	
$n - \text{Tr}: (V_{GS} \leq V_{Tn}) I_{DS} = \beta_n(n-1)\phi_t^2 e^{(V_{GS}-V_{Tn})/\phi_t n}(1 - e^{-V_{DS}/\phi_t})$	
$p - \text{Tr}: (V_{GS} \geq V_{Tp}) I_{DS} = -\beta_p(n-1)\phi_t^2 e^{-(V_{GS}-V_{Tp})/\phi_t n}(1 - e^{V_{DS}/\phi_t})$	
$n = 1 + C'_j/C'_{ox}$	

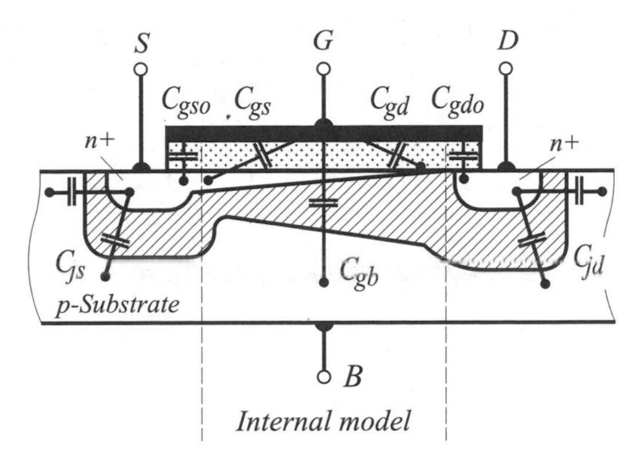

Figure 4.63 Cross-section of a MOS transistor with internal capacitances

The transistor frame is unchanged, compared to Figure 4.61, except for the omission of the diodes. The internal transistor includes three capacitances which are strongly bias dependent. This is shown in Figure 4.64 for the case that the V_{DS} voltage is constant. In the cut-off region with $V_{GS} < V_{Tn}$, the gate-substrate capacitance has a value of $C_{gb} = C_{ox}$. Increasing the V_{GS} voltage leads the transistor into current saturation. In this condition a channel exists, causing the gate-bulk capacitance C_{gb} to approach zero and the gate-source capacitance C_{gs} a value of $\frac{2}{3}\, C_{ox}$. If the V_{GS} voltage is increased further, the transistor moves into the resistive region. In this situation a continuous channel between source and drain exists. This leads to a gate-drain capacitance C_{gd} with a value of

$$C_{gs} = C_{gd} = \frac{1}{2} C_{ox} \tag{4.110}$$

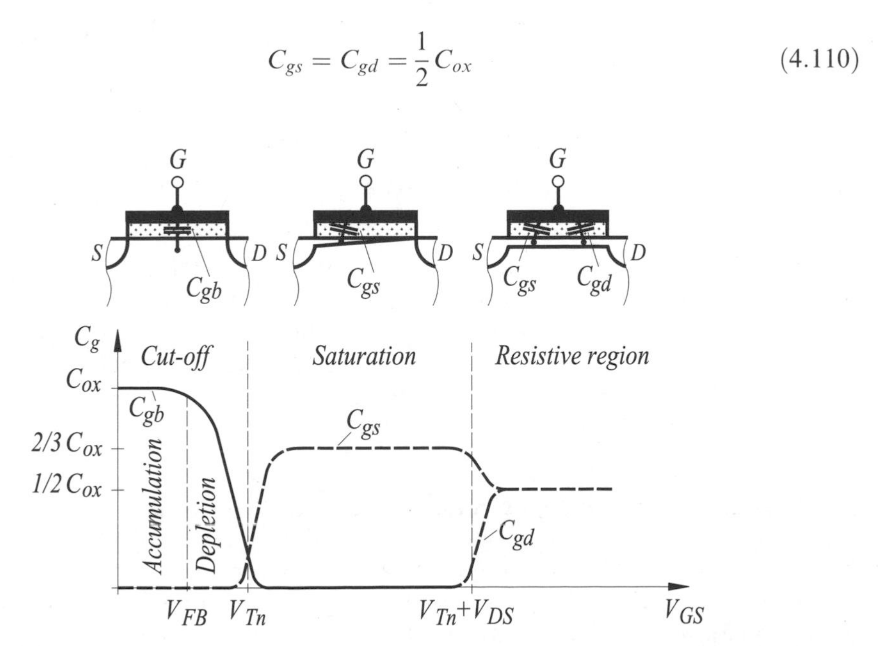

Figure 4.64 Qualitative behavior of the internal transistor capacitances

Previously it was mentioned that the gate-source capacitance C_{gs} approaches a value of $\frac{2}{3}\,C_{ox}$. Why this happens is discussed in the following.

The local dependence of the channel voltage (Equation 4.56) in saturation is given by

$$\phi_C(x) = (V_{GS} - V_{Tn})\left(1 - \sqrt{1 - \frac{x}{l}}\right) \tag{4.111}$$

where $V_{DS} = V_{GS} - V_{Tn}$ is used to describe the saturation condition. This yields a local charge distribution of

$$\begin{aligned} \sigma_n(x) &= -C'_{ox}(V_{GS} - V_{Tn} - \phi_C(x)) \\ &= -C'_{ox}(V_{GS} - V_{Tn})\sqrt{1 - \frac{x}{l}} \end{aligned} \tag{4.112}$$

and after integration

$$w\int_o^l \sigma_n(x)\,\mathrm{d}x = -wC'_{ox}(V_{GS} - V_{Tn})\int_o^l \sqrt{\left(1 - \frac{x}{l}\right)}\,\mathrm{d}x \tag{4.113}$$

a total inversion layer charge of

$$Q_n = -\frac{2}{3}C_{ox}(V_{GS} - V_{Tn}) \tag{4.114}$$

Using the definition of the small-signal capacitance (Equation 2.45) results in a gate source capacitance of

$$\boxed{C_{gs} = \left|\frac{\mathrm{d}Q_n}{\mathrm{d}U_{GS}}\right| = \frac{2}{3}C_{ox}} \tag{4.115}$$

The absolute value is used, since in comparison to Equation (2.45) the negative inversion layer charge is used in the derivation. Otherwise an uncommon negative capacitance value results. As is obvious from the derivation, the factor $\frac{2}{3}$ in the above equation results from the non-uniform charge distribution, which approaches a value close to zero at the drain under saturation condition.

4.7.3 Transistor Model for Small-Signal Calculations

For the derivation of this model the same procedure is used as in the case of the bipolar transistor. The starting point is the small-signal excitation of the MOS transistor (Figure 4.65). Its effect can be described by three small-signal conductance parameters at a defined operation point (A).

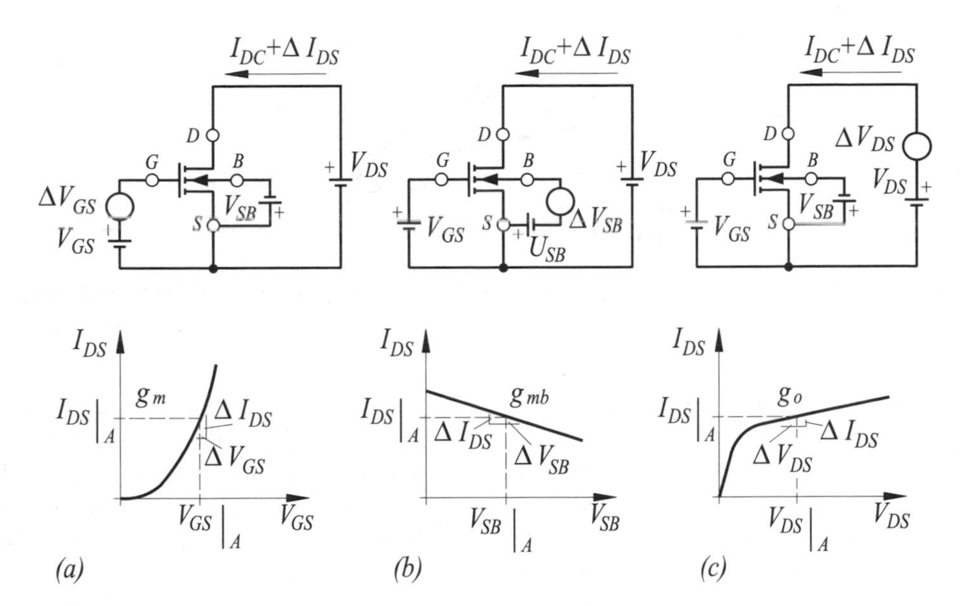

Figure 4.65 Small-signal excitation of a MOS transistor: (a) $I_{DS}\ (V_{GS})$, (b) $I_{DS}\ (V_{SB})$ and (c) $I_{DS}\ (V_{DS})$

Transconductance of the gate:

$$g_m = \frac{\partial I_{DS}}{\partial V_{GS}} \tag{4.116}$$

Transconductance of the substrate:

$$g_{mb} = \frac{\partial I_{DS}}{\partial V_{SB}} \tag{4.117}$$

Output conductance:

$$g_o = \frac{\partial I_{DS}}{\partial V_{DS}} \tag{4.118}$$

If all small-signal voltages – all referenced to the source – are changed simultaneously, the following drain current change occurs.

$$\begin{aligned}\Delta I_{DS} &= \frac{\partial I_{DS}}{\partial V_{GS}} \cdot \Delta V_{GS} + \frac{\partial I_{DS}}{\partial V_{SB}} \cdot \Delta V_{SB} + \frac{\partial I_{DS}}{\partial V_{DS}} \cdot \Delta V_{DS} \\ &= g_m \Delta V_{GS} + g_{mb} \Delta V_{SB} + g_o \Delta V_{DS}\end{aligned} \tag{4.119}$$

This relationship is represented by the small-signal equivalent circuit of Figure 4.66, except for the added small-signal capacitances from Figure 4.63.

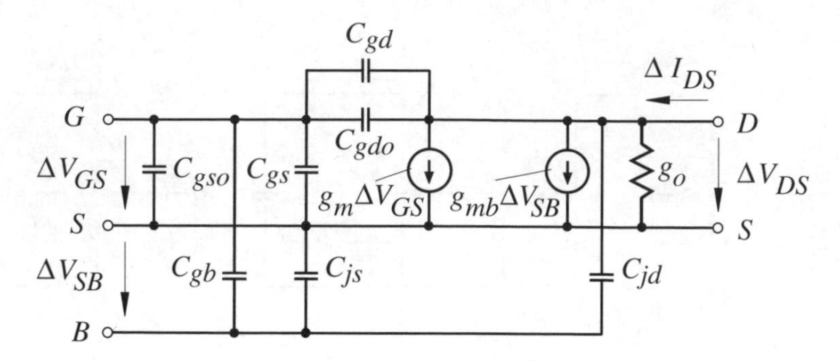

Figure 4.66 Small-signal equivalent circuit of the MOS transistor

The small-signal equivalent circuit is also valid when, instead of infinitesimal voltage and current changes, sinusoidal excitations are present. As a reminder: *in order to avoid confusion, large letter symbols are used throughout the text to indicate static parameters, whereas small letter symbols or indices are used for small-signal components.*

Transconductance of the gate

Differentiating the current–voltage equation, which includes the channel length modulation (Equation 4.82), results in a transconductance value of

$$g_m = \beta_n(V_{GS} - V_{Tn})(1 + \lambda V_{DS})$$

$$\boxed{g_m = \sqrt{2I_{DS}\beta_n(1 + \lambda V_{DS})}} \tag{4.120}$$

The transconductance has a square law behavior with respect to the drain current. This is a substantial drawback compared to the bipolar transistor, where the transconductance is proportional to the collector current (Equation 3.102). This is discussed in detail in Section 10.4.2, where the small-signal voltage gain of the MOS transistor and bipolar transistor are compared.

Transconductance of the substrate

In some analog circuits, a voltage change ΔV_{SB} between source and bulk occurs (Figure 4.65(b)). This leads usually to an undesirable behavior of the circuit, as is discussed in Chapter 8. The reason for this behavior results from the body effect, causing a change in threshold voltage and consequently in the drain current, when the V_{SB} voltage varies (Section 4.3.3). Differentiating Equation (4.82) yields

$$g_{mb} = \sqrt{2I_{DS}\beta_n(1 + \lambda V_{DS})}\left(-\frac{\partial V_{Tn}}{\partial V_{SB}}\right) \tag{4.121}$$

The quotient $\partial V_{Tn}/\partial V_{SB}$ can be found by differentiating the threshold equation (Equation 4.36)

$$V_{Tn} = V_{Ton} + \gamma\left(\sqrt{2\phi_F + V_{SB}} - \sqrt{2\phi_F}\right)$$

which leads to the following transconductance description

$$g_{mb} = \frac{-\gamma}{2}\sqrt{\frac{2I_{DS}\beta_n(1+\lambda V_{DS})}{2\phi_F + V_{SB}}} \tag{4.122}$$

Compared to the gate transconductance, the substrate transconductance has a negative sign and therefore opposes the gate excitation $g_m \cdot \Delta V_{GS}$ (Figure 4.66).

Output conductance

The channel length modulation is responsible for a finite output conductance (Section 4.5.2). Differentiating Equation (4.82) leads directly to

$$g_o = \frac{I_{DS}\lambda}{1+\lambda V_{DS}} \approx I_{DS}\lambda \tag{4.123}$$

In this case, $1/g_o$ describes the gradient of the output characteristic (Figure 4.65(c)).

In the derivation of the small-signal components, the transistor is assumed to be biased always into current saturation. The reason – as discussed in Chapter 8 – is that only under this condition will the transistor yield an acceptable voltage gain in an amplifier configuration.

Summary of the most important results

A MOS structure is used to analyze the characteristic conditions accumulation, depletion, and inversion, and the influence of different work functions on the flat-band voltage. For the case of strong inversion it is possible to describe the surface voltage by $\phi_S = 2\phi_F + V_{SB}$. This voltage leads to the definition of a threshold voltage, which can be varied by the V_{SB} voltage. A measure of this effect is the body-effect parameter γ.

The current–voltage equation of the transistor is derived under the conditions of strong inversion and small V_{DS} voltage. As it turns out, the electron speed is highest and the charge density lowest at the drain side of the transistor channel, when the V_{DS} voltage is raised. This leads to a phenomenological description, when the transistor is biased into the current saturation region and behaves like a voltage controlled current generator. With $V_{GS} \leq V_{Tn}$ the transistor is biased into weak inversion and shows an exponential current–voltage relationship. A very useful expression or, better, a figure of merit is the sub-threshold swing S. That is the gate voltage swing needed to change the current by one decade. When one assumes the best case, a sub-threshold swing at room temperature of 60 mV per decade results. Contrary to the operation in strong inversion, in the weak inversion region the transistor shows a pronounced positive temperature dependence.

Second-order effects, e.g. mobility degradation and channel length modulation, are considered. The mobility changes under the influence of the vertical and horizontal electrical field of the MOS transistor. This leads to a substantial reduction in drain current, particularly at small dimensions. In current saturation a shift of the pinch-off point toward the source can be observed when the V_{DS} voltage is increased. This effect can be interpreted as a reduction in effective channel length responsible for an increase in drain current. This so-called channel length modulation plays an impotent role in analog circuits.

Problems

4.1 The onset of strong inversion is given when the charge density at the semiconductor surface n_s has the same value as the majority carrier – that is, the holes – of the substrate. In this case, the surface voltage ϕ_S has a value of $\phi_S(SI)$ and remains approximately constant even when the gate voltage, and in conjunction the inversion layer charge, is increased.

Sketch the band diagram and show that the surface voltage does not change by more than 60 mV, when the gate voltage, and therefore the inversion layer charge n_s, is increased by a factor of 10.

4.2 Shown below is the small-signal behavior of a MOS structure.

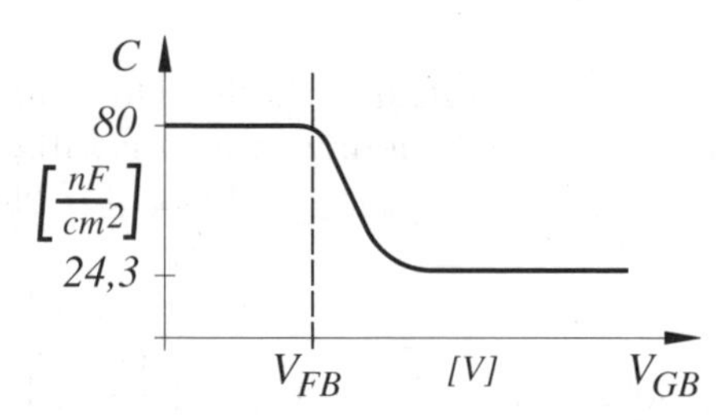

The substrate has a doping concentration of $N_A = 10^{16}\,\text{cm}^{-3}$. Calculate:

(a) The width of the depletion region at strong inversion.

(b) The surface voltage at the onset of strong inversion.

(c) The threshold voltage of the transistor, when the flat band voltage has a value of $V_{FB} = -0.1\,\text{V}$ and $V_{SB} = 0\,\text{V}$.

(d) The threshold voltage at $V_{SB} = 5\,\text{V}$.

4.3 Shown below is the cross-section sketch of a one-transistor DRAM cell (Section 7.5). The n$^+$-poly of the capacitor can be charged either to 1.8 V or to 0 V. This charge may cause an unexpected large leakage current to flow through the vertical parasitic n-channel transistor. Calculate or show:

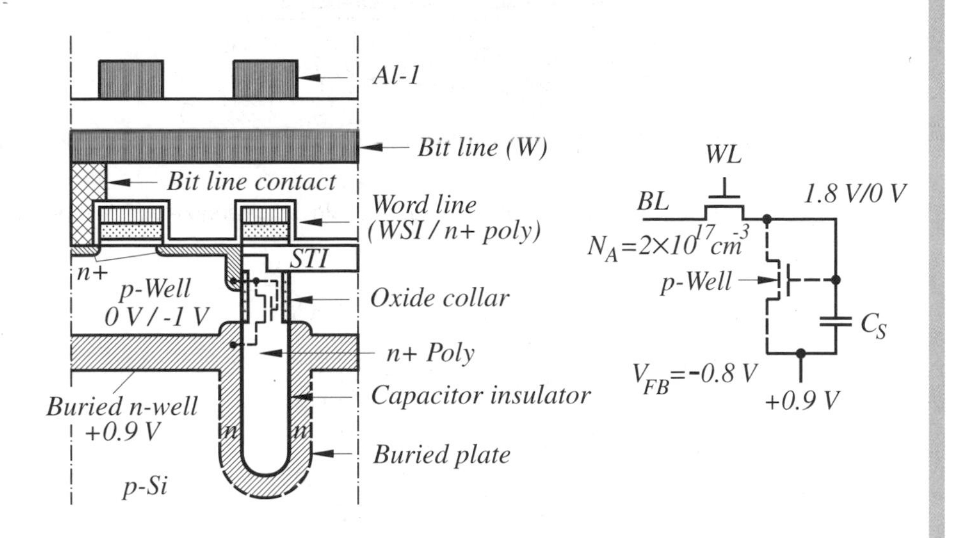

(a) The worst case voltage condition.

(b) The threshold voltage of the parasitic transistor at $d_{ox} = 7\,\text{nm}$ (oxide collar).

(c) The threshold voltage, when d_{ox} is increased to 30 nm, and when a voltage of -1 V is applied to the p-well.

4.4 If a metal layer crosses, e.g., two diffusion layers via a thick oxide region, this leads to a field oxide transistor (Section 4.3.3). In this case the metal layer acts as gate and the diffusion regions as source and drain.

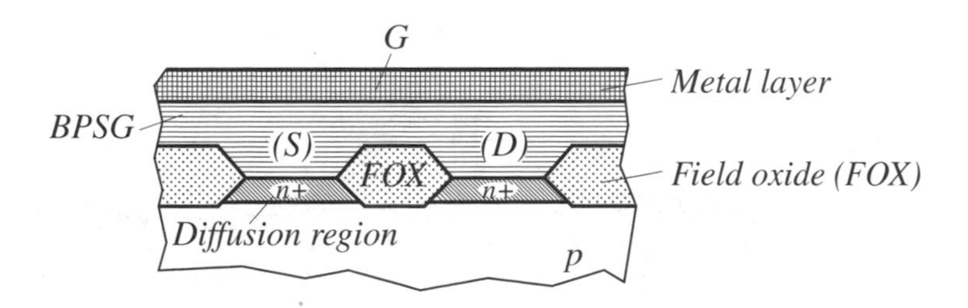

Determine the field threshold voltage V_{FT}, when the BPSG and FOX layers are together 100 nm or 200 nm. How do the threshold values change when a source bulk voltage of $V_{SB} = 1.0\,\text{V}$ is applied? Data: $N_A = 10^{17}\,\text{cm}^{-3}$, $V_{FB} = -0.4\,\text{V}$, effective dielectric constant $\varepsilon_r \approx 4.1$.

4.5 Shown below is a cross-section of a MOS transistor.

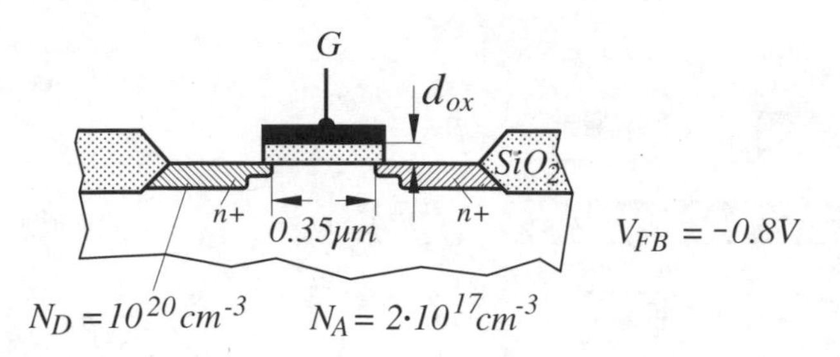

Determine at room temperature:

(a) The threshold voltage at $V_{SB} = 0$, when d_{ox} is 9.5 nm.

(b) The threshold voltage change when the oxide thickness in production varies between 9.5 nm and 10.5 nm.

(c) Do you expect a change in threshold voltage due to a short channel length effect (calculate the depletion region width of source and drain)?

4.6 An n-channel MOS transistor is used as a voltage-controlled resistor. Determine:

(a) The over-drive voltage, that is the voltage difference between the gate voltage and the threshold voltage, under the condition that the resistance between drain and source is 2.5 kΩ and $V_{DS} \rightarrow 0\,\text{V}$.

(b) The charge density σ_n under the same conditions.

Data: $w/l = 1.5$, $d_{ox} = 5\,\text{nm}$, $\varepsilon_{ox} = 3.9$, $\mu_n = 600\,\text{cm}^2/\text{Vs}$, $V_{SB} = 0\,\text{V}$.

4.7 In an n-channel transistor a current of 300 μA can be observed. The data are: $k_n = 120\,\mu\text{A/V}^2$, $w/l = 5$, $V_{Ton} = 0.5\,\text{V}$, $V_{GS} = 3\,\text{V}$. Determine the V_{DS} voltage.

4.8 Shown below is a MOS transistor connected in a so-called MOS diode configuration.

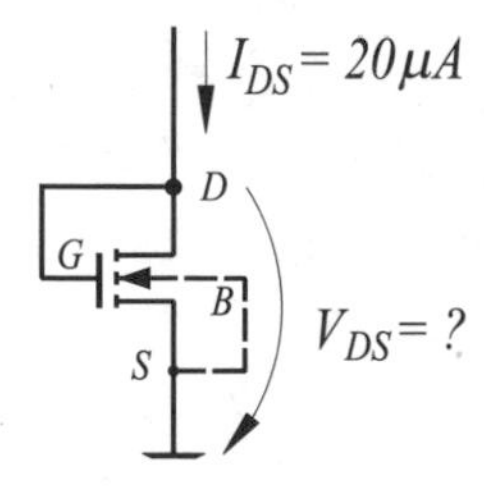

Determine the V_{DS} voltage at room temperature for the following cases: (a) $w/l = 5$, (b) $w/l = 500$, and (c) $w/l \gg 500$. Data: $k_n = 120\,\mu\text{A/V}^2$, $V_{Ton} = 0.8\,\text{V}$, $n = 2$.

4.9 Determine the I_{DS} current at room temperature, when the shown transistor enters the sub-threshold region.

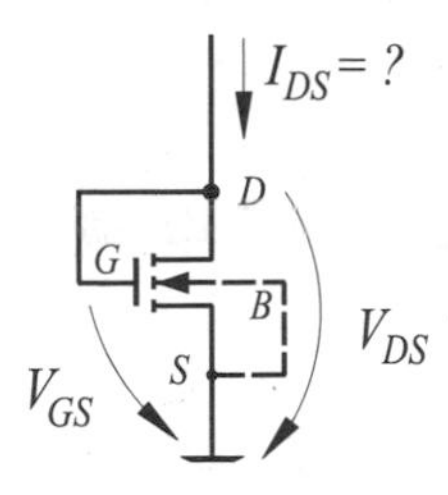

Data: $k_n = 120\,\mu A/V^2$, $w/l = 15$, $n = 2$, $V_{Ton} = 0.6\,V$.

4.10 Shown below is the cross-section of a one-transistor DRAM cell (Section 7.5) together with its equivalent circuit representation and data. Determine:

(a) The sub-threshold current under worst-case condition of the cut-off transistor.

(b) The refresh time under worst case condition. It can be assumed here that a reduced high level of 1.5 V and an increased low level of 0.3 V is still acceptable from a sense amplifier point of view.

(c) By how much can the refresh time be improved, when −1 V is applied to the p-well?

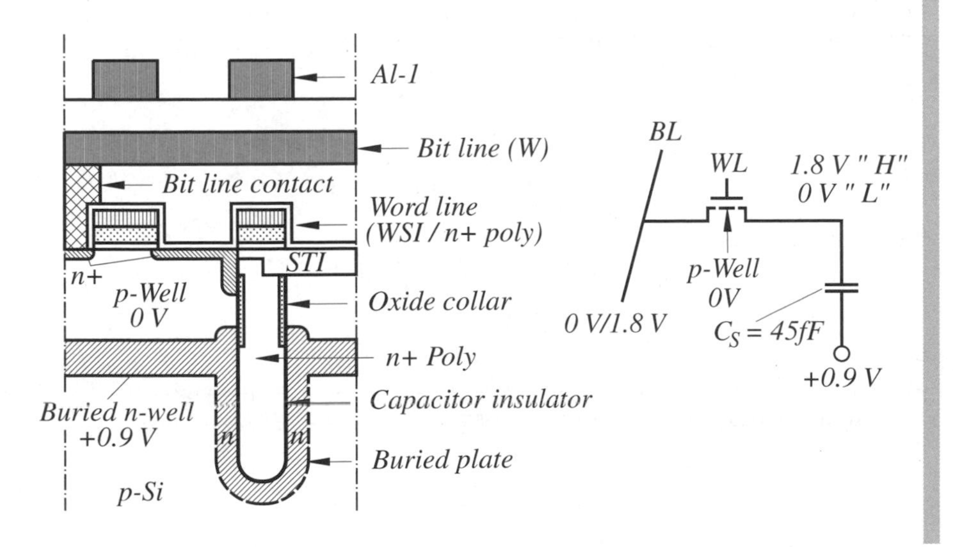

Data of n-channel transistor:

	27 °C	90 °C
I_{DS} (measured at $V_{GS} = 0.6\,V$)	250 pA	4.8 nA
V_{Ton} (p-well 0 V)	1.0 V	0.87 V
γ	$0.3\sqrt{V}$	$0.3\sqrt{V}$
S	120 mV/dec	145 mV/dec
$2\phi_F$	0.82 V	0.76 V
$n = 1 + \frac{C'_j}{C'_{ox}}$	2	2

4.11 Determine the built-in voltage (Equation 4.5) (a) by assuming that drift and diffusion currents compensate each other and (b) via different work functions. Both approaches must deliver the same result.

The solutions to the problems can be found under: *www.unibw-muenchen.de/campus/ET4/index.html*

APPENDIX A

Current–Voltage Equation of the MOS Transistor under Weak Inversion Condition

The derivation of the charge equations of the MOS structure (Section 4.3.1) is simplified, since the charge-sheet approximation and strong inversion with a constant surface voltage of $\phi_S(SI) = 2\phi_F$ are assumed (Figure 4.16). Under weak inversion conditions these prerequisites are not valid. This leads to more complex equations, as the following derivation illustrates. The starting point is – similar to Section 4.3.2 – Poisson's equation

$$\frac{d^2\phi}{dx^2} = -\frac{\rho}{\varepsilon_0\varepsilon_{Si}} = \frac{q}{\varepsilon_0\varepsilon_{Si}}(N_A + n(x)) \tag{A.1}$$

where the electron distribution (Equation 2.11) is given by

$$n(x) = n_i e^{\frac{E_{Fn} - E_i(x)}{kT}} \tag{A.2}$$

According to Figure 4.18(b), repeated for convenience in Figure A.1, the voltage of the semiconductor has a local distribution of

$$\phi(x) = \frac{E_i(x) - E_i(x_d)}{-q} \tag{A.3}$$

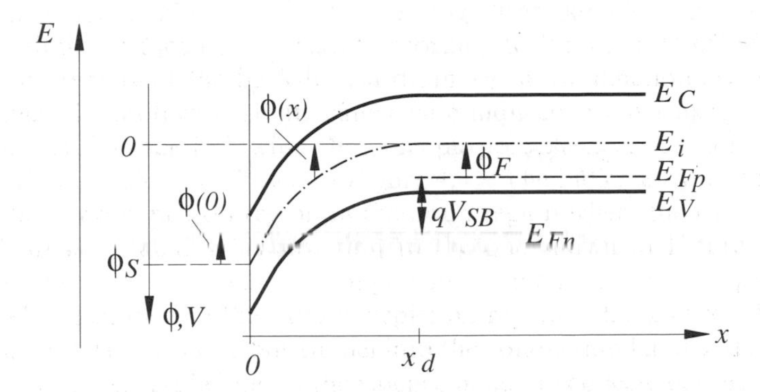

Figure A.1 Band diagram of the MOS structure under the influence of a V_{SB} voltage

This leads to the following local electron density distribution (Equation A.2) as a function of $\phi(x)$

$$n(x) = n_i e^{\frac{E_{Fn} - E_i(x_d) + q\phi(x)}{kT}} \tag{A.4}$$

Since furthermore

$$\frac{E_{Fn} - E_i(x_d)}{-q} - \phi_F + V_{SB} \tag{A.5}$$

is valid, an electron density of

$$n(x) = n_i e^{(\phi(x) - \phi_F - V_{SB})/\phi_t} \tag{A.6}$$

results. Poisson's equation can thus be rewritten in the form

$$\frac{\mathrm{d}^2\phi}{\mathrm{d}x^2} = \frac{q}{\varepsilon_0 \varepsilon_{Si}} \left(N_A + n_i e^{(\phi(x) - \phi_F - V_{SB})/\phi_t}\right) \tag{A.7}$$

Using the identity

$$\frac{1}{2}\frac{\mathrm{d}}{\mathrm{d}x}\left(\frac{\mathrm{d}\phi}{\mathrm{d}x}\right)^2 = \left(\frac{\mathrm{d}\phi}{\mathrm{d}x}\right)\left(\frac{\mathrm{d}^2\phi}{\mathrm{d}x^2}\right) \tag{A.8}$$

and the integration border with $\mathcal{E}_{Si}(x_d) = 0$, a simplified solution results

$$\begin{aligned} \left[\frac{1}{2}\left(\frac{\mathrm{d}\phi}{\mathrm{d}x}\right)^2\right]_o^{x_d} &= \frac{q}{\varepsilon_0 \varepsilon_{Si}} \left[N_A \phi(x) + \phi_t n_i e^{(\phi(x) - \phi_F - V_{SB})/\phi_t}\right]_{\phi_S}^{o} \\ \mathcal{E}_{Si}^2(0) &= \frac{2q}{\varepsilon_0 \varepsilon_{Si}} \left(N_A \phi_S + \phi_t n_i e^{(\phi_S - \phi_F - V_{SB})/\phi_t}\right) \end{aligned} \tag{A.9}$$

Gauss' law leads to the relationship (Equation 4.15)

$$-\varepsilon_0\varepsilon_{Si}\mathcal{E}_{Si}(0) = \sigma_n + \sigma_d \tag{A.10}$$

Joining Equations (A.9) and (A.10) results in

$$\sigma_n + \sigma_d = -(2\varepsilon_0\varepsilon_{Si}qN_A)^{1/2}\left(\phi_S + \phi_t e^{(\phi_S - 2\phi_F - V_{SB})/\phi_t}\right)^{1/2} \tag{A.11}$$

where (Equation 4.26)

$$p(x_d) = N_A = n_i e^{\phi_F/\phi_t}$$

is used to describe the hole density in the substrate. Since the width of the inversion layer d_i is much smaller than the width of the depletion region x_d (Figure 4.13(b)), the depletion region charge can be approximated by Equation (4.13)

$$\sigma_d = -qN_A x_d = -(qN_A 2\varepsilon_0\varepsilon_{Si}\phi_S)^{1/2}$$

With this assumption Equation (A.11) yields the following inversion layer charge

$$\sigma_n = -(2\varepsilon_0\varepsilon_{Si}qN_A)^{1/2}\left[\left(\phi_S + \phi_t e^{(\phi_S - 2\phi_F - V_{SB})/\phi_t}\right)^{1/2} - (\phi_S)^{1/2}\right] \tag{A.12}$$

What is still needed is a relationship between the applied V_{GB} voltage and the surface voltage ϕ_S, in order to derive an equation relating the inversion layer charge to the applied V_{GB} voltage. Using Equations (4.15) and (4.18) results in

$$D_{ox} = C'_{ox}(V_{GB} - V_{FB} - \phi_S) = -(\sigma_n + \sigma_d) \tag{A.13}$$

and in conjunction with Equation (A.11) leads finally to

$$V_{GB} = V_{FB} + \phi_S + \frac{1}{C'_{ox}}\left[2\varepsilon_0\varepsilon_{Si}qN_A\left(\phi_S + \phi_t e^{(\phi_S - 2\phi_F - U_{SB})/\phi_t}\right)\right]^{1/2} \tag{A.14}$$

$$V_{GS} = V_{FB} + \phi_S - V_{SB} + \gamma\left[\phi_S + \phi_t e^{(\phi_S - 2\phi_F - V_{SB})/\phi_t}\right]^{1/2} \tag{A.15}$$

where the definition of the body-effect parameter (Equation 4.20) is used and V_{GB} is substituted by $V_{GS} + V_{SB}$. Equations (A.15) and (A.12) are implicit relationships, which can be solved numerically or by appropriate approximations. For this purpose Equation (A.15) is plotted in Figure A.2.

For the weak inversion case with $V_{GS} \leq V_{Tn}$, the surface voltage $\phi_S(V_{GS})$ can be approximated by

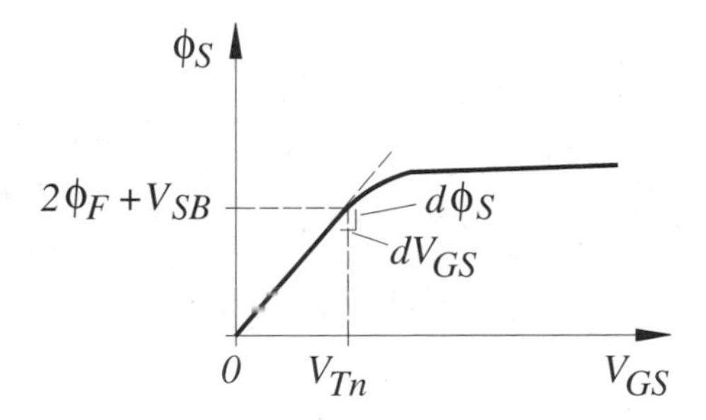

Figure A.2 Surface voltage as a function of the V_{GS} voltage

$$\begin{aligned}\phi_S &\approx \phi_S(V_{GS} = V_{Tn}) + \frac{\mathrm{d}\phi_S(V_{Tn})}{\mathrm{d}V_{GS}}(V_{GS} - V_{Tn}) \\ &\approx 2\phi_F + V_{SB} + \frac{C'_{ox}}{C'_{ox} + C'_j}(V_{GS} - V_{Tn}) \\ &\approx 2\phi_F + V_{SB} + \frac{1}{n}(V_{GS} - V_{Tn})\end{aligned} \tag{A.16}$$

where

$$\boxed{n = 1 + \frac{C'_j}{C'_{ox}}} \tag{A.17}$$

is used to describe a capacitance ratio. How it comes to this capacitance ratio can be understood by considering the expanded differential quotient

$$\begin{aligned}\frac{\mathrm{d}\phi_S}{\mathrm{d}V_{GS}} &= \frac{\mathrm{d}Q}{\mathrm{d}V_{GS}}\frac{\mathrm{d}\phi_S}{\mathrm{d}Q} \\ \frac{\mathrm{d}\phi_S}{\mathrm{d}V_{GS}} &= C'_{ox}\frac{1}{C'_j + C'_{ox}}\end{aligned} \tag{A.18}$$

The individual terms can be expressed by small-signal capacitances. Figure A.3 illustrates the situation. The term $\mathrm{d}Q/\mathrm{d}V_{GS}$ is described by C_{ox} (Figure A.3(a)) and the term $\mathrm{d}Q/\mathrm{d}\phi_S$ by the sum of $C_{ox} + C_j$ (Figure A.3(b)).

The equation of the inversion layer charge (Equation A.12) can be simplified for small ϕ_S values – which is the case in weak inversion – by the approximation (valid for small x values)

$$(1 + x)^{1/2} \approx 1 + x/2 \tag{A.19}$$

$$\tau_n \approx -(2\varepsilon_0\varepsilon_{Si}qN_A)^{1/2}\left(\frac{1}{2}\phi_t\frac{1}{\sqrt{\phi_S}}e^{(\phi_S - 2\phi_F - V_{SB})/\phi_t}\right) \tag{A.20}$$

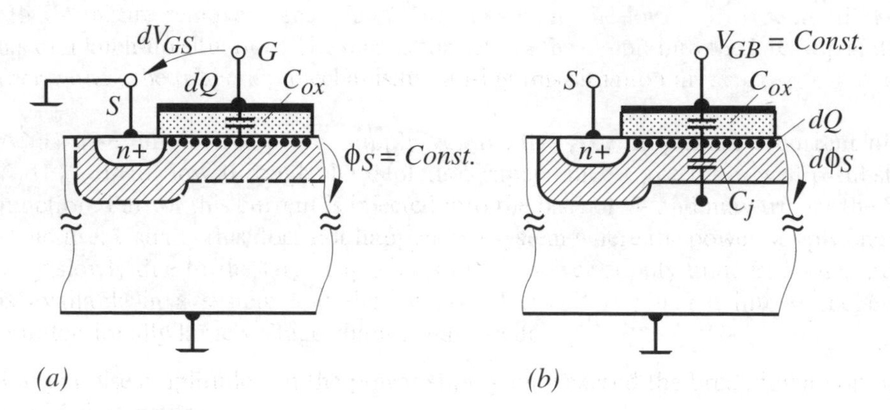

Figure A.3 Capacitance presentation: (a) ϕ_S is constant and (b) V_{GB} is constant

Since the depletion capacitance (Equations 4.3 and 4.12) can be described by

$$C_j = \frac{\varepsilon_0 \varepsilon_{Si}}{x_d} = \frac{(2\varepsilon_0 \varepsilon_{Si} q N_A)^{1/2}}{2\sqrt{\phi_S}} \tag{A.21}$$

the following simplified charge equation results

$$\sigma_n \approx -C_j \phi_t e^{(\phi_S - 2\phi_F - V_{SB})/\phi_t} \tag{A.22}$$

As is discussed in Section 4.4, at small gate voltages, a diffusion current (Equation 4.52) dominates the current–voltage relationship. This equation leads to a current of

$$\begin{aligned} \int_{y=0}^{y=1} I_{DS}\, \mathrm{d}x &= w\mu_n \phi_t \int_{\tau_n(\mathrm{Source})}^{\tau_n(\mathrm{Drain})} d\sigma_n \\ I_{DS} &= \frac{w}{l}\mu_n \phi_t [\sigma_n(\mathrm{Drain}) - \sigma_n(\mathrm{Source})] \end{aligned} \tag{A.23}$$

The charge on the source side of the channel can be found directly from Equation (A.20) and the one at the drain side, by exchanging in Equation (A.20) V_{SB} by V_{DB}. With the surface voltage direct from Equation (A.16) and the relationship $V_{DB} = V_{DS} + V_{SB}$, the following sub-threshold current results

$$\begin{aligned} I_{DS} &= \frac{w}{l}\mu_n C_j' \phi_t^2 e^{(V_{GS} - V_{Tn})/\phi_t n}\left(1 - e^{-V_{DS}/\phi_t}\right) \\ I_{DS} &= \frac{w}{l}\mu_n C_{ox}' \frac{C_j'}{C_{ox}'} \phi_t^2 e^{(V_{GS} - V_{Tn})/\phi_t n}\left(1 - e^{-V_{DS}/\phi_t}\right) \end{aligned}$$

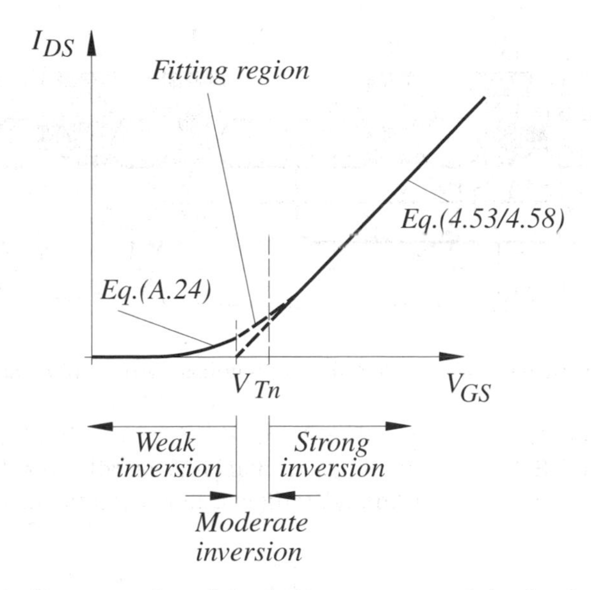

Figure A.4 Representation of the drain current around the threshold voltage

$$I_{DS} = \beta_n(n-1)\phi_t^2 e^{(V_{GS}-V_{Tn})/\phi_t n}\left(1 - e^{-V_{DS}/\phi_t}\right) \tag{A.24}$$

This leads to the situation illustrated in Figure A.4.

Due to the various approximations used at strong inversion and weak inversion, a discontinuity around the threshold voltage exists. To avoid problems with the discontinuity, in some CAD models the so-called moderate inversion region is separately described or fitted to the measured data.

REFERENCES

A. Amerasekera, L. van Roozendaal, J. Bruines and F. Kuper (1991) Characterization and modeling of second breakdown in NMOST's for the extraction of ESD-related processes and design parameters. *IEEE Trans. Electron. Devices*, **38**(9), 2161–8.

M. Bagheri *et al.* (1985) The need for an explicit model describing MOS transistors in moderate inversion. *Electron. Lett.*, **21**(12), 873–4.

G. Baum *et al.* (1970) Drift velocity saturation in MOS transistors. *IEEE Trans. Electron. Devices*, **ED-17**, 481–2.

J.R. Brews, (1978) A charge-sheet model of the MOSFET. *Solid-State Electron.*, **21**, 345–55.

D. Caughy and R. Thomas (1967) Carrier mobilities in silicon empirically related to doping and field. *Proc. IEEE*, **55**, 2192–3.

T.E. Chang, C. Huang and T. Wang (1995) Mechanisms of interface trap-induced drain leakage current in off-state n-MOSFETs. *IEEE Trans. Electron. Devices*, **42**(4), 738–43.

Y.H. Cheng and C. Hu (1999) *MOSFET Modelling and BSIM3 User's Guide*. Boston, MA: Kluwer Academic.

I.M. Filanovsky (2000) Voltage reference using mutual compensation of mobility and threshold voltage temperature effects. In ISCAS 2000; IEEE International Symposium on Circuits and Systems, pp. 197–200.

D. Foty (1997) *MOSFET Modeling with SPICE*. Upper Saddle River, NJ: Prentice Hall.

J. Hu Genda (1984) A better understanding of CMOS latch-up. *IEEE Trans. Electron. Devices*, **ED-31**(1), 62–7.

E. Gondro, O. Kowarik, G. Knoblinger and P. Klein (2001) When do we need non-quasistatic CMOS RF-models? In *Proceedings of the IEEE 2001 Custom Integrated Circuit Conference*, pp. 377–80.

C. Hu *et al.* (1982) Second breakdown of vertical power MOSFETs. *IEEE Trans. Electron. Devices*, **ED-29**(8), 1287–93.

M. Koyanagi, H. Sunami, N. Hashimoto and M. Ashikawa (1987) Subbreakdown drain leakage current in MOSFETs. *IEEE Electron. Device Lett.*, **EDL-8**, 515–17.

R. Kraus and H.J. Mattausch (1998) Status and trend of power semiconductor device models for circuit simulation. *IEEE Trans. Power Electron.*, **13**(3), 452–65.

C.H. Kreuzer, N. Krsichke and N. Nance (1996) Physically based description of quasi-saturation region of vertical DMOS power transistors. *IEDM 96 Technical Digest*, pp. 489–92.

Y. Leblebici, Y. Leblebici and S.-M. Kang (1993) *Hot-Carrier Reliability of MOS VLSI Circuits*. Boston, MA: Kluwer Academic.

Medici User's Manual, (1997) Technology Modeling Associates, Inc., Sunnyvale, CA.

J.E. Meyer (1971) MOS models and circuit simulation. *RCA Rev.*, **32**, 42–63.

A. Preussger, E. Glenz, K. Heift, K. Malek, W. Schwetlick, K. Wiesinger and W.M. Werner (1991) SPT – a new smart power technology with a fully self aligned DMOS cell. In *IEEE International Symposium on Power Semiconductor Devices and ICs, ISPSD'91*, 195–7.

J. Schrietter (1955) Effective carrier mobility in surfaces space charge layers. *Phys. Rev.*, **97**, 641–6.

A. Schwerin, W. Haensch and W. Weber (1987) The relationship between oxide charge and device degradation: a comparative study of n- and p-channel MOSFETs. *IEEE Trans. Electron. Devices*, **ED-34**, 2493–9.

H. Shichman and D.A. Hodges (1968) Modelling and simulation of insulated gate-field-effect transistor switching circuits. *IEEE J. Solid-State Circuits*, **SC-3**(3), 285–9.

T. Skotnicki, G. Merckel and T. Pedron (1986) A new approach to threshold voltage modelling of short-channel MOSFETs. *Solid-State Electron.*, **29**(11), 1115–27.

S. Tanaka (1995) Theory of the drain leakage current in silicon MOSFETs. *Solid-State Electron.*, **38**(3), 683–91.

J. Tihany *et al.* (1977) DIMOS – a novel IC technology with submicron effective channel MOSFETs. *IEEE International Electron Devices Meeting Digest of Technical Papers*, pp. 399–401.

Y. Tsividis (1987) *Operation and Modeling of the MOS Transistor*. Boston, MA: McGraw-Hill.

C. Turchetti (1983) Relationship for the drift and diffusion components of the drain current in an MOS transistor. *Electron. Lett.*, **19**(23), 960–61.

A. Vladimirescu and S. Liu (1980) The simulation of MOS integrated circuits using Spice 2, memorandum no. UCB/ERL M80/7, University of California, Berkeley.

D.E. Ward and R.W. Dutton (1978) A charge-oriented model for MOS transistor capacitances. *IEEE J. Solid-State Circuits*, **SC-13**(5), 703–8.

R.A. Wunderlich and P.K. Ghosh (1994) Modeling the gate more accurately for power MOSFETs. *IEEE Trans. Power Electron.*, **9**(1), 105–11.

L.D. Yau, (1974) A simple theory to predict the threshold voltage of short-channel IGFETs. *Solid-State Electron.*, **17**, 1059.

FURTHER READING

H. Ballan and M. Declercq (1999) *High Voltage Devices and Circuits in Standard CMOS Technologies*. Boston, MA: Kluwer Academic.

C.Y. Chang and S.M. Sze (1966) *ULSI Technology*. New York: McGraw-Hill.

E.H. Nicollian and J.R. Brews (1982) *MOS (Metal Oxide Semiconductor) Physics and Technology*. New York: John Wiley & Sons.

Y. Tsividis (1987) *Operation and Modeling of the MOS Transistor*. Boston, MA: McGraw-Hill.

5

Basic Digital CMOS Circuits

Before starting with the analysis and the design of basic digital circuits, it is useful to begin this chapter with a general discussion about geometric and electrical design rules. These design rules are prerequisite for all designs, independent of whether they are used for digital or analog applications. Inverters and buffers are chosen for the discussion of basic sizing rules. Depending on the technology and design style used, four basic inverter configurations can be distinguished. Of interest are the voltage levels, the power dissipation, and the switching performance. Buffer stages are presented and input output circuits analyzed. In conjunction with these circuits the ESD protection is discussed. The knowledge gained is then applied to more complex circuits and memories in following chapters.

5.1 GEOMETRIC DESIGN RULES

In the final design stage of an integrated circuit the so-called layout is drawn. This layout includes all geometrical structures needed for mask generation. The masks are then used in production to transmit the designed geometries onto the silicon wafer (see Figure 4.2). The layout can therefore be considered to be the connecting link between the circuit design and the manufacturing process. An example is used to demonstrate how the individual layout levels and accordingly mask levels are put together in a manufacturing process, as described in Section 4.1. The individual steps are (Figure 5.1):

1. Definition of n-wells (mask 1).
2. Determination of the active areas (mask 2). These are all gate areas and diffusion areas. The remaining ones are covered with field oxide (FOX) (Figure 5.1(a)). If dimensions below approximately 0.2 µm are used, trench insulations are employed in these areas.
3. Definition of the polysilicon gates and polysilicon interconnection layers (mask 3) (Figure 5.1(b)).

System Integration: From Transistor Design to Large Scale Integrated Circuits. Kurt Hoffmann.
 ISBN: 0-470-85407-3

4. Definition of n^+-implantation areas (mask 4). These are all source and drain regions of the n-channel transistors, diffused interconnection layers, and n-well contacts (Figure 5.1(c)).
5. Definition of p^+-implantation areas (mask 5). Similar to mask 4 but this time used for p-channel transistors (Figure 5.1(d)).
6. Contact definition (mask 6). Usually one minimum contact size is used only, in order to optimize the lithographic process. If the resistance of one contact is too large, parallel connected ones can be used to ease the situation (Figure 5.1(e)).
7. Metal layer 1 definition (mask 7).
8. Connecting metal layer 1 (via) with metal layer 2 (mask 8).
9. Definition of metal layer 2 (mask 9).

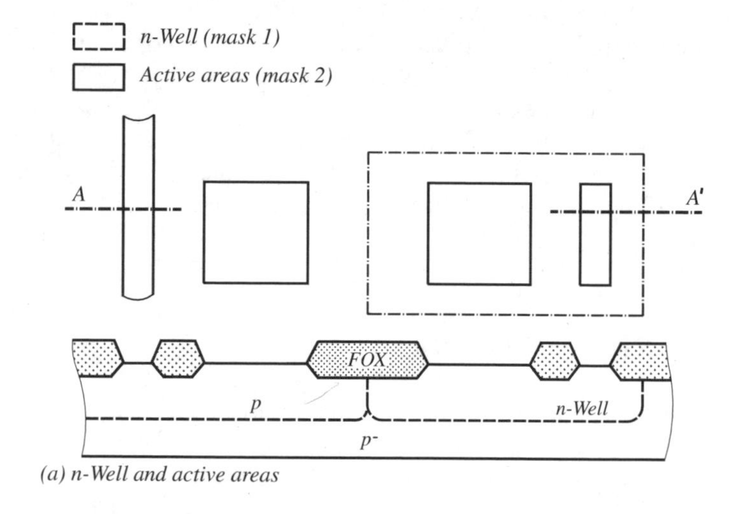

(a) n-Well and active areas

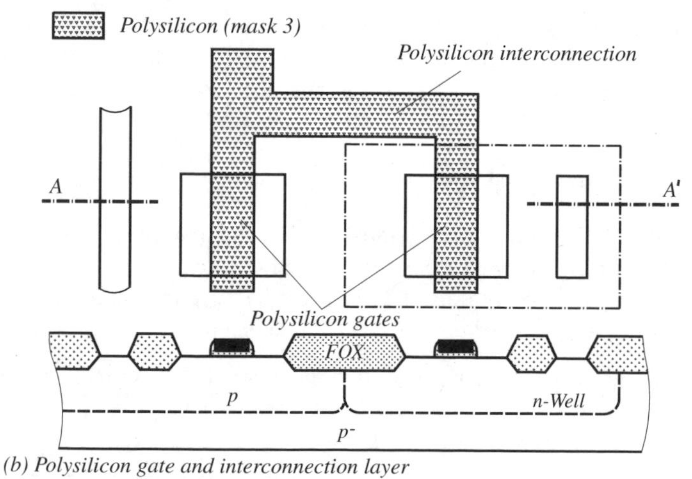

(b) Polysilicon gate and interconnection layer

Figure 5.1 (Continued)

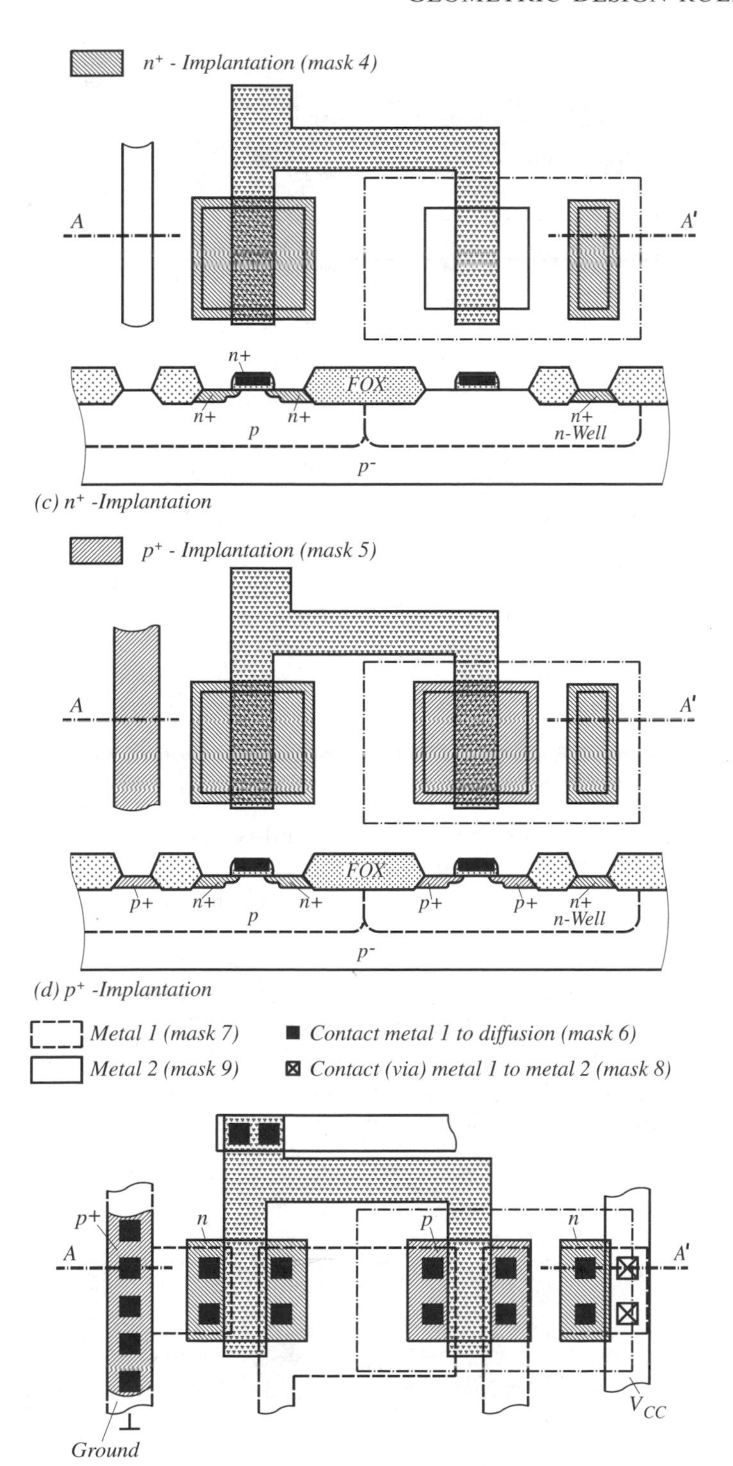

Figure 5.1 (Continued)

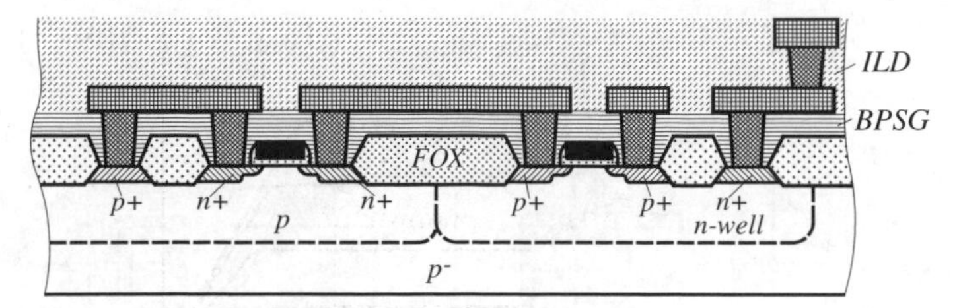

(e) Contact metal 1 to diffusion, metal layer 1, contact (via) metal layer 1 to metal layer 2 (not all steps shown)

Figure 5.1 Layout of a partial CMOS circuit with technology cross-section A–A'

10. Connecting metal layer 2 (via) with metal layer 3 (mask 10) not shown.
11. Definition of metal layer 3 (mask 11) not shown, and so on.

This example demonstrates that permissible design rules are required. These are, e.g., minimum spacings and dimensions of diffusion regions, polysilicon layers, contact sizes, metal layers and so on. Only when these geometrical design rules are not violated, a high yield during manufacturing can be guaranteed. Yield describes the ratio of good dies or chips with respect to the total number of chips available on a wafer. If larger design rule dimensions are used, in order to be on the safe side, this reduces the number of chips available on the wafer and consequently the production cost is increased. Additionally to the minimum allowed design rules, other requirements have to be taken into consideration. Examples are: the spacing between adjacent diffusion layers has to be in accordance with the required breakdown voltage, or the metal layers have to meet the electromigration requirement, or the voltage drop across polysilicon, diffusion or metal layers has to be taken into account. Furthermore, the design rules have to include the tolerance between mask and wafer alignment. Two examples illustrate the situation (Figure 5.2). In the first case (Figure 5.2(a)), the contact is not properly aligned, causing a short between metal layer 1 and the substrate. In the second case (Figure 5.2(b)), a bridge between source and drain of a transistor exists. Under consideration of all manufacturing influences design rules are generated. As a simplified example a summary of some important design rules of a CMOS process with typical 0.3 μm dimensions is shown in Table 5.1.

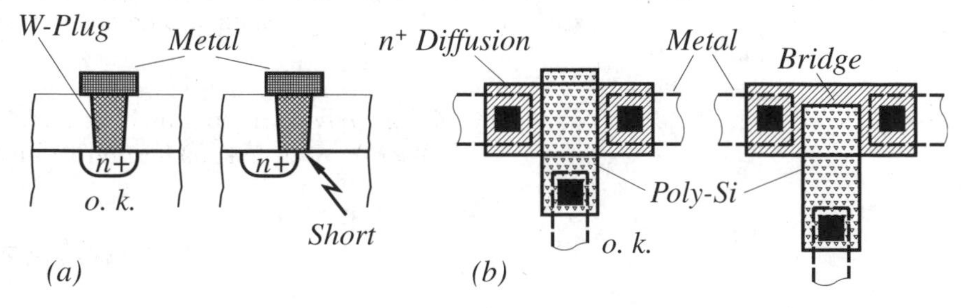

Figure 5.2 Influence of mask and wafer alignment: (a) between metal 1 and diffusion and (b) between polysilicon and diffusion

Table 5.1 Minimum geometrical design rules of a 0.3 μm CMOS process. Measurements in μm (not to scale)

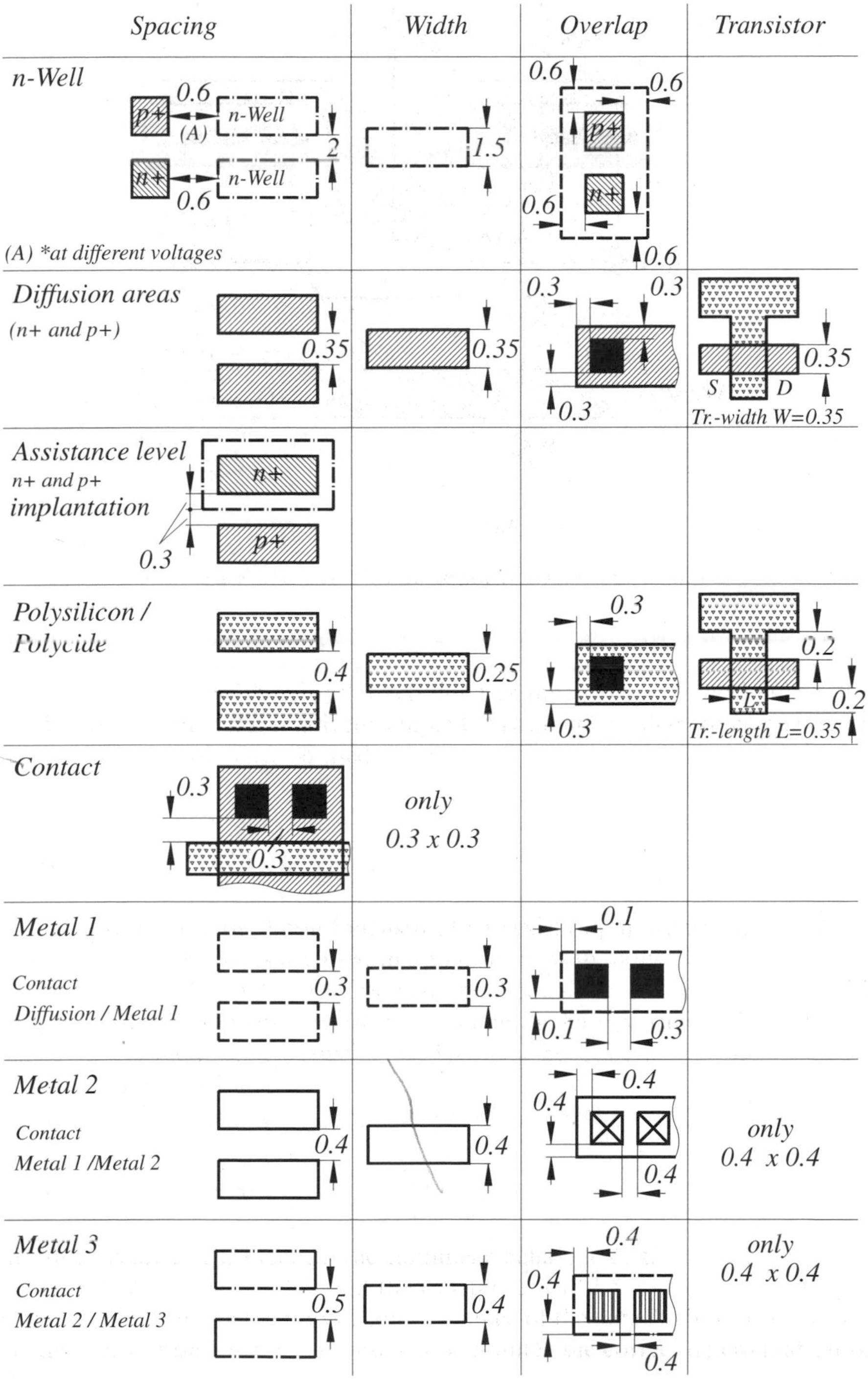

	Spacing	*Width*	*Overlap*	*Transistor*
n-Well (A) *at different voltages	p+ 0.6 (A) n-Well n+ 0.6 n-Well 2	1.5	0.6 0.6 p+ n+ 0.6 0.6	
Diffusion areas (n+ and p+)	0.35	0.35	0.3 0.3 0.3	0.35 S D Tr.-width W=0.35
Assistance level n+ and p+ *implantation*	n+ p+ 0.3			
Polysilicon / Polycide	0.4	0.25	0.3 0.3	0.2 L 0.2 Tr.-length L=0.35
Contact	0.3 0.3	only 0.3 x 0.3		
Metal 1 Contact Diffusion / Metal 1	0.3	0.3	0.1 0.1 0.3	
Metal 2 Contact Metal 1 /Metal 2	0.4	0.4	0.4 0.4 0.4	only 0.4 x 0.4
Metal 3 Contact Metal 2 / Metal 3	0.5	0.4	0.4 0.4 0.4	only 0.4 x 0.4

Relationship between drawn and final measurements

The geometric design rules are used for the generation of a layout and are given in drawn dimensions. These dimensions are then transmitted via lithographic processes to the wafer. In due process there exists a difference between the drawn dimensions and the ones implemented on the wafer. As an example, Figure 5.3 shows cross-sections through a MOS transistor. As the figure illustrates, the final channel length of the transistor is

$$l = L - 2LD \tag{5.1}$$

and the final width is

$$w = W - 2\Delta W \tag{5.2}$$

In the case of the trench insulation basically no difference exists between drawn and final width geometries. For the process presented in Section 4.1, the total lateral diffusion LD is about 0.05 µm. This leads to a minimum final or effective channel length of about $l = 0.35\,\mu\text{m} - 2 \cdot 0.05\,\mu\text{m} = 0.25\,\mu\text{m}$.

The design rules presented are of a standard three-layer metal CMOS process. Additional process steps lead to a larger variety of design features, but at the expense of additional manufacturing cost. Examples are:

(a) The cut-out of silicide film at ESD structures (Section 5.6.3) in order to increase the resistance of source, drain, and gate regions.

(b) A second polysilicon layer for the implementation of precise capacitors.

(c) An additional implantation step for the generation of precise resistors.

(d) Modified transistors able to withstand an extended voltage range (Section 4.6).

(e) Additional metal layers in order to implement improved inductors.

(f) Special elements to store information (Chapter 7).

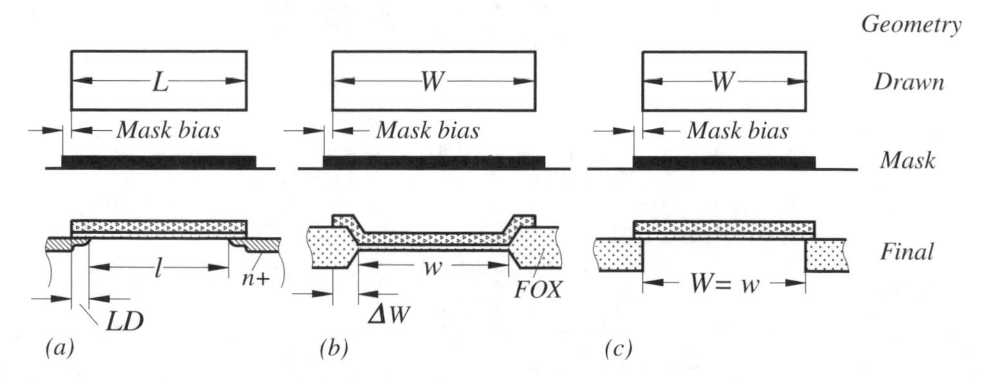

Figure 5.3 Cross-section through a MOS transistor: (a) gate length, (b) gate width using LOCOS insulation and (c) gate width using trench insulation $\Delta W = 0$

5.2 ELECTRICAL DESIGN RULES

Additionally to geometric design rules, electrical ones are required. These are, e.g., capacitance and resistance values of diffusion, polysilicon, and metal layers, and the parameters of the transistors.

Transistor parameters

The starting point is a test chip with a variety of test structures and transistors. These are evaluated and the C (V) and I (V) parameters are determined. Basically two parameter extraction methods are in use. With the first method the transistor is measured and simulated at identical voltages and currents. The discrepancy between the simulation and the measurement is then minimized with an algorithm by changing the parameters during simulation until the desired result is available. A disadvantage of this scheme is that the parameters do not necessarily have a physical meaning anymore, even when a physically based transistor model is used.

The second extraction method is also based on a physical transistor model, but the transistor parameters are generated with the aid of extraction routines. For example, the threshold voltage is determined at transistors with different gate lengths and widths. Then the parameters dealing with the short channel and width effect are extracted. This method is not as precise as the first one, but the physical nature of the parameters remains maintained.

Resistance values

In Table 5.2 typical resistance and sheet resistance values are shown for the process of Section 4.1. The contact resistances are given in Ω for the dimensions shown in Table 5.1,

Table 5.2 Typical resistance and sheet resistance values of a 0.3 μm CMOS process

n-Well	1 kΩ/□
n^+/p^+-S/D silicide	3 Ω/□
n^+-S/D without silicide	80 Ω/□
p^+-S/D without silicide	100 Ω/□
Polysilicon with silicide	3 Ω/□
n^+-Polysilicon without silicide	250 Ω/□
p^+-Polysilicon without silicide	200 Ω/□
Metal-1	100 mΩ/□
Metal-2	80 mΩ/□
Metal-3	60 mΩ/□
Metal-1 contact with n^+- or p^+-regions	4 Ω
Metal-1 contact with polysilicon	3 Ω
Metal-1 contact with metal 2 (Via)	1 Ω
Metal-3 contact with metal 2 (Via)	1 Ω

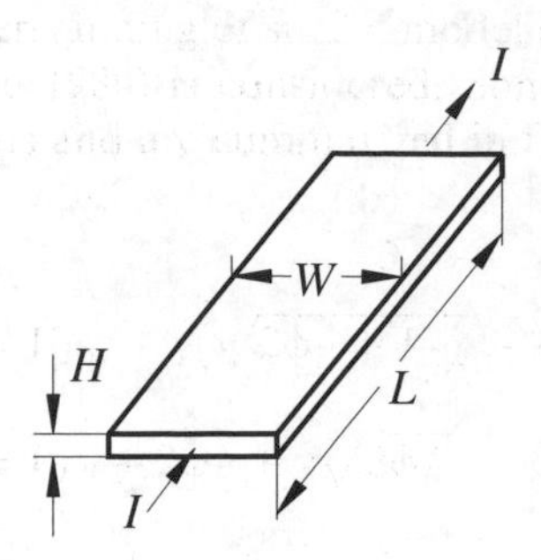

Figure 5.4 Sketch of an interconnection layer

whereas the sheet resistances in Ω/□ apply to interconnection layers. For example, the sheet resistance of the interconnection layer (Figure 5.4) can be determined by

$$R = \rho \cdot \frac{L}{H \cdot W} = \frac{\rho}{H}\frac{L}{W} \tag{5.3}$$

where ρ is the resistivity in Ωm of the material and H, W, and L are the dimensions of the layer. If the resistivity is divided by the thickness H of the layer this results in

$$\boxed{R = R_S \frac{L}{W}} \tag{5.4}$$

where R_S is called sheet resistance. If the sheet resistance is known, the resistance of an interconnection layer can simply be found by multiplying the L/W ratio of the layer with the sheet resistance. The sheet resistance has the unit Ω but is usually given in Ω/□, in order to indicate that the resistance of an interconnection layer can be found from the product of the number of squares – that is, the L/W ratio – times the sheet resistance.

Example

The diffused interconnection layers shown in Figure 5.5 are given. Structure (a) has 9 squares. This results in an interconnection resistance of $R = 3\Omega/\square \cdot 9\square = 27\Omega$ when

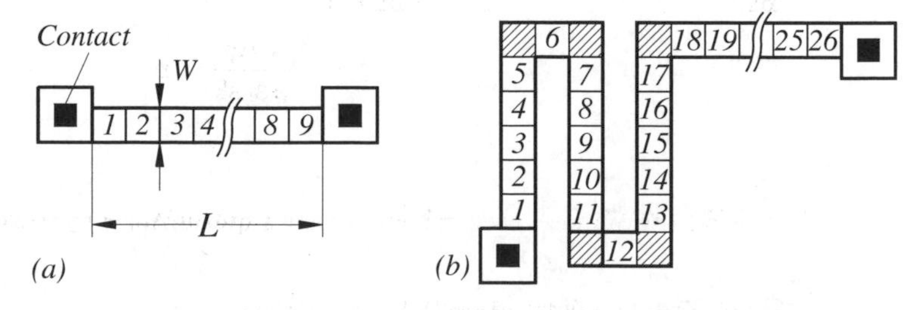

Figure 5.5 Top view of diffused interconnection layers

the contact resistance is neglected. The meander structure (b) has 26 squares and 5 corners with a sheet resistance of $0.55 \cdot R_S$ (Walton *et al.* 1985). This results in a total resistance of

$$R = 3\Omega/\square(26 + 0.55 \cdot 5)\square = 86\Omega$$

The example illustrates how the calculation of interconnection layers can be simplified by more or less just inspecting the structure. That is the reason why in integrated circuit design, sheet resistances are used exclusively.

Electro-migration

A very important parameter of metal layers is the lifetime, expressed as mean time between failures (MTBF). The lifetime is limited by electro-migration. This effect is caused by moving electrons which collide with metal ions. This gives rise to metal ions moving in the direction of the electrons. The resulting material transport leads to a rupture of the interconnection layer and a failure of the integrated circuit. The lifetime can be calculated by using the Arrhenius relationship

$$\boxed{\mathrm{MTBF} \sim J^{-2} e^{E_A/kT}} \tag{5.5}$$

where J is the current density and E_A an activation energy with a typical value of about 0.65 eV. For example, with a current density of $2\,\mathrm{mA}/\mu\mathrm{m}^2$ and a 80 °C chip temperature, a lifetime of MTBF $\approx$ 80 years results. This is an exceptionally good value, but one has to keep in mind that the lifetime deteriorates exponentially with an increase in temperature (Problem 5.1). Electro-migration can be observed not only at metal layers but also at metal–silicon contacts. Particularly at the contact corners large currents occur, causing silicon atoms to move into the connecting metal. This limits the current per contact to about 0.1 mA.

From the preceding discussion it is obvious that before the design of an integrated circuit is started, the power-supply distribution network has to be planned. As examples two networks are shown in Figure 5.6. Both networks have the advantage that no metal lines are crossing, so bridges and therefore resistive connections are avoided.

Capacitance values

The electrical design rules include the values of parasitic capacitances also. For the process described in Section 4.1, some important ones are listed in Table 5.3.

Cad tools for the physically based design

In the preceding sections geometric and electrical design rules are presented. These documents, together with the appropriate parameters of the transistor model, are used for the design of an integrated circuit. How this is done and supported by computer aided design (CAD) tools is considered in the following (Figure 5.7).

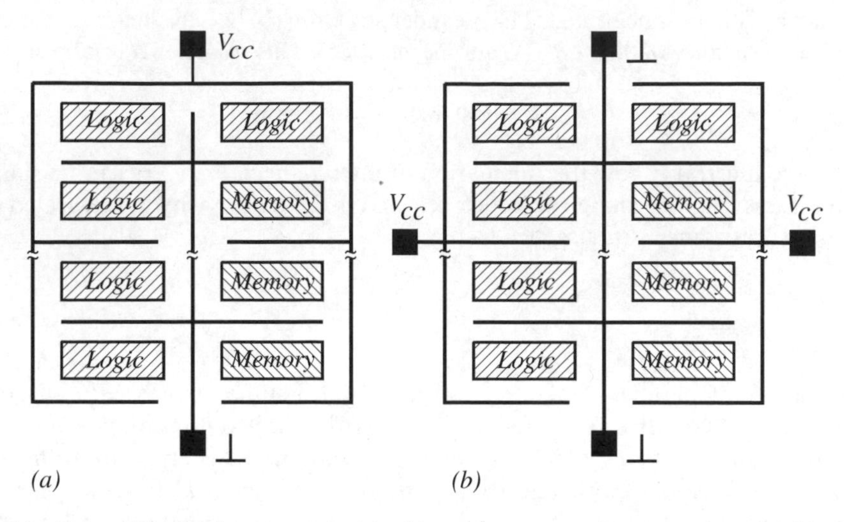

Figure 5.6 Power distribution network: (a) with one power-supply connection and (b) with two power-supply connections

Table 5.3 Typical capacitance values of a 0.3 μm CMOS process

Gate/substrate	$C'_{gb} = 4\,\text{fF}/\mu\text{m}^2$	
Poly-FOX-substrate	$C'_p = 0.15\,\text{fF}/\mu\text{m}^2$	
Metal-BPSG-FOX-substrate	$C'_{BPS} = 0.03\,\text{fF}/\mu\text{m}^2$	
Metal 2-ILD metal 1	$C'_{M2} = 0.05\,\text{fF}/\mu\text{m}^2$	
Meta 3-ILD metal 2	$C'_{M3} = 0.05\,\text{fF}/\mu\text{m}^2$	
n^+/p-Substrate	$C'_j = 1\,\text{fF}/\mu\text{m}^2$	$M \approx 0.3$
	$C^*_j = 0.12\,\text{fF}/\mu\text{m}$	$M \approx 0.22$
	$\phi_i = 0.72\,\text{V}$	
p^+/n-Well	$C'_j = 1.1\,\text{fF}/\mu\text{m}^2$	$M \approx 0.35$
	$C^*_j = 0.09\,\text{fF}/\mu\text{m}$	$M \approx 0.18$
	$\phi_i = 0.76\,\text{V}$	

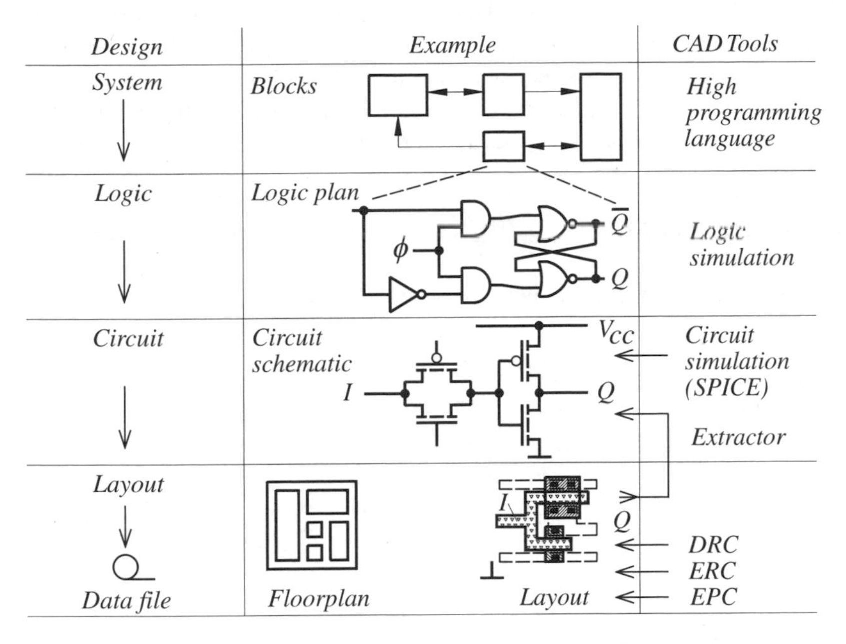

Figure 5.7 Design flow with CAD tools

In general, the design procedure can be organized into the categories system, logic, circuit, and layout design. The first two blocks are the logical design, whereas the second two blocks are the physical design. Since the last block is part of the topic of this book it is looked at in more detail.

The starting point is the system specification, which leads to a system architecture consisting of memories, programmable logic arrays, and controllers, in order to name a few. The logic design follows, yielding a logic plan. This plan is broken down into various circuit design tasks. At this level the circuit simulation is used for the verification and optimization of the circuits. The layout phase can be started when the circuit is more or less known. A data file will then be generated once the layout is finished. This data file is used at a pattern generator in order to generate or, better, write the individual masks. But before this happens, elaborate tests have to be performed to guarantee the validity of the design and layout. For this purpose the following CAD tools are used.

Design rule check (DRC)

This program checks whether all geometric design rules, e.g. minimum width, spacing, overlapping, and so on, are observed (Baker and Terman 1980).

Circuit extraction

Connecting errors in circuit design cannot be uncovered with the previous program. For this purpose circuit extraction programs are available (Trimberger 1987). These programs extract the circuit schematic from the layout and compare it with the original schematic. A discrepancy between the two corresponds to a connecting error. Additionally, parasitic resistances and capacitances are extracted from the layout and transmitted to the circuit simulator. At this point a realistic circuit simulation can be performed. If the result is not adequate, a redesign of the circuit and layout has to be started.

Electrical rule check (ERC)

This check tests, e.g., for the following rule violations:

(a) shorts between V_{CC} and ground

(b) connections which never lead to a V_{CC} or ground

(c) open connections

(d) shorts between drain and source of the transistors.

Electrical parameter check (EPC)

This check deals with the influence the layout has on some electrical properties. This verification includes, e.g.:

(a) width and length of each transistor

(b) high resistive interconnection layers

(c) circuit notes with large capacitance loading.

The briefly presented design procedure is lengthy and thus expensive. This leads to a variety of design approaches, to shorten and automate the design process. These are, e.g., array-based designs and standard cell designs. In the latter case a library is provided that contains a wide selection of circuit elements such as gates, flip-flops, adder decoder, amplifiers, and so on. Then the task of the designer is to select and wire together the appropriate components to a system.

In the following sections basic digital circuits are presented and analyzed, based on the geometric and electrical design rules discussed.

5.3 MOS INVERTER

This inverter is the simplest circuit which will be used to cover more or less all electrical effects of digital MOS circuits. These are e.g., the w/l sizing, the voltage reduction caused by the threshold voltage, the influence of the body effect, the power dissipation, and the switching performance. The goal of the following various inverter analysis is the derivation of easy to use equations, useful for the understanding of circuit characteristics in general and for a first estimate of circuit performance. The final evaluation of a circuit is only possible with an adequate circuit simulation tool anyway. A simple

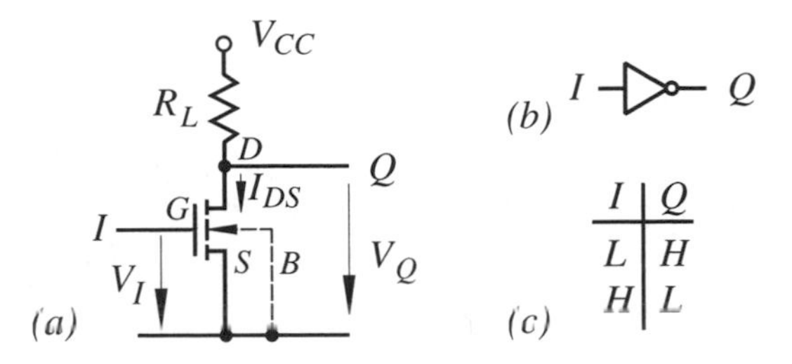

Figure 5.8 (a) MOS inverter with resistive load, (b) logic symbol and (c) truth table

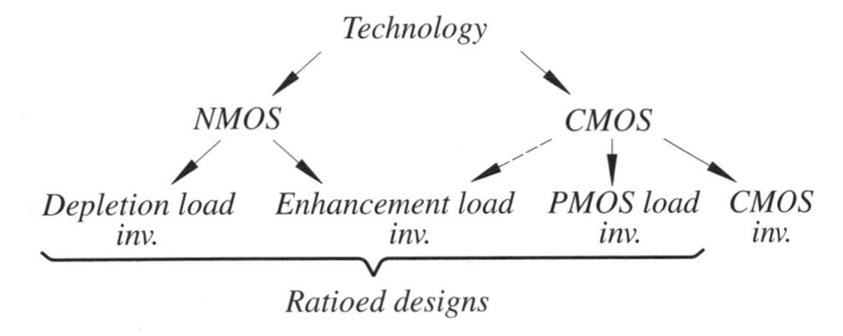

Figure 5.9 Overview of inverter realizations

inverter with a resistive load is shown in Figure 5.8. With a signal of $V_I < V_{Tn}$ applied to the input I, the transistor is cut off. Since no drain current flows – the sub-threshold current is neglected – the output signal has a value of $V_Q = V_{CC}$. Applying an input signal of $V_I \gg V_{Tn}$, the transistor is conducting. A relatively large current flows, causing a voltage drop at the resistive load R_L and consequently a small output voltage. When binary states are assigned, 'low (L)' for voltages $V < V_{Tn}$ and 'high (H)' for voltages $V \gg V_{Tn}$, then the truth table of Figure 5.8(c) results. The resistor may be implemented with a long meander structure of diffusion or polysilicon material. This approach would cause an inappropriate consumption of silicon area in large scale integrated circuits. To avoid this, the resistor is replaced by various transistor loads, depending on the technology available and design style used (Figure 5.9).

The NMOS process uses n-channel transistors only and was phased out by the more complex CMOS process starting as early as 1980. Today, the NMOS technique is used only in some small-scale integrated circuits such as smart power designs, where high voltage requirements dominate.

5.3.1 Depletion Load Inverter

This is the most common implementation in an NMOS technique. As load device, a depletion n-channel transistor T_2 is used, with gate and source short-circuited (Figure 5.10). With an additional implantation step during processing, the threshold voltage of transistor T_2 is shifted from a positive value to a negative one, usually in the order of -3.5 V (Figure 5.11). The depletion transistor is therefore still conducting (current $I_{DS,1}$

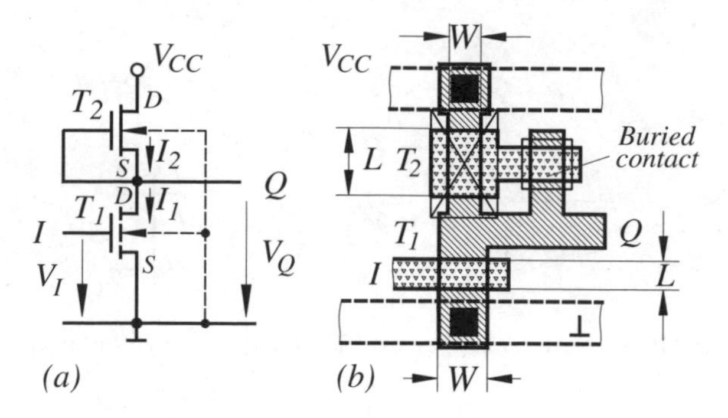

Figure 5.10 (a) Depletion load inverter and (b) layout

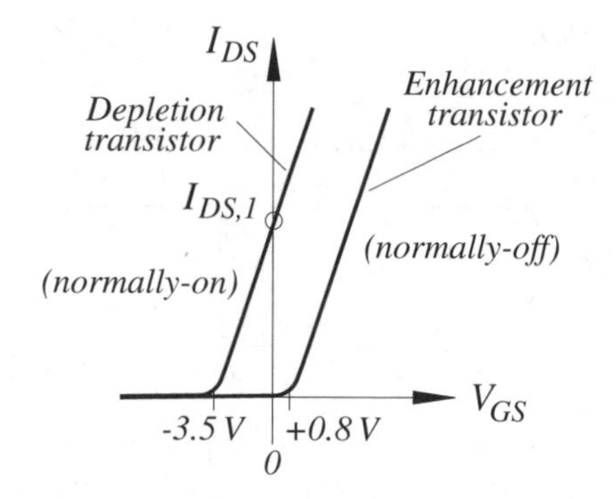

Figure 5.11 Comparison between enhancement and depletion n-channel transistors

in Figure 5.11), when gate and source are short-circuited. The layout is very area-effective since a buried contact between the polysilicon gate and the n^+-diffusion region is used. The substrate for both n-channel transistors is common and is tied to 0 V.

Note: In all layouts shown in this book the mask layer symbols shown in Table 5.1 are used.

The output voltage of a MOS inverter in the low state V_{QL} has to be smaller than the threshold voltage V_{Tn}. This guarantees that the low state is properly interpreted, when the output is connected to other MOS circuits. Usually a value of

$$\boxed{V_{QL} = 1/2 V_{Tn}} \tag{5.6}$$

is used in order to have a sufficient signal to noise ratio. Under this condition and with an input high signal of $V_{IH} = V_{CC}$, the geometry ratio of the transistors can be determined, or in other words, the transistors can be sized.

The load transistor T_2 is, with a threshold voltage of $V_{Tn,2} = -3.5\,\text{V}$, in saturation since $(V_{GS,2} - V_{Tn,2})$ is equal to $(0 - V_{Tn,2}) \leq V_{DS,2}$. The drain voltage $V_{DS,2}$ has, with $V_{CC} - V_{QL}$, a value of 4.6 V. Hereby it is assumed that $V_{CC} = 5\,\text{V}$ and $V_{QL} = 0.4\,\text{V}$.

The pull-down transistor T_1 is in the resistive region since $(V_{IH} - V_{Tn,1}) > V_{QL}$. This leads, with transistor Equations (4.53) and (4.58), to

$$I_2 = I_1$$

$$\frac{\beta_{n,2}}{2}(-V_{Tn,2})^2 = \beta_{n,1}\left[(V_{IH} - V_{Tn,1})V_{QL} - \frac{V_{QL}^2}{2}\right] \tag{5.7}$$

and under consideration of Equation (5.6) to a so-called inverter sizing parameter of

$$Z = \frac{\beta_{n,1}}{\beta_{n,2}} = \frac{(-V_{Tn,2})^2}{(V_{IH} - V_{Tn,1})V_{Tn,1} - (V_{Tn,1}/2)^2}$$

$$\boxed{Z = \frac{\beta_{n,1}}{\beta_{n,2}} = \frac{k_{n,1}\,(w/l)_1}{k_{n,2}\,(w/l)_2}} \tag{5.8}$$

This ratio is strongly dependent on the input signal value V_{IH}. In most cases this value is equal to the power-supply V_{CC}.

It is interesting to know what the value of a typical inverter sizing parameter is. The data of a high voltage NMOS technique are: $k_{n,1} = 30 \cdot 10^{-6}\,\text{A/V}^2$, $k_{n,2} = 25 \cdot 10^{-6}\,\text{A/V}^2$, $V_{Tn,2} = -3.5\,\text{V}$, $V_{Tn,1} = 0.8\,\text{V}$, and $V_{IH} = V_{CC} = 5\,\text{V}$. This leads to a sizing parameter of $Z = 3.8$. Due to the additional implantation of the depletion transistor T_2, the mobility of this transistor is slightly reduced compared to the pull-down transistor T_1 and consequently the gain factor $k_{n,2} < k_{n,1}$. This leads to a geometry ratio according to Equation (5.8) of $(w/l)_1/(w/l)_2 = 3.2$. This is shown in the layout (drawn dimensions) of Figure 5.10(b).

Up to now it is assumed that the threshold voltage of the transistors is constant. This is only true for T_1, where source and bulk are tied together. At transistor T_2, the situation is different. Between source and bulk exists the output voltage (Figure 5.10(a)). Therefore $V_{SB} = V_Q$, causing an increase in the threshold voltage via the body effect (Problem 5.4). This condition is discussed in more detail in the following section.

Power dissipation

In general, the power dissipation can be divided into a static P_{stat} and dynamic P_{dyn} contribution. The latter is caused by the charging and discharging of capacitances. At this type of inverter, the static power dissipation is usually by far dominating and can be calculated by

$$\begin{aligned} P_{\text{stat}} &= I_2 V_{CC} S \\ &= \frac{\beta_{n,2}}{2}(-V_{Tn,2})^2 V_{CC} S \end{aligned} \tag{5.9}$$

since T_2 is in saturation when T_1 is conducting. The factor S describes the on/off ratio of T_1, usually about 0.5 or 50%.

5.3.2 Enhancement Load Inverter

As indicated in Figure 5.9, this inverter type can be implemented in an NMOS or CMOS technology, since only n-channel transistors are used (Figure 5.12). The analysis will show that this inverter is inferior to all the other ones. The inverter is therefore usually not used in this form. But from a didactical point of view the analysis is useful, since the observed effects can be transferred to other MOS circuits. The sizing of the inverter is similar to the one presented in the preceding case. Transistor T_2 has a threshold voltage of, e.g., $V_{Tn,2} = 0.45\,\text{V}$ and in this configuration is always in saturation since $(V_{GS,2} - V_{Tn,2}) < V_{DS,2} = V_{GS,2}$. This leads to

$$I_2 = I_1$$

$$\frac{\beta_{n,2}}{2}(V_{GS,2} - V_{Tn,2})^2 = \beta_{n,1}\left[(V_{IH} - V_{Tn,1})V_{QL} - \frac{V_{QL}^2}{2}\right] \tag{5.10}$$

and with

$$V_{GS,2} = V_{CC} - V_{QL}$$

$$V_{QL} = V_{Tn}/2$$

$$V_{Tn,1} = V_{Tn,2} = V_{Tn}$$

$$k_{n,1} = k_{n,2}$$

to an inverter sizing of

$$Z = \frac{\beta_{n,1}}{\beta_{n,2}} = \frac{(w/l)_1}{(w/l)_2} = \frac{(V_{CC} - (3/2)V_{Tn})^2}{(V_{IH} - V_{Tn})V_{Tn} - (V_{Tn}/2)^2} \tag{5.11}$$

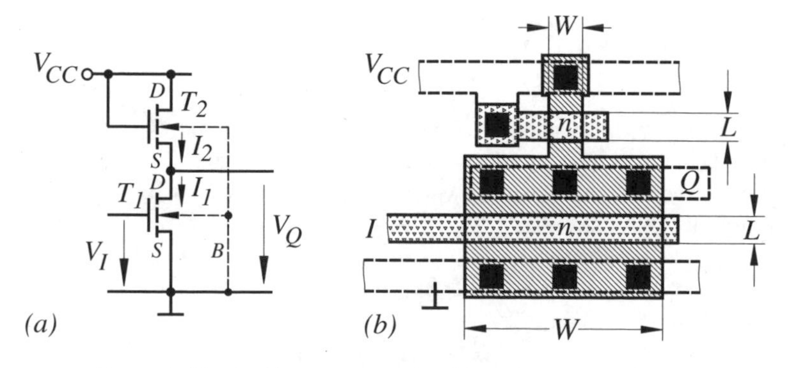

Figure 5.12 (a) Enhancement load inverter and (b) layout

With the parameters $V_{CC} = 3\,\text{V}$, $V_{IH} = V_{CC}$, and $V_{Tn} = 0.45\,\text{V}$ an inverter sizing parameter of $Z = 4.9$ results.

A considerable disadvantage of the enhancement load inverter is that the output voltage does not reach the full level of V_{CC} when T_1 is cut off, but instead a value reduced by the threshold voltage of the load transistor

$$V_{QH} = V_{CC} - V_{GS,2} = V_{CC} - V_{Tn,2} \tag{5.12}$$

This is obvious when one considers that T_2 is non-conducting or cut off – sub-threshold current neglected – when $V_{GS,2} = V_{Tn,2}$ (Figure 5.13).

Between source and bulk of T_2 the output voltage V_{QH} exists. With $V_{SB} = V_{QH}$ an increase in the threshold voltage of T_2 via the body effect occurs, which leads to a further decrease in the V_{QH} voltage. According to Equation (4.36) the threshold voltage has, with $V_{SB} = V_{QH}$, a value of

$$V_{Tn,2} = V_{Ton,2} + \gamma\left(\sqrt{2\phi_F + V_{QH}} - \sqrt{2\phi_F}\right) \tag{5.13}$$

Substituting this equation in Equation (5.12) gives

$$V_{QH} = V_{CC} - V_{Ton,2} - \gamma\left(\sqrt{2\phi_F + V_{QH}} - \sqrt{2\phi_F}\right)$$

$$\boxed{V_{QH} = V_N + \gamma^2/2 - \gamma\sqrt{V_N + 2\phi_F + \gamma^2/4}} \tag{5.14}$$

where

$$\boxed{V_N = V_{CC} - V_{Ton,2} + \gamma\sqrt{2\phi_F}}$$

is used as an abbreviation.

In order to develop a feeling for the influence the body effect has on the V_{QH} voltage, the following example is used.

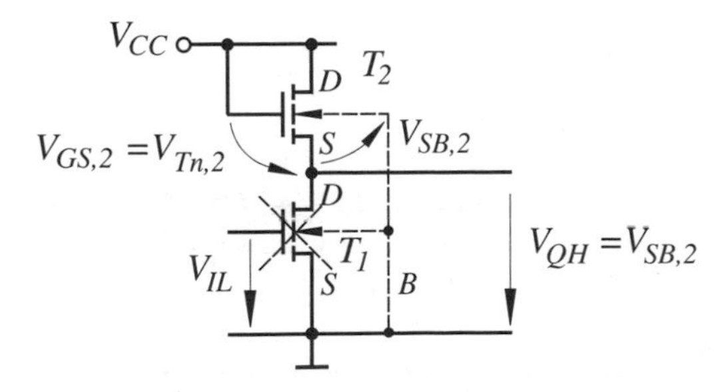

Figure 5.13 Enhancement inverter in the *H* state

Example

An enhancement load inverter is given according to Figure 5.13. Transistor T_1 is cut off. The values $V_{Ton,2} = 0.45\,\text{V}$, $2\phi_F = 0.6\,\text{V}$, $\gamma = 0.4\,\text{V}^{1/2}$, and $V_{CCmin} = 3\,\text{V}$ substituted into Equation 5.14 result in a V_{QH} voltage of 2.19 V. Without considering the body effect the output voltage would be $V_{QH} = V_{CC} - V_{Ton,2} = 2.55\,\text{V}$ and the error about 16%.

When this reduced high signal is used to drive another MOS circuit, this leads to a large sizing factor of the following stage or to an undesirable DC current path. This situation is covered in more detail in Chapter 6.

Power dissipation

The static power dissipation of the enhancement load inverter is with $V_{Tn,2} \approx V_{Tn,1} = V_{Tn}$.

$$
\begin{aligned}
P_{\text{stat}} &= I_2 V_{CC} S \\
&= \frac{\beta_{n,2}}{2}(V_{GS,2} - V_{Tn})^2 V_{CC} S \\
&= \frac{\beta_{n,2}}{2}(V_{CC} - V_{QL} - V_{Tn})^2 V_{CC} S \\
&= \frac{\beta_{n,2}}{2}\left(V_{CC} - \frac{3}{2}V_{Tn}\right)^2 V_{CC} S
\end{aligned}
\tag{5.15}
$$

5.3.3 PMOS Load Inverter

Before this inverter type is analyzed, it is useful to consider first the definition of source and drain at a p-channel transistor (Figure 5.14). All pn-junctions are back-biased and the source terminal of each transistor is connected to the respective bulk terminal B: that is, the p-substrate of the n-channel transistor and the n-substrate – that is, the n-well – of the p-channel transistor. Contrary to the ground-connected source of the n-channel transistor is the one of the p-channel transistor connected to V_{CC}. With respect to this source terminal, the other voltages V_{GS} and V_{DS} are then negative, whereas in the case

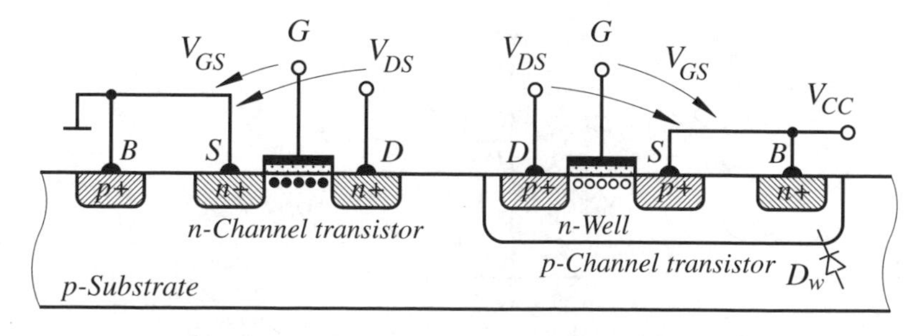

Figure 5.14 Cross-section sketch of a CMOS process

of the n-channel transistor these voltages are positive. Sometimes MOS transistors are connected as pass-transistors (Section 6.1.3). In this case it is not always obvious what is source and what is drain of the symmetrical transistor. An easy way to remember is that *the negatively charged electrons of the n-channel transistor move to the most positive terminal, the drain. In the case of the p-channel transistor the positively charged holes move to the most negative terminal, namely the drain.*

Using a CMOS process with an n-well has the advantage that different V_{CC} voltages can be used in an integrated circuit.

The PMOS load inverter is shown in Figure 5.15. The p-channel transistor is connected as load device and always conducting. Therefore, the behavior of this inverter type is similar to the discussed NMOS versions, thus it is sometimes called a quasi-NMOS inverter. With an H signal applied to the input, transistor T_n is in the resistive region and transistor T_p is in saturation, since $|V_{GS,p} - V_{Tp}| < |V_{DS,p}|$. This leads to an inverter sizing factor of

$$-I_p = I_n$$

$$\frac{\beta_p}{2}(-V_{CC} - V_{Tp})^2 = \beta_n\left[(V_{IH} - V_{Tn})V_{QL} - \frac{V_{QL}^2}{2}\right] \quad (5.16)$$

$$Z = \frac{\beta_n}{\beta_p} = \frac{k_n(w/l)_n}{k_p(w/l)_p} = \frac{(-V_{CC} - V_{Tp})^2}{(V_{IH} - V_{Tn})V_{Tn} - (V_{Tn}/2)^2}$$

where $V_{QL} - V_{Tn}/2$ is used.

With the following typical values: $V_{Tn} = 0.45\,\text{V}, V_{Tp} = -0.45\,\text{V}$, and $V_{IH} = V_{CCmin} = 3\,\text{V}$, a sizing factor of $Z = 5.9$ results. This yields with $k_n = 120\,\mu\text{A/V}^2$ and $k_p = 50\,\mu\text{A/V}^2$ a geometry ratio of $(w/l)_n/(w/l)_p = 2.5$. This is shown in the layout of Figure 5.15(b) for drawn dimensions.

In order to ease the figure presentations in the following sections and chapters, the *bulk terminals are usually omitted in the circuit schematics, and in the layouts the n-well*

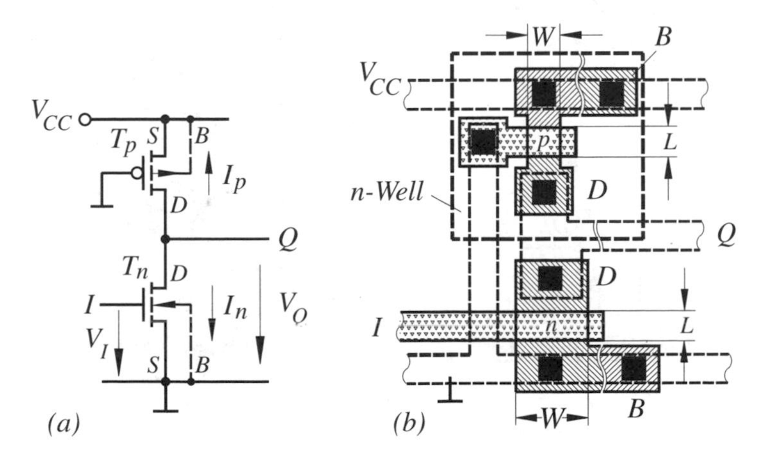

Figure 5.15 (a) PMOS load inverter and (b) layout

and the bulk connections are usually not drawn. Deviation from this simplification occurs only in cases of particular importance.

Power dissipation

The static power dissipation of this inverter is

$$P_{\text{stat}} = I_p V_{CC} S$$
$$= \frac{\beta_p}{2}(-V_{CC} - V_{Tp})^2 V_{CC} S \tag{5.17}$$

This static power dissipation can be almost completely neglected when the gates of the p-channel and n-channel transistors are connected together, as is discussed in the following section in some detail. Then the question arises, what is the use of the presented PMOS load inverter? When a large number of p-channel transistors are connected in series, as is the case in some static logic implementations (Section 6.1.2), it may be necessary to replace the series-connected transistors by one PMOS load in order to cope with the rise time requirement.

5.3.4 CMOS Inverter

The inverters so far looked at are ratioed designs where the logic levels are determined by the geometry ratio between the load and switching transistor. These designs have the drawback of a relatively large static power dissipation. This is not the case for the CMOS inverter which can be designed ratioless and is therefore the most important inverter type (Figure 5.16).

If an *H* signal of V_{CC} is applied to the input of the inverter, this causes the n-channel transistor to be turned on and the p-channel transistor to be turned off, since the voltage between gate and source is zero. An output signal of 0 V results. If an *L* signal is applied to the input the situation reverses. The n-channel transistor is cut off and the p-channel

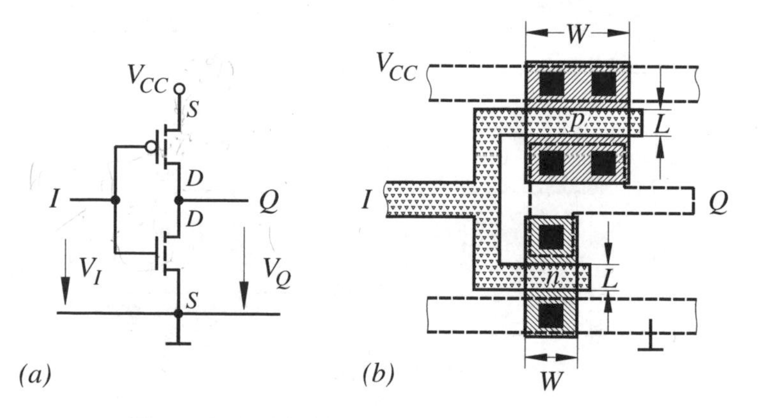

Figure 5.16 (a) CMOS inverter and (b) layout

transistor is turned on, causing an output signal of V_{CC}. Since the output has either an L signal of 0 V or an H signal of V_{CC}, one speaks of an output with a rail to rail swing. This swing is independent of the transistor geometry and therefore ratioless. A further advantage of this inverter is that no body effect influences the performance of the circuit, since source and bulk in each case are connected together. Finally, the input resistance is, as in all the preceding cases, extremely high, as the gates of the MOS transistors are an almost perfect insulator.

The described advantages are the prime reasons why the CMOS technique superseded the NMOS technique, starting as early as 1980.

The CMOS inverter can be designed ratioless. A tremendous freedom in the choice of the transistor w/l dimensions results. This feature can be used, e.g., to achieve a high switching speed or an optimized signal to noise ratio. The latter is a common approach and is considered next. For this purpose the voltage transfer characteristic of a CMOS inverter is considered (Figure 5.17).

If the input signal is increased, starting from 0 V, or decreased, starting from V_{CC}, voltage changes at the output of $\mathrm{d}V_Q/\mathrm{d}V_I = -1$ can be observed at the switching points V_{S1} and V_{S2}. When a symmetrical signal to noise ratio is chosen as a design criterion, this leads to the requirement that $V_{S1} = V_{CC} - V_{S2}$. In this case it is possible to superimpose onto the L signal a noise signal of V_{S1}, before the output changes markedly. An analogous situation exists, when a noise signal of $V_{S2} - V_{CC}$ is superimposed onto the H signal. The choice of a symmetrical signal to noise ratio requires that the current I_p and I_n and therefore the gain factors β_n and β_p are equal, assuming $V_{Tn} \approx -V_{Tp}$. This leads to a geometry relationship of

$$\left(\frac{w}{l}\right)_p = \frac{k_n}{k_p}\left(\frac{w}{l}\right)_n = \frac{\mu_n C'_{ox}}{\mu_p C'_{ox}}\left(\frac{w}{l}\right)_n \tag{5.18}$$

Due to the 2–3 times larger mobility of the electrons compared to the holes, this results in a geometry ratio of

$$\boxed{\left(\frac{w}{l}\right)_p = 2 \text{ to } 3\left(\frac{w}{l}\right)_n} \tag{5.19}$$

which is illustrated in Figure 5.16(b).

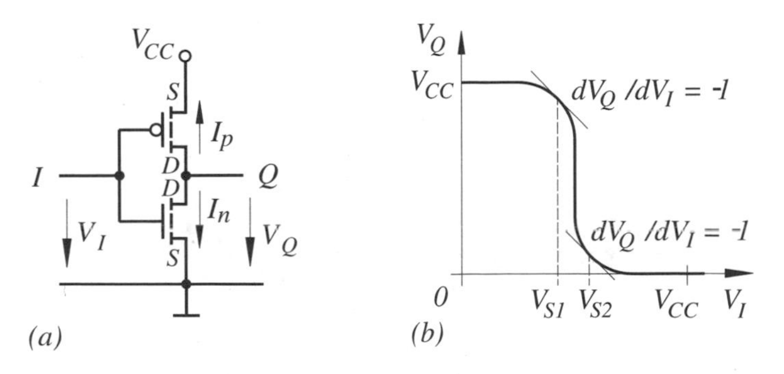

Figure 5.17 (a) CMOS inverter and (b) voltage transfer characteristic

Power dissipation

For this analysis it is useful to divide the total power dissipation into the following three contributions: static, transient, and dynamic power dissipation.

Static power dissipation

If a signal of $V_{IL} = 0\,\text{V}$ is applied to the input of the CMOS inverter (Figure 5.18), then the n-channel transistor is cut off and the p-channel transistor is turned on. With the voltage condition $V_{IL} = V_{GS} = 0\,\text{V}$, a sub-threshold current of (Equation 4.65)

$$I_{DS,n} = \beta_n(n-1)\phi_t^2 e^{(-V_{Tn})/\phi_t n}(1 - e^{-V_{DS}/\phi_t}) \tag{5.20}$$

flows through the n-channel transistor. The smallest current – that is, the sub-threshold current of the n-channel transistor – determines the total current from V_{CC} to ground.

In Figure 5.18(b) the situation is reversed and the p-channel transistor is responsible for the sub-threshold current in analogy to the proceeding equation. This leads to a total static power dissipation of

$$P_{\text{stat}} = V_{CC}[I_{DS,n}S + I_{DS,p}(1 - S)] \tag{5.21}$$

where S describes the on/off ratio of the transistors. If the sub-threshold currents are extremely small, it may be necessary to add the leakage currents or saturation currents (Equation 2.28) of the back-biased junctions to the static power dissipation.

In some applications, such as static MOS memories (Chapter 7), even a small sub-threshold current may still be too large. With the millions of memory cells on one chip, this may add up to an unacceptably high stand-by current. One obvious way to decrease the sub-threshold current is to increase the threshold voltage of critical transistors (Equation 5.20). When, e.g., the threshold voltage is increased by 100 mV, this leads with a sub-threshold swing of 100 mV/decade (Equation 4.70) to a current reduction by a factor of ten. Another way to reduce the sub-threshold current is to use a circuit solution, shown in Figure 5.19. The changed power-supply voltages are responsible for the fact that in the cut-off condition either a $V_{GS,n}$ voltage of $-0.15\,\text{V}$ for the n-channel

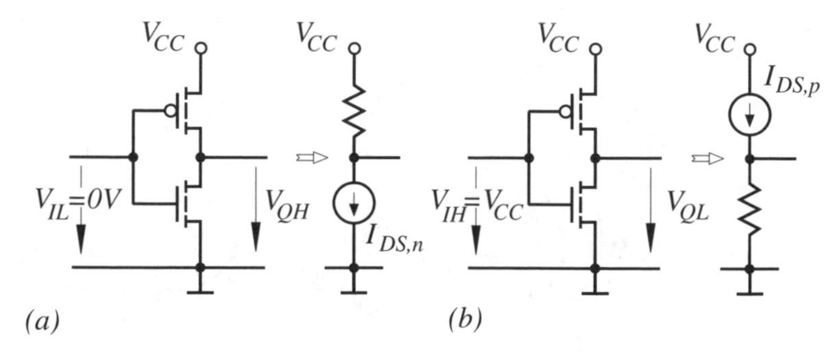

Figure 5.18 CMOS inverter with equivalent circuit: (a) L state with $V_{IL} = 0\,\text{V}$ and (b) H state with $V_{IH} = V_{CC}$

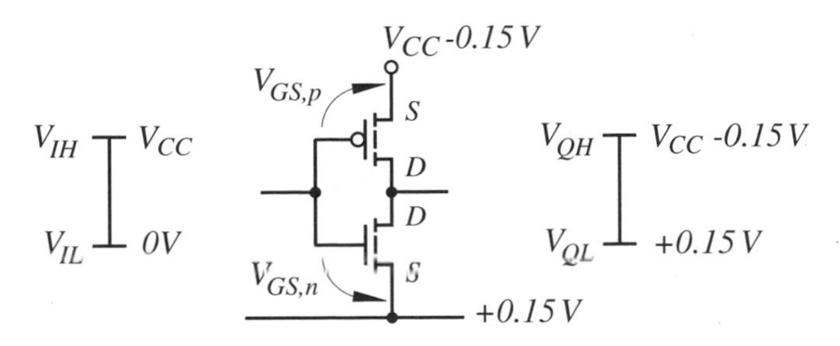

Figure 5.19 CMOS inverter with reduced sub-threshold currents

transistor or a $V_{GS,p}$ voltage of +0.15 V for the p-channel transistor is available. This results in a sub-threshold current reduction by a factor of more than ten. A drawback is the deterioration of the output levels, which is the reason why this approach is only useful in particular situations.

Transient power dissipation

This power dissipation is caused during the transition of the input signal from high to low or vice versa, since both transistors are simultaneously conducting during the transition (Figure 5.20). With $V_I = 0\,\text{V}$ the p-channel transistor is in the resistive region and conducting and the n-channel transistor is cut off. The input signal and the current of the n-channel transistor increase during the time interval t_1 to t_2, whereas the one through the p-channel transistor reduces. In this time interval is $V_Q > (V_I - V_{Tn})$ and the n-channel transistor is in saturation and controls with its smaller current value the total current from V_{CC} to ground

$$I(t) = \frac{\beta_n}{2}(V_I(t) - V_{Tn})^2 \tag{5.22}$$

With the assumption that the inverter is designed symmetrically, that is $\beta_n = \beta_p = \beta$ and $V_{Tn} = -V_{Tp}$, the current approaches its maximum value when the input signal V_I assumes a value of $V_{CC}/2$ (Figure 5.21). Both transistors are in saturation. If the input voltage is increased further the situation reverses. The p-channel transistor remains in saturation, whereas the n-channel transistor moves into the resistive region. In this case

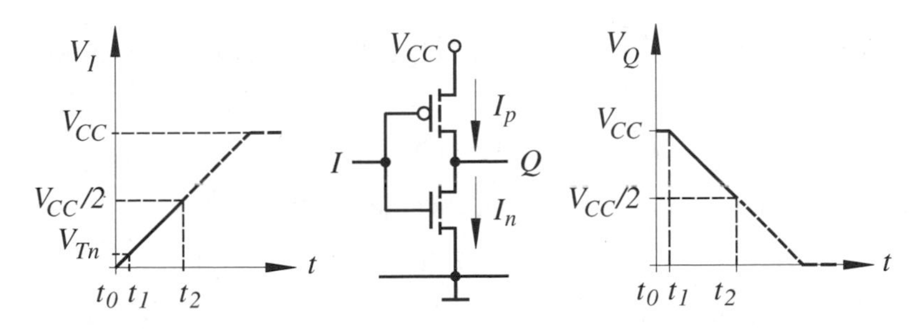

Figure 5.20 *L* to *H* transition at the input of a CMOS inverter

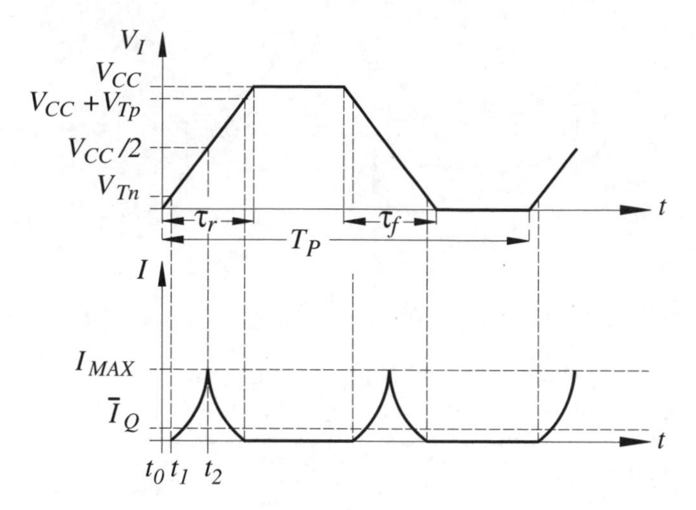

Figure 5.21 Transient current in a CMOS inverter

the reduced current through the p-channel transistor determines the total current from V_{CC} to ground. These transitions yield, during the time period T_P, an average current of

$$\bar{I} = 4\frac{1}{T_P}\int_{t_1}^{t_2} I(t)\,\mathrm{d}t = \frac{2\beta}{T_P}\int_{t_1}^{t_2}(V_I(t) - V_{Tn})^2\,\mathrm{d}t \tag{5.23}$$

Assuming a symmetrical linear input signal with $\tau_r = \tau_f = \tau$

$$V_I(t) = \frac{V_{CC}}{\tau}t \tag{5.24}$$

a current of

$$\begin{aligned}\bar{I} &= \frac{2\beta}{T_P}\int_{t_1}^{t_2}\left(\frac{V_{CC}}{\tau}t - V_{Tn}\right)^2 \mathrm{d}t \\ &= \frac{1}{12}\frac{\beta}{V_{CC}}(V_{CC} - 2V_{Tn})^3\frac{\tau}{T_P}\end{aligned} \tag{5.25}$$

results, where $t_1 = (V_{Tn}/V_{CC})\tau$ and $t_2 = \tau/2$. This leads to a transient power dissipation of

$$\boxed{P_{\mathrm{tr}} = \bar{I}V_{CC} = \frac{\beta}{12}(V_{CC} - 2V_{Tn})^3\frac{\tau}{T_P}} \tag{5.26}$$

This is probably not an unexpected result. It states that the smallest transient power results, when the transition times are short and the gain factors are small.

If the gain factors are very large, as is the case in output driver stages, which are discussed in Section 5.6.2, this leads to transient currents, which are extremely troublesome to cope with.

Dynamic power dissipation

What happens when the output of the CMOS inverter drives a capacitance load? In this case the capacitance is charged by the current of the p-channel transistor and discharged by that of the n-channel transistor (Figure 5.22). These charging currents cause an average power dissipation in the n-channel transistor and p-channel transistor of

$$P_{\text{dyn}} = \frac{1}{T_P}\left[\int_0^{T_P/2} I_n V_Q \,\mathrm{d}t + \int_{T_P/2}^{T_P} I_p (V_{CC} - V_Q)\,\mathrm{d}t\right] \tag{5.27}$$

Hereby it is assumed that the rise and fall times τ_r and τ_f at the output are much smaller than the time period T_P. From this assumption it follows that the capacitance at the output is always completely charged or discharged during a time period. With $I = C_L\,\mathrm{d}V_Q/\mathrm{d}t$ the following currents in the respective time intervals $I_n = -C_L\,\mathrm{d}V_Q/\mathrm{d}t$ and $I_P = C_L\,\mathrm{d}V_Q/\mathrm{d}t$ result. This gives, after solving the integral, a dynamic power dissipation of

$$P_{\text{dyn}} = \frac{C_L}{T_P}\left[-\int_{V_{CC}}^{0} V_Q\,\mathrm{d}V_Q + \int_0^{V_{CC}} (V_{CC} - V_Q)\,\mathrm{d}V_Q\right]$$

$$P_{\text{dyn}} = \frac{C_L}{T_P} V_{CC}^2$$

$$\boxed{P_{\text{dyn}} = C_L f V_{CC}^2} \tag{5.28}$$

where $f = 1/T_P$ describes the clock frequency.

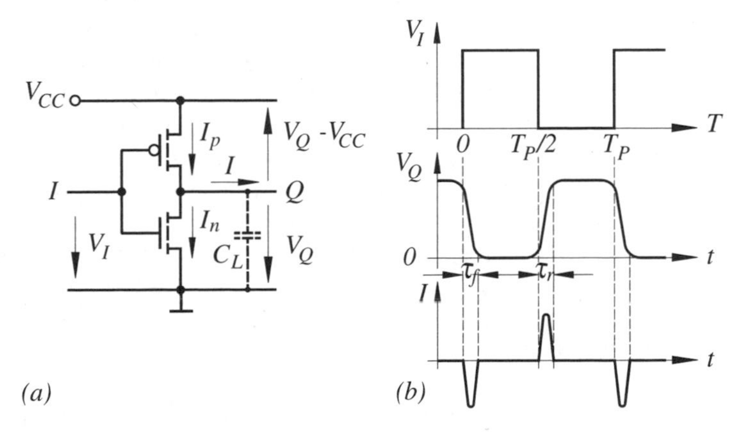

Figure 5.22 (a) CMOS inverter with capacitance loading and (b) time behavior

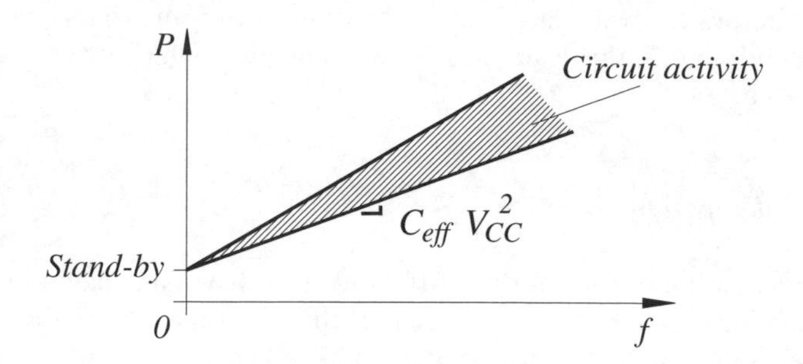

Figure 5.23 Power dissipation of an integrated CMOS circuit

An important observation is that the power dissipation does not depend on any transistor parameter. The reason for this behavior is that the load capacitance is always completely charged and discharged during a time period and therefore the time constants of the transistors have no influence.

The total power dissipation of the CMOS inverter is the sum of the individual contributions

$$P = P_{\text{stat}} + P_{\text{tr}} + P_{\text{dyn}} \tag{5.29}$$

In most designs the dynamic power dissipation is usually dominating as long as the rise and fall times of the input and output signals are comparable.

Note: In the preceding analysis the transient power dissipation is determined as if no capacitance loading exists. In reality this is not the case and the capacitance has to be included in the analysis (Veendrick 1984). The derived transient power dissipation can therefore be considered as a rough estimate only. For an accurate power dissipation prediction a circuit simulation is therefore the most effective way.

Usually, integrated CMOS circuits or systems show a power dissipation as illustrated in Figure 5.23. At zero frequency the system is in stand-by. Only circuits which are absolutely necessary are active, causing a static power dissipation. An increase in frequency causes a rise in power dissipation, due dominantly to the dynamic contribution. Its value depends very much on the number of activated gates in a system. Usually one can assume that not more than 25% of the gates available in an integrated circuit or system are active at a time. The gradient in Figure 5.23 can therefore be considered to represent the effective capacitance C_{eff} switched every clock cycle in a system. With the trend to larger integrated systems and higher clock frequencies, the existing power dissipation problem will worsen in the future. One obvious way to cope with the problem is the reduction of the power-supply voltage, due to its quadratic influence on the power dissipation. Since this leads to a worsening in switching performance, one finds some integrated circuits using two or more different supply voltages to deal with different circuit tasks.

5.3.5 Ratioed Design Issues

In logic circuits transistors are connected in series, e.g. as a *NAND* gate, and in parallel, e.g. as a *NOR* gate. In the following it is analyzed how the sizing of series or parallel connected transistors is done in order to guarantee an appropriate L signal at the output of the circuit configurations. The results of this analysis can then be used in Chapter 6, dealing with logic design issues.

In Figure 5.24 examples of ratioed circuits are shown. The load can be implemented by any one of the presented devices (Figure 5.9). Since the low voltage output condition $V_{QL} \leq V_{Tn}/2$ has to be met in all three configurations, the output resistance of the pull-down transistors or network has to be the same in each case, which simplifies the analysis. The resistance of the pull-down transistor or switching transistor T_S of the inverter is

$$R_S = \frac{V_{QL}}{\beta_S[(V_{IH} - V_{Tn})V_{QL} - V_{QL}^2/2]} \approx \frac{1}{\beta_S[V_{IH} - V_{Tn}]} \tag{5.30}$$

where β_S is the gain factor of the transistor and V_{QL} is assumed to be very small for simplicity reasons. By analogy, the resistance of the m in series-connected transistors (Figure 5.24(b)) is

$$\Sigma R \approx \frac{m}{\beta_n[V_{IH} - V_{Tn}]} \tag{5.31}$$

Since, as mentioned before, in either case the pull-down resistance has to be the same in order to guarantee an appropriate L output signal, the sizing rule for the series-connected transistors of the *NAND* gate is

$$\beta_n = m\beta_S$$

$$\boxed{(w/l)_n = m(w/l)_S} \tag{5.32}$$

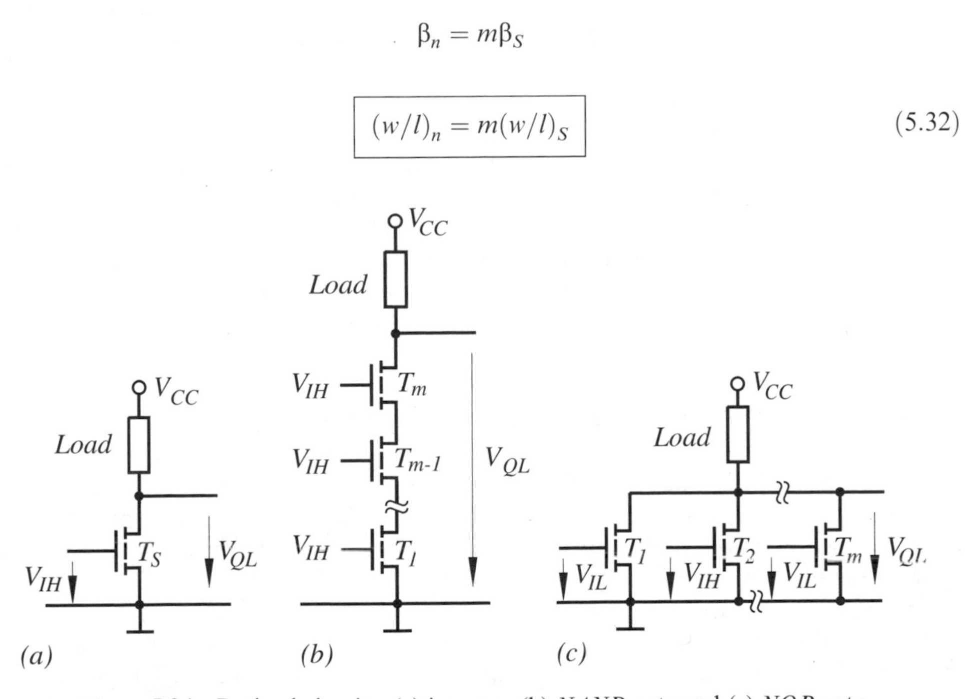

Figure 5.24 Ratioed circuits: (a) inverter, (b) *NAND* gate and (c) *NOR* gate

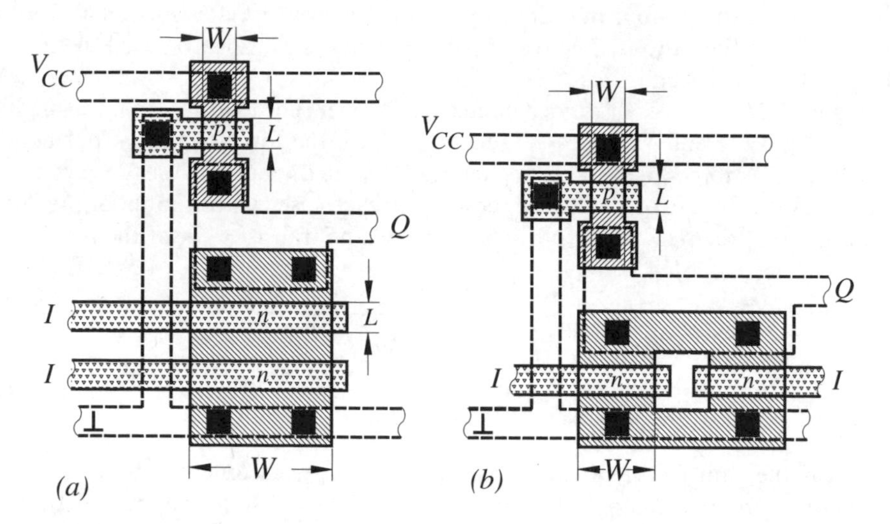

Figure 5.25 Layouts with PMOS loads: (a) *NAND* gate and (b) *NOR* gate

In digital circuits usually the same minimum channel length is used for the transistors. When m transistors are connected in series each individual transistor width must therefore be increased by a factor m, in comparison to the sizing of the switching transistor of the inverter.

Contrary to this situation is that of the parallel connected transistors in the *NOR* gate, which is much more favorable. The worst case exists when one pull-down transistor is conducting only. Therefore the parallel-connected transistors have to be sized like an inverter. The layouts of the analyzed gates using PMOS load devices are shown in Figure 5.25. These layouts illustrate the unfavorable situation of the *NAND* gate. The parasitic capacitances and particularly the gate capacitances increase with the number of transistors connected in series. Therefore in real designs it is very seldom encountered that more than five transistors are connected in series. When this is required by the logic, circuit techniques, such as clocked designs or PMOS loads, may be used (Section 6.1). So far issues of series- and parallel-connected transistors in a ratioed situation are treated. How this is done in ratioless cases is discussed in Section 6.2.5, after the switching performance and some other topics of the inverters are considered.

5.4 SWITCHING PERFORMANCE OF THE INVERTERS

In this section the switching performance and the delay time of the inverters are analyzed. These results, in conjunction with the ones already obtained, can be considered as rules of thumb useful for a first estimate of circuit performance before CAD tools are used.

In general, an output of an inverter drives a capacitance load, since the gate of a following MOS transistor stage is an almost perfect insulator. This is illustrated in the example of Figure 5.26, where two CMOS inverters are connected in series. The effective capacitance at the output Q of the first inverter results from contributions of

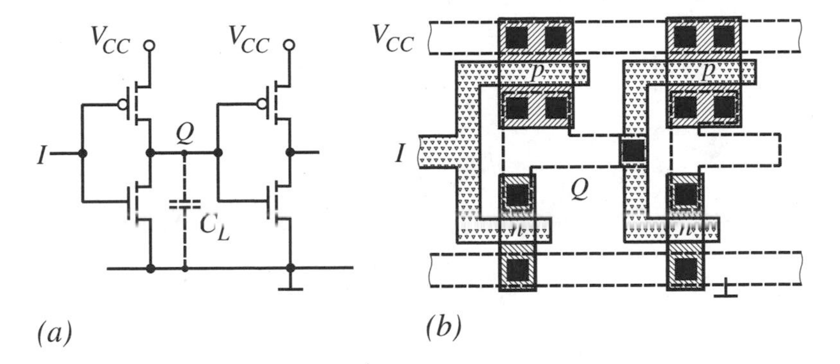

Figure 5.26 (a) Series connected CMOS inverters and (b) corresponding layout

interconnections, overlap areas, and transistor gates. In most designs the nonlinear gate capacitance is dominating. Since, furthermore, the transistor equations are nonlinear, an accurate switching performance prediction of a circuit is only time effective with the aid of a circuit simulation program.

The equations to be derived are approximate ones and not very accurate but, for the following reasons, useful:

(a) to aid in understanding the circuit behavior

(b) to develop a feeling about the switching and delay times of MOS circuits

(c) to come up with a first design estimate before the layout is drawn.

The equations should be easy to use. In order to fulfill this requirement the following assumptions are made:

(a) C_L is an average voltage-independent load capacitance

(b) a step function is applied to the input

(c) resistances in interconnections are neglected.

Load capacitance estimation

In order to come up with a first estimate of the load capacitance, it is useful to revisit the CMOS inverter (Figure 5.27). The input voltage of the inverter is switched from an L to an H state and the effective capacitances of the n-channel transistor are considered first. To ease the derivation and in order to come up with a worst-case estimate, it is assumed that the transistor remains in the resistive region during switching. Under this condition the gate capacitance can be divided into source and drain contributions of 50% each, as shown in Figure 4.64 of Chapter 4. At time $t = 0$ the n-channel transistor is cut off, its gate source capacitance $C_{gs} = C_{ox}/2$ is charged to 0 V and its gate drain capacitance $C_{gd} = C_{ox}/2$ to $V_{GD} = -V_{CC}$ via the p-channel transistor. If the n-channel transistor is

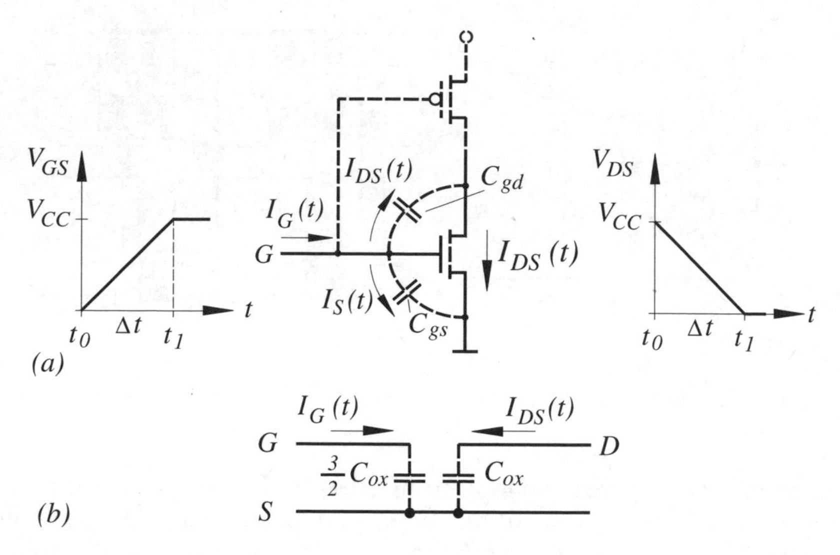

Figure 5.27 (a) Switching behavior of a CMOS inverter and (b) equivalent capacitance model of an n-channel transistor

turned on causes this the C_{gs} capacitance to be charged from 0 V to V_{CC} and the C_{gd} capacitance to be charged to a *different polarity* from $-V_{CC}$ to $+V_{CC}$. The resulting charging currents are for C_{gs}

$$I_S(t) = \frac{C_{ox}}{2}\frac{\mathrm{d}V_{GS}}{\mathrm{d}t} \approx \frac{C_{ox}}{2}\frac{V_{CC}}{\Delta t} \tag{5.33}$$

and for C_{gd}

$$\begin{aligned} I_{DS}(t) &= \frac{C_{ox}}{2}\left(\frac{\mathrm{d}V_{GS}}{\mathrm{d}t} - \frac{\mathrm{d}V_{DS}}{\mathrm{d}t}\right) \approx \frac{C_{ox}}{2}\left[\frac{V_{CC}}{\Delta t} - \left(-\frac{V_{CC}}{\Delta t}\right)\right] \\ &\approx C_{ox}\frac{V_{CC}}{\Delta t} \end{aligned} \tag{5.34}$$

This leads to a total charging current at the gate of

$$I_G(t) = I_S(t) + I_{DS}(t) \approx \frac{3}{2}C_{ox}\frac{V_{CC}}{\Delta t} \tag{5.35}$$

and accordingly to the equivalent capacitance model of the n-channel transistor shown in Figure 5.27(b).

Is this model, used for the series-connected CMOS inverters, a loading situation as shown in Figure 5.28 results where the effective loading at note Q is

$$C_L \approx \frac{5}{2}(C_{ox,n} + C_{ox,p}) \tag{5.36}$$

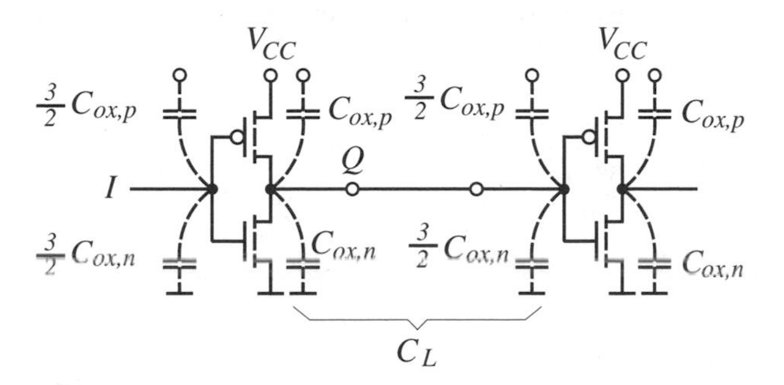

Figure 5.28 Effective capacitance loading of series-connected CMOS inverters

If the p-channel transistor geometry is chosen to be twice as large as that of the n-channel transistor (Equation 5.19), a capacitance loading of

$$\boxed{C_L \approx 7.5C_{ox}} \tag{5.37}$$

results.

Switching performance of the CMOS inverter

The switching behavior is illustrated in Figure 5.29. If the input signal V_I changes abruptly from 0 V to V_{CC}, this causes the p-channel transistor to be cut off and the n-channel transistor to be conducting, discharging capacitance C_L. At a value of 0.1 V_{CC} it can be assumed that a following stage interprets this value as an L signal. Up to a voltage of $V_{DS} = V_{CC} - V_{Tn}$ the n-channel transistor is in the saturation and subsequently in the resistive region. Because of these two operation conditions the fall time t_f has to be divided into two time intervals. With the current of the

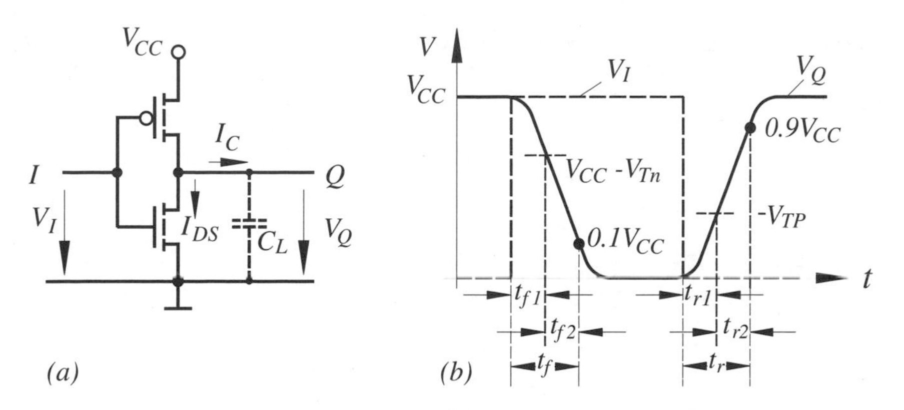

Figure 5.29 (a) CMOS inverter and (b) switching behavior

n-channel transistor I_{DS} being equal to that discharging the capacitance I_C, the first time interval

$$I_C = -I_{DS}$$
$$C_L \frac{dV_Q}{dt} = -\frac{\beta_n}{2}(V_{CC} - V_{Tn})^2$$
$$\int dt = \int_{V_{CC}}^{V_{CC}-V_{Tn}} \frac{2C_L}{-\beta_n(V_{CC} - V_{Tn})^2} dV_Q$$
$$t_{f1} = \frac{2C_L V_{Tn}}{\beta_n(V_{CC} - V_{Tn})^2} \tag{5.38}$$

can be found. A similar derivation yields the second time interval

$$C_L \frac{dV_Q}{dt} = -\beta_n[(V_{CC} - V_{Tn})V_Q - V_Q^2/2]$$
$$\int dt = -\frac{C_L}{\beta_n} \int_{V_{CC}-V_{Tn}}^{0.1V_{CC}} \frac{1}{(V_{CC} - V_{Tn})V_Q - V_Q^2/2} dV_Q$$
$$t_{f2} = \frac{C_L}{\beta_n} \frac{1}{V_{CC} - V_{Tn}} \ln \frac{1.9V_{CC} - 2V_{Tn}}{0.1V_{CC}} \tag{5.39}$$

Adding the two, a total fall time of

$$t_f = t_{f1} + t_{f2} = \frac{C_L}{\beta_n} \frac{1}{V_{CC} - V_{Tn}} \left(\frac{2V_{Tn}}{V_{CC} - V_{Tn}} + \ln \frac{1.9V_{CC} - 2V_{Tn}}{0.1V_{CC}} \right) \tag{5.40}$$

results. In analogy to this derivation, the rise time determined by the p-channel transistor can be derived

$$t_r = t_{r1} + t_{r2} = \frac{C_L}{\beta_p} \frac{1}{V_{CC} + V_{Tp}} \left(\frac{-2V_{Tp}}{V_{CC} + V_{Tp}} + \ln \frac{1.9V_{CC} + 2V_{Tp}}{0.1V_{CC}} \right) \tag{5.41}$$

These equations can be modified into easy to use ones, when the typical values of $V_{CC} = 3\,\text{V}$ and $V_{Tn} = -V_{Tp} = 0.45\,\text{V}$ are substituted. This leads to the simplifications

$$\boxed{t_f = \frac{C_L}{\beta_n} 1.2(1/V)} \tag{5.42}$$

and

$$\boxed{t_r = \frac{C_L}{\beta_p} 1.2(1/V)} \tag{5.43}$$

These equations are probably not surprising. They state that the rise and fall times are proportional to the capacitance loading and that these times can be reduced by increasing the gain factors of the transistors.

Delay time of the CMOS inverter

The delay time of the inverter is caused by the charging and discharging of all parasitic capacitances. In the preceding analysis a step function is applied to the input of the inverter. In reality the input is driven by a signal which behaves similarly to that considered at the output. The delay between input and output signal can be approximated for the charging and discharging case by $t_{dr} \approx t_r/2$ and $t_{df} \approx t_f/2$. This leads to the following average delay time of the CMOS inverter

$$t_d \approx \frac{1}{2}\left(t_{dr} + t_{df}\right) \approx \frac{1}{4}(t_r + t_f) \tag{5.44}$$

useful for first hand estimations. With respect to the already used values of $V_{Tn} = -V_{Tp} = 0.45\,\mathrm{V}$ and $V_{CC} = 3\,\mathrm{V}$, a delay time of

$$t_d \approx C_L\left(\frac{1}{\beta_p} + \frac{1}{\beta_n}\right)0.3(1/V) \tag{5.45}$$

results. This equation shows an interesting feature, when the load capacitance is expressed by the transistor geometries (Equation 5.36)

$$C_L \approx \frac{5}{2}((w \cdot l)_n + (w \cdot l)_p)C'_{ox} \tag{5.46}$$

Substituting Equation (5.46) into Equation (5.45) results in an average delay of

$$t_d \approx \frac{5}{2}((w \cdot l)_n + (w \cdot l)_p)\left(\frac{1}{(w/l)_p\,\mu_p} + \frac{1}{(w/l)_n\,\mu_n}\right)0.3(1/V) \tag{5.47}$$

or in one of

$$t_d \approx 2l^2\left(\frac{1}{2\mu_p} + \frac{1}{\mu_n}\right)(1/V) \tag{5.48}$$

when a geometry ratio of $(w/l)_p = 2 \cdot (w/l)_n$ is used. This interesting result states that the delay time is independent of the oxide capacitance C'_{ox} and the gate width. This makes sense when one keeps in mind that an increase in oxide capacitance or gate width results in an increased current and simultaneously in an increased load capacitance.

A prerequisite for this statement is that the gate capacitances are dominating. A further important observation is that the delay time is proportional to the square of the channel length. This is one reason why designers of digital circuits usually use a minimum channel length.

Switching performance of the PMOS load inverter

The charging of the capacitance is caused, as in the previous case, by the current $I_C = I_p$ of the p-channel transistor (Figure 5.30) and the discharging of the capacitance by the difference between the currents $I_C = I_p - I_n$. Since, according to Equation (5.16) $\beta_n = Z \cdot \beta_p$ it is possible to neglect the current of the p-channel transistor in comparison to that of the n-channel transistor. This results in a discharging current of $I_C \approx -I_n$. With this assumption the switching behavior of the PMOS load inverter is similar to that of the CMOS inverter. If it is further assumed that $V_{Tn}/2 \approx 0.1 \cdot V_{CC}$ the results of Equations (5.42) and (5.43)

$$t_f = \frac{C_L}{\beta_n} 1.2(1/V)$$

$$t_r = \frac{C_L}{\beta_p} 1.2(1/V)$$

can be applied directly to the PMOS load inverter. This gives a ratio between rise and fall time of

$$\boxed{\frac{t_r}{t_f} = \frac{\beta_n}{\beta_p} = Z} \tag{5.49}$$

since $\beta_n = Z \cdot \beta_p$. This means that the rise time is slower than the fall time by the sizing parameter factor Z.

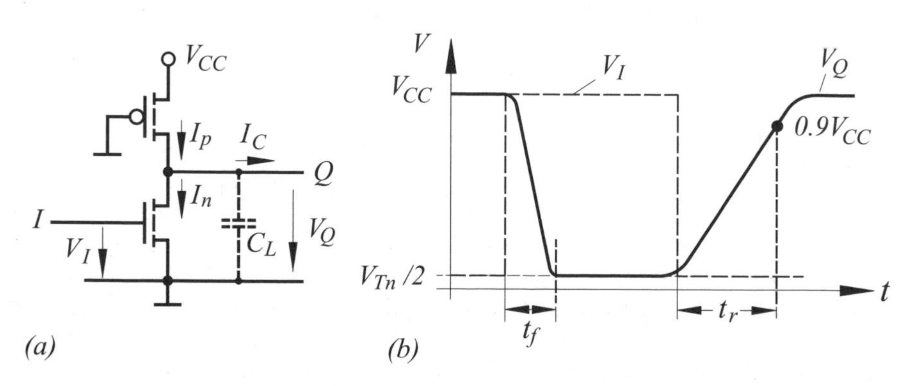

Figure 5.30 (a) PMOS load inverter and (b) switching behavior

Switching performance of the enhancement load inverter

This inverter is, just as in the preceding case, a ratioed one (Figure 5.31). The discharging occurs by the current difference of $I_C = I_2 - I_1$. Since in this ratioed inverter $\beta_1 = Z \cdot \beta_2$ is applicable (Equation 5.11), the discharging current can be approximated by $I_C \approx -I_1$. When one disregards the difference in the H levels, the results of Equation (5.40) or (5.42) can be used directly. But a considerable difference exists in the charging of the capacitance. The n-channel enhancement transistor is always in saturation. This leads to

$$I_C = I_2$$
$$C_L \frac{\mathrm{d}V_Q}{\mathrm{d}t} = \frac{\beta_{n,2}}{2}(V_{CC} - V_Q - V_{Tn,2})^2 \tag{5.50}$$

and after separating the variables and integration to a rise time of

$$t_r = \frac{2C_L}{\beta_{n,2}}\left[\frac{1}{0.1(V_{CC} - V_{Tn})} - \frac{1}{V_{CC} - (3/2)V_{Tn}}\right] \tag{5.51}$$

where $V_{Tn,1} = V_{Tn,2} = V_{Tn}$ is used, and as integration limits, the voltages $V_{Tn}/2$ and $V_Q = 0.9 \cdot (V_{CC} - V_{Tn})$ are applied. At this point it is useful to remember that the H level is reduced by a threshold voltage below V_{CC} (Equation 5.12). With the typical values of $V_{CC} = 3\,\mathrm{V}$ and $V_{Tn} = 0.45\,\mathrm{V}$, a rise time of

$$\boxed{t_r = \frac{C_L}{\beta_{n,2}} 7(1/V)} \tag{5.52}$$

results. The rise time is substantially slower compared to all the previous cases. The reason for this poor performance is that the gate source voltage of the load transistor $V_{GS} = V_{CC} - V_Q(t)$ is not constant, but reduces during charging until the transistor cuts off at $V_{GS} = V_{Tn,2}$. The enhancement inverter thus has two drawbacks, namely a reduced H voltage level and a substantially increased rise time. This inverter is therefore,

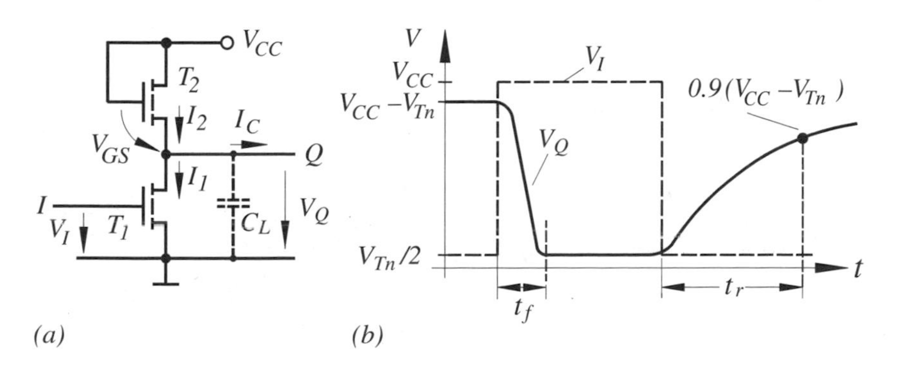

Figure 5.31 (a) Enhancement load inverter and (b) switching behavior

in general, not used in designs. But the results of the inverter analysis can be applied to similar situations useful in the following chapters.

Switching performance of the depletion load inverter

From the beginning of the rise time up to a voltage of $V_{DS,2} = V_{CC} - V_Q = -V_{Tn,2}$, the depletion transistor remains in saturation and shifts at a voltage of $V_Q = V_{CC} + V_{Tn,2}$ into the resistive region (Figure 5.32). Since the charging current I_2 is equal to that of the capacitance I_C, the following relationships lead to the first time interval

$$C_L \frac{dV_Q}{dt} = \frac{\beta_{n,2}}{2}(-V_{Tn,2})^2$$

$$\int dt = \frac{2C_L}{\beta_{n,2}(-V_{Tn,2})^2} \int_{V_{Tn,1}/2}^{V_{CC}+V_{Tn,2}} dV_Q$$

$$t_{r1} = \frac{2C_L}{\beta_{n,2}(-V_{Tn,2})^2}(V_{CC} + V_{Tn,2} - V_{Tn,1}/2) \tag{5.53}$$

In the second time interval

$$C_L \frac{dV_Q}{dt} = \beta_{n,2}\left[(-V_{Tn,2})(V_{CC} - V_Q) - \frac{(V_{CC} - V_Q)^2}{2}\right] \tag{5.54}$$

is valid, which results after separation of variables and integration in a time of

$$t_{r2} = \frac{C_L}{\beta_{n,2} V_{Tn,2}} \ln\left[\frac{0.1 V_{CC}}{-0.1 V_{CC} - 2V_{Tn,2}}\right] \tag{5.55}$$

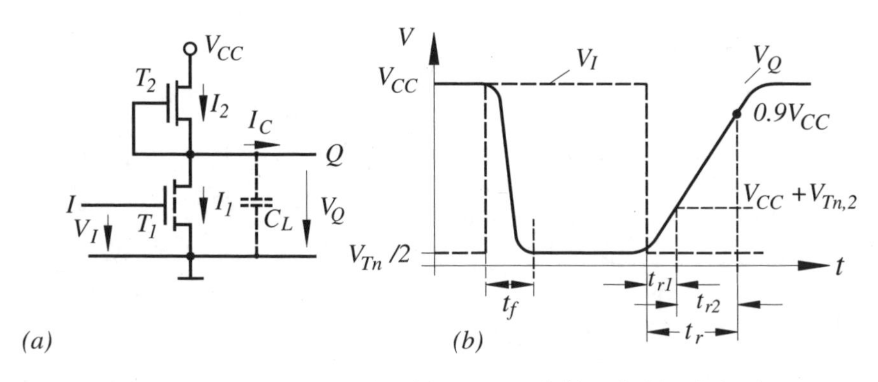

Figure 5.32 (a) Depletion load inverter and (b) switching behavior

Table 5.4 Comparison of characteristic MOS inverter data

Inverter	Inverter ratio Z	Rise time	Fall time
CMOS	–	$\frac{C_L}{\beta_p}1.2(1/V)$	$\frac{C_L}{\beta_n}1.2(1/V)$
P-load	5.9	$\frac{C_L}{\beta_p}1.2(1/V)$	$\frac{C_L}{\beta_n}1.2(1/V)$
Enhancement load	4.9	$\frac{C_L}{\beta_{n,2}}7(1/V)$	$\frac{C_L}{\beta_{n,1}}1.2(1/V)$

Conditions: $V_{CC} = 3\,\text{V}$; $V_{IH} = 3\,\text{V}$; $V_{Tn} = 0.45\,\text{V}$; $V_{Tp} = -0.45\,\text{V}$.

The total rise time is therefore

$$t_r = \frac{C_L}{\beta_{n,2}}\left[\frac{2(V_{CC} + V_{Tn,2} - V_{Tn,1}/2)}{(-V_{Tn,2})^2} + \frac{1}{V_{Tn,2}}\ln\frac{0.1V_{CC}}{-0.1V_{CC} - 2V_{Tn,2}}\right] \tag{5.56}$$

With the typical values for an NMOS depletion technique with $V_{Tn,1} = 0.8\,\text{V}$, $V_{Tn,2} = -3.5\,\text{V}$, and $V_{CC} = 5\,\text{V}$, the following simplified rise time expression results

$$\boxed{t_r = \frac{C_L}{\beta_{n,2}}0.9(1/V)} \tag{5.57}$$

The fall time of the inverter is comparable to that of the PMOS load inverter or CMOS inverter described by Equation (5.42).

A summary of the most important results of the MOS inverter analysis is shown in Table 5.4, with the exception of the NMOS inverter, which is used in special applications only.

5.5 BUFFER STAGES

Buffer stages are used in integrated circuits to drive relatively large parasitic on-chip capacitances, which are present in conjunction, e.g., with clock and data lines or output stages. The buffers or drivers which are analyzed can be divided into super buffers and bootstrap ones.

5.5.1 Super Buffer

The CMOS inverter shown in Figure 5.33(a) is supposed to drive a large capacitance C_L. This results in a substantial delay time of the circuit, which might not be acceptable from a system point of view. Increasing the current gain of the transistors by making the geometry ratios $(w/l)_n$ and $(w/l)_p$ larger may not necessarily lead to the required

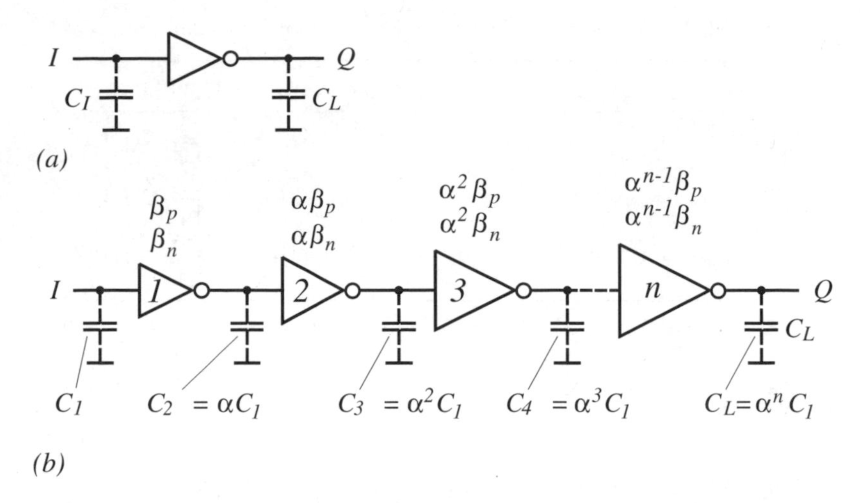

Figure 5.33 (a) Single CMOS inverter and (b) cascaded CMOS inverters

reduction in the delay time, since the input capacitance of the inverter (Equation 5.36) increases also. In the extreme case the value of the input capacitance may even be of the some order as the load capacitance. Cascaded CMOS inverters (Figure 5.33(b)) with staggered geometry ratios, called super buffers, are a solution (Deschacht *et al.* 1988; Lin 1975). The first inverter has a relatively small input capacitance of C_1, which corresponds to the geometry dimensions of the input transistors. This inverter drives a second one with a (w/l) ratio of the transistors α times larger than that of the first inverter. This results in an increased input capacitance of $C_2 = \alpha \cdot C_1$. The second inverter in turn drives a third one with also an α times larger geometry ratio and larger input capacitance of $C_3 = \alpha \cdot C_2 = \alpha^2 \cdot C_1$, and so on until the nth inverter drives the load capacitance C_L. The question arises as to how many inverters are needed and what capacitance ratio

$$\alpha = \frac{C_{(N+1)}}{C_{(N)}} \tag{5.58}$$

is required in order to achieve a minimum delay time. If identical inverters are cascaded, as shown in Figure 5.26, each inverter has an identical delay time of t_d. If staggered geometry ratios are used, the delay of each inverter increases to

$$t'_d = \alpha t_d \tag{5.59}$$

This results in a total inverter chain delay time of

$$T_d = nt'_d = n\alpha t_d \tag{5.60}$$

With the load capacitance (Figure 5.33(b)) given by

$$C_L = \alpha^n C_1 \tag{5.61}$$

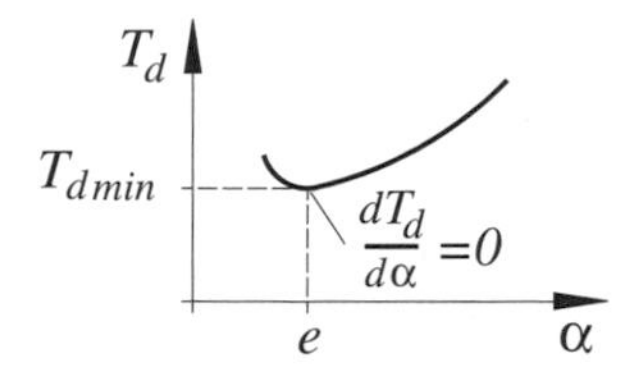

Figure 5.34 Delay time of cascaded inverters as a function of the capacitance ratio α

this leads to the relationship of

$$T_d = \frac{\alpha}{\ln \alpha} t_d \ln \frac{C_L}{C_1} \tag{5.62}$$

which is sketched in Figure 5.34.

A minimum delay time of

$$T_{d\,\min} = e t_d \ln \frac{C_L}{C_1} \tag{5.63}$$

results at $\mathrm{d}T_d/\mathrm{d}\alpha = 0$ with $\alpha = e$, the Euler number. In this case the required number of inverters

$$\boxed{n = \ln \frac{C_L}{C_1}} \tag{5.64}$$

can be found directly from Equation (5.61).

An example is used to demonstrate the improvement when cascaded inverters are employed.

Example

A buffer has to drive a clock line of $C_L = 5\,\text{pF}$ and the input capacitance of the buffer should not be larger than $C_1 = 100\,\text{fF}$. Determine the delay times: (a) when one inverter is used only and (b) when cascaded inverters are employed.

The data are: $C'_{ox} = 4.3\,\text{fF}/\mu\text{m}^2$, $k_n = 100\,\mu\text{A/V}^2$, $k_p = 40\,\mu\text{A/V}^2$, channel length $l = 0.35\,\mu\text{m}$.

One inverter only

The starting point is the input capacitance of a CMOS inverter, which has, according to Figure 5.28, a value of

$$C_1 = \frac{3}{2}(C_{ox,n} + C_{ox,p}) = \frac{3}{2}C'_{ox}\left((w \cdot l)_n + (w \cdot l)_p\right)$$

This equation simplifies to

$$C_1 = 4.5C'_{ox}(w \cdot l)_n$$

when a geometry ratio of $(w/l)_p = 2 \cdot (w/l)_n$ is used. Since the channel length is fixed at 0.35 μm, this leads to the following width of the transistor

$$w = \frac{100\,\text{fF}}{4.5 \cdot 4.3 \cdot \text{fF}/\mu\text{m}^2 \cdot 0.35\,\mu\text{m}} = 14.76\,\mu\text{m}$$

and in conjunction to a current gain factor of

$$\beta_n = \left(\frac{w}{l}\right)_n k_n = \frac{14.76\,\mu\text{m}}{0.35\,\mu\text{m}} 100\,\mu\text{A}/\text{V}^2 = 4.2\,\text{mA}/\text{V}^2$$

The p-channel transistor with twice as large a width has a gain factor of

$$\beta_p = \left(2\frac{w}{l}\right)_p k_p = 2 \cdot \frac{14.76\,\mu\text{m}}{0.35\,\mu\text{m}} \cdot 40\,\mu\text{A}/\text{V}^2 = 3.4\,\text{mA}/\text{V}^2$$

With a total load capacitance of 5 pf, this results in an inverter delay (Equation 5.45) of

$$t_d \approx C_L\left(\frac{1}{\beta_p} + \frac{1}{\beta_n}\right)0.3(1/V) = 800\,\text{ps}$$

Super buffer

The number of required inverter stages needed can be determined directly from Equation (5.64)

$$n = \ln\frac{C_L}{C_1} = \ln\frac{5000\,\text{fF}}{100\,\text{fF}} \approx 4$$

This leads to a delay time (Equation 5.60) of

$$T_{d\,\min} = net_d$$

To solve this equation, the delay time t_d of cascaded identical stages is needed. According to Figure 5.28 and Equation (5.36), the load capacitance has a value of

$$C_L = \frac{5}{2}(C_{ox,n} + C_{ox,p})$$

The input capacitance is known

$$C_1 = \frac{3}{2}(C_{ox,n} + C_{ox,p}) = 100\,\text{fF}$$

and therefore the load capacitance of 166 fF also. This results in a delay time, according to Equation (5.45), of

$$t_d \approx C_L\left(\frac{1}{\beta_p}+\frac{1}{\beta_n}\right)0.3(1/V) = 26.5\,\text{ps}$$

This value applied to the super buffer leads to a total delay time of

$$T_{d\,\min} = e \cdot t_d \cdot n = 288\,\text{ps}$$

As the numbers demonstrate, the super buffer shows an approximately three times shorter delay time in comparison to the single inverter.

5.5.2 Bootstrap Buffer

The bootstrap technique enables a voltage multiplication on chip (Figure 5.35), which can be used advantageously in the realization of various bootstrap buffers. With a clock signal of $\phi = 0\,\text{V}$, the capacitor is charged to $V_{A1} = V_{CC} - V_{Tn}$. This voltage increases to $V_{A2} = 2V_{CC} - V_{Tn}$ when the clock signal changes from 0 V to V_{CC}. Since in this time interval the V_{A2} voltage is larger than V_{CC}, the drain and source functions of the transistor are reversed (indicated in brackets), causing the transistor with $V_{GS} = 0\,\text{V}$ to cut off. As a result, the capacitor remains charged when the small sub-threshold current of the transistor is neglected. Usually, a parasitic capacitance C_P exists at the output of the circuit, causing a charge sharing

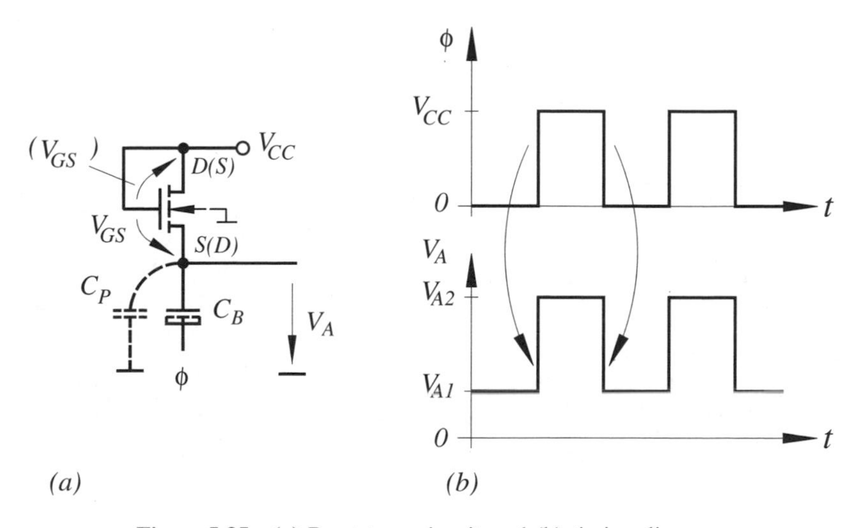

Figure 5.35 (a) Bootstrap circuit and (b) timing diagram

between C_B and C_P, when the clock signal is raised. This reduces the bootstrap voltage to

$$\begin{aligned} V_{A2} &= V_{CC} - V_{Tn} + V_{CC}\frac{C_B}{C_B + C_P} \\ &= V_{CC}\left(1 + \frac{C_B}{C_B + C_P}\right) - V_{Tn} \end{aligned} \tag{5.65}$$

The capacitor C_B is very often implemented by a MOS transistor (Figure 5.35), with source and drain tied together. In this configuration, the relatively large gate capacitance C_{ox} (Figure 4.55) is effective.

As an example a buffer is shown using the bootstrap principle (Figure 5.36). In this buffer the transistor T_3 and the capacitor C_B implement the bootstrap function and the transistors T_1 and T_2 act as output buffer. At time $t < t_1$, the input signal has a voltage of $V_I = V_{CC}$, causing the output voltage V_Q to have a value close to zero. The capacitor is charged via transistor T_3 to a value of $V_{CC} - V_{Tn}$. If the input signal at $t > t_1$ changes from V_{CC} to 0 V, this causes transistor T_1 to be cut off and the capacitance C_L to be charged. The output voltage V_Q rises and with it the voltage V_A at *(1)*. The V_A voltage approaches a maximum value of $2 \cdot V_{CC} - V_{Tn}$. The capacitor, in effect, acts as a built-in voltage source, with a value of $V_{CC} - V_{Tn}$ and therefore as a constant gate source voltage of transistor T_2. As a result, the driving capability of T_2 is improved, charging the load capacitance C_L to the full voltage level of V_{CC}. In order to reduce the charge sharing effect between C_P and C_B, the capacitance value of C_B should be much larger than the parasitic capacitance value of C_P. This is usually not difficult to attain, since a large part of the bootstrap capacitance is made up of the gate source capacitance of transistor T_2.

The presented bootstrap buffer is not ideal, since a static power dissipation occurs, when an input *H* signal is applied. But the buffer is a good example of the application of the bootstrap principle. Improved versions will be discussed in conjunction with word line buffers or drivers in Section 7.5.2.

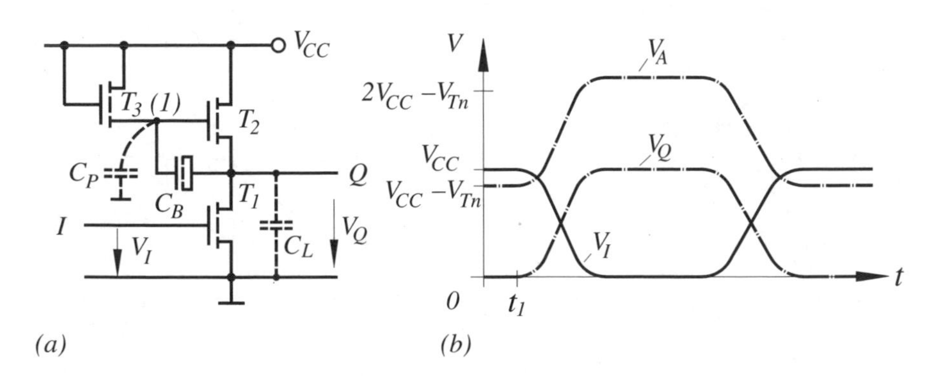

Figure 5.36 (a) Bootstrap buffer and (b) timing behavior

5.6 INPUT/OUTPUT STAGE

All integrated circuits have input and output stages. What requirements these circuits have to meet and what problems are encountered in the design is the topic of this section.

It is useful to start with a discussion about some common interface specifications. As an example the specifications of the JEDEC Standard No. 8 – A for the LVTTL (Low Voltage Transistor Transistor Logic) and for the LVCMOS (Low Voltage CMOS) interfaces are used (Figure 5.37).

5.6.1 Input Stage

The input stages have to be designed in such a way that they operate with LVCMOS or LVTTL levels. In the first case this is not a problem, since the logic levels have an almost rail to rail swing, causing the input stage to operate in a push–pull mode without a static current (Figure 5.38(a)). This situation is completely different when LVTTL levels are used. A V_{IHMIN} signal of 2.0 V causes a static current (Figure 5.38(a))

$$I_0 = \frac{\beta_p}{2}(V_{GS,p} - V_{Tp})^2 = \frac{\beta_P}{2}(V_{\mathrm{IHMIN}} - V_{\mathrm{CCMAX}} - V_{Tp})^2 \tag{5.66}$$

to flow, since both transistors are conducting. This results in an undesirable increase in power dissipation when many input stages are available on a chip.

In order to guarantee an appropriate output level for this case, a ratioed design is required, similar to that of the PMOS load inverter. A drawback of a ratioed design is the relatively large transistor width required, leading to larger parasitic capacitances and therefore to a slower switching performance.

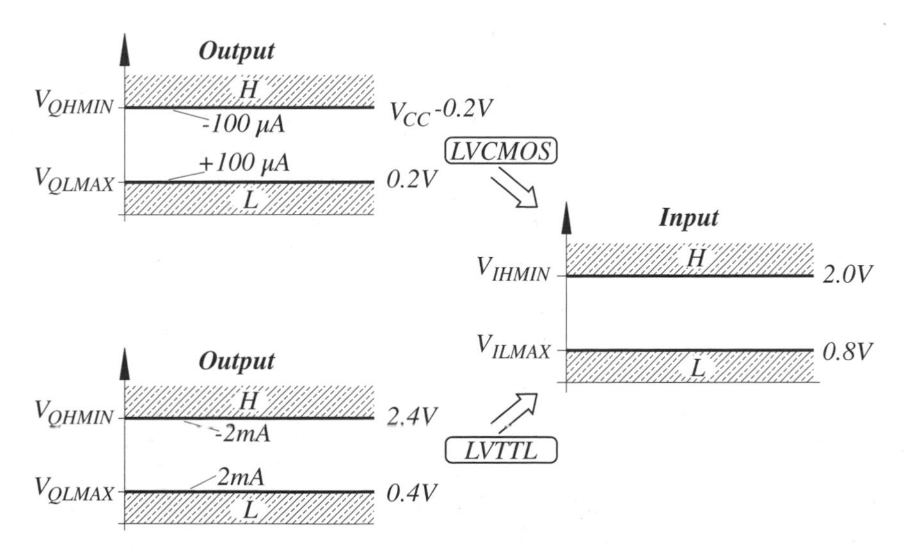

Figure 5.37 LVCMOS and LVTTL specification for $V_{CC} = 3.3\,\mathrm{V} \pm 10\%$

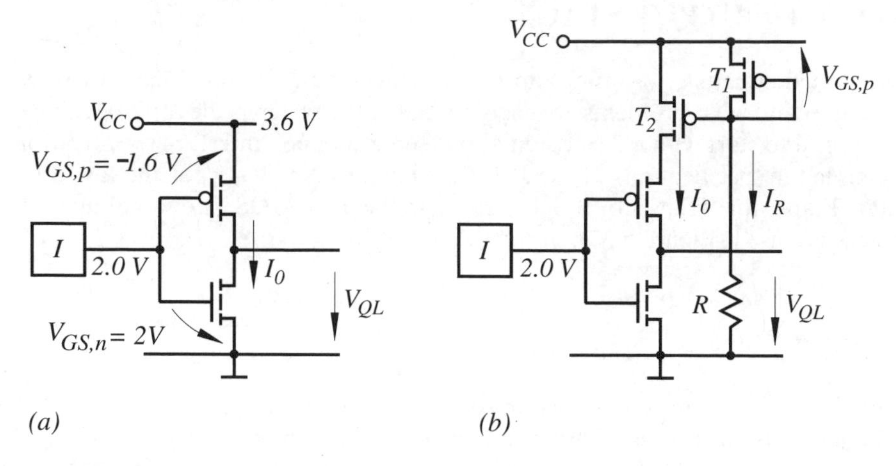

Figure 5.38 Input stages: (a) CMOS inverter and (b) CMOS inverter with current mirror

In some applications it is necessary to limit the I_0 current variation caused by the power-supply voltage and the H input level. This can be achieved with the input stage of Figure 5.38(b). Transistors T_1 and T_2 are in saturation, causing the following currents to flow when the channel length modulation (Section 4.5.2) is neglected

$$I_R = \frac{\beta_{p,1}}{2}(V_{GS,p} - V_{Tp})^2 \tag{5.67}$$

$$I_0 = \frac{\beta_{p,2}}{2}(V_{GS,p} - V_{Tp})^2 \tag{5.68}$$

These two equations lead to a current ratio or a so-called current mirroring of

$$\frac{I_0}{I_R} = \frac{(w/l)_{p,2}}{(w/l)_{p,1}} \tag{5.69}$$

which depends on the geometry ratio only. The current I_R has a value of

$$I_R = \frac{V_{CC} - V_{GS,p}}{R} = \frac{V_{CC} - \sqrt{\frac{2I_R}{\beta_{p,1}}} + V_{Tp}}{R} \tag{5.70}$$

How good the circuit is in reality depends mainly on the manufacturing tolerance of the resistor R.

Schmitt trigger

This is a very useful input circuit for cases where slow varying input signals in a noisy environment are present (Figure 5.39). When the input signal, starting from an L level, approaches the switching threshold voltage V_{SH}, this causes the output voltage to

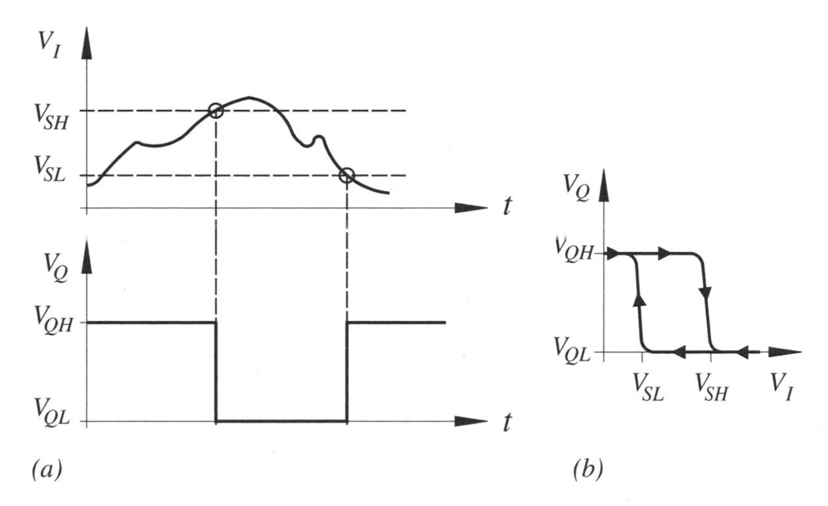

Figure 5.39 (a) Input and output signal of a Schmitt trigger and (b) voltage transfer characteristic

change from H to L. Contrary is the situation when the input signal starting at an H level approaches the switching threshold voltage V_{SL}, which causes the output voltage to change from L to H. A circuit with this performance has to have a voltage hysteresis characteristic defined by the voltage difference between the two switching threshold voltages (Figure 5.39(b)). The aim of the Schmitt trigger is thus to suppress noise and, at the same time, cause a fast change in logic states at the output. A circuit with this performance is shown in Figure 5.40 (Ohtomo and Nogawa 1995). Basically the circuit consists of a CMOS inverter with an additional n-channel and p-channel transistor connected in series. The voltage between transistors T_1 and T_3 and between T_6 and T_4 is supplied during switching by the transistors T_2 and T_5. In order to explain the circuit,

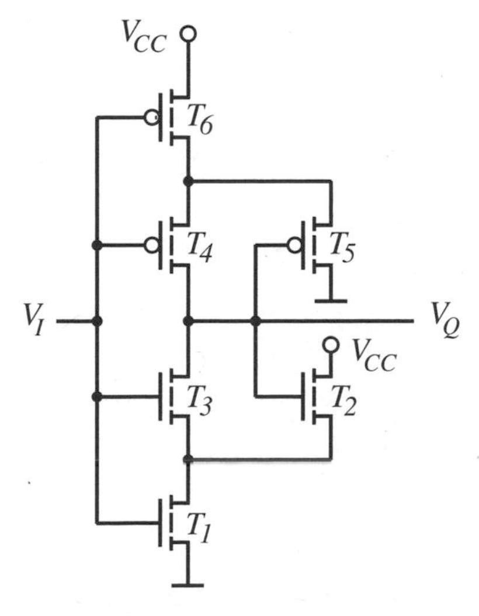

Figure 5.40 Schmitt trigger circuit

the illustration of Figure 5.41(a) is used. The V_I voltage has a value of 0 V and the output one of $V_Q = V_{CC}$. Then the V_I voltage starts to rise. As long as this voltage is smaller than the threshold voltage $V_{Tn,1}$ of transistor T_1 the source of transistor T_2 has a value of $V_S = V_{CC} - V_{Tn,2}$. If the V_I voltage increases further, T_1 starts to conduct, which causes the V_S voltage to decrease. The switching threshold V_{SH} is approached when T_3 is at the onset of conduction. In this case the input voltage has a value of

$$V_I = V_{SH} = V_S + V_{Tn,3} \tag{5.71}$$

Increasing V_I further causes T_3 to conduct and the output voltage to decrease. This enhances the decrease of the V_S voltage, since T_2 becomes less conducting. Finally T_2 is cut off and T_1 and T_3 are turned on.

If the V_I voltage approaches the switching threshold of Equation (5.71), T_3 is at the onset of conduction and approximately the same current flows through T_2 and T_1. This condition, together with Equation (5.71), can be used to determine the sizing of transistors T_1 and T_2.

$$\begin{aligned} I_1 &= I_2 \\ \frac{\beta_1}{2}(V_{SH} - V_{Tn,1})^2 &= \frac{\beta_2}{2}(V_{CC} - V_S - V_{Tn,2})^2 \\ \frac{\beta_1}{\beta_2} = \frac{(w/l)_1}{(w/l)_2} &= \left(\frac{V_{CC} - V_{SH}}{V_{SH} - V_{Tn,1}}\right)^2 \end{aligned} \tag{5.72}$$

In this equation it is considered that T_2 and T_3 have the same threshold voltage, since the sources of both transistors are connected to a common note.

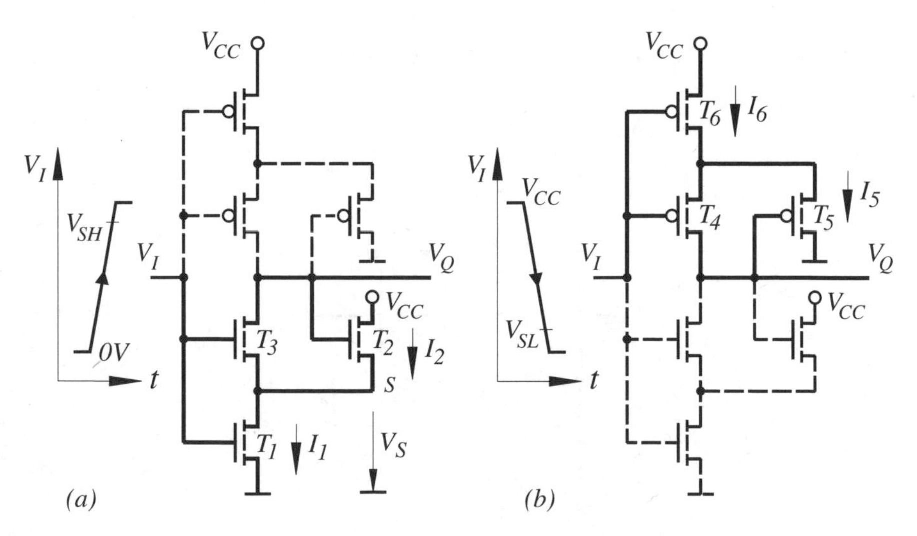

Figure 5.41 Schmitt trigger: (a) input voltage approaches V_{SH} and (b) input voltage approaches V_{SL}

A similar analysis can be performed for the switching threshold V_{SL}. The p-channel transistors of the circuit (Figure 5.41(b)) are responsible for this threshold. The input signal has a value of V_{CC}, causing an output voltage of $V_Q = 0\,\text{V}$. If the input voltage is reduced to the value of the switching threshold V_{SL}, an almost identical current flows through T_5 and T_6. This condition leads to a gain and geometry ratio for these transistors of

$$\frac{\beta_6}{\beta_5} = \frac{(w/l)_6}{(w/l)_5} = \left(\frac{V_{SL}}{V_{CC} - V_{SL} + V_{Tp,6}}\right)^2 \tag{5.73}$$

Since the transistors T_3 and T_4 operate basically as switching devices, their gain factors should be in the order of

$$\begin{aligned} \beta_3 &\geq 8\beta_1 \\ \beta_4 &\geq 8\beta_6 \end{aligned} \tag{5.74}$$

5.6.2 Output Stage

Buses are used in all kinds of systems. A bus connects sending and receiving integrated circuits, such as memories, processors, and controllers, on a printed circuit board. In order to guarantee the data integrity, only one integrated circuit is allowed to send at a time. This can be achieved by putting the outputs of the remaining circuits via a chip select signal CS in a high impedance state called tri-state.

If a data bus is designed for an LVTTL interface, as shown in Figure 5.42, the output and input stages of the integrated circuits have to fulfill the requirement of the specification, indicated in Figure 5.37. Additionally, the capacitance of the data bus, of up to 30 pF, has to be charged or discharged in a particular time. This leads to maximum

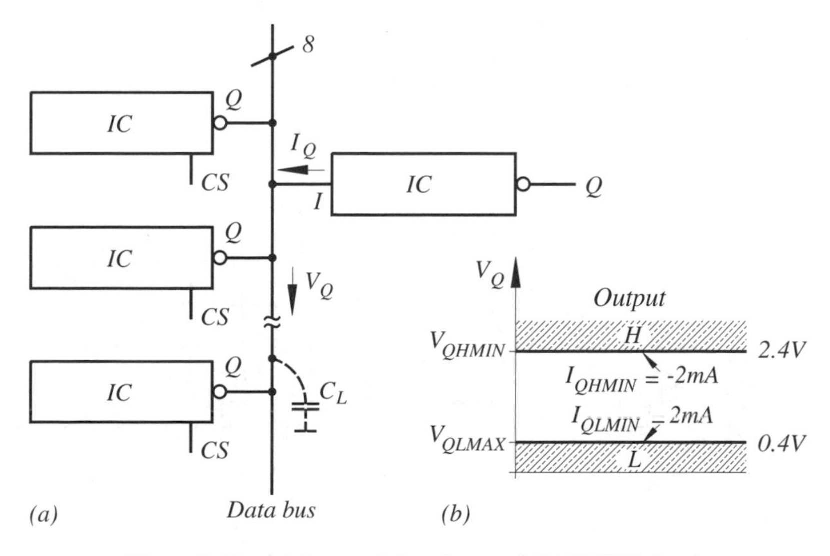

Figure 5.42 (a) External data bus and (b) LVTTL levels

clock frequencies in the order of 100 MHz, where the wave character of the data bus may still be neglected. As a consequence of the mentioned requirements, the output buffers usually have very large (w/l) ratios of about 800 (Figure 5.43). The diffusion areas are connected periodically to metal lines in order to reduce the source and drain resistances.

In Figure 5.44 a tri-state output buffer is shown. For this buffer it is not only important that output changes occur very fast, but also that the chip select signal CS leads the buffer in and out of the high impedance state very quickly. The chip select signal is in the H state and the buffer is selected. Transistors T_5 and T_4 are conducting, whereas T_3 and T_6 are cut off. In this situation, the gates of T_2 and T_1 are connected as well as the drains of T_7 and T_8. The circuit behaves like two in-series-connected CMOS inverters. The contrary situation is when the chip select signal is in the L state (Figure 5.44(b)). Transistors T_3 and T_6 are conducting and T_4 and T_5 are cut off. This causes the gates of the output transistors T_1 and T_2 to be connected to ground and V_{CC}, respectively, independent of the information applied to the input I. The output transistors are in a high resistive state.

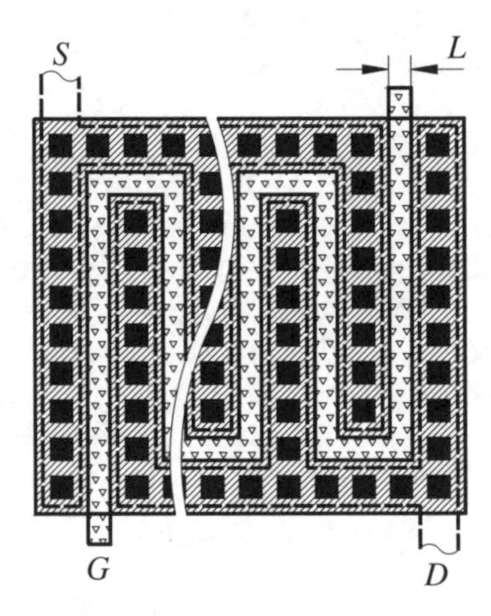

Figure 5.43 Layout of a transistor with large geometry ratio (W average meanderwidth)

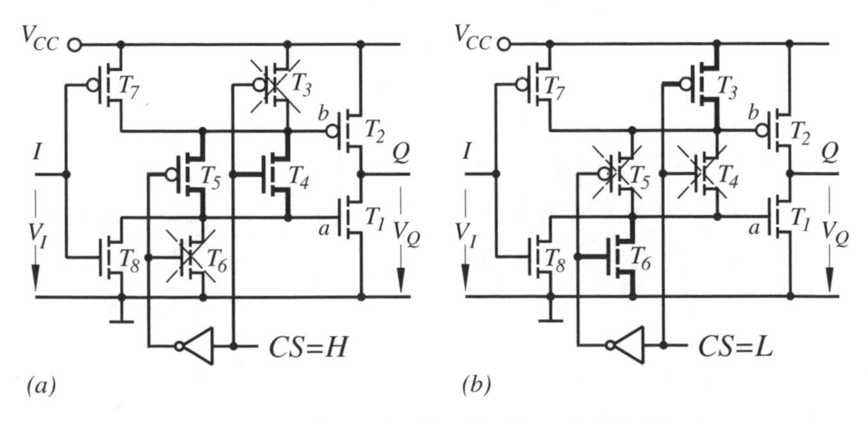

Figure 5.44 Tri-state buffer: (a) $CS = H$, (b) $CS = L$

High resistive state means that the transistors are in the sub-threshold region, and a sub-threshold current occurs, which is largest at elevated temperature (Section 4.4.3). This current is usually not allowed to be larger than 10 μA. This seems to be high, but one has to keep in mind that the output transistors have a very large (w/l) ratio. In order to meet this specification it is recommended not to use a minimum channel length for these transistors, due to short channel length effect (Figure 4.39). This makes sure that an unintended reduction in threshold voltage, and with it an increase in sub-threshold current, does not occur with process variations.

How those currents can be measured is shown in Figure 5.45. In case (a) the voltage across the p-channel transistor T_2 is zero and the transistor neutralized. Thus, only the sub-threshold current through the n-channel transistor is measured by the current meter. In case (b) the situation is reversed.

The charging and discharging of a large output capacitance C_L causes large currents to flow. These may be able to generate such large switching noise on the supply line or at the data outputs that the performance of the total circuit is severely impaired. In order to avoid such a scenario, rise and fall times at the output are not allowed to fall below a particular switching time (Shoji 1986). In order to explain the situation, the charging and discharging of a capacitance C_L is considered in more detail (Figure 5.46). Included in this illustration are the bond wires between the integrated circuit pads and the package pins and the resistance of the power-supply lines. This configuration leads to an undesirable relationship between the current change $\mathrm{d}I/\mathrm{d}t$ at the output and the switching noise. To analyze this situation, the current behavior at the output during switching is considered (Figure 5.47).

During the rise time t_r the current increases first and approaches a maximum I_{MAX} and then starts to decrease, when the capacitance is substantially charged. A similar situation exists during the fall time t_f. If one assumes that the current change, shown in Figure 5.47, can be described by

$$\left|\frac{\mathrm{d}I}{\mathrm{d}t}\right|_{\mathrm{MAX}} \approx \frac{I_{\mathrm{MAX}}}{t_S/2} \tag{5.75}$$

with

$$Q_L \approx I_{\mathrm{MAX}} \cdot t_S/2 = C_L V_{CC} \tag{5.76}$$

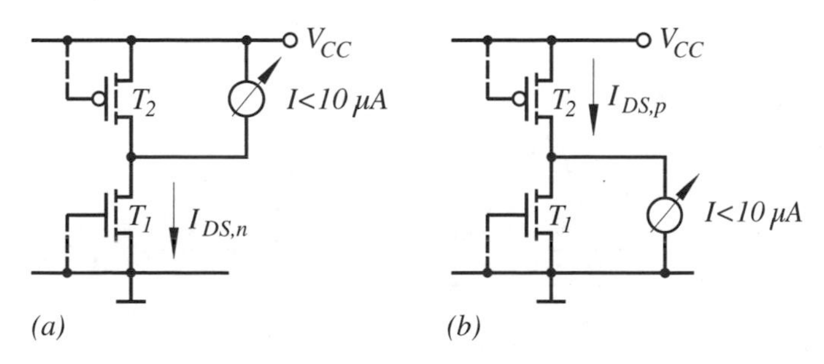

Figure 5.45 Output buffer transistors in tri-state: (a) $I_{DS,n}$ measurement and (b) $I_{DS,p}$ measurement

this leads to the relationship

$$\boxed{\left|\frac{\mathrm{d}I}{\mathrm{d}t}\right|_{\mathrm{MAX}} \approx \frac{4C_L V_{CC}}{t_S^2}} \tag{5.77}$$

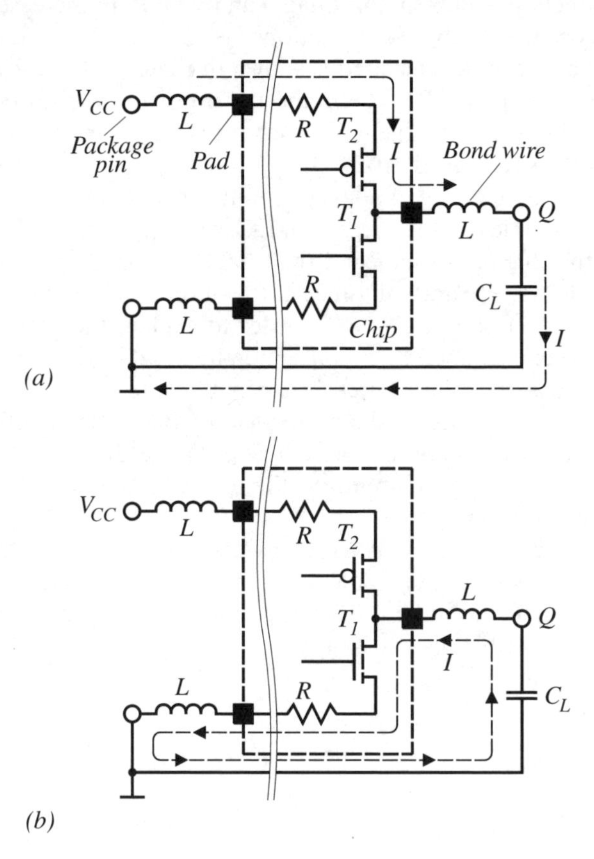

Figure 5.46 Equivalent circuit of an *IC* after contacting: (a) charging of C_L and (b) discharging of C_L

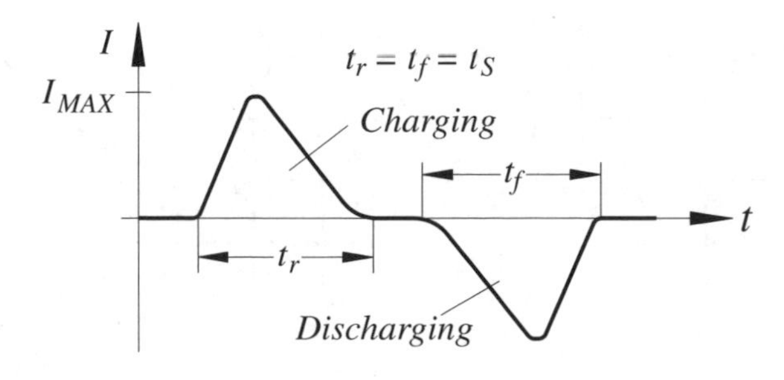

Figure 5.47 Current behavior at the output during charging and discharging of C_L

This current change causes switching noise voltages at the inductance of the bond wires $V_L = L \cdot dI/dt$ and the resistance of the supply lines $V = R \cdot I$. The problem is that the current change is inversely proportional to the square of the switching time t_S, which limits the rise and fall time of the output stage severely, as is demonstrated in the following example.

Example

The data of the schematic, shown in Figure 5.46, are: $C_L = 30\,\text{pF}$, $t_S = 2\,\text{ns}$, $V_{CC} = 3.3\,\text{V}$, and a bond wire inductance L of 5 nH. This results during switching in a voltage across one wire only of 0.5 V.

In the preceding analysis it is assumed that the output buffer has no transient current caused by the simultaneous conduction of the output transistors during switching (Section 5.3.4). Unfortunately this effect leads to additional switching noise, particularly when large output transistors are involved. One efficient way to control the transient current or slew rate is to split the output buffer into different parts and connect the individual circuits by resistors (Figure 5.48). The slew rate can then be adjusted by an appropriate choice of resistor values. Compared to the standard solution a reduction of the switching noise by about 50% is possible at the expense of a slight increase in delay time (Senthinathan and Prince 1993).

The described situation is aggravated when many output buffers are available on a chip and are switched simultaneously. With the following compromises an easing of the problem is possible:

(a) extending the slew rate of all buffers

(b) adopting different delays to each buffer so that a simultaneous switching is not possible

(c) providing each buffer with its own V_{CC} and ground pad

(d) using a novel interface with reduced voltage swing (Nakase *et al.* 1999).

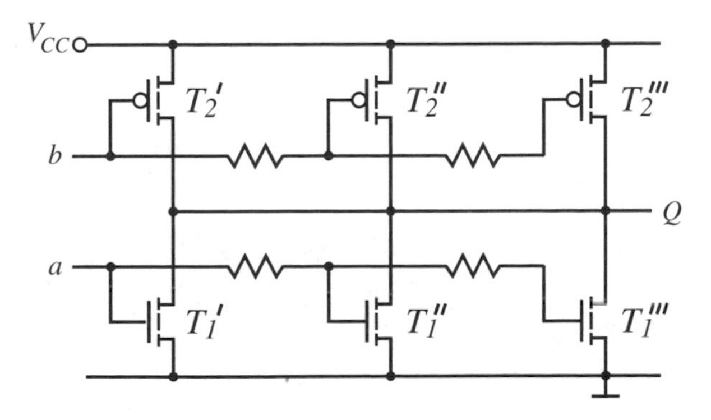

Figure 5.48 Split output buffer transistors with adjusted slew rate (*a* and *b* connections shown in Figure 5.44)

Additionally to the previously suggested proposals, the packages have to have a minimum of parasitic inductances. Some examples of available plastic packages are shown in Figure 5.49.

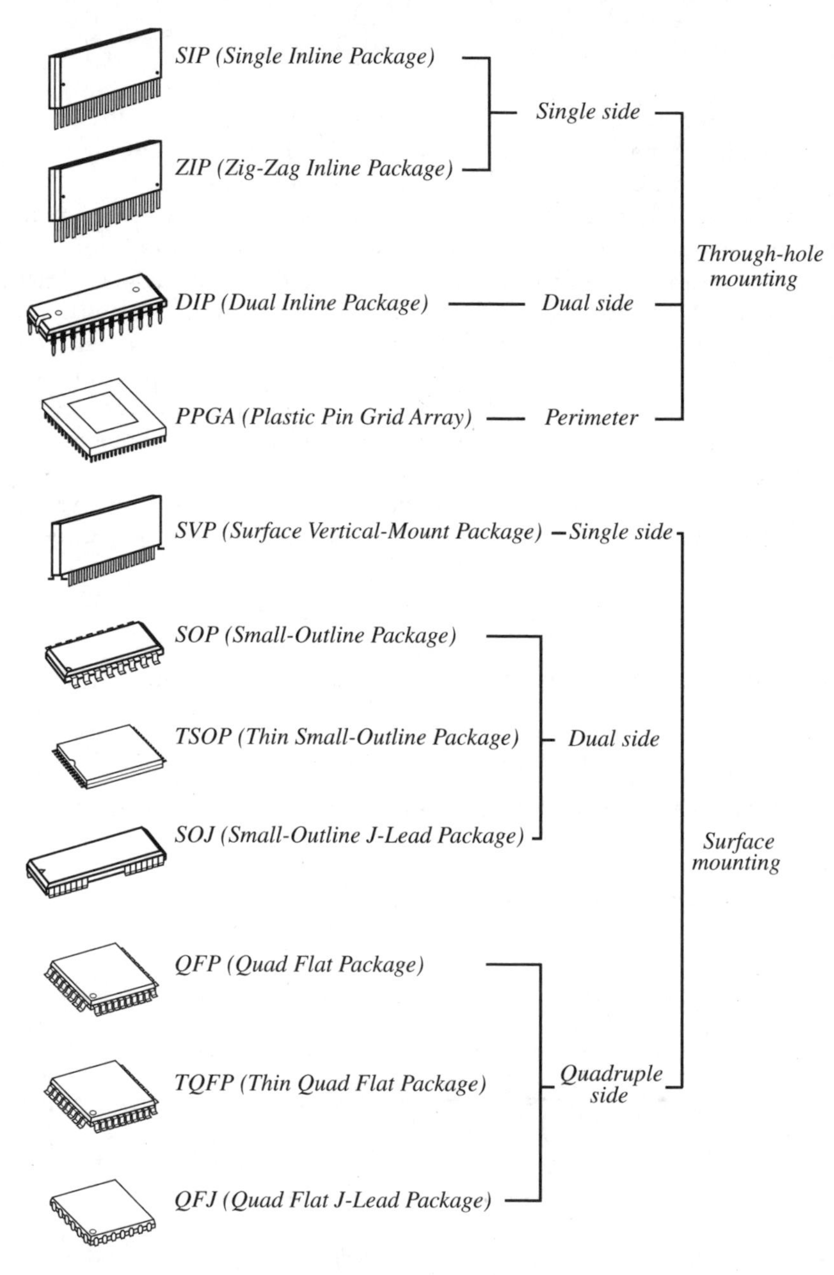

Figure 5.49 Examples of plastic packages

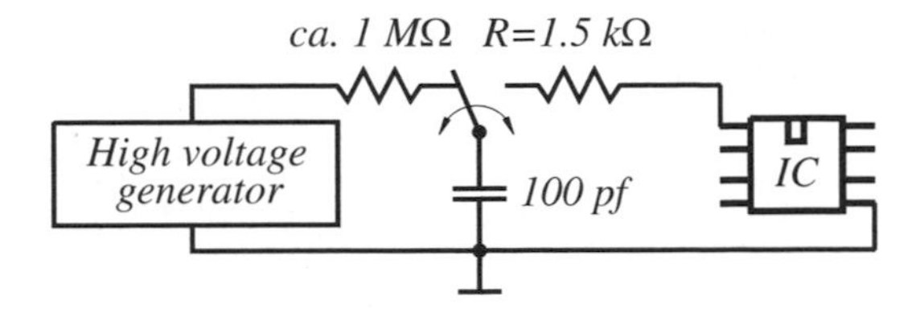

Figure 5.50 Sketch of an HBM test set-up

5.6.3 ESD Protection

Electro-static discharge (ESD) at a semiconductor can lead, particularly at input and output stages, to reliability problems or, even worse, to damage and failure of the integrated circuit. Three models simulate what can happen in reality.

Human body model (HBM)

This model simulates the discharge of a semiconductor by a human (Figure 5.50). The charge of the capacitor corresponds to that of a charged person. If the switch is connected to the circuit, this causes a discharge of the capacitor via the integrated circuit (IC). The resistor of 1.5 kΩ simulates the finger of a person touching the IC. The high voltage value, in the following called ESD voltage, which the IC can withstand without damage is a measure of the ESD protection. In practice, current changes from 5 to 20 A in a time of 50–200 ns have been observed (Tucker 1968).

Machine model (MM)

Additionally to the discharge of an IC by a person, this can happen when the IC comes into contact with a machine, e.g. at the printed circuit board (PCB) assembly. In this case, the possibility of damaging the IC is even larger since no body resistance is involved. Current changes of up to 10 A in 8 ns have been observed. This can be simulated by the HBM set-up, except that the resistor has a value close to zero and the capacitor one of about 200 pf.

Charged device model (CDM)

The HBM and MM models simulate situations when a person or machine is charged and discharging takes place via an IC. The charged device model accounts for a different situation, where the device is charged. This can happen, e.g., by the removal of a device from the packaging or wrapping. The following contact of the device with a grounded object can then lead to an undesirable discharge. Observed current changes are in the order of about 60 A in 0.25 ns. A possible test set-up is shown in Figure 5.51, where the sketched capacitances indicate the charge storage of the device with respect to the environment.

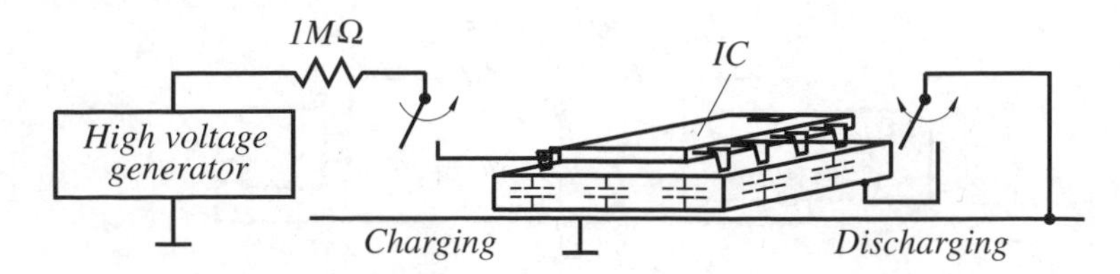

Figure 5.51 Sketch of a possible CDM test set-up

For protection against ESD, the IC needs protection circuits at the pads. It is necessary to distinguish between those pads that are connected to gates and thus to input stages, and those connected to diffusion regions, like output buffers and power-supply lines. Whereas the ones connected to the gates definitely require a protection circuit, for the rest of the terminals this is only necessary where the diffusion region cannot withstand the ESD stress. A typical input protection circuit is shown in Figure 5.52. If a negative ESD voltage is applied between the input pad I and ground pad (GND), this causes a discharge via the forward biased diodes. Since the voltage drop across the diodes is very small, and thus the dissipated power, a critical situation does not exist. This is different when a positive ESD voltage is applied between the terminals. In this case the ESD protection has to be such that the positive logic levels are not affected, but a reliable discharge is guaranteed. This can be achieved with a two-stage protection. The first one uses, e.g., a FOX device similar to the one shown in Figure 4.22. Due to the thick oxide this device is robust and moves up a particular voltage into the snap-back region (Section 4.5.6). In this condition the voltage V_{SP} (Figure 4.49) across the device is smaller than in the case of an avalanche break-down. But the voltage is still too large to guarantee a safe protection of the input stage. For this reason, a series resistor R_S is provided in conjunction with a second voltage limitation. For this limitation a transistor with standard gate oxide is used, which is also driven into snap-back. As an alternative a transistor with a small gate length, operating in the punch-through region (Section 4.5.6), may be employed.

So far only the ESD between the input and GND terminals is considered. But in order to guarantee sufficient ESD protection, it has to be possible to apply the ESD voltage between any two terminals. What can happen is illustrated in Figure 5.53. At the input pad I a positive ESD voltage with respect to the V_{CC} pad is applied. This causes

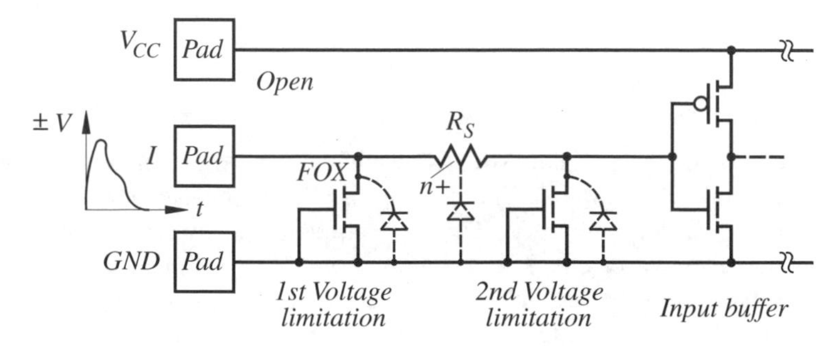

Figure 5.52 Two-stage input protection circuit

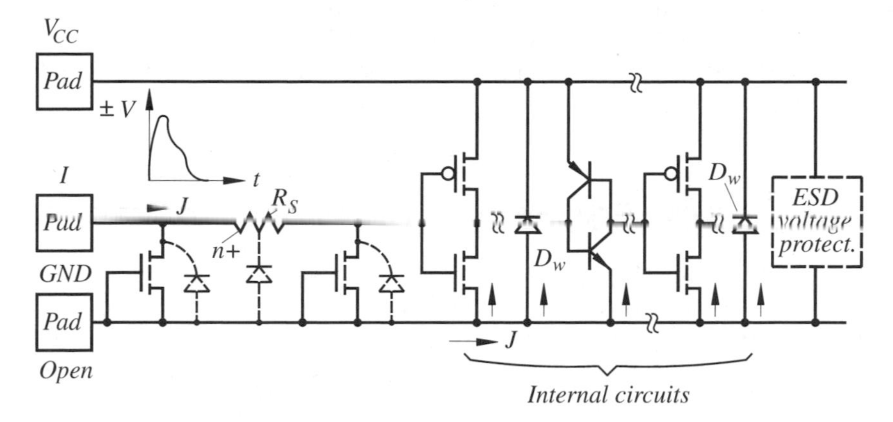

Figure 5.53 Possible ESD current paths in an integrated circuit

a discharge via the input protection circuit and the forward biased diode D_W of the n-well junction (Figure 5.14). Since the dissipated power at the diode is small, the ESD constellation is not critical. The situation is very much different when a negative ESD voltage is applied between the input pad I and the V_{CC} pad. The diodes of the input protection circuit are forward biased and the D_W diode backward. Thus the voltage across this diode is large and may lead to undesirable discharging paths in the circuit (arrows in Figure 5.53) and possible destruction of the circuit. Even a latch-up effect (Section 4.5.7) may occur and damage the circuit. The probability of destruction can be reduced when the ESD voltage across the power-supply is limited by a protection circuit to an acceptable level (Figures 5.53 and 5.54).

A negative ESD voltage with respect to the V_{CC} pad is applied either to GND or to the input pad, or a positive ESD voltage with respect to GND or to the input pad is applied to the V_{CC} pad. This voltage causes transistors T_2 and T_3 to be conducting, since the capacitor C is initially not charged. Since T_3 has a very large (w/l) ratio of about 3000, the ESD voltage will be limited by this transistor. Simultaneously the capacitor is charged via the resistor R. Finally, the voltage at the capacitor approaches a value where T_1 is conducting and T_3 is cut off. The RC time constant is chosen on the one hand to such a value that the ESD protection is guaranteed during the ESD voltage

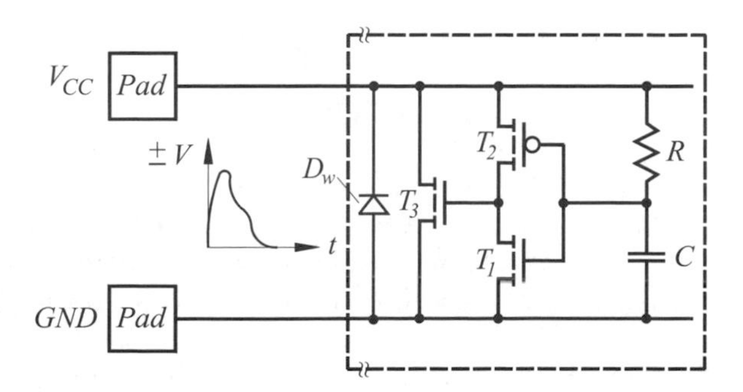

Figure 5.54 ESD voltage protection circuit between power-supply rails (after Ker *et al.* 2000; © 2000 IEEE)

decay of $t < 200$ ns, and, on the other hand, not to interfere with the turn-on of the power-supply voltage $t \gg 1$ μs. According to Ker *et al.* (2000) the ESD voltages can be raised to 6000–8000 V for the human body model and to approximately 400 V for the machine model.

Summary of the most important results

At the beginning of this chapter geometric and electrical design rules are presented. These design rules are prerequisite for all designs. Inverters are chosen for the discussion of basic sizing rules. In order to cover all possible constellations inverters with depletion, enhancement, and PMOS loads are considered. These inverters require a ratioed design style, since a direct current path exists when an H signal is applied to the input. This is different at the CMOS inverter, where the sizing of the transistors is optimized for a maximum signal to noise ratio. That is, the (w/l) ratio of the p-channel transistor is about two times as large as that of the n-channel transistor. In most practical cases, the dominant power dissipation of the CMOS inverter is $P = CV^2f$ and caused by the charging and discharging of capacitances. The analysis of the switching performance showed that the enhancement load inverter has a roughly six times slower rise time than all other inverter types. Furthermore, the enhancement inverter showed a reduced H level at the output of about one threshold voltage.

Two buffer circuits are presented. A super buffer uses an inverter chain with a cascading (w/l) ratio of $\alpha = e$. The improved driving capability of the bootstrap buffer is caused by a voltage multiplication via a capacitor.

Input and output stages are considered. It turns out that the square of the switching time t_S is inversely proportional to the current change $\mathrm{d}I/\mathrm{d}t$ at the output stage. As a consequence, severe switching noise at outputs and power-supply lines may occur.

ESD at a device can lead, particularly at input and output stages, to reliability problems or, even worse, to damage and failure of the integrated circuit. Three models simulate what can happen in reality. In the human body model, an IC has to withstand current changes of about 5 to 30 A in a time of 50–200 ns without being damaged.

Problems

5.1 The average lifetime (MTBF) of a metal supply line is 75 years under the following conditions: current density 1 mA/μm^2, chip temperature 80 °C, activation energy 0.65 eV. To what value does the lifetime reduce when the average chip temperature is raised to 160 °C?

5.2 An IC is used in an automobile. Under operation a current density of 2 mA/μm^2 exists in a metal power-supply line. The lifetime (MTBF) of the supply line is about 60 years at 70 °C. Calculate whether the lifetime is sufficient to guarantee a car operation of 10 000 hours, when the chip temperature (self-heating and outside temperature) is on average 150 °C. The activation energy is 0.65 eV.

5.3 The inverter shown below has a resistive load.

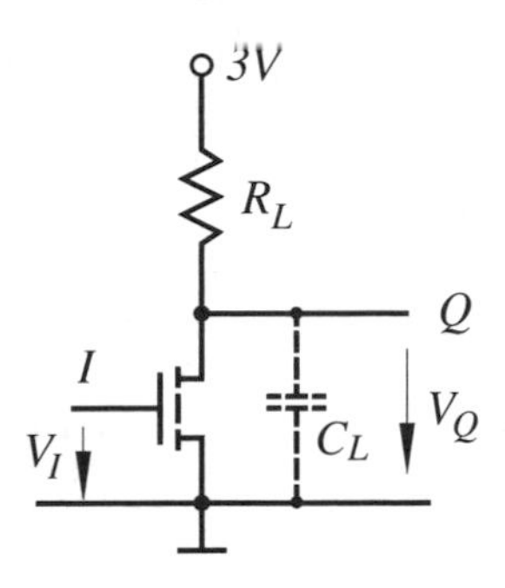

The data are: $V_{Ton} = 0.6\,\text{V}$, $k_n = 120\,\mu\text{A/V}^2$. Calculate:

(a) The resistor value causing the capacitance to be charged in 3 ns to 86% (two time constants) to its final value.

(b) The (w/l) ratio of the transistor in order to guarantee an L signal of 0.3 V at the output, when an H signal of 3 V is applied to the input.

(c) The (w/l) ratio required when the input signal is reduced to 2 V.

5.4 The depletion load inverter shown below is used in a power application. Determine the output voltage when the input voltage is zero. The process data are: $V_{Tn,2} = -3.5\,\text{V}$, $\phi_F = 0.3\,\text{V}$, and $\gamma = 0.7\,\text{V}^{1/2}$.

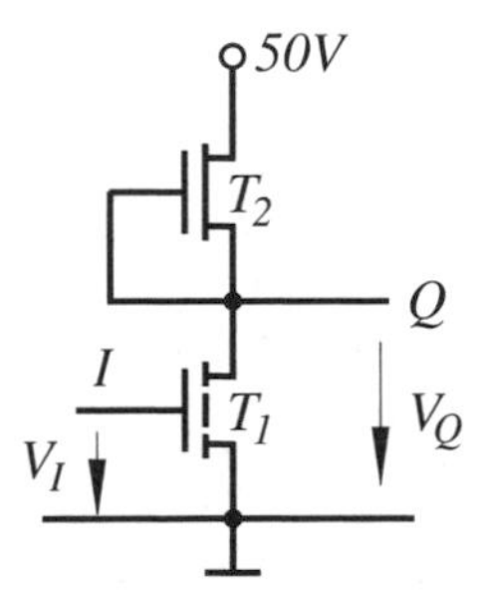

5.5 An enhancement load inverter is shown below. The input voltage is zero. Does the load transistor operate in the resistive or saturation region? Determine the output H level under worst case conditions. The process data are: $V_{Tn,2} = 0.6\,\text{V}$, $\phi_F = 0.35\,\text{V}$, and $\gamma = 0.6\,\text{V}^{1/2}$.

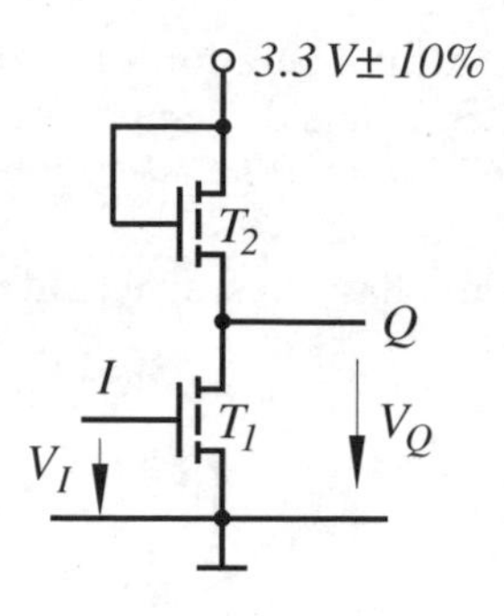

5.6 A CMOS inverter clocked with a symmetrical signal with $T_P = 10\,\text{ns}$ and a rise and fall time of 1 ns is shown below. Calculate the transient power dissipation and compare the result with the dynamic one. The data are: $\beta_n = \beta_p = 120\,\mu\text{A/V}^2$, $V_{Ton} = -V_{Top} = 0.45\,\text{V}$, and $C_L = 0.1\,\text{pF}$.

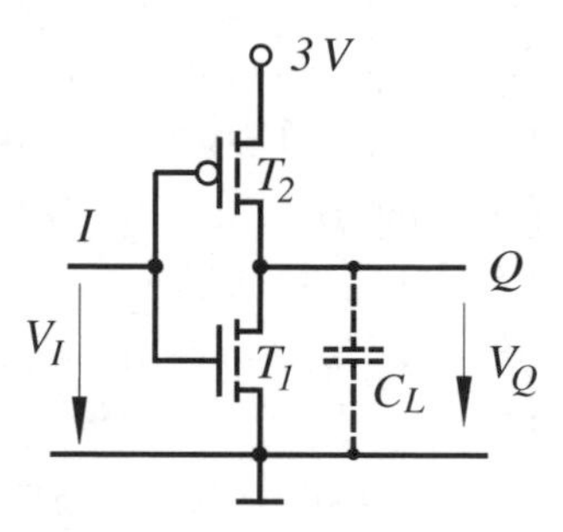

5.7 Shown below is part of a shift register. If a clock signal of $\phi = V_{CC}$ is applied to the circuit, then the parasitic capacitance C_L is charged to the inverse of the logic state of the input D.

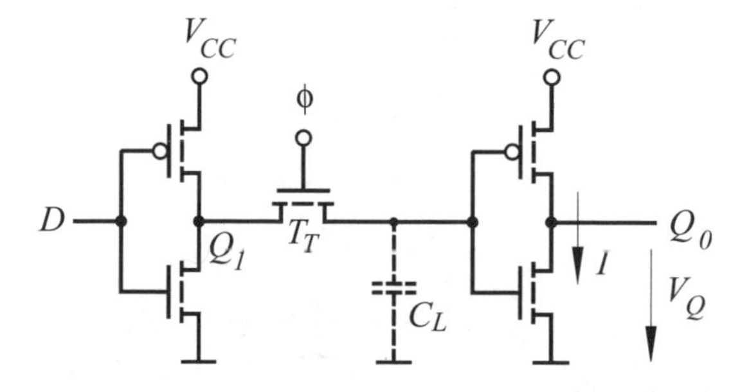

With the clock signal at $\phi = 0\,\text{V}$ transistor T_T is cut off and the logic function stored as different charge at C_L. Calculate the voltage at C_L for the case of a stored H signal and L signal. Is it possible that at a particular logic state a current I passes through the output transistors? If so, how large is the current? Data for all n-channel transistors are: $V_{Ton} = 0.5\,\text{V}$, $\gamma = 0.4\,\text{V}^{1/2}$, $\beta_n = 300\,\mu\text{A/V}^2$, $2\phi_F = 0.65\,\text{V}$. Data for all p-channel transistors are $V_{Top} = -0.5\,\text{V}$, $\gamma = 0.4\,\text{V}^{1/2}$, $\beta_p = 300\,\mu\text{A/V}^2$, $2\phi_F = 0.65\,\text{V}$. The power-supply has a voltage of $V_{CC} = 3\,\text{V}$.

5.8 Shown below is a multi-function gate. Determine the width of the transistors under the condition that the load capacitance has to be charged in 1 ns. The data are: $k_n = 100\,\mu\text{A/V}^2$, $k_p = 40\,\mu\text{A/V}^2$, and $l_{min} = 0.3\,\mu\text{m}$.

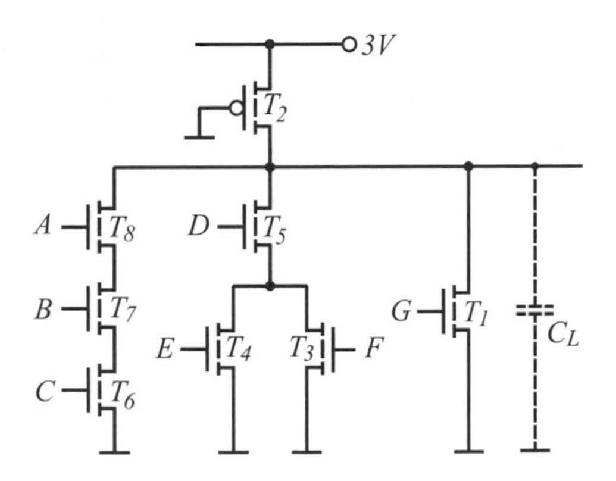

5.9 An integrated circuit has four data output buffers, which cause an unacceptably large switching noise at the power-supply lines and at the output, when switched simultaneously. In order to reduce the simultaneous switching noise, would it be better to increase the switching time of all buffers or would it be preferable to use staggered delays?

The solutions to the problems can be found under: *www.unibw-muenchen.de/campus/ET4/index.html*

REFERENCES

C.M. Baker and C. Terman (1980) Tools for verifying integrated circuit designs. *Lambda*, **1**(3), 22–30.

D. Deschacht, M. Robert and D. Auvergne (1988) Explicit formulation of delays in CMOS data paths. *IEEE J. Solid-State Circuits*, **23**(5), 1257–64.

M.D. Ker, T.Y. Chen, C.Y. Wu and H.H. Chang (2000) ESD protection design on analog pin with very low input capacitance for high-frequency or current-mode applications. *IEEE J. Solid-State Circuits*, **35**(8), 1194–9.

H.C. Lin (1975) An optimized output stage for MOS integrated circuits. *IEEE J. Solid-State Circuits*, **SC-10**(2), pp. 106–9.

Y. Nakase, Y. Morooka, D.J. Perlman, D.J. Kolor, J.M. Choi, H.J. Shin, T. Yoshimora, T. Watanabe, Y. Matsuda, M. Kumanoya and M. Yamada (1999) Source synchronization and timing vernier techniques for 1.2-GB/s SLDRAM interface. *IEEE J. Solid-State Circuits*, **34**(4), 494–501.

Y. Ohtomo and Nogawa (1995) Low power Gb/s CMOS interface. *Symposium on VLSI Circuits, Digest of Technical Papers*, pp. 29–30.

R. Senthinathan and J.L. Prince (1993) Application specific CMOS output driver circuit design techniques to reduce simultaneous switching noise. *IEEE J. Solid-State Circuits*, **28**(12), 1383–8.

M. Shoji (1986) Reliable chip design method in high performance CMOS VLSI. In *Proceedings of IEEE International Conference on Computer Design*, pp. 389–92.

S.M. Trimberger (1987) *An Introduction to CAD for VLSI.* Boston, MA: Kluwer.

T.J. Tucker (1968) Spark Initiation Requirements. *Annals of the New York Academy of Science*, Vol. 152, Art 1. Oct. 28.

H.J.M. Veendrick (1984) Short-circuit dissipation of static CMOS circuitry and Its impact on the design of buffer circuits. *IEEE J. Solid-State Circuits*, **SC-19**(4), 468–73.

A.J. Walton, R.J. Holwill and J.M. Robertson (1985) Numerical simulation of resistive interconnects for integrated circuits. *IEEE J. Solid-State Circuits*, **SC-20**(6), 1252–8.

6

Combinational and Sequential CMOS Circuits

In general, a digital system consists of combinational and sequential circuits. The combinational circuits have the property that at any point in time, the outputs of the circuits are related directly to the input signals by Boolean expressions. Examples are *NOR*, *NAND*, and *XOR* combinations. The situation in the case of sequential circuits is different. In sequential circuits the outputs are not only a function of the input data, but also of previously stored data. Examples are counter, shift registers, and, to some extent, programmable logic arrays. Combinational and sequential CMOS circuits can be implemented in a variety of circuit techniques with different trade-offs.

6.1 STATIC COMBINATIONAL CIRCUITS

One way to organize the different circuit techniques for the realization of the mentioned system components is to divide the implementations into static or clocked versions. An overview of the resulting different circuit design styles is shown in Figure 6.1 and treated in this chapter, starting with complementary circuits.

6.1.1 Complementary Circuits

The most important feature of these circuits is that they behave like a CMOS inverter (Section 5.3.4): that is, they exhibit rail to rail swing, no static power dissipation (ignoring the sub-threshold current), and a ratioless design style. To maintain these advantages in complementary logic circuits, a particular configuration of series and parallel connected transistors has to be followed. As an example a *NAND* gate is shown in Figure 6.2. A current pass from V_{CC} to ground is possible only when both n-channel transistors are conducting. This is the case when *H* signals are applied to both inputs. But the *H* signals turn off the p-channel transistors, guaranteeing that no current path is generated. A similar situation exists at the *NOR* gate (Figure 6.3).

System Integration: From Transistor Design to Large Scale Integrated Circuits. Kurt Hoffmann.
 ISBN: 0-470-85407-3

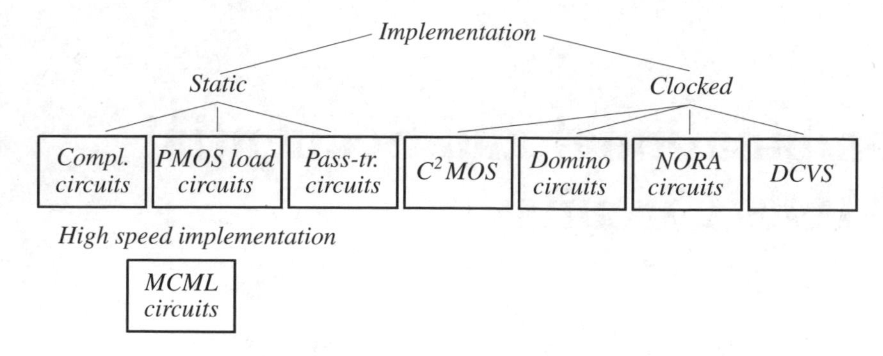

Figure 6.1 CMOS circuit design styles

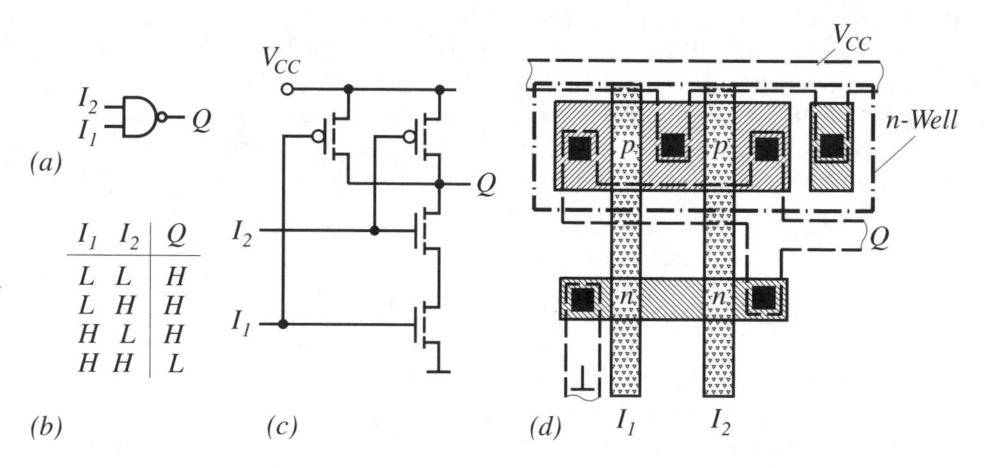

Figure 6.2 Two-input *NAND* gate: (a) logic symbol, (b) truth table, (c) circuit and (d) layout

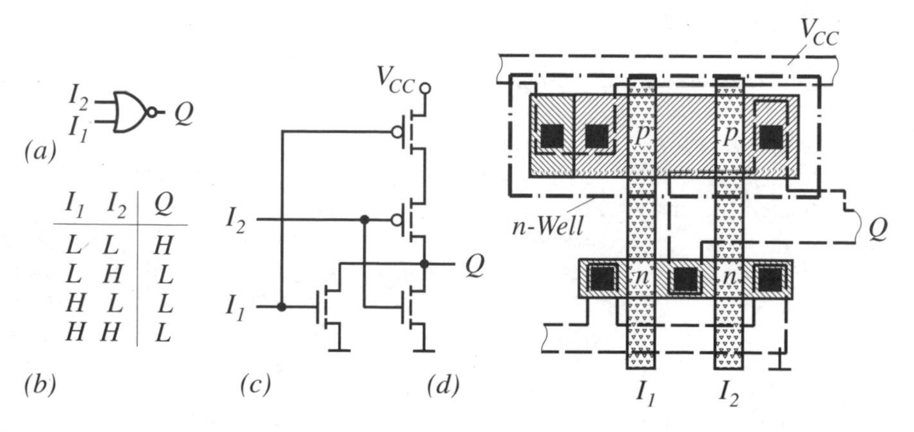

Figure 6.3 Two-input *NOR* gate: (a) logic symbol, (b) truth table, (c) circuit and (d) layout

If both input signals are in an L state, both p-channel transistors are turned on and the n-channel transistors are cut off. Thus, no current pass between V_{CC} and ground exists. From this brief consideration the following findings result:

(a) All p-channel transistors can be placed in a common n-well, saving layout area.

(b) The output is always in a defined H or L state.

(c) The H state is guaranteed by the p-channel transistors and the L state by the n-channel transistors.

(d) The implemented logical function by the n-channel transistors is realized in complementary form by the p-channel transistors.

How the last statement can be understood and transferred to more complex logic gates follows next. For this purpose, a general complex complementary network is considered where $Q(I)$ describes the H state and $\overline{Q}(I)$ the L state at the output (Figure 6.4).

The complement of a logic function can be found by using DeMorgan's theorem, which states

$$Q = \overline{I_1 + I_2 + I_3 + I_4 \ldots} = \bar{I}_1 \cdot \bar{I}_2 \cdot \bar{I}_3 \cdot \bar{I}_4 \ldots\ldots$$
$$\text{and} \qquad (6.1)$$
$$Q = \overline{I_1 \cdot I_2 \cdot I_3 \cdot I_4 \ldots\ldots} = \bar{I}_1 + \bar{I}_2 + \bar{I}_3 + \bar{I}_4 \ldots\ldots$$

This theorem can be expressed in simplified wording: 'break the line change the sign.' If, e.g., the following function $\overline{Q} = I_1 \cdot [I_2 + I_3] + I_4 \cdot I_5 + I_6$ is given, then the connections of the n-channel transistors are defined and, after applying DeMorgans's theorem, those of the p-channel transistors $Q = [\bar{I}_1 + \bar{I}_2 \cdot \bar{I}_3] \cdot [\bar{I}_4 + \bar{I}_5] \cdot \bar{I}_6$ also (Figure 6.5). Since p-channel transistors change an L state at the input to an H state at the output an inversion of the input signals at these transistors is not required.

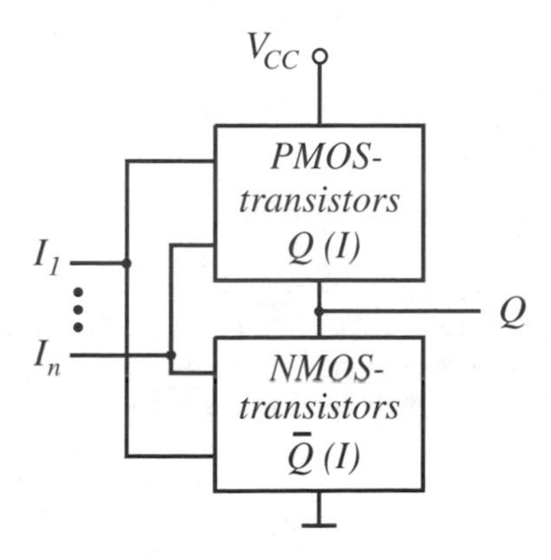

Figure 6.4 General complementary network

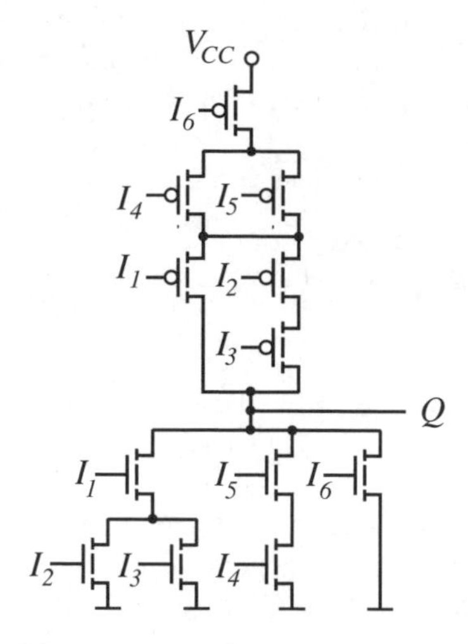

Figure 6.5 Example of a complementary circuit

Layout of complementary circuits

The layout of these circuits is particularly advantageous, when the polysilicon or polycide lines are placed orthogonally to the metal power-supply lines. This is shown for the preceding circuit in Figure 6.6(a).

If a polysilicon or polycide line crosses an n-diffusion or p-diffusion region, this leads to a transistor. These transistors are individually interconnected by metal lines according to the circuit configuration. The layout area can be minimized when it is possible to join adjacent diffusion areas in one diffusion strip, interrupted by polysilicon lines only

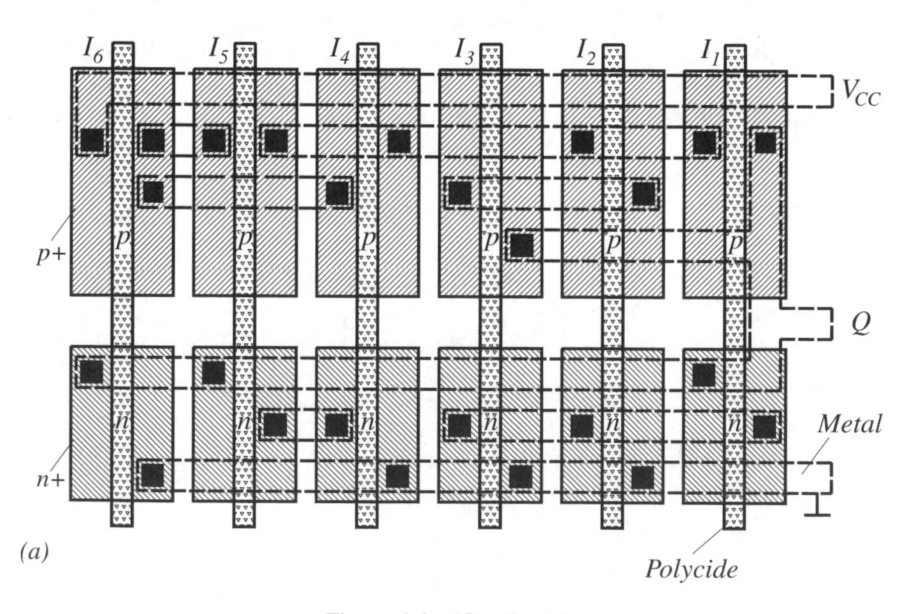

Figure 6.6 (Continued)

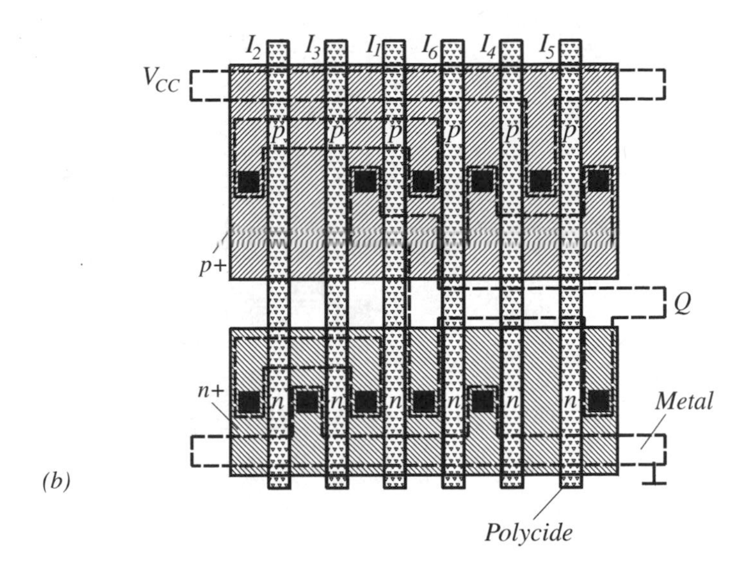

Figure 6.6 Layout of the complementary circuit of Figure 6.5: (a) individual transistor connections and (b) joined transistor connections

(Figure 6.6(b)). This is possible from a circuit point of view when the polysilicon lines – these are the transistors – are rearranged. For this purpose the complementary circuit is represented by a graph. How this can be accomplished is demonstrated with the following function $Q = \overline{I_1 \cdot I_2 + I_3 \cdot I_4}$. The graph consists of two parts, one representing the n-channel transistors and the other the p-channel transistors (Figure 6.7). The vertices are source and drain connections and the edges in the graph are transistors that connect source and drain vertices. The graph mirrors the connections of the

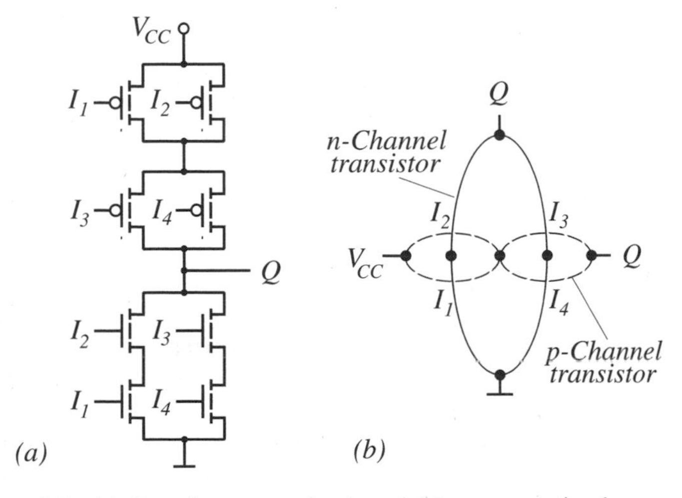

Figure 6.7 (a) Complementary circuit and (b) representation by a graph

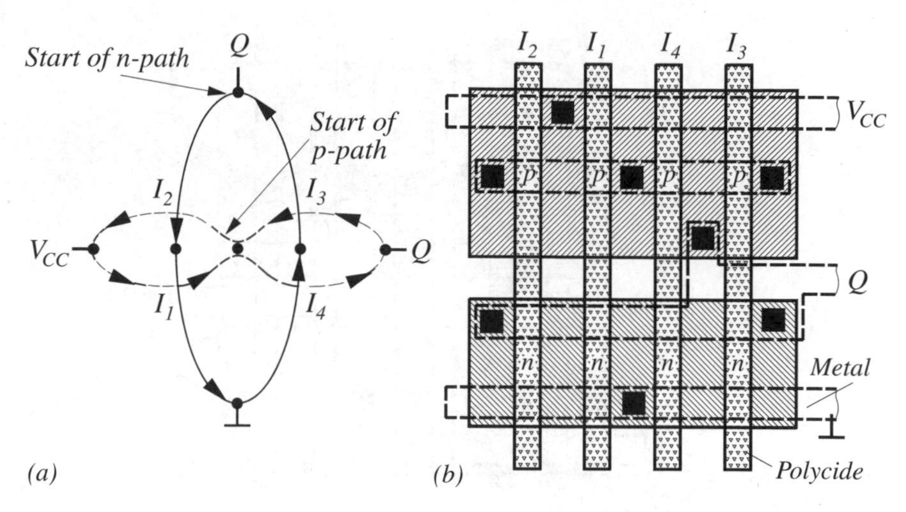

Figure 6.8 (a) Graph of Figure 6.7 with Euler path and (b) layout

transistors in the circuit. Each edge is named with the gate signal name for the transistor. The graphs are dual of each other, just as the n- and p-channel transistors are. The p-graph overlays the n-graph to illustrate the point. If there exists a sequence of edges in the n-graph and p-graph that have identical labeling, then the diffusion regions may be joined in one diffusion strip, interrupted by polysilicon lines only. Such a sequence is known as an Euler path.

In other words, all n-channel and p-channel transistors must be rearranged and connected in an identical sequence; that is, the source of one device is the drain of the following transistor or vice versa. No transistor (edge) is allowed to be omitted or accounted for more than once.

For the example a consistent Euler path I_2, I_1, I_4, I_3, is shown in Figure 6.8. This path is not necessarily unique since many different solutions might exist.

The advantage of using this layout style is obvious: not only is silicon area saved but also the delay times of the shortened lines and strips are reduced. A further advantage is that this layout technique can easily be automated (Uehara and vanCleemput 1981).

6.1.2 PMOS Load Circuits

The described complementary circuits have the advantage that they are very robust against power-supply variations and noise injections. A disadvantage is the relative complexity of these gates in large arrays due to the complementary function. This causes a relatively large silicon area consumption and in the extreme may lead to circuits which cannot be implemented in a meaningful way using this technique. Examples are 1024 input *NOR* and *NAND* gates used, e.g., in decoders. These gates have either 1024 n-channel or p-channel transistors connected in series: not a thrilling solution for fast rise or fall times. One solution is to replace the total p-transistor part, e.g., of Figure 6.5 by one PMOS load device (Figure 6.9).

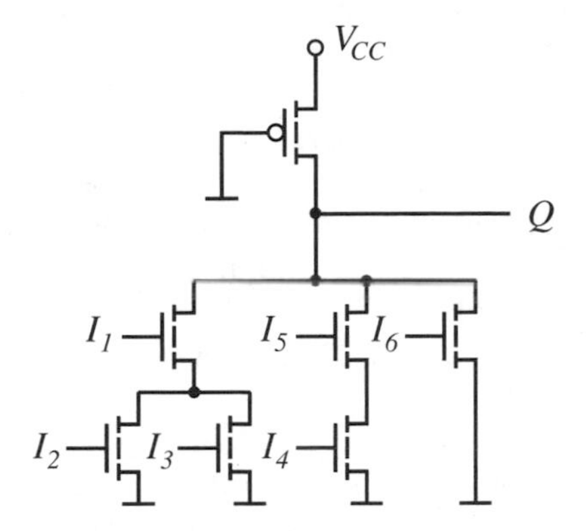

Figure 6.9 Example of PMOS load circuit

A disadvantage of this circuit configuration is that, depending on the input variables, a static power dissipation arises. This requires a ratioed sizing strategy, as presented in Sections 5.3.3 and 5.3.5.

6.1.3 Pass-Transistor Circuits

The basic element of these circuits is the pass transistor needed to form a controlled connection between two circuit notes or terminals (Figure 6.10). Applying a clock signal ϕ to the gate of an n-channel transistor, the connection between the terminals A and B is open or closed, that is, the transistor is non-conducting with $\phi = 0\,\text{V}$ or conducting with $\phi = V_{CC}$ (Figure 6.10(a)). In the last case, the maximum voltage to be transferred between the terminals is one threshold voltage below the clock voltage $V_{CC} - V_{Tn}$. This is a comparable situation to the one described at the enhancement inverter, where the V_{QH} level is reduced by a threshold voltage (Equation 5.14). If this reduced voltage is applied to a following CMOS inverter, an unintentional current path may occur. This situation is discussed in some detail after the pass-transistor analysis is completed. The

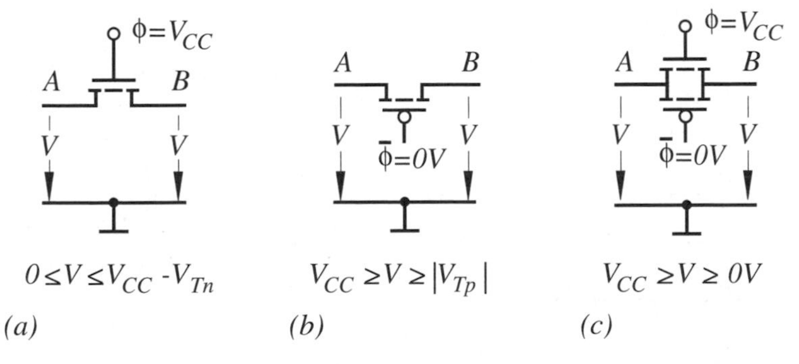

Figure 6.10 Pass transistor: (a) n-channel transistor, (b) p-channel transistor and (c) n-channel and p-channel transistors in parallel

reduced voltage level can be avoided when an increased clock voltage of $\phi > V_{CC} + V_{Tn}$ is applied. But this solution is only meaningful when many pass transistors are clocked due to the added complexity of the clock driver circuit. When a p-channel transistor is used as a pass transistor the situation is reversed. Only a voltage transfer of $V \geq |V_{Tp}|$ is possible when the transistor is turned on with $\overline{\phi} = 0\,\text{V}$ (Figure 6.10(b)). By comparing the two pass-transistor implementations it is obvious that the full voltage swing $V_{CC} \geq V \geq 0\,\text{V}$ can be transferred when n-channel and p-channel transistors are connected in parallel (Figure 6.10(c)).

In some applications a pass transistor is connected to the input of a CMOS inverter (Figure 6.11).

The n-channel transistor is usually preferred, since its current gain is approximately twice as large as that of the p-channel transistor (Equation 5.19). If a high signal V_{IH} is applied to the input, this leads to a reduced input voltage at the inverter by one threshold voltage below V_{CC} and thus to a V_{GS} voltage of the p-channel transistor of $-V_{Tn}(V_{SB})$ (Figure 6.11(a)). This causes a current I to flow through the CMOS inverter. One has to be aware that the threshold voltage of the pass transistor T_T is increased due to its body effect (Equation 5.14). One way to eliminate this current I is to apply a level restore transistor to the circuit. With $V_{QL} \approx 0\,\text{V}$ this transistor is turned on, pulling the input of the CMOS inverter to V_{CC} and thus tuning off transistor T_2. A change of the input signal to $V_{IL} = 0\,\text{V}$ initially causes a current I_T to flow until the circuit is switched to an output signal of $V_{QH} = V_{CC}$ (Figure 6.11(b)). In order to guarantee a safe switching of this circuit, the (w/l) ratio of the restore transistor should be smaller than that of the pass transistor by at least a factor of two, when a symmetrical CMOS inverter is assumed.

The described pass transistor can be used to implement all kinds of logic circuits. A popular one is a multiplexer (Figure 6.12). The path selector selects one of the four inputs and passes the information of the data inputs to the output Q. Logically the output of the circuit has a function as shown in Figure 6.12. Since only n-channel transistors are used as pass transistors, the output H voltage is reduced by a threshold voltage. This leads, as mentioned before, to an unintentional current path in the following CMOS inverter or circuit, which can be avoided by using the described level restore transistor.

Another example is the multipexer/demultiplexer shown in Figure 6.13. In this implementation parallel-connected n-channel and p-channel pass-transistor branches

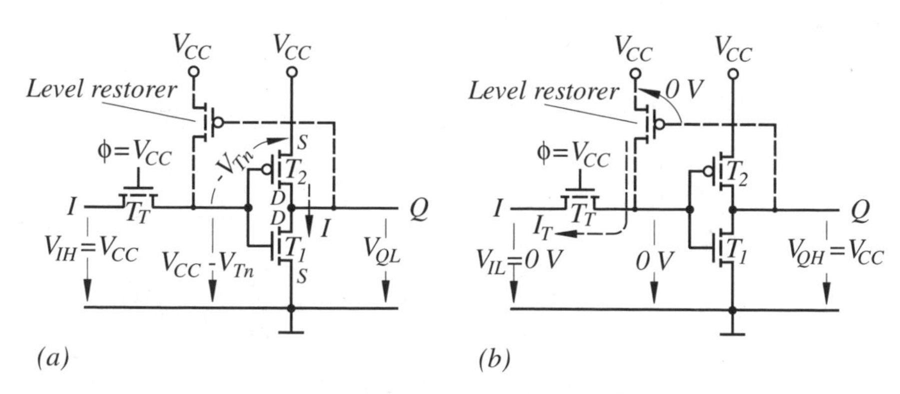

Figure 6.11 Pass transistor with level restorer: (a) $V_{IH} = V_{CC}$ and (b) $V_L = 0\,\text{V}$

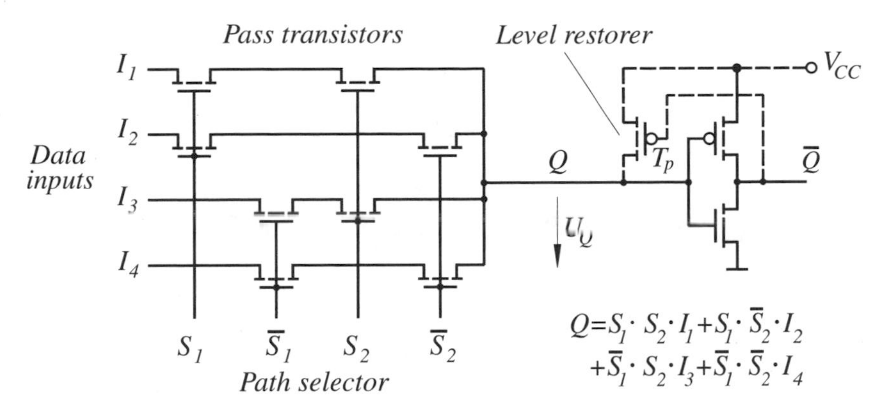

Figure 6.12 Circuit of a 4 to 1 multiplexer with level restorer

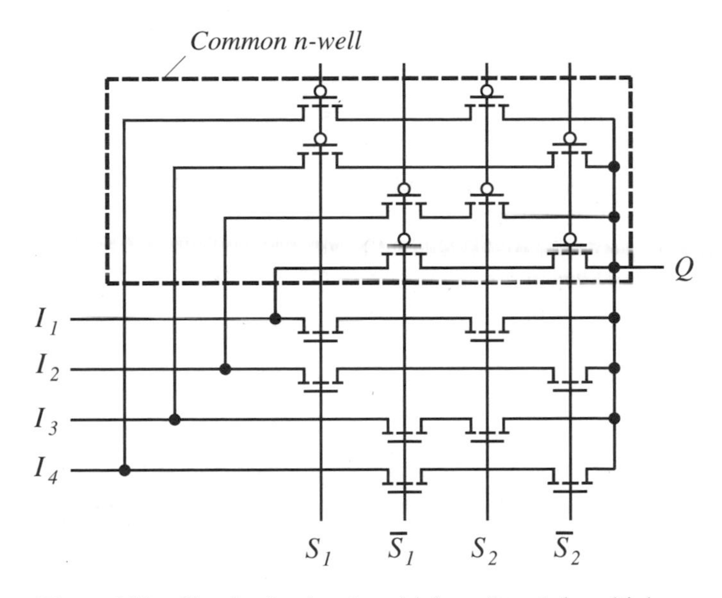

Figure 6.13 Circuit of a 4 to 1 multiplexer/1 to 4 demultiplexer

are used. These branches guarantee that the full voltage swing is transferred from the inputs to output. This circuit is bi-directional; that is, it can be used as a multiplexer or as a demultiplexer by exchanging inputs and outputs. Another solution would be to use parallel-connected n-channel and p-channel transistors, as shown in Figure 6.10(c), for the implementation. From an electrical point of view this would be identical to the preceding solution, but the layout would be more complex.

A further example is the use of pass transistors in *XOR* circuits (Figure 6.14). Regardless of the logic values of I_1 and I_2, the output is always connected either to an *H* signal or to an *L* signal, as can easily be proved.

These examples demonstrate the diversity of the pass-transistor logic. The pass-transistor logic is not limited to the relatively simple examples presented, but can be

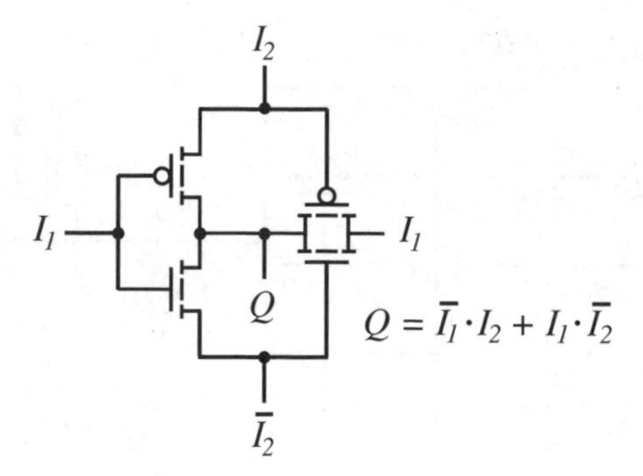

Figure 6.14 *XOR* implementation with pass transistors

used for the implementation of general logic functions, as is described in some detail in Whitaker (1983) and Radhakrishnan *et al.* (1985).

6.2 CLOCKED COMBINATIONAL CIRCUITS

A drawback of the preceding circuits is that they are relatively slow, caused to a great extent by series-connected transistors or by the implementation of the complementary function. The PMOS load circuits consume static power and are therefore not ideal either. Clocked circuits ease these restraints, as the following section reveals.

6.2.1 Clocked CMOS Circuits (C^2MOS)

The basic idea behind the clocked circuit technique, sometimes called dynamic circuits is illustrated in Figure 6.15. With a clock signal of $\phi = L$ applied to the circuit, the p-channel transistor is conducting and the n-channel one is cut off. The parasitic

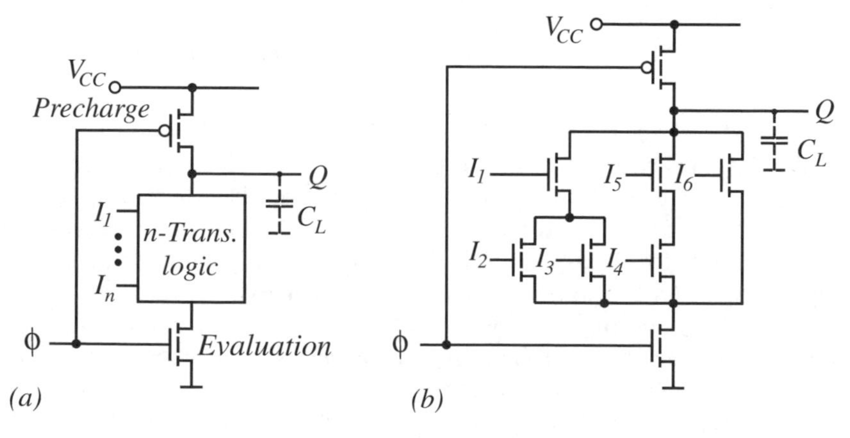

Figure 6.15 (a) Basic C^2MOS circuit and (b) C^2MOS logic

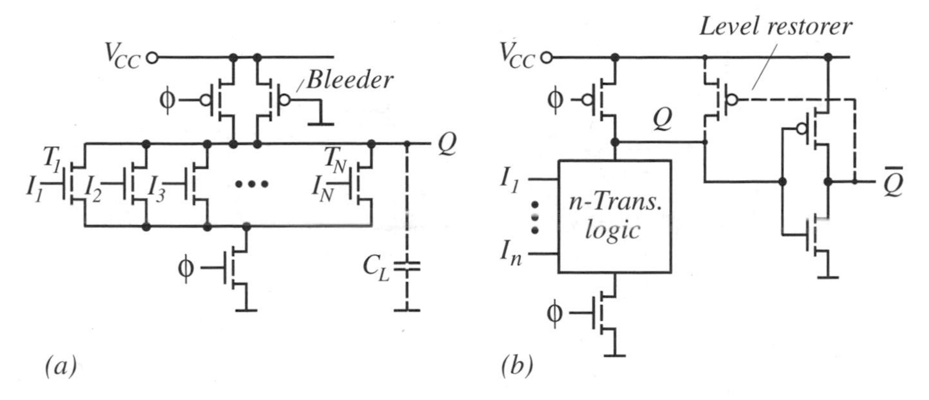

Figure 6.16 (a) C^2MOS *NOR* gate and (b) C^2MOS logic with level restore circuit

capacitance C_L at the output is precharged to V_{CC}. A change in the clock signal from L to H causes the p-channel transistor to cut off and the n-channel transistor to be conducting. The evaluation of the logic begins. Depending on the input information, the capacitance remains charged or is discharged. As an example, the logic function of Figure 6.9 is used. This demonstrates that the circuit can be implemented without a PMOS load or a complementary function in a ratioless design fashion. A further example is shown in Figure 6.16(a) and illustrates how an N input *NOR* gate can be built, even when N is very large, e.g. 1024.

The clocked circuit designs are implemented in a ratioless design fashion. Usually, in this design technique, transistors with very small dimensions are used. This has the advantage that the parasitic capacitances are very small and the dynamic power dissipation (Equation 5.28) also. Further benefits are the fast rise and fall times. This circuit technique is thus a very good contender when high performance is required. But two drawbacks exist also. These are leakage currents and charge redistribution problems.

Leakage currents

The high signal at the output of the C^2MOS circuits is stored as a particular amount of charge at the capacitance C_L. In this condition the n-channel transistors in the logic part of the circuit are cut off and operate in the sub-threshold region. As a consequence leakage currents – mainly sub-threshold currents – occur (Section 4.4.3), leading to a discharge of the stored charge and to a limited time the charge can be guaranteed.

Charge redistribution

This occurs when differently charged capacitances are connected, causing unwanted voltage changes. The following example demonstrates such a situation (Figure 6.17). At time t_0 capacitance C_L is charged to $V_Q = V_{CC}$ and at time t_1 the parasitic capacitance C_A is charged to a voltage of $V_A = 0\,\text{V}$, since I_4 is in an H state. The voltages at the capacitances remain unchanged, when at time t_2 the input I_4 changes from an H state to

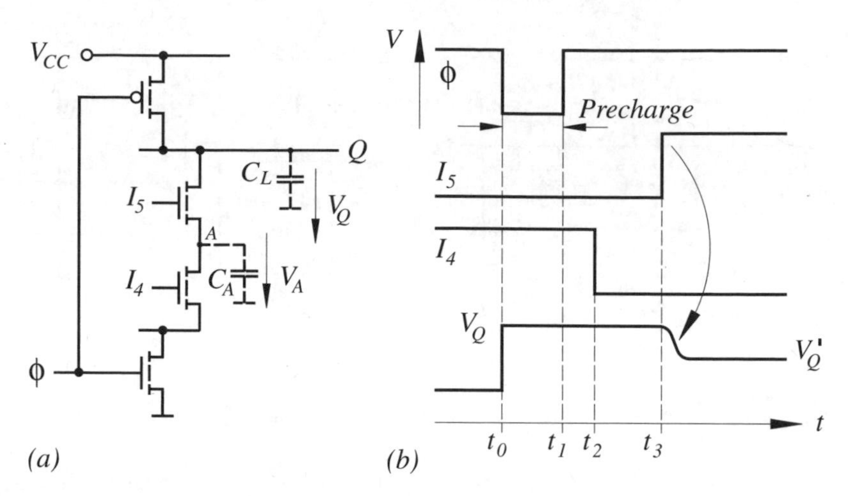

Figure 6.17 (a) Partial circuit of Figure 6.15(b) and (b) charge redistribution at the output

the L state. But if I_5 now changes from the L to the H state, the charged capacitances are connected together. This causes a charge redistribution and a reduced voltage at the output Q of

$$
\begin{aligned}
V_Q' &= \frac{Q_G}{C_G} = \frac{C_L V_Q + C_A V_A}{C_A + C_L} \\
&= \frac{C_L}{C_A + C_L} V_{CC}
\end{aligned}
\tag{6.2}
$$

where Q_G is the total charge and C_G the total capacitance. The charge redistribution effect can be minimized or eliminated when the change of the inputs is limited to the precharge time or the layout is optimized in such a way that $C_A < C_L/10$.

An effective way to cope with leakage currents is to include in the circuit a so-called bleeder transistor (Figure 6.16(a)). The purpose of this transistor is to compensate with its current the sub-threshold currents of the cut-off transistors. Unfortunately, this transistor introduces a static power dissipation when an L signal exists at the output. In order to reduce this effect, the bleeder transistor should have a relatively long channel length and a small width. But this design has an adverse effect on the chip area. Another solution exists, when a non-inverted output signal – such as in the following domino circuits – is required. In this case, a level restorer (Figure 6.16(b)) – described in conjunction with the pass transistor – can be used to maintain the integrity of the H level.

6.2.2 Domino Circuits

Cascading of clocked CMOS circuits is only possible with additional circuit components. Why this is the case is demonstrated with the following example (Figure 6.18). The capacitances C_L are charged via the p-channel transistors to a voltage of V_{CC}. If the clock changes from L to H, both logic circuits are activated simultaneously. Depending on the

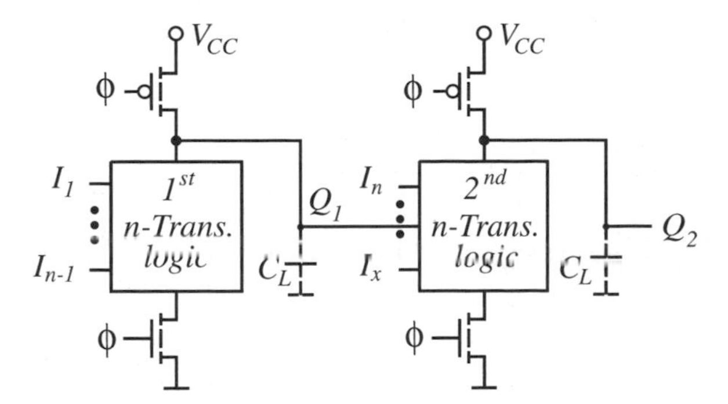

Figure 6.18 Illustration of raise condition at cascading C^2MOS circuits

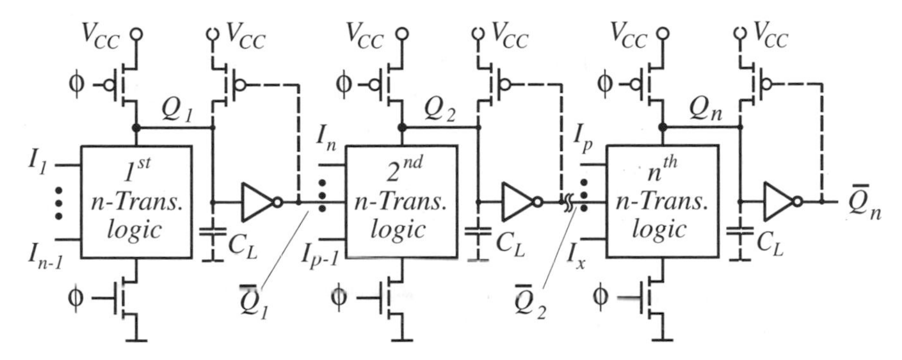

Figure 6.19 Domino principle

input conditions, the output Q_1 may change from H to L. Since this can only happen with a particular delay time, e.g. of one ns, it is possible that the second logic circuit may interpret this erroneously as an H signal, causing the capacitance at output Q_2 partly or totally to be discharged. To avoid this problem the clock to the second gate could be delayed or the domino technique (Krambeck *et al.* 1982) could be used (Figure 6.19).

In this case, only non-inverted signals are cascaded by using additional inverters. The signals are delayed through each stage after activating, just as before, but with the difference that the input variables are only able to change delayed from L to H. A proper signal propagation is guaranteed since an L signal at the input does not lead to an erroneous condition at the n-channel logic. The advantage of the domino technique is that only one clock signal is required and thus timing problems are avoided. With the use of level restorer, the leakage problem is circumvented. A slight disadvantage is that only non-inverting logic blocks are applicable.

6.2.3 NORA Circuits

A modification of the domino principle with fewer components is the NO RAise (NORA) implementation (Goncalves and DeMan 1983; Lee and Szeto 1986), sometimes called zipper logic (Figure 6.20). In this case, alternating n-channel and p-channel

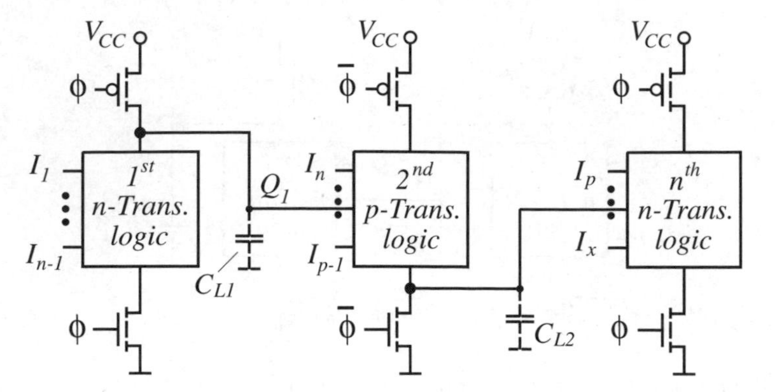

Figure 6.20 NORA principle

logic blocks are cascaded and additionally an inverted clock signal is applied. During precharge with $\phi = L$ and $\overline{\phi} = H$, C_{L1} is charged to V_{CC} and C_{L2} to 0 V. This guarantees that during the activation of the logic gates, when the clock signal changes, a proper signal propagation is guaranteed, since an L signal at the input of the n-channel logic does not lead to an erroneous condition and an H signal to one at the p-channel logic block. An example, illustrating the effectiveness of this circuit technique is a full adder with the logic function shown in Figure 6.21.

$$S_N = [A_N \oplus B_N] \oplus C_{N-1} \quad \text{and}$$

$$C_N = [A_N \oplus B_N]C_{N-1} + A_N B_N \tag{6.3}$$

All presented circuits have in common that during precharge, the outputs are either at a defined L state or H state (Figure 6.22(a)). If this is undesirable for further data

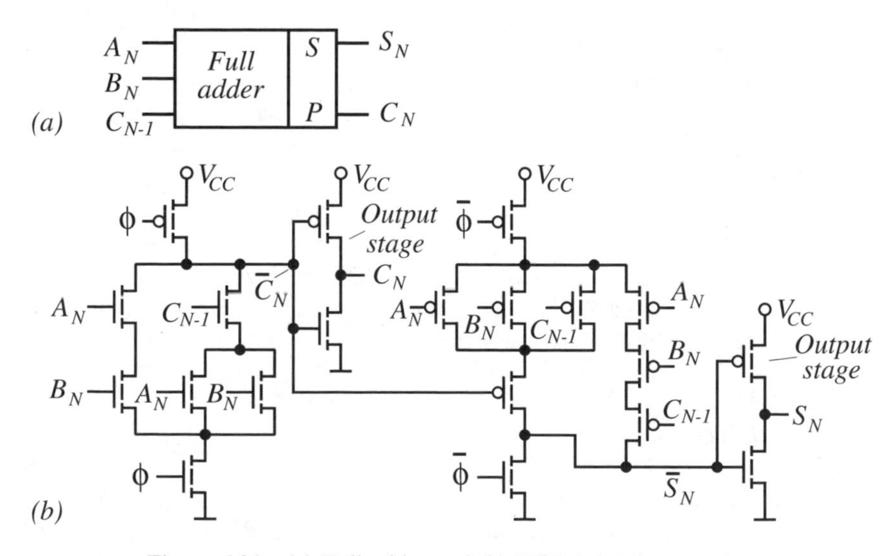

Figure 6.21 (a) Full adder and (b) NORA implementation

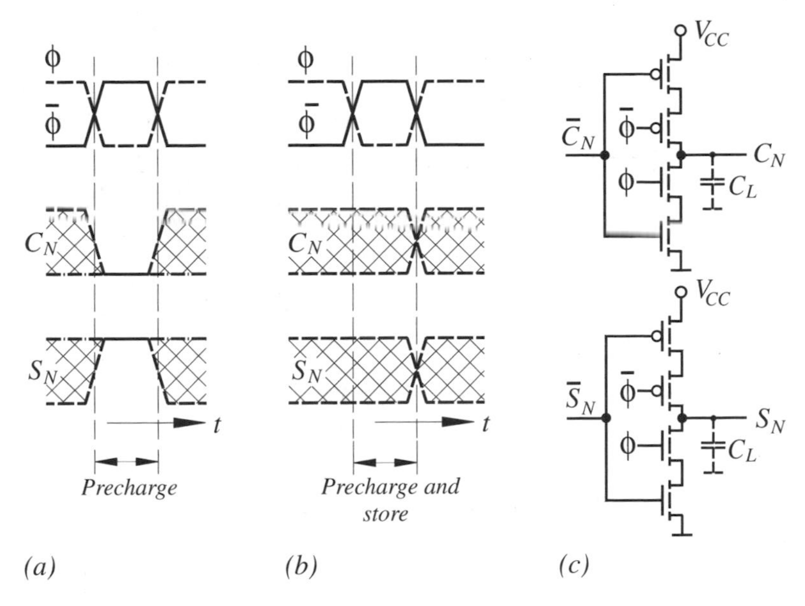

Figure 6.22 (a) Timing diagram of the full adder of Figure 6.21, (b) timing diagram employing intermediate data storage and (c) modified output stage

processing, the information can be stored during the precharge time at the parasitic capacitance C_L of the modified output stages of the full adder (Figure 6.22(c)). During precharge with $\phi = L$ and $\overline{\phi} = H$, the output transistors are non-conducting and the capacitances remain charged to the previous logic state (Figure 6.22(b)). A change in clock signals connects the capacitances to the output stages, and the circuit behaves like its predecessor.

6.2.4 Differential Cascaded Voltage Switch Circuits (DCVS)

This technique is based on the C^2MOS principle but requires a complement to each input signal (Figure 6.23). If the clock signal ϕ changes after precharging from L to H, this causes either the output Q or $\overline{Q}$ to change to an L state, whereas the other one remains at an H state, depending on the input variables. To demonstrate the principle an XOR gate is used (Figure 6.23(b)). The transistor connection for output Q is given by the logic function $Q = I_1 \cdot \overline{I}_2 + \overline{I}_1 \cdot I_2$ and for the output $\overline{Q}$ by $\overline{Q} = I_1 \cdot I_2 + \overline{I}_1 \cdot \overline{I}_2$. As this example demonstrates, standard logic design methods can be used for the realization of such circuits (Chu and Pulfrey 1987). To ease further data processing the information can be stored during precharge at the parasitic capacitances C_L, as described previously (Figure 6.24).

With $\phi = L$ the outputs $\overline{Q'}$ and Q' are in the H state. Simultaneously, the p-channel output transistors are non-conducting and thus the capacitances remain charged to the previous logic state. Only a clock signal change from L to H is able to modify the output conditions.

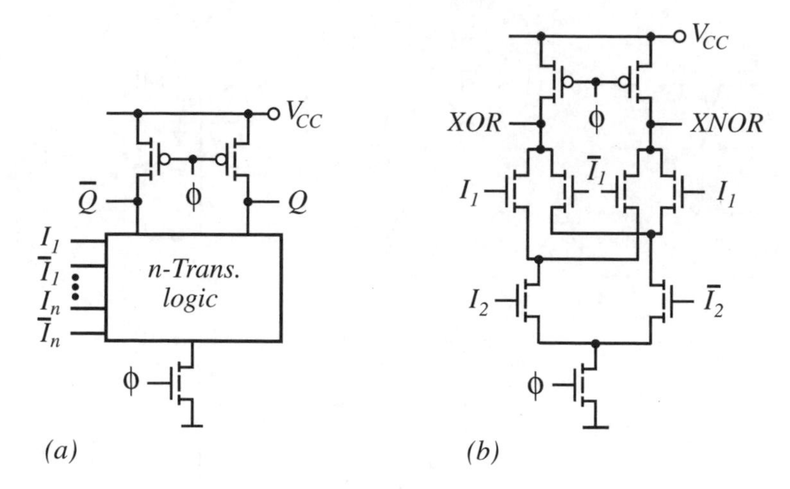

Figure 6.23 (a) DCVS principle and (b) example of a *XOR*/*XNOR* implementation

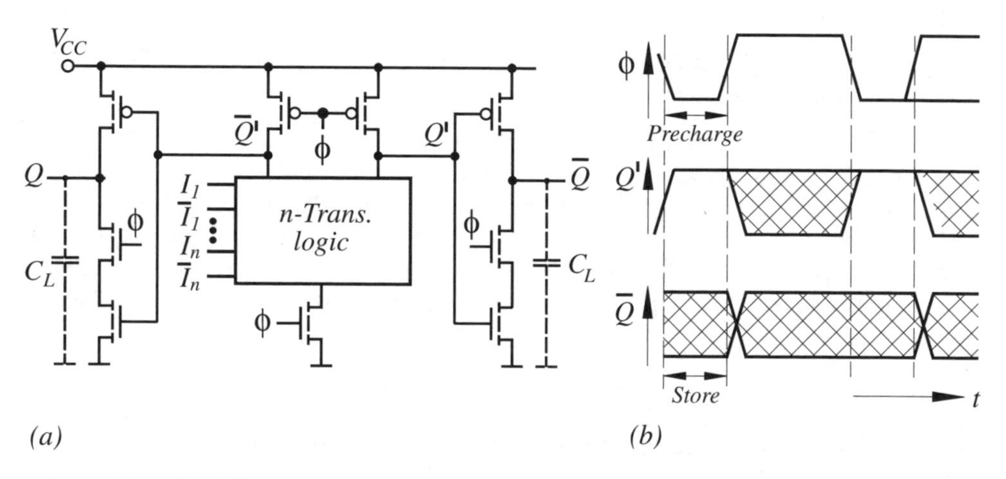

Figure 6.24 (a) DCVS logic circuit with intermediate data storage and (b) timing diagram

Leakage currents are able to change an *H* state into an *L* state, just as is the case in C^2MOS and domino circuits. This can be avoided by adding two p-channel transistors to the circuit as level restorer, similar to Figure 6.11. Since they are cross-coupled, no static current path exists (Figure 6.25).

When one compares the DCVS technique with the other circuit techniques, a distinct speed advantage results, due to the differential nature of these circuits (Ng *et al.* 1996).

6.2.5 Switching Performance of Ratioless Logic

The switching time of the ratioless CMOS inverters is treated in Section 5.4. These results can be transferred to logic circuits and used for a first estimate of rise and fall

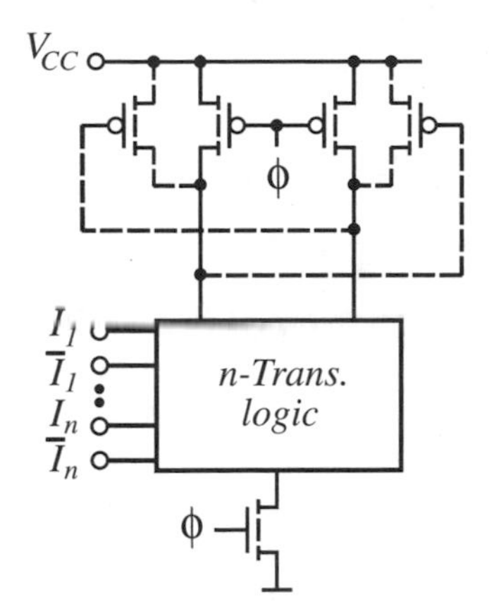

Figure 6.25 Modified DCVS logic circuit

times. If, e.g., m n-channel transistors are parallel-connected as in a *NOR* gate (Figure 6.3), this leads, according to Equation (5.42), to a fall time of

$$\frac{1}{m}\frac{C_L}{\beta_n}1.2(1/V) \leq t_f \leq \frac{C_L}{\beta_n}1.2(1/V) \tag{6.4}$$

which depends on the number of transistors switched simultaneously. If the same number of transistors are connected in series, as is the case in a *NAND* gate (Figure 6.2), this results in an increased fall time of

$$t_f = m\frac{C_L}{\beta_n}1.2(1/V) \tag{6.5}$$

This is the best case, as the following example of a four-input *NAND* gate demonstrates (Figure 6.26).

At time t_0 the input variables are in the logic states shown in Figure 6.26(b). As a result transistors T_1, T_2, and T_3 are turned on and T_4 is cut off. This leads to the shown note voltages V_Q to V_4. At time t_1 the inputs I_1 to I_3 change into an L state. The voltages V_Q and V_4 remain unchanged. At time t_2 all inputs change from an L state to an H state, discharging all note capacitances. But this does not take place simultaneously since the signal ripples through the circuit. At first capacitance C_{L4} is discharged before transistor T_3 starts to discharge capacitance C_{L3} and so on, until finally C_{L1} is discharged to zero volts. This propagation effect results in an increased fall time. In order to reduce the propagation delay the n-channel transistors are progressively sized (Figure 6.27). The one to start the propagation delay T_4 has the largest width, whereas the last transistor in the propagation chain T_1 has the smallest one. The progressive sizing leads to a reduction in the fall time of between 15 and 25% (Shoji 1985).

So far the behavior of n-channel transistors is considered only. But it is obvious from the discussion that the gained results can be applied to the p-channel transistors and configuration in an analogous way.

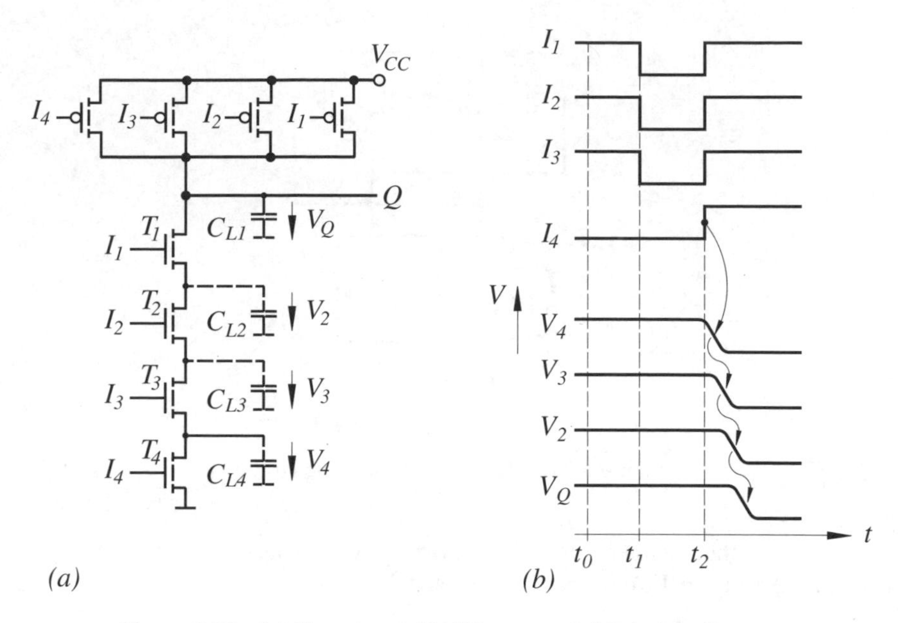

Figure 6.26 (a) Four-input *NAND* gate and (b) timing diagram

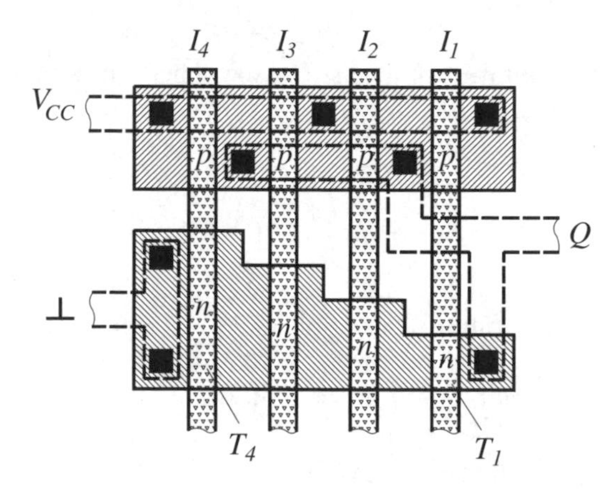

Figure 6.27 Layout of a four-input *NAND* gate with progressively sized n-channel transistors

6.3 IIIGH SPEED CIRCUITS

Communication systems and fiber optic applications require data rates far beyond 1 Gb/s. Circuits for these applications are a domain of bipolar transistor circuits, such as current mode logic (CML) and emitter couple logic (ECL), treated in Section 10.1.

With the trend to ever smaller device dimensions, MOS transistors became so fast that from a cost point of view, MOS current mode logic (MCML) became an interesting alternative to bipolar solutions (Yamashina and Yamada 1992).

Basically the speed performance is achieved by reducing the logic swing $\Delta V = V_{IH} - V_{IL}$ which is, for the circuits considered so far, the full rail to rail swing. The reduced logic swing leads to an improved switching time of

$$\Delta t \approx C\frac{\Delta V}{I} \tag{6.6}$$

The main element of the MCML circuits is the current switch, which can be used as an inverter operated in a differential manner (Figure 6.28). The current of the current source I_K is divided between the transistors T_1 and T_2. If the input voltage $V_I > V_{\bar{I}}$ then $I_{DS,1} \gg I_{DS,2}$. This causes a voltage drop ΔV across transistor T_3 and almost none across T_4. Thus the output voltages $V_{\bar{Q}L} = V_{CC} - \Delta V$ and $V_{QH} \approx V_{CC}$ result. A reverse situation exists when $V_I < V_{\bar{I}}$. Values for ΔV are between 0.2 V and 0.4 V. At the smallest value, the robustness of the circuit starts to deteriorate due to threshold variation and noise coupling.

The value of the voltage difference $\Delta V = I_K \cdot R$ across the transistors T_3 or T_4 can be adjusted by the voltage V_{Rp}. In this case, R is the resistance of the transistors which operate in the resistive region.

Instead of p-channel transistors, resistors may be used, when they are available in the manufacturing process. This can be an advantage. When transistors with small resistance values are required, this leads to relatively large (w/l) ratios and in conjunction to large capacitances at the transistors, which slow down the circuit.

Transistor T_S – the so-called current source (Section 8.1) – operates in current saturation and is controlled by the voltage V_{Rn}. The voltage at note (k) in Figure 6.28 remains almost constant due to the differential input signals (Section 8.3.2). This voltage does not even change when the data frequency is raised. Therefore, the power dissipation of the MCML technique is almost independent of the frequency (Tanabe *et al.* 2001) (Figure 6.28(b)). This is contrary to the CMOS logic, where the dissipated power (Equation 5.28) increases with frequency.

$$P = CV_{CC}^2 f \tag{6.7}$$

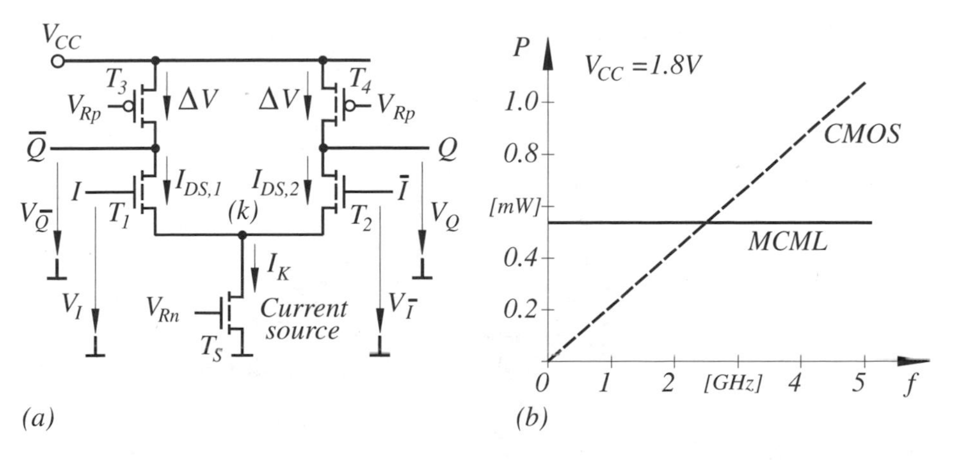

Figure 6.28 (a) Current switch and (b) power dissipation of CMOS and MCML as a function of the frequency (after Tanabe *et al.* 2001; © 2001 IEEE)

Therefore, the MCML circuits are very well suited to high speed applications.

A further advantage is that the constant current I_K is switched either by T_1 or T_2 to the left or right branch of the circuit. As a consequence, an almost constant current in the power-supply lines exists, causing only a very small switching noise in comparison to the one generated by CMOS gates. An additional advantage is that common noise acting on both inputs is suppressed. The reason is that the common noise has no influence on the source current I_K and the ratio of the currents between transistors T_1 and T_2 (Section 8.3.2). And last but not least, the power-supply voltage, and thus the power dissipation $P = I_K \cdot V_{CC}$, can be reduced substantially without adversely influencing the switching time (Equation 6.6) of the circuit.

The minimum supply voltage can be determined by analyzing the circuit of Figure 6.29. Transistors T_S, T_1, and T_2 operate under current saturation condition in order to achieve a high voltage gain (Section 8.3). The minimum V_{DS} voltage of the transistors to guarantee this operation condition is, according to Equation (4.54)

$$V_{DSsat} = V_{GS} - V_{Tn} \tag{6.8}$$

This leads to a current (Equation 4.58) of

$$I_{DS} = I_K = \frac{\beta_n}{2}(V_{GS} - V_{Tn})^2 \tag{6.9}$$

and therefore to a saturation voltage of

$$V_{DSsat} = \sqrt{\frac{2I_K}{\beta_n}} \tag{6.10}$$

This yields a minimum supply voltage, which is the sum of the saturation voltages at T_S, T_1, and the voltage at T_3, of

$$\boxed{V_{CC}(\min) = \sqrt{\frac{2I_K}{\beta_{n,s}}} + \sqrt{\frac{2I_K}{\beta_{n,1}}} + \Delta V} \tag{6.11}$$

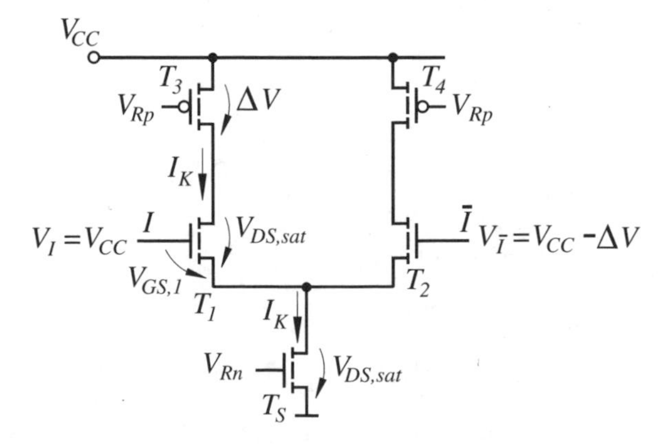

Figure 6.29 Current switch at minimum supply voltage

But there exists an additional criterion for the minimum supply voltage. If a voltage of V_{CC} is applied to the input I and one of $V_{CC} - \Delta V$ at input $\bar{I}$, then the supply voltage is not allowed to be below

$$V_{CC}(\min) = V_{DS,sat} + V_{GS,1}$$

$$\boxed{V_{CC}(\min) = \sqrt{\frac{2I_K}{\beta_{n,s}}} + \sqrt{\frac{2I_K}{\beta_{n,1}}} + V_{Tn}} \quad (6.12)$$

This analysis shows that it is advantageous to have transistors with small threshold voltages, around 0.2 V, available (mizuno *et al.* 1996). This leads to an improved switching performance at a reduced supply voltage. With the typical values of $V_{Dssat} = 0.4$ V for transistors T_S and T_1 and $\Delta V = 0.3$ V, Equation (6.11) yields a minimum supply voltage of 1.1 V.

How logic implementations in the MCML technique can be realized is illustrated with a few examples in Figure 6.30. In the examples, the transistors are connected in such a way that the source current I_K is either switched to the left or right branch of the circuit, but never to both branches simultaneously. This results in series-connected transistors, called series gating, and to additional voltage drops across these transistors. The result is a small increase in supply voltage. A further drawback of the series gating is that some transistors move into the resistive region, causing a reduction in voltage gain. An example illustrates the situation (Figure 6.31).

Via MCML gates two H levels of V_{CC} are applied to the transistor inputs T_1 and T_2 and two L levels of $V_{CC} - \Delta V$ to the differential counter parts. This leads to a situation where two in-series-connected transistors T_1 and T_2 have the same gate voltage of V_{CC} (Figure 6.31(b)). Since $V_{GS,1} > V_{GS,2}$ and $\beta_{n,1} = \beta_{n,2}$ an output characteristic, as shown in Figure 6.31(c), results. Since the current I_K flows through both transistors, an operation point A must exist where one transistor, that is T_1, is in the resistive region and the other, that is T_2, is in the current saturation region.

A further important example, the D flip-flop, is considered in conjunction with sequential circuits in Section 6.5.1.

The switching performance of the MCML circuits depends on the source current I_K and on the voltage swing ΔV. In order to keep these parameters independent of technology variation and operation conditions, the V_{Rn} voltage can be controlled by the circuit shown in Figure 6.32 (Mizuno *et al.* 1996).

At the differential inputs of the MCML circuit, the voltages $V_{CC} - \Delta V$ and V_{CC} are permanently applied. This results in currents of $I_1 \approx 0$ A and $I_2 \approx I_K$. If the voltage across T_{p2} is smaller or larger than $V_{CC} - \Delta V$, this causes an increase or decrease of the V_{Rn} voltage at the output of the amplifier and thus, a change in the current I_K until the voltage across the amplifier input approaches zero. Under this condition the V_{Rn} voltage has a value that guarantees that at the note (1) in the circuit a voltage of $V_{CC} - \Delta V$ exists. The generated V_{Rn} voltage can then be used as a reference voltage for other MCML circuits. A prerequisite is that they use the same transistor sizing.

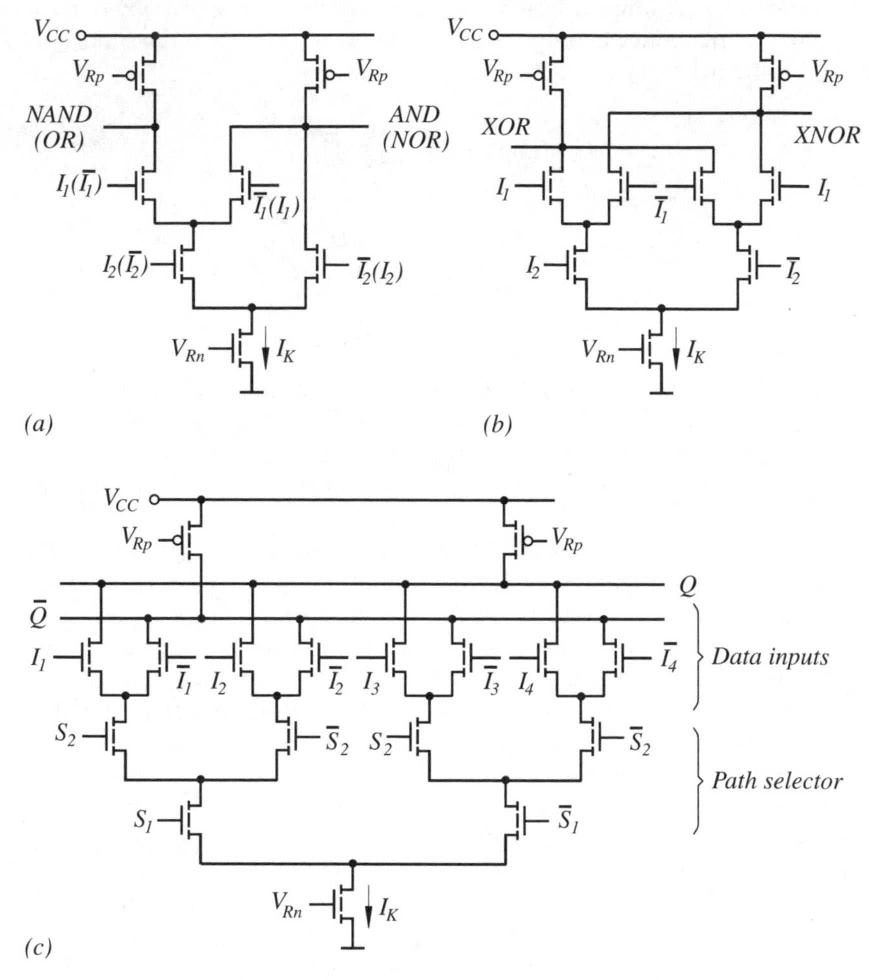

Figure 6.30 MCML gates: (a) *NAND*(*OR*)/*AND*(*NOR*), (b) *XOR*/*XNOR* and (c) 4 to 1 multiplexer

Offset voltage

A considerable disadvantage of all MCML circuits is their sensitivity towards offset voltages. If, e.g., a voltage of $V_{CC} - \Delta V$ is applied to both inputs of the current switch, the outputs have a voltage of V_M and the voltage difference between the outputs is $V_{off} = 0\,\text{V}$ (Figure 6.33).

This changes when an offset voltage exists (Figure 6.33(b)). If this voltage is too large, it causes a malfunction of the circuit. The offset voltage between the outputs is caused by asymmetries between the transistors. Usually the difference between the two threshold voltages of the n-channel and the p-channel transistors is dominant (Section 10.4.1). In order to analyze the influence of an asymmetry between the n-channel transistors, a small-signal analysis is useful. To simplify the situation, the p-channel transistors are replaced by identical resistors (Figure 6.34).

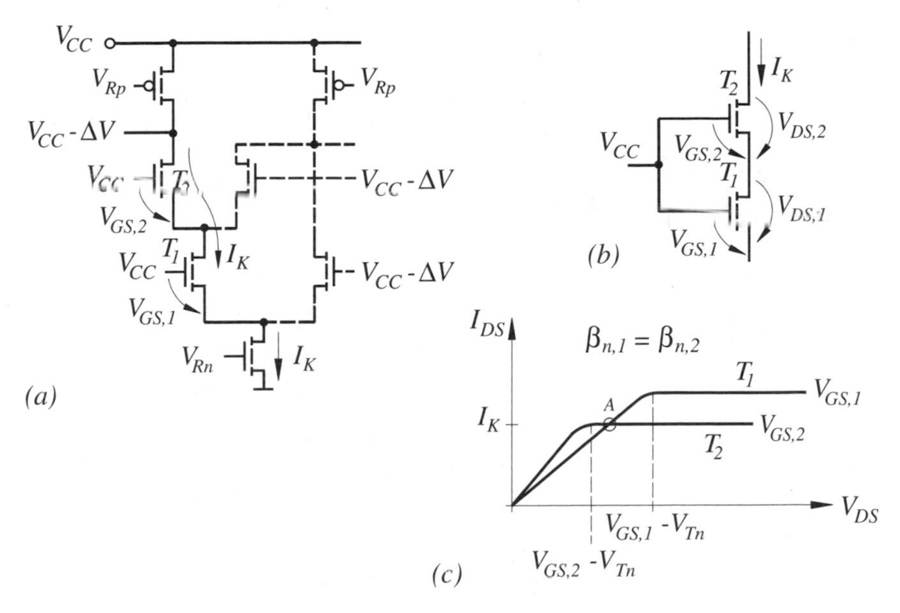

Figure 6.31 (a) Series-gated MCML gate, (b) voltage assignment and (c) output characteristic of T_1 and T_2

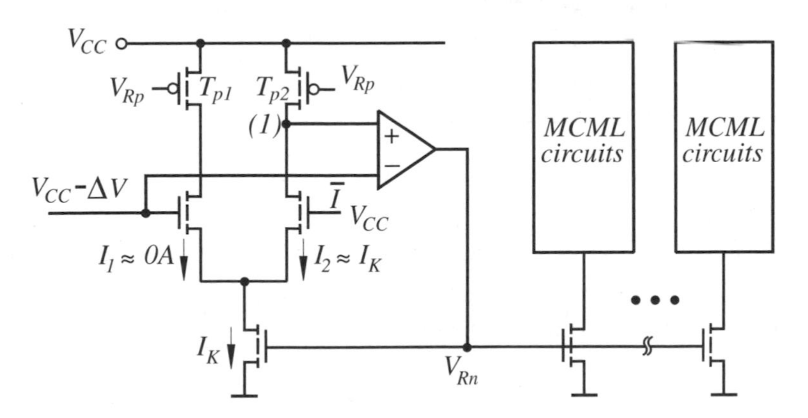

Figure 6.32 V_{Rn} voltage control circuit

According to Equation (8.34) of Section 8.3.2, the voltage gain of the circuit at low frequency – not considering transistors T_R – is

$$a_{dm}(0) = \frac{v_o}{v_i} = -g_m R_L \tag{6.13}$$

The transconductance has a value of (Equation 6.9)

$$g_m = \frac{\partial I_{DS}}{\partial V_{GS}} = \beta_n (V_{GS} - V_{Tn}) \tag{6.14}$$

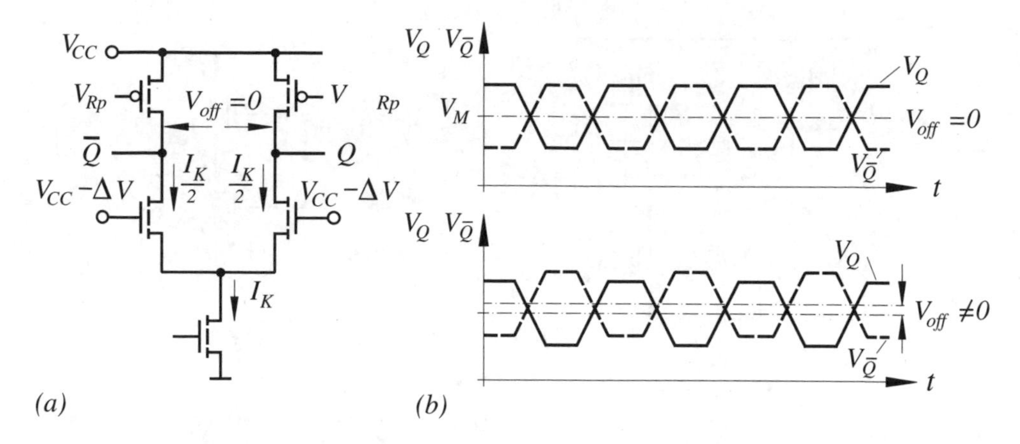

Figure 6.33 (a) Current switch with short circuited inputs and (b) influence of an offset voltage on the outputs

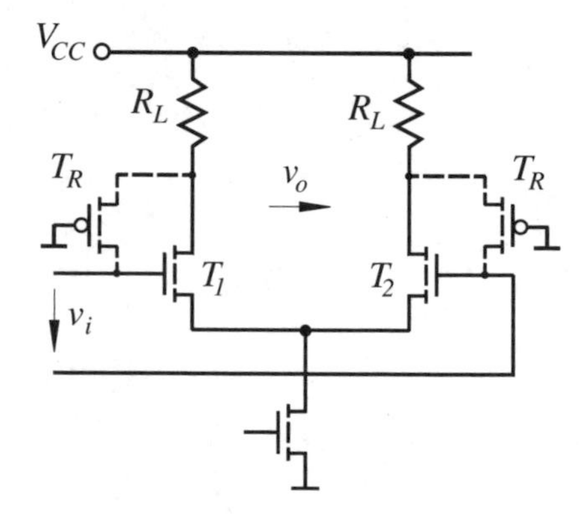

Figure 6.34 Current switch with small-signal excitation (after Tanabe *et al.* 2001; © 2001 IEEE)

and the offset voltage can be determined by

$$V_{off} = \Delta I_{DS} R_L \tag{6.15}$$

where ΔI_{DS} describes the difference between the n-channel transistor currents caused by a difference in threshold voltage. Expanding this equation yields

$$V_{off} \approx \frac{\partial I_{DS}}{\partial V_{Tn}} \Delta V_{Tn} R_L \tag{6.16}$$

and differentiation of Equation (6.9) results in

$$\frac{\partial I_{DS}}{\partial V_{Tn}} = -g_m \tag{6.17}$$

Finally, combining these equations leads to an offset voltage of

$$V_{off} \approx a_{dm}(0)\Delta V_{Tn} \tag{6.18}$$

This is probably not a very surprising result. It states that the offset voltage of the circuit increases with the voltage gain. To cope with the offset voltage, the gain $a_{dm}(0)$ is usually limited to about 1.5. In Tanabe *et al.* (2001) a different solution is proposed. The gain of the circuit is controlled by feedback transistors T_R, shown in Figure 6.34.

6.4 LOGIC ARRAYS

In the preceding chapters various logic circuits are discussed. These logic circuits can be combined to form more complex systems with an increased silicon area consumption per gate, due to the various interconnections. This dependence can be avoided and chip area more effectively used by employing logic arrays (Figure 6.35). A basic component of all logic arrays is the decoder. Its power dissipation, propagation delay, and area consumption determine to a large extent the performance of the logic array or, as is shown in Chapter 7, the performance of memory circuits. This is the reason why this circuit is considered in some detail in the following section.

6.4.1 Decoder

Decoders are components which translate an N-bit input word into an M-bit output word and where $M = 2^N$. Each output word has, e.g., only one H signal out of the remaining L ones. In order to demonstrate the principle a simple PMOS load NOR decoder is used (Figure 6.36).

The input consists of a 3-bit input word and its complement and the output of a $2^3 = 8$-bit word. Output Y_0 can only be in an H state when A, B, and C are in an L state. Y_1 can only be in an H state when $\overline{A}$, B, and C are in an L state, and so on. The decoding is always defined since only one determined transistor combination is able to cut off the transistors in a row. When one considers the transistor arrangement in the decoder one recognizes that the least significant bit (LSB) leads to an alternating transistor

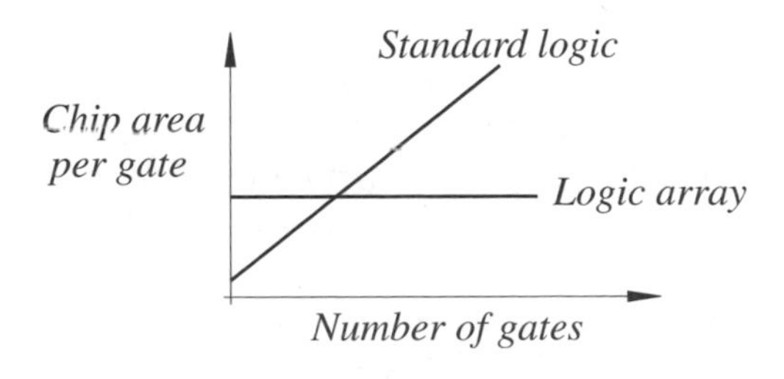

Figure 6.35 Comparison of different chip areas per gate

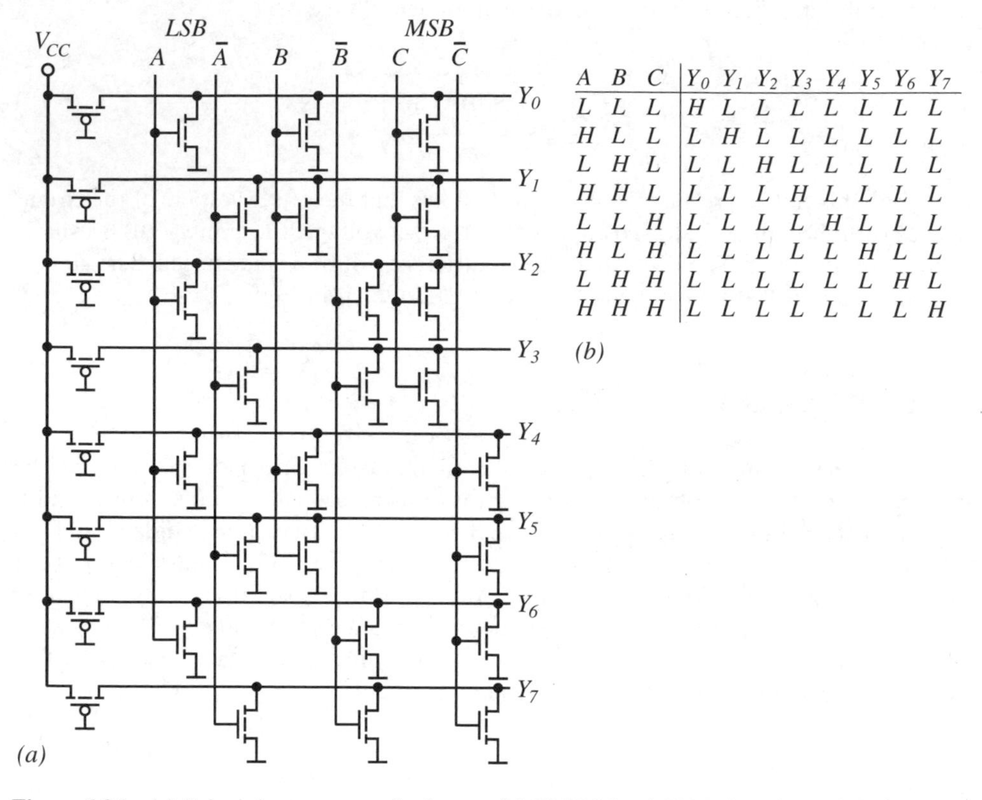

A	B	C	Y_0	Y_1	Y_2	Y_3	Y_4	Y_5	Y_6	Y_7
L	L	L	H	L	L	L	L	L	L	L
H	L	L	L	H	L	L	L	L	L	L
L	H	L	L	L	H	L	L	L	L	L
H	H	L	L	L	L	H	L	L	L	L
L	L	H	L	L	L	L	H	L	L	L
H	L	H	L	L	L	L	L	H	L	L
L	H	H	L	L	L	L	L	L	H	L
H	H	H	L	L	L	L	L	L	L	H

Figure 6.36 (a) Principle structure of a 1-out-of-8 PMOS load *NOR* decoder and (b) truth table

connection, whereas the most significant bit (MSB) changes the transistor in a quadruple manner. The circuit reveals that the transistor placements are a representation of the binary code. The PMOS load decoder is an example, and a very bad one with respect to power dissipation, since each L state – and there are $2^N - 1$ – consumes static power. Therefore only ratioless designs are considered in the next sections.

Complementary decoder

An example of such a decoder is shown in Figure 6.37 implementing the truth table of Figure 6.36(b). The n-channel transistors implement – exactly as in Figure 6.36 – the *NOR* function, whereas the complementary one is realized by the p-channel transistors. A further example is a complementary *NAND* decoder (Figure 6.38). Contrary to the preceding cases, this decoder selects one L signal out of the remaining H signals. The n-channel transistors implement the *NAND* function and the p-channel transistors the complementary one.

To reduce the number of interconnections or the transistor count, two modifications of the n-channel implementation of the 1-out-of-8 *NOR* and *NAND* decoder are shown as examples in Figure 6.39.

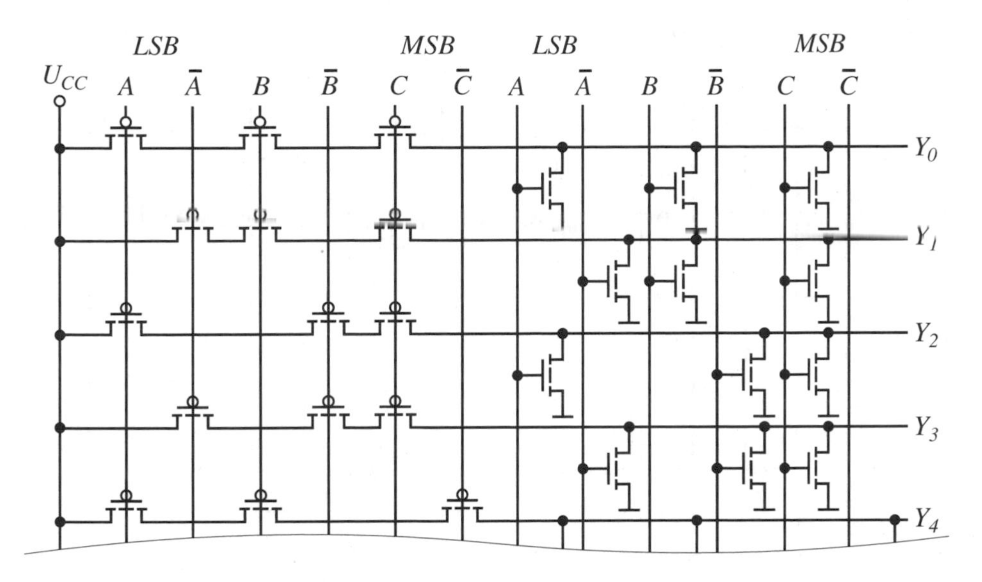

Figure 6.37 Complementary 1-out-of-8 *NOR* decoder

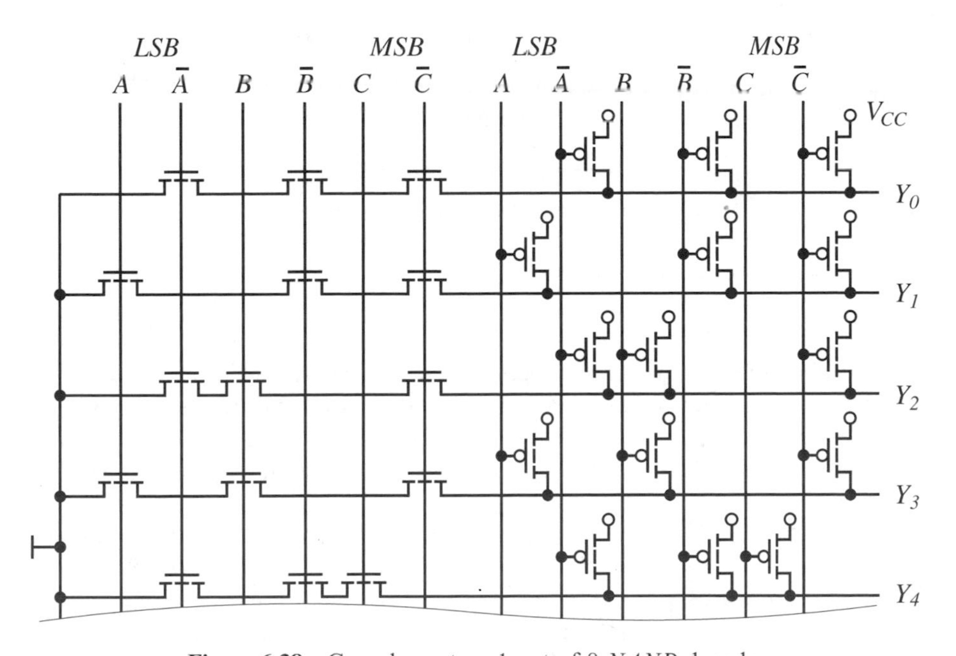

Figure 6.38 Complementary 1-out-of-8 *NAND* decoder

As is illustrated in Figure 6.37, each individual n-channel transistor must be connected to ground. This individual ground connection can be avoided with a so-called virtual ground scheme (Figure 6.39(a)). The even or odd outputs of the decoder are always connected via the LSB to ground. These grounded output lines can therefore be used by the remaining transistors as ground connection.

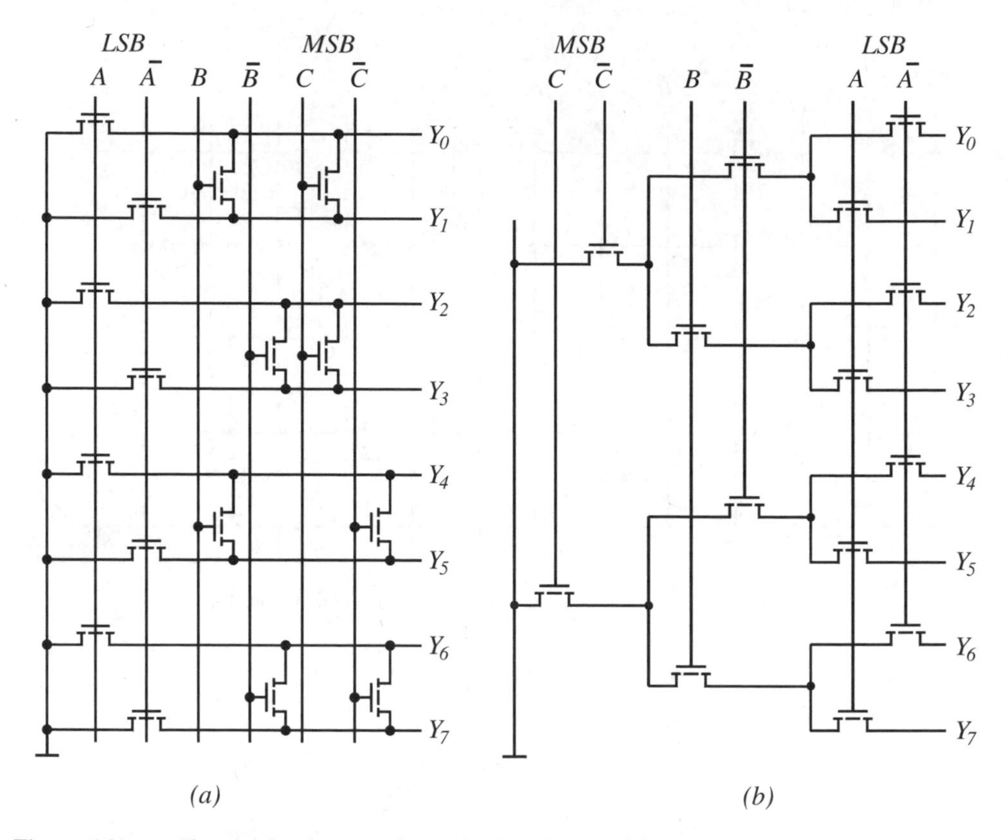

Figure 6.39 n-Channel implementation: (a) virtual ground-based *NOR* decoder and (b) tree-based *NAND* decoder

Another efficient implementation is possible with a tree decoder. This decoder uses a binary reduction scheme as shown in Figure 6.39(b). It is obvious that the number of devices is substantially reduced. The schemes of Figure 6.39 are not only applicable to the n-channel transistors but of course to the complementary function realized in p-channel transistors also.

A disadvantage of all complementary decoders is that either n-channel transistors or p-channel transistors are connected in series. When more than five transistors are connected in series a substantial increase in rise or fall time results. To ease this situation a progressive transistor sizing (Section 6.2.5) is useful and makes sense, particularly in the case of a tree decoder. A way around this problem is to cascade decoders. The principle is shown in Figure 6.40. Altered CMOS inverters are used, where the sources of the p-channel transistors are controlled by the *NOR* decoder and the gates by the *NAND* decoder. As the example illustrates, an *H* signal of the *NOR* decoder can only propagate to the output Z_0, when the *NAND* decoder supplies an *L* signal (consult Problem 6.3 for worst case condition).

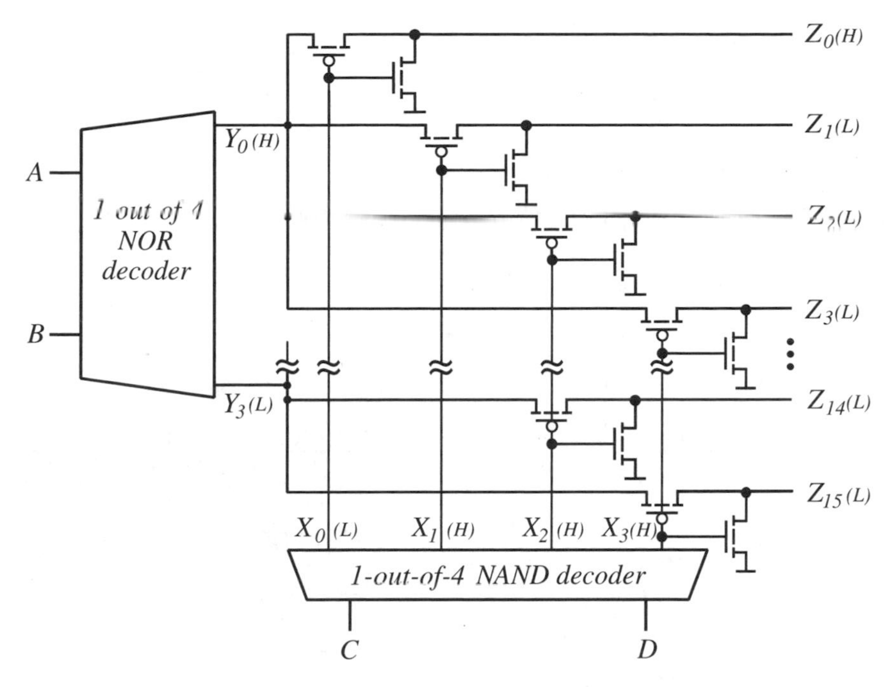

Figure 6.40 Cascaded 1-out-of-16 decoder

Clocked decoders

The principle of the clocked CMOS circuits of Section 6.2.1 can be used advantageously in decoder implementations in order to avoid series-connected transistors. An example of a *NOR* decoder is shown in Figure 6.41. With $\phi = L$ and $\overline{\phi} = H$ all outputs of the *NOR* gates are in an *L* state, causing the transistors in the decoding array to be cut off during precharge. The p-channel transistors are conducting and all parasitic capacitances of the output lines *y* are charged to V_{CC}. When the clock signal changes, the capacitances of the output lines are discharged to 0 V, except for one in accordance with the input information.

Two decoder layouts are shown. The one in Figure 6.41(b) uses as input polycide lines and for the output metal lines. The ground connections of the n-channel transistors are implemented with diffusion stripes. Since these stripes have a relatively high resistance of $3\,\Omega/\square$ even when silicide is processed, they have to be periodically connected to ground. The advantage of this layout is the small pitch between the output lines. This is particularly important in memory applications, where decoders control very small memory cells. An alternative layout is shown in Figure 6.41(c). The inputs are metal lines and the outputs are diffusion stripes. This yields a large pitch between the output lines. But this scheme has the advantage that the propagation delay of the input lines is reduced, whereas that of the diffusion strips is not significant, since its dimensions are usually much smaller than in the y-direction.

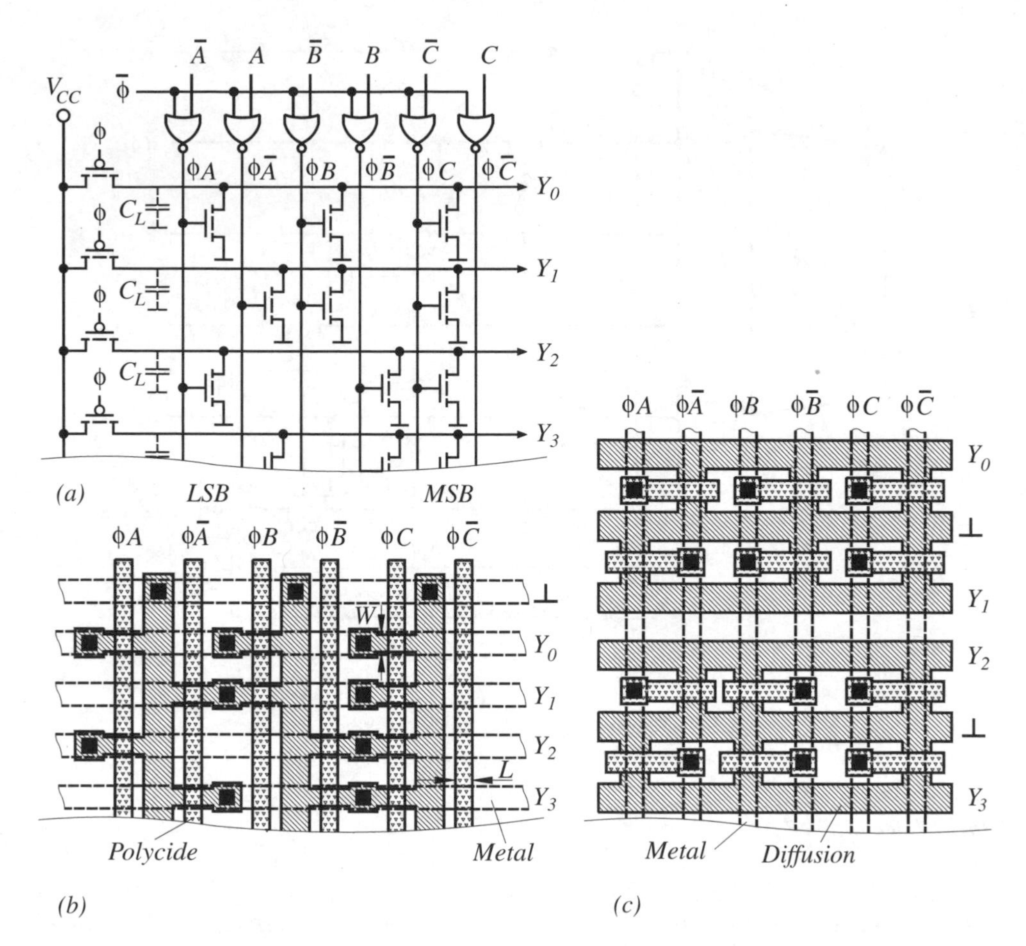

Figure 6.41 (a) Clocked 1-out-of-8 *NOR* decoder, (b) layout with polycide input lines and (c) layout with metal input lines

6.4.2 Programmable Logic Array

A combinational logic circuit with N inputs relates to a maximum of 2^N output words. This logic can be implemented in one of the logic circuit styles presented or by the combination of transistors in an array forming a programmable logic array (PLA). This yields a more effective use of chip area, as is mentioned in the introduction to this section. In these arrays the sums of binary product terms such as

$$Q_1 = A \cdot \overline{B} \cdot \overline{C} + \overline{A} \cdot B \cdot \overline{C}; \quad Q_2 = \overline{A} \cdot \overline{B} \cdot \overline{C} + A \cdot \overline{B} \cdot \overline{C} + \overline{A} \cdot \overline{B} \cdot C + \overline{A} \cdot B \cdot C;$$
$$Q_3 = \overline{A} \cdot B \cdot \overline{C} + A \cdot B \cdot \overline{C} + A \cdot \overline{B} \cdot C; \quad Q_4 = A \cdot B \cdot \overline{C} + \overline{A} \cdot B \cdot C,$$

are implemented. In the following example *NOR*/*NOR* arrays are used to convey the idea. For simplicity reasons PMOS load implementations are employed (Figure 6.42).

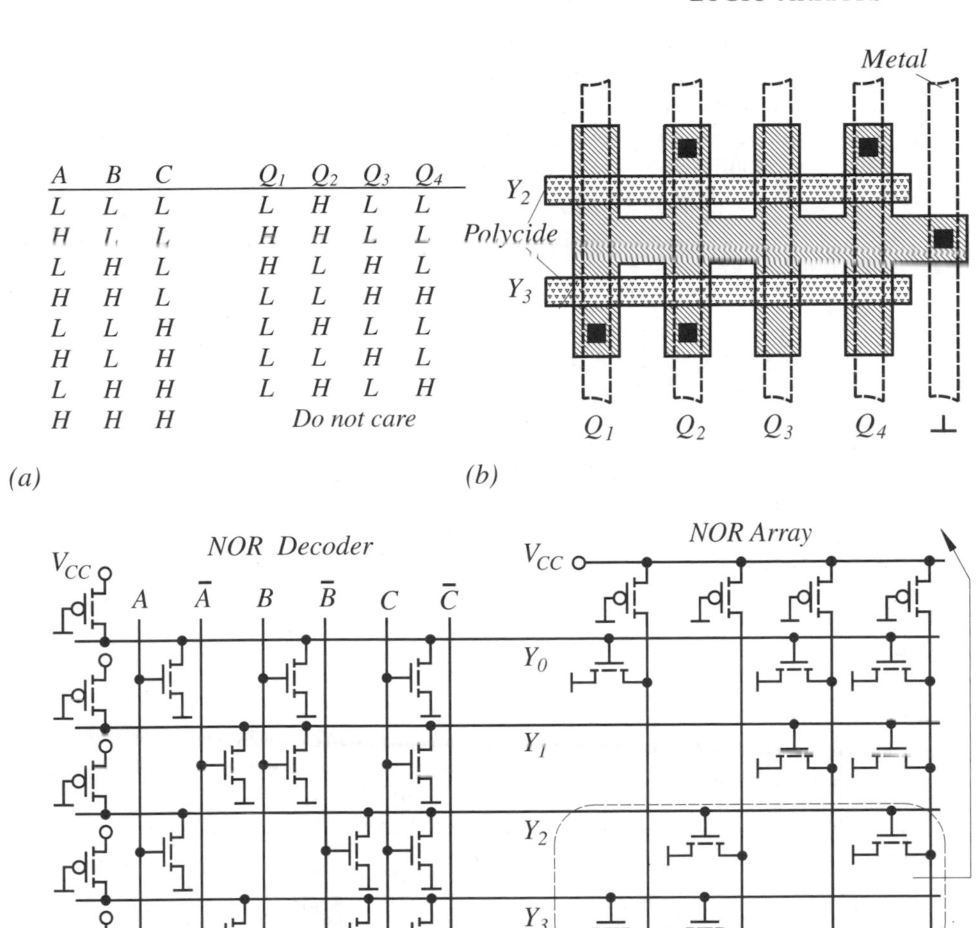

A	B	C	Q_1	Q_2	Q_3	Q_4
L	L	L	L	H	L	L
H	L	L	H	H	L	L
L	H	L	H	L	H	L
H	H	L	L	L	H	H
L	L	H	L	H	L	L
H	L	H	L	L	H	L
L	H	H	L	H	L	H
H	H	H	Do not care			

Figure 6.42 PLA with *NOR*/*NOR* arrays: (a) truth table, (b) layout detail and (c) circuit implementation

The starting point is a *NOR* decoder. Its outputs are the inputs to a following *NOR* array or matrix with the outputs Q_1 to Q_4. The connections in this matrix are made by contacts between the metal output lines and the drains of the transistors, according to the truth table of Figure 6.42(a).

An even more effective layout with respect to silicon area can be achieved by the use of *NAND/NAND* arrays. This is demonstrated with an example where a clocked design approach is used and the truth table of the preceding PLA is implemented (Figure 6.43).

The programming of the *NAND* array or matrix is done with an additional lithographic step, followed by ion implantation. This changes enhancement transistors with a threshold voltage of, e.g., 0.8 V to depletion transistors with a threshold

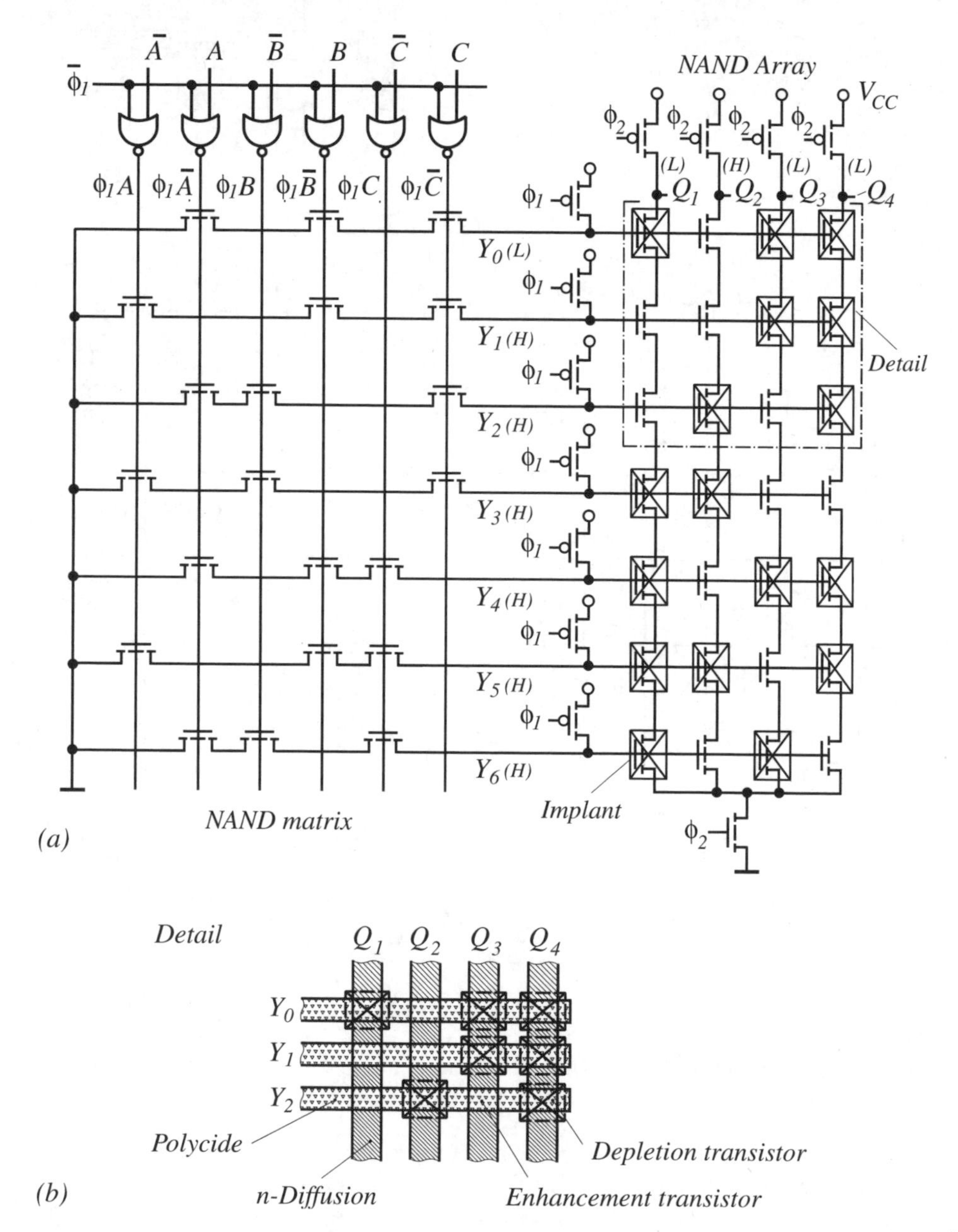

Figure 6.43 (a) *NAND/NAND* PLA implementation and (b) layout detail

voltage of −3.5 V (Figure 5.11). These transistors cannot be cut off when an *L* signal is applied by the decoder output. A *NAND* decoder selects one *L* signal out of all *H* signals. Thus all transistors in the array are turned on except the ones connected to a selected row. In this row it depends on the implantation whether a transistor is conducting or non-conducting. If the transistor conducts, the total transistor chain is conducting and able to discharge the precharged output capacitance. The saving in layout area is obvious, since no contact zones are required (Figure 6.43(b)). A disadvantage of this PLA concept is the large number of series-connected transistors, which lead to slow fall times. Therefore, this scheme is usually used in applications, such as dictionaries or speech recording, where the slow switching performance is of no concern.

6.5 SEQUENTIAL CIRCUITS

At the beginning of this chapter it is mentioned that a digital system consists of combinational and sequential circuits. The sequential circuits not only depend on the input data but are also a function of previously stored data. Examples are flip-flops and registers, which are treated in this section.

6.5.1 Flip-flop

In digital circuits the set – reset (SR) flip-flop and the delay (*D*) flip-flop are the most important temporary storage mediums. These flip-flops can be grouped into static and dynamic ones, as is described in the following.

Static SR flip-flops

A flip-flop can be implemented by two cross-coupled *NOR* gates (Figure 6.44(a)). The inputs are named set and reset. If an *H* signal is applied to the set input *S*, this causes the

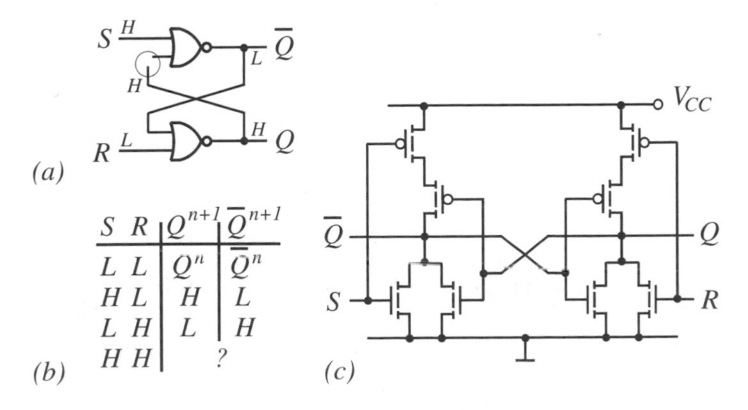

S	R	Q^{n+1}	$\overline{Q}^{n+1}$
L	L	Q^n	$\overline{Q}^n$
H	L	H	L
L	H	L	H
H	H	?	

Figure 6.44 (a) SR flip-flop, (b) truth table and (c) circuit implementation

output $\overline{Q}$ to be in an L state. Since this output is connected to the input of a second NOR gate and an L signal exists at the reset input R, its output Q must be in an H state. If the output Q is connected to the second input of the first NOR gate (circle in Figure 6.44(a)), the cross-coupled NOR gates remain in these states, even when the set input is changed from the H to the L state. If an H signal is applied to the reset input, then the flip-flop changes into the opposite direction. The flip-flop thus has two stable states into which it can be switched, either by applying an H signal to the set or an H signal to the reset input. If $R = S = L$ no output change occurs. This is shown in the truth table by the previous output Q^n. If in comparison $R = S = H$, then both outputs are in an L state. This is not an allowed condition, since the flip-flop is in an undefined state when released.

The complementary circuit implementation is based on the two-input NOR gate of Figure 6.3. In order to use this flip-flop in sequential circuits, it has to be controlled by a clock signal. This is achieved in the realization of Figure 6.45 by the added $NAND$ gates.

The disadvantage of the undefined state with $R = S = H$ still exists. This problem can be circumvented with a D flip-flop.

Static D flip-flop

For this purpose the reset input is controlled via an inverter by the set input (Figure 6.46(a)). It is obvious that for the circuit realization the implementation of Figure 6.45(b) plus an additional CMOS inverter can be used. In order to reduce the transistor count of this circuit an alternative approach with pass transistors (Section 6.1.3) and an additional inverted clock signal is preferred. Information is stored when the output Q is fed back by the turned on pass transistors *P2* to the input of the two inverters. Information is transmitted to the flip-flop when the pass transistors *P2* are cut off and the pass transistors *P1* are turned on. The cut-off of the pass transistors *P2* is necessary to avoid the input signal D of the flip-flop having to work against the low impedance output.

In Section 6.3 MCML gates are described. How a D flip-flop can be implemented in this technique is shown in Figure 6.47. With $\phi = H$ and $\overline{\phi} = L$ the information at inputs $D, \overline{D}$ is transmitted to the outputs. If the clock signal changes to $\phi = L$ and $\overline{\phi} = H$ the

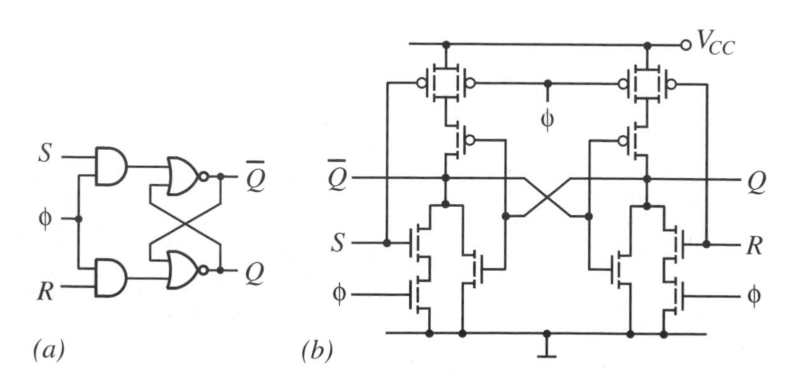

Figure 6.45 Clocked SR flip-flop and (b) circuit implementation

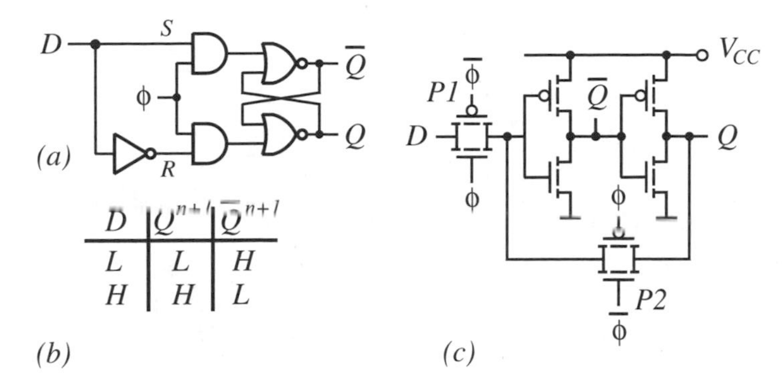

D	Q^{n+1}	$\bar{Q}^{n+1}$
L	L	H
H	H	L

Figure 6.46 (a) Clocked D flip-flop, (b) truth table and (c) circuit implementation

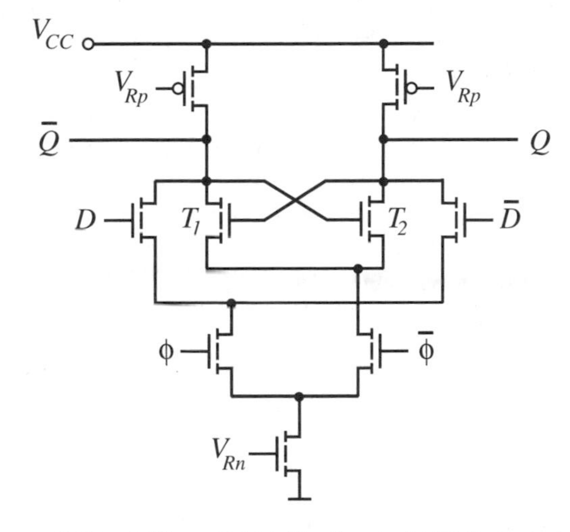

Figure 6.47 Differential D flip-flop in MCML technique

input transistors are cut off and the information is stored statically by the feedback transistors T_1 and T_2.

Dynamic D flip-flop

The flip-flops considered so far are static ones. That is, the information is stored by the cross-coupled inverters or *NOR* gates and remains there for as long as the power-supply voltage is available. Contrary to this kind of data storage is the one where the information is stored as different amounts of charge. The advantage of such information storage is that the implementation requires a minimum of components, and thus of silicon area. The best-known dynamic D flip-flops are considered next. Figure 6.48 shows a first example.

With a clock signal of $\phi = H$, the n-channel pass transistor is turned on and the capacitance C_L is charged to the voltage applied by the input. With $\phi = L$ the transistor is cut off and the information is stored as a different amount of charge at C_L.

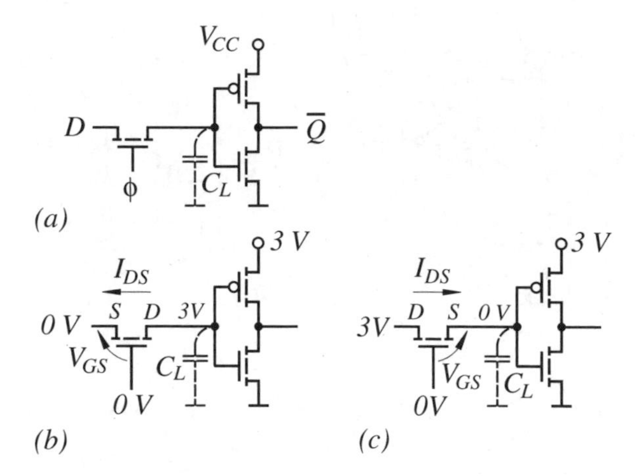

Figure 6.48 (a) Dynamic D flip-flop, (b) sub-threshold current at stored H level and (c) sub-threshold current at stored L level

The capacitance consists mainly of the gate capacitance of inverter transistors. This implementation needs the smallest number of components but has the disadvantage that the full H level is only transmitted via the pass transistor, when a clock signal of $\phi > V_{CC} + V_{Tn}(V_{SB})$ is applied (Section 6.1.3).

A disadvantage of all dynamic memories is that, due to leakage currents, the charge can only be guaranteed for a particular time. That is the reason why these memories are called dynamic flip-flops.

Refresh time determination

The leakage current discharging or charging the capacitance is usually the sub-threshold current of the transistor (Section 4.4.3). This current can be determined, according to Equation (4.65), by

$$I_{DS} = \beta_n(n-1)\phi_t^2 e^{(V_{GS}-V_{Tn})/\phi_t n} \tag{6.19}$$

where V_{DS} is assumed to be larger than 100 mV. To analyze the situation in more detail the voltage conditions of the pass transistor of Figure 6.48(b) are repeated in Figure 6.49. In this example the capacitance C_L is charged to a voltage of, e.g., 3 V, and the transistor is, with $\phi = 0$ V, cut-off. The voltage at input D changes to 0 V. The V_{GS} voltage is thus zero. This leads, according to Equation (6.19), to a sub-threshold current discharging the charged capacitor. The opposite situation exists when the capacitance is charged to 0 V.

The pass transistor is cut off and the input signal changes to 3 V (Figure 6.48(c)). In this case, the capacitance is charged by the sub-threshold current. But there is a distinct difference. In the first case, the V_{GS} voltage remains zero and in the second a negative V_{GS} voltage develops, opposing the charging of the capacitance and in effect self-limiting the current.

For example, when the n-channel transistor has a sub-threshold swing of $S = 120$ mV/decade and the voltage at C_L raises to 120 mV, this leads to a reduction in

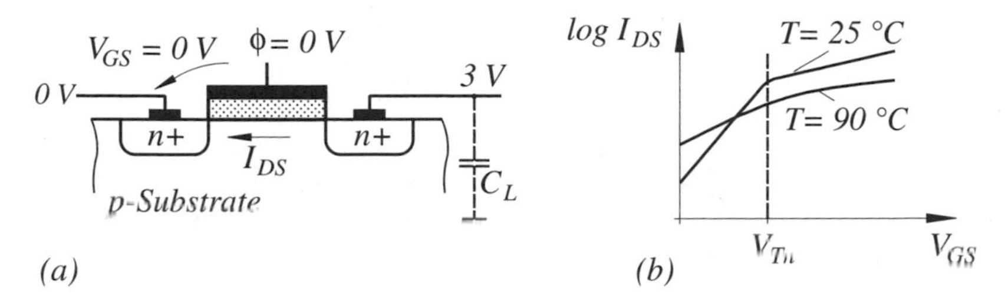

Figure 6.49 (a) Sub-threshold currents of the cut-off pass transistor and (b) sub-threshold behavior as a function of temperature

sub-threshold current by one decade. Thus, the situation with the voltage conditions of Figure 6.48(b) or Figure 6.49(a) represents the worst case conditions for the sub-threshold current. Unfortunately, the sub-threshold current increases substantially with temperature (Figure 6.49(b)), due to its positive temperature coefficient (Section 4.4.4). The maximum possible storage time, called refresh time, can be estimated by

$$\Delta t \approx C_L \frac{\Delta V}{I_{DS}} \tag{6.20}$$

In this equation ΔV is an acceptable voltage change, which does not lead to a circuit malfunction.

The circuit of Figure 6.48(a) requires an elevated clock level for charging the capacitance to the full H level of the D input. If an elevated clock level is not desirable, a level restorer (Section 6.1.3) can be used (Figure 6.50(a)). With a level restorer one

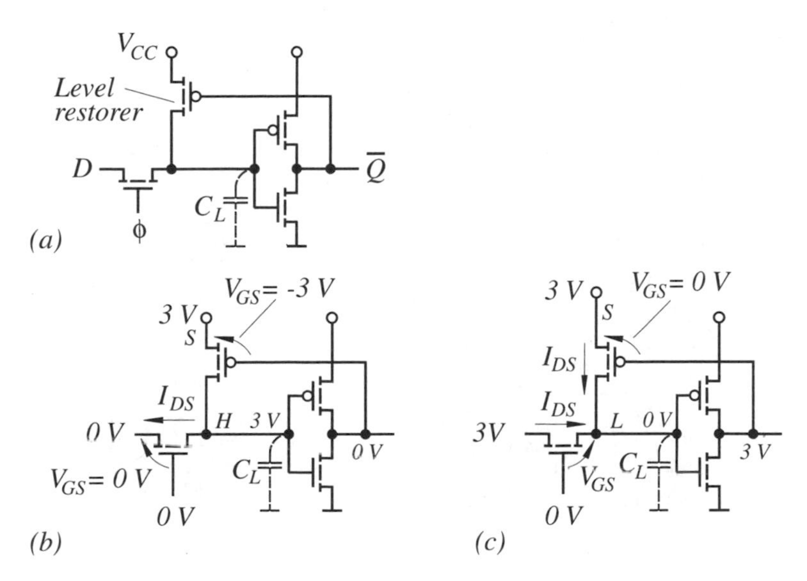

Figure 6.50 (a) Dynamic D flip-flop with level restorer, (b) sub-threshold current at stored H level and (c) sub-threshold current at stored L level

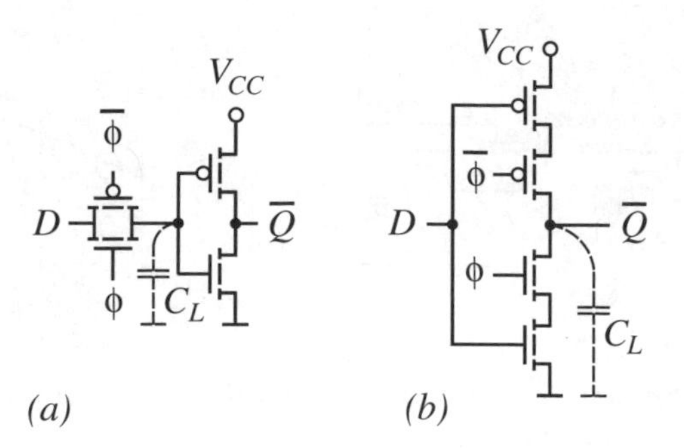

Figure 6.51 Dynamic D flip-flop: (a) with pass transistors at the input and (b) with pass transistors at the output

could get the idea that the sub-threshold current (Figure 6.50(b)) is of no concern anymore, since this transistor is conducting. This is correct. But unfortunately a situation exists (Figure 6.50(c)) with some unpleasant surprises. The current of the pass transistor is self-limiting, just as in the preceding example, but the level restorer has a constant V_{GS} voltage of zero volts and is thus responsible for the charging of the capacitance and the time the L level can be guaranteed.

Two further examples of dynamic D flip-flops are shown in Figure 6.51. Both circuits require additionally an inverted clock signal. In Figure 6.51(a), the information is stored, as in the preceding examples, at the gate capacitances, whereas in Figure 6.51(b), the output capacitance is used. With respect to the sub-threshold current the situation is comparable to the previous cases. There always exists one transistor with a worst case V_{GS} voltage of zero. Thus, these implementations show no improved storage time behavior. But a comparison of the layouts (Figure 6.52) reveals that the one with the pass transistors at the output is preferable due to the smaller number of interconnections.

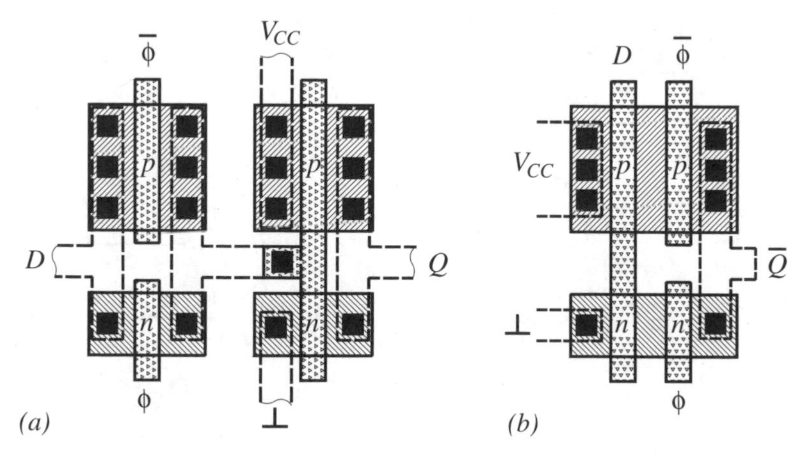

Figure 6.52 Layout of dynamic D flip-flops: (a) according to Figure 6.51(a) and (b) according to Figure 6.51(b)

Master–slave principle

Master–slave flip-flops are built by cascading two flip-flops. The first one is called master and the second one slave (Figure 6.53). The purpose of this scheme is to guarantee a controlled propagation of logic signals through a register or counter at any time. The master is activated by clock ϕ_1, whereas the slave is activated by ϕ_2. These clock signals are non-overlapping in order to make sure that only one flip-flop is able to accept information, whereas the other one stores the data. This scheme guarantees that the input D of the master–slave flip-flop is never connected to the output at any time. The flip-flop is edge-triggered, as the following discussion of the timing diagram shows. At time t_1 the clock signal ϕ_1 changes to an H state, activating the master. At the outputs Q and $\overline{Q}$ the data remain unchanged since ϕ_2 is in an L state. In order to guarantee proper functioning of the circuit, the input data must be stable before the negative edge of ϕ_1 disconnects the master from the input D. This time is called set-up time t_S. Its value corresponds approximately to the delay time of the master. A further characteristic timing parameter is the maximum time t_{FF} required, until the outputs Q and $\overline{Q}$ change into the appropriate state after the negative edge of clock signal ϕ_1.

6.5.2 Two-Phase Clocked Register

A register consists of consecutively connected master–slave flip-flops with common clock lines. The input and output of data can be sequential or parallel. These registers

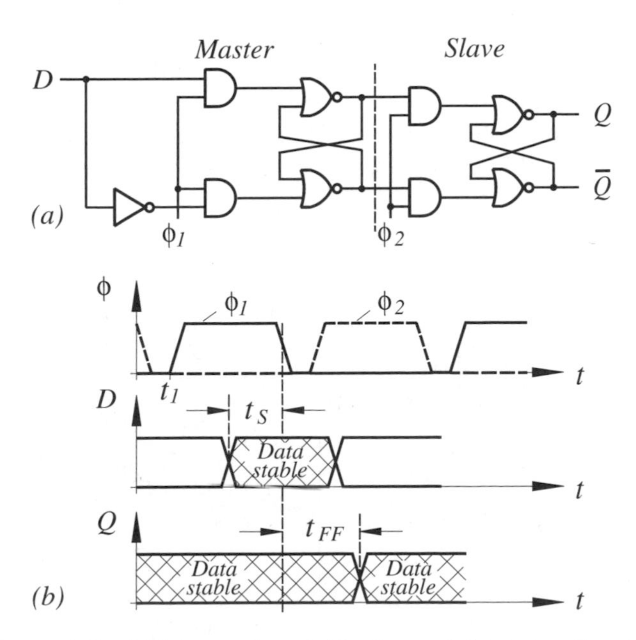

Figure 6.53 (a) Master–slave D flip-flop and (b) timing diagram

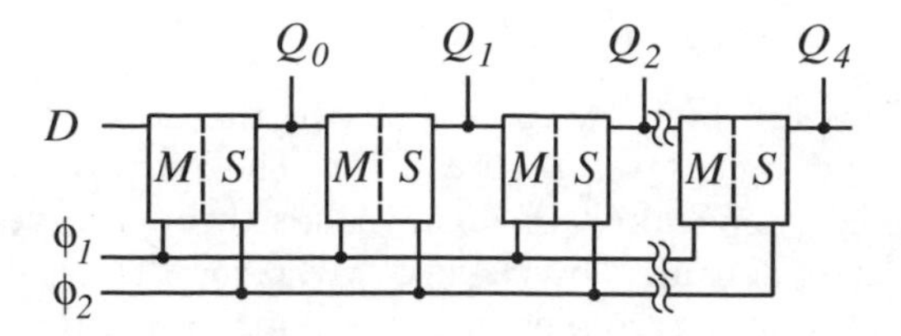

Figure 6.54 Series–parallel converter with mater–slave D flip-flops

can therefore be used as a series–parallel or parallel–series converter. An example of a series–parallel register is shown in Figure 6.54.

In the following, examples of such register implementations are considered. These realizations can be grouped into static, quasi-static, and dynamic solutions.

Static master–slave register

This implementation uses the D flip-flop of Figure 6.46 for the implementation of the master and the slave part of the register. This leads to a relatively large transistor count, as shown for the first stage of the register in Figure 6.55.

Quasi-static master–slave register

The number of transistors can be reduced when a dynamic flip-flop is used as master and a static one as slave (Figure 6.56).

Pass transistors *P1* are activated by ϕ_1 and $\overline{\phi}_1$, and *P2* and *P3* are deactivated by ϕ_2 and $\overline{\phi}_2$. In this condition, the information is stored as different charge levels at C_S, whereas the charge at C_M can be altered by the input data up to the negative edge of ϕ_1. The output data remain unchanged. If *P1* is deactivated and *P2* and *P3* activated, the information shifts from C_M to the output and is statically stored by the feedback via *P2*.

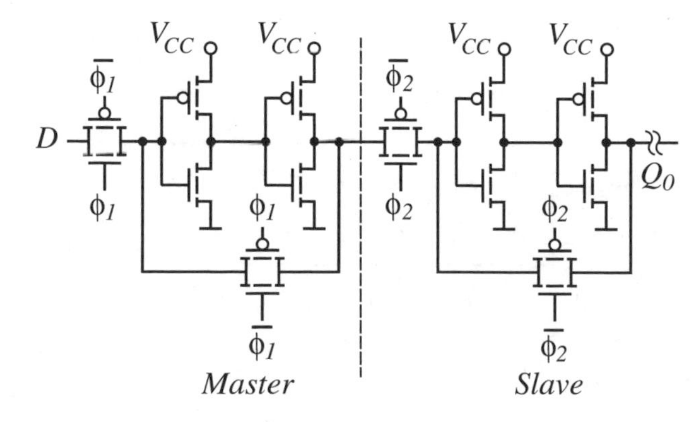

Figure 6.55 Series–parallel register with static master–slave D flip-flops

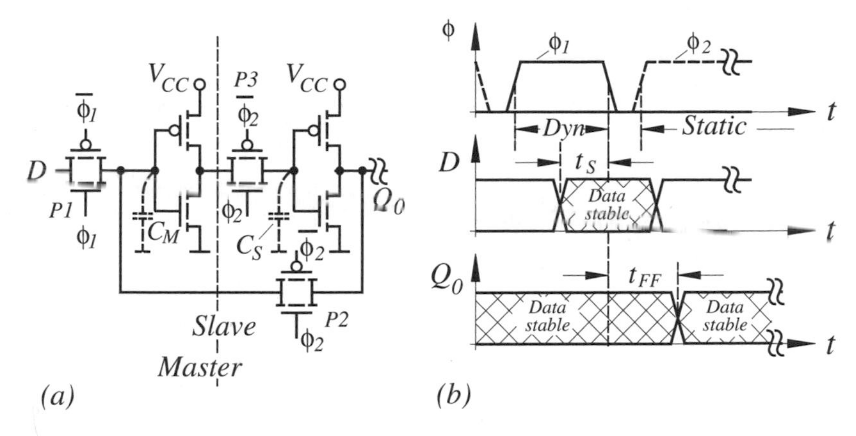

Figure 6.56 (a) Series–parallel register with quasi-static master–slave D flip-flops and (b) timing diagram

It is obvious that no storage time limitation exists, as long as the information is stored in the static slave stage.

Dynamic master–slave register

The number of components can be reduced even further when dynamic flip-flops are used for the master and slave implementation of the register (Figure 6.57). In this register the dynamic flip-flop of Figure 6.51(a) is employed. As in the previous case, a prerequisite for the proper functioning of the circuit is that non-overlapping clocks are available.

If the dynamic flip-flop of Figure 6.51(b) is used it is even possible to control the register with one clock and its complement only (Suzuki *et al.* 1973). This circuit implementation is called clocked CMOS (C^2MOS) master–slave flip-flop (Figure 6.58).

If $\phi = H$ and $\overline{\phi} = L$, the information is stored as different charge levels at C_S, whereas the charge at C_M can be changed by the input data up to the negative edge of ϕ. With $\phi = L$ and $\overline{\phi} = H$, the data reach the output Q_0. An important feature of this flip-flop is that it is insensitive against an overlapping of the ϕ and $\overline{\phi}$ clocks. This is illustrated in Figure 6.59.

The output is always disconnected from the input when the signals are $\phi = \overline{\phi} = H$ or $\phi = \overline{\phi} = L$. This is so, since either the n-channel or the p-channel pass transistors are

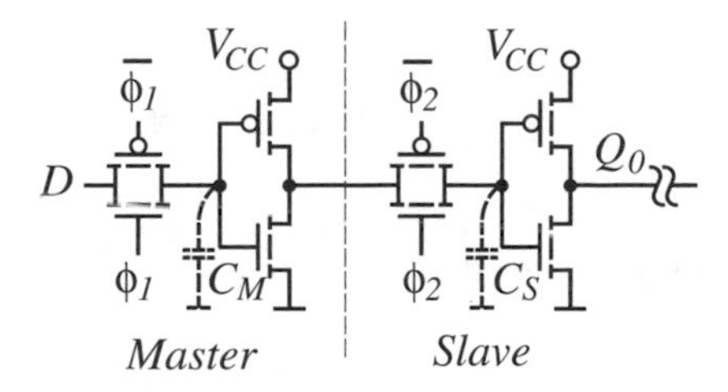

Figure 6.57 Series–parallel register with dynamic master–slave flip-flops (timing diagram according to Figure 6.53(b))

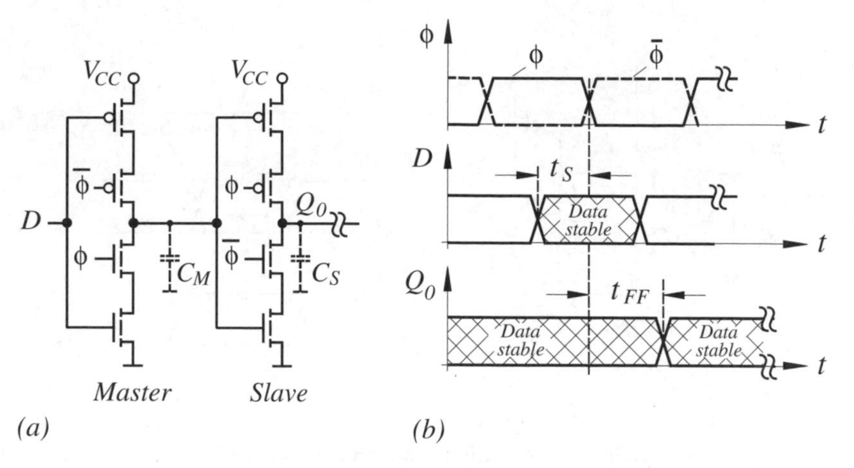

Figure 6.58 (a) Series–parallel register with C²MOS master–slave D flip-flops and (b) timing diagram

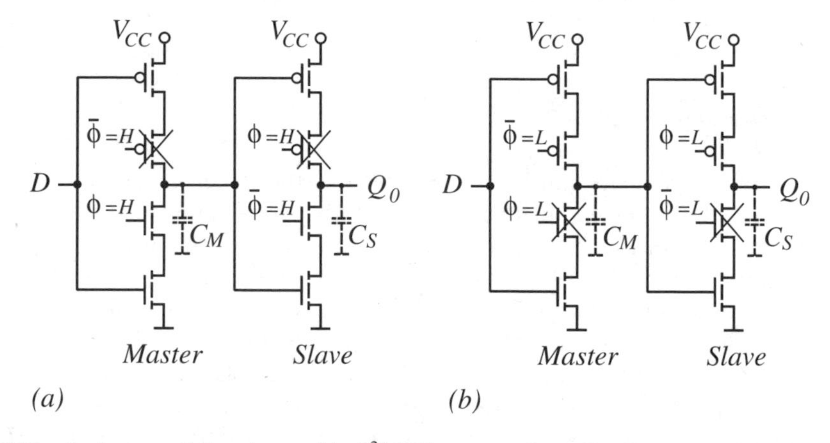

Figure 6.59 Series–parallel register with C²MOS master–slave flip-flops: (a) overlapping clocks $\phi = \overline{\phi} = H$ and (b) overlapping clocks $\phi = \overline{\phi} = L$

non-conducting and a signal propagation from the input to output is possible only when the flip-flop acts as two cascaded inverters. This is the case when the n-channel and p-channel pass transistors are conducting simultaneously. This situation may occur during relatively slow clock transition times. To avoid this situation the rise and fall times of the clock signals should be substantially shorter than the delay time of the flip-flop stages.

6.5.3 One-Phase Clocked Register

Two-phase clocked registers and other two-phase clocked components have the disadvantage that they need a sophisticated clock distribution system in order to guarantee the correct circuit function over a wide range of parameter spread and temperature. The situation eases obviously when one-phase clocked circuits are used (Yuan and Svensson 1989). The basic element is a dynamic flip-flop (Figure 6.60), consisting of two inverters with n-channel pass-transistors. Note: This is not a master–slave flip-flop even if it may look like it at a first glance. If $\phi = H$, then the pass transistors are turned on and the circuit behaves like two in-series-connected inverters (Figure 6.60(a)). If $\phi = L$, the data are stored as different charge levels. No logic state at input D is able to change the stored charge at C_2. This can be examined by considering the following two situations:

(a) The capacitances are charged during the transmit condition by an H input signal at the input to $C_1 = L$ and $C_2 = H$. Then with $\phi = L$, the pass transistors are turned off. Next, the input signal changes from an H state to an L state. This causes C_1 to be charged to an H level. It is important that the output H state at C_2 remains unchanged by this transition.

(b) The capacitances are charged during the transmit condition by the L input signal at the input to $C_1 = H$ and $C_2 = L$. Then, with $\phi = L$, the pass transistors are turned off. Next, the input signal changes from an L state to an H state. In this condition, neither the charge at capacitance C_1 nor the one at C_2 can be altered.

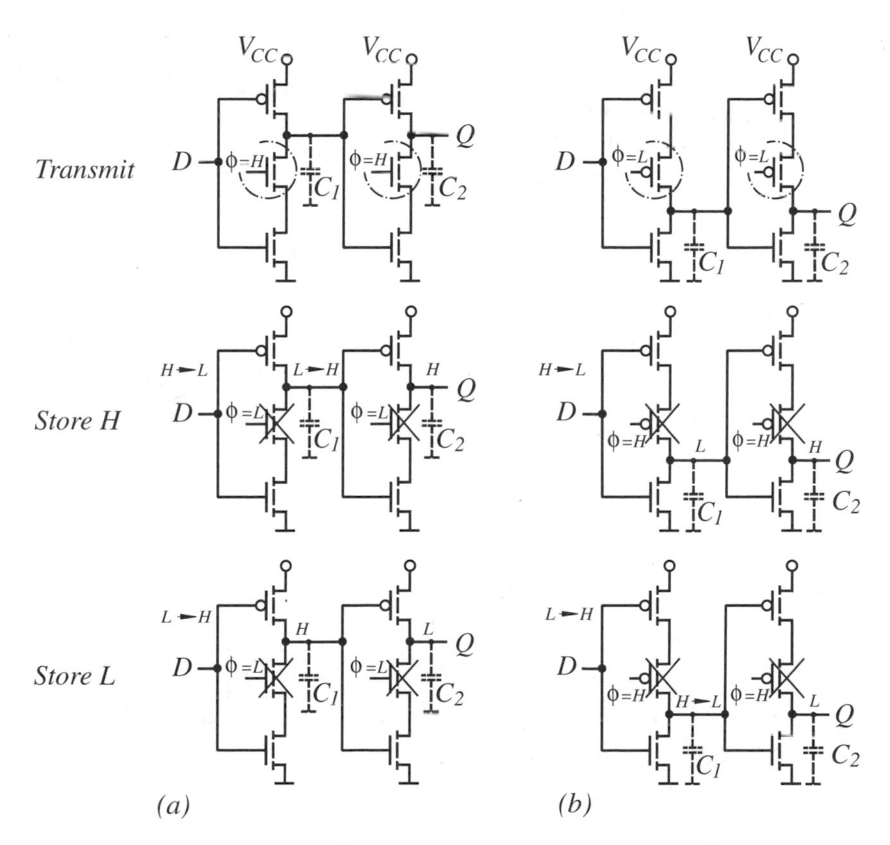

Figure 6.60 (a) Doubled n-C²MOS flip-flop and (b) doubled p-C²MOS flip-flop

Thus, the output Q and the input D are separated whenever $\phi = L$. A similar circuit can be designed, using p-channel pass transistors (Figure 6.60(b)). But in this case the output Q and input D are separated whenever $\phi = H$.

Connecting the two C^2MOS flip-flops in series, a negatively edge-triggered one-phase clocked master–slave flip-flop or the first stage of a series–parallel register results (Figure 6.61).

The so-called doubled n-C^2MOS or p-C^2MOS flip-flop can be simplified to some extent, when only the first inverter is clocked. The circuit schematic, called split-output flip-flop, is shown in Figure 6.62.

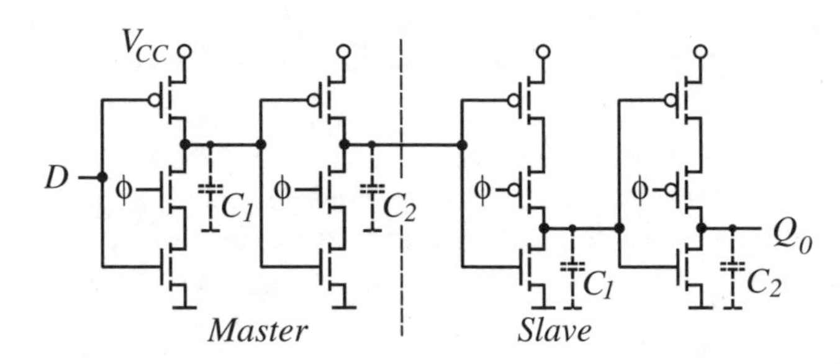

Figure 6.61 One-phase clocked series–parallel register

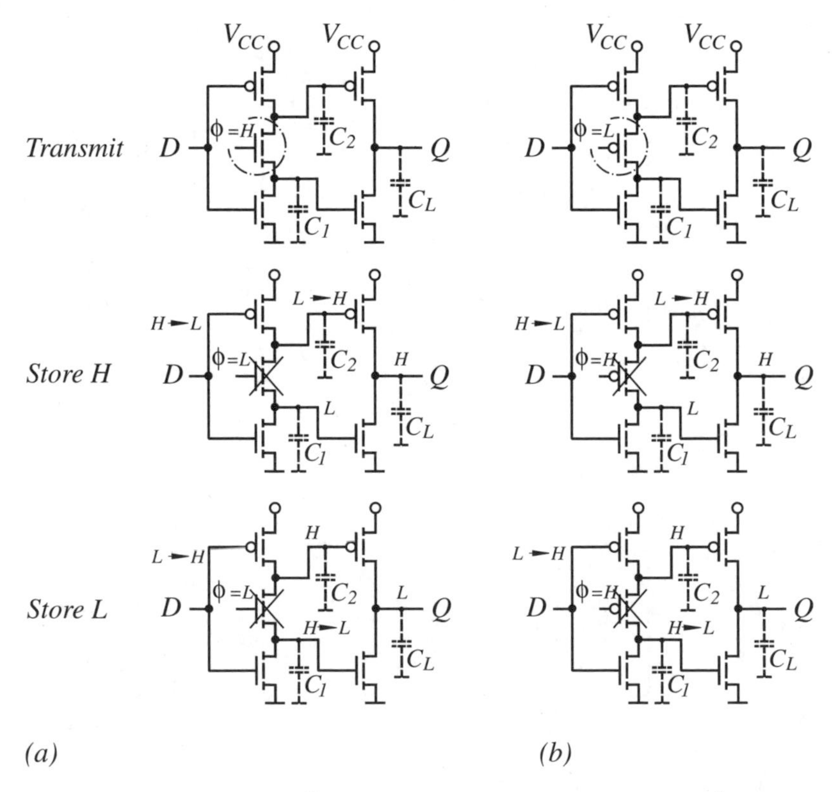

Figure 6.62 (a) Split-output n-C^2MOS flip-flop and (b) split-output p-C^2MOS flip-flop

If $\phi = H$, then the n-channel pass transistor is turned on and the circuit behaves like two in-series-connected inverters (Figure 6.62(a)). If $\phi = L$, the output is separated from the input and the information is stored as different charge levels at C_L. A data change at the input has no influence on the stored output data, as illustrated in Figure 6.62(a). In the implementation of Figure 6.62(b), a p-channel transistor is used as pass transistor. In this case, the input and output are separated when $\phi = H$. The advantage of this scheme is that the loading of the clock line is reduced by one half. A disadvantage is that in the case of Figure 6.62(a), C_1 is charged to a reduced H level of $V_{CC} - V_{Tn}$ and in the implementation of Figure 6.62(b), C_2 is discharged to an increased L level of $|V_{Tp}|$. This leads to a reduced operation speed of the circuit. A master–slave flip-flop using this technique is shown in Figure 6.63.

The one-phase clocked scheme is not only useful for registers but also for pipeline structures used in microprocessors and memories (Figure 6.64), where the different takes are handled as in an assembly line.

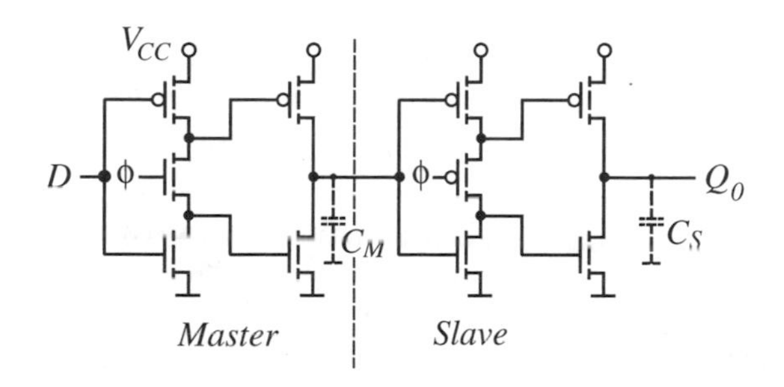

Figure 6.63 One-phase clocked series–parallel register with reduced clock loading

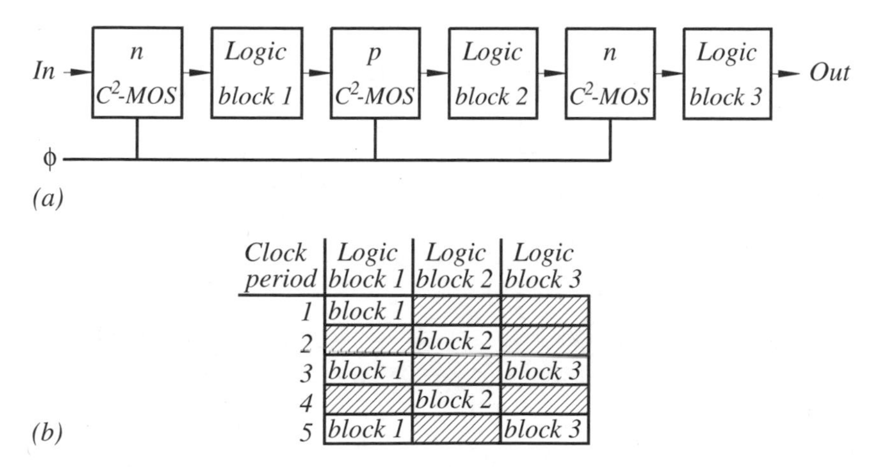

Clock period	Logic block 1	Logic block 2	Logic block 3
1	block 1		
2		block 2	
3	block 1		block 3
4		block 2	
5	block 1		block 3

Figure 6.64 (a) Principle of a one-phase pipeline structure and (b) timing sequences

6.5.4 Clock Distribution and Generation

From the preceding discussion it is obvious that only a sophisticated clock distribution system can guarantee the correct function of clocked circuits over a wide range of parameter spread and temperature. In these distribution systems it is not so important how large the propagation delay time is, but rather how large the delay or skew between communicating circuits is. Parameters influencing the design of the clock distribution system are the material of the distributing network and the loading by the respective circuits. If the propagation delay of the network is in the order of the rise and fall time of the clock buffer, a transmission line model has to be used for circuit simulations. In all the other cases a distributed *RC* model is sufficient. A network that causes a minimum of skew between the individual circuit components is the *H*-tree network, with the same line distance between the central clock buffers and all individual logic blocks *B* (Figure 6.65).

Of more practical importance is the distributed buffer approach, where the delay time between the individual logic blocks can be adjusted to some extent by the decentralized buffers.

The requirements on the clock buffer depend very much on the capacitance loading and the rise and fall times required. The super buffer, described in Section 5.5.1, is a good candidate for this task.

The one-phase clock system is obviously the easiest to implement and most appropriate for high speed operation. More complex is the situation when non-overlapping clocking schemes are used. An example illustrates the case, where the complement of a clock signal is generated by one buffer (Figure 6.66). An overlapping of the clocks results due to the associated propagation delay of the buffer. The clocking skews (hatched areas) may cause severe problems in high speed circuits. The situation can be eased by using the altered circuit of Figure 6.67.

In this case, besides the two identical buffers, additional pass transistors *P* are used. The purpose of these transistors is to compensate the propagation delay caused by inverter *I*. A sizing rule is that the pass transistors should duplicate the geometries of the buffer transistors.

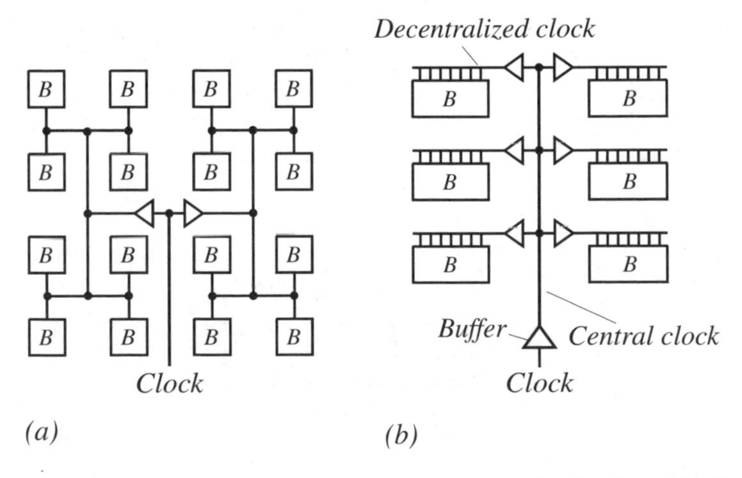

Figure 6.65 Clock distribution systems: (a) *H*-tree network and (b) distributed buffering network

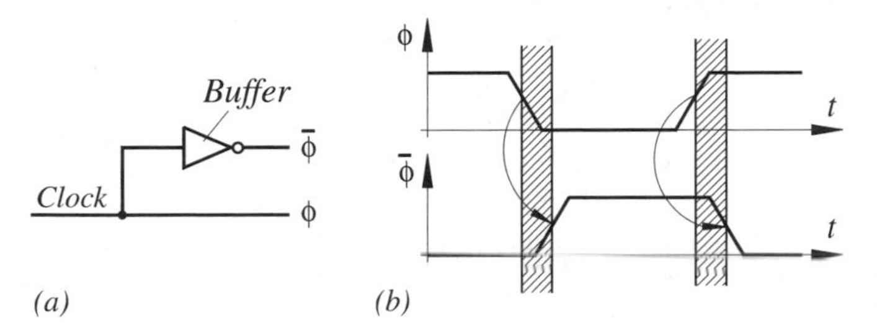

Figure 6.66 (a) Two-phase clocking scheme with overlapping clocks and (b) timing diagram

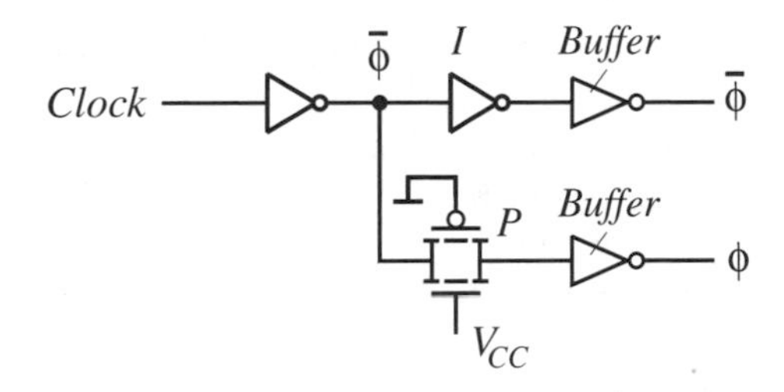

Figure 6.67 Two-phase clocking scheme with reduced clocking skew

A truly non-overlapping clocking scheme can be designed, using the implementation shown in Figure 6.68. The time during which both clock signals ϕ_1 and ϕ_2 are simultaneously in the L state is basically controlled by the propagation delay of the gates and the two following inverters. When the clock signal changes into an H state, this causes the gate *G1* to shift into an H state, which leads to clock signals of $\phi_1 = L$ and delayed via *G2* and the two inverters to one of $\phi_2 = H$. When the clock changes from H to L a reverse situation exists. If the time duration in which both clock signals are in the L state is not sufficient, then additional inverter pairs may be added to the circuit.

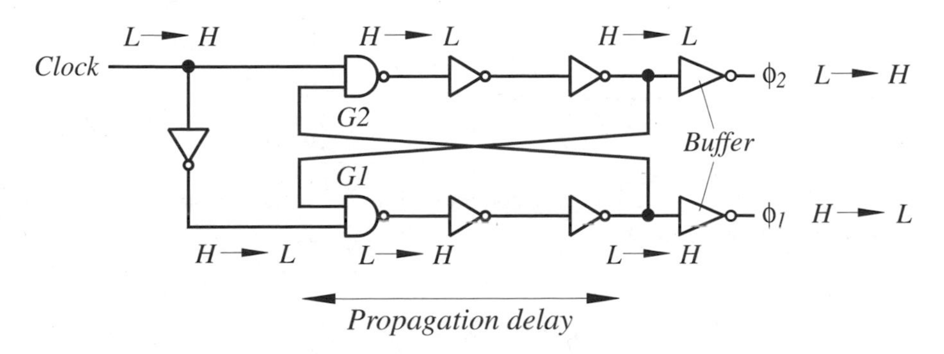

Figure 6.68 Two-phase clocking scheme with non-overlapping clocks

Summary of the most important results

Static and clocked CMOS circuits are considered. The most robust ones with respect to noise and power-supply variations are the complementary static circuits. The layout of these circuits can be optimized by substituting the circuit by a graph and by finding an Euler path. If such a path exists, then the diffusion regions may be joined in one diffusion strip, interrupted by polysilicon lines only. The realization of the complementary function by p-channel transistors may lead to situations where many transistors are connected in series. This usually results in an unacceptably slow rise time. In this case, the total p-channel implementation network can be substituted by a single PMOS load device. Unfortunately, this device causes a static power dissipation. A way out of this situation is the use of clocked circuits, which usually require the least amount of power and silicon area.

The MCML technique allows gates to operate at frequencies far into the GHz region. This is possible by using a current switching approach, allowing the logic swing to be reduced below 0.4 V.

From a particular number of connected logic gates it is advantageous to use programmable logic arrays. A basic element of these arrays is the decoder. Complementary *NOR* and *NAND* implementations have the disadvantage that they are usually only applicable to small array sizes, due to the large number of series-connected transistors. A solution is offered by clocked versions, where no series-connected transistors are required.

Static and dynamic flip-flops are compared and used to implement series–parallel registers employing the master–slave principle. One-phase clocked solutions are the preferred option in order to ease the requirements on the clock distribution network.

Problems

6.1 Implement the logic function $Q = \overline{I_1 \cdot I_2 + (I_3 + I_4) \cdot (I_5 + I_6)}$ with a complementary CMOS circuit and sketch the layout. Use the graph theory (Section 6.1.1) to determine the Euler path for the layout.

6.2 Determine the logic functions which can be implemented by the pass transistor circuit shown below, when the input variables assume the values shown in the table.

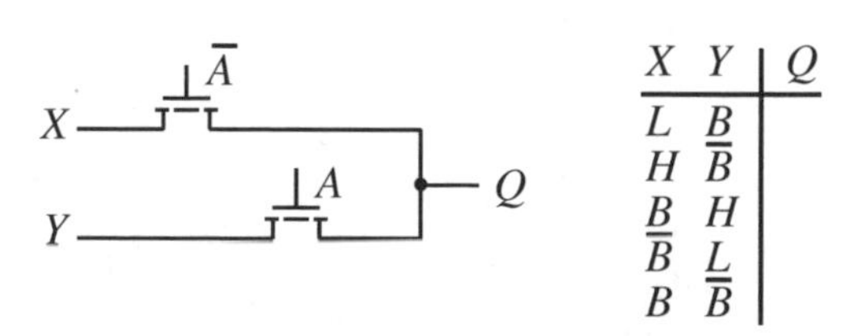

X	Y	Q
L	B	
H	$\overline{B}$	
B	H	
$\overline{B}$	L	
B	$\overline{B}$	

6.3 In Figure 6.40 a cascading of decoders is shown. At the outputs Z a deteriorated output level occurs.

(a) Is this the case for an L level or H level?

(b) Determine the deteriorated voltage value.

(c) How can the situation be improved?

The parameters are: $V_{Top} = -0.45\,\text{V}$, $\gamma = 0.3\,\text{V}^{1/2}$, $V_{CC} = 3\,\text{V}$.

6.4 Draw the circuit of a PLA implementing the truth table shown below and determine the function described by the truth table.

A	B	C	Q1	Q2
L	L	L	L	L
H	L	L	H	L
L	H	L	H	L
H	H	L	L	H
L	L	H	H	L
H	L	H	L	H
L	H	H	L	H
H	H	H	H	H

6.5 Shown below are two clocked C^2MOS master–slave D flip-flops. Which implementation should be preferred? Determine for the solution with a particular drawback, the worst case H voltage and L voltage at the parasitic capacitance C_L. It is assumed that $C_L = 2C_A$.

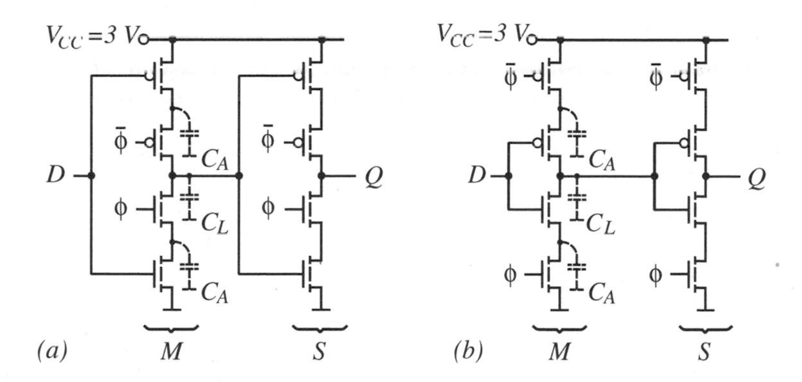

6.6 At the MCML circuit shown below, the input voltages of 1.3 V and 0.9 V are applied. Determine approximately the currents I_1 and I_2. The data are: $V_{Tn} = 0.4\,\text{V}$, $(wl)_1 = (w/l)_2 = 3$, $k_n = 150\,\mu\text{A/V}^2$.

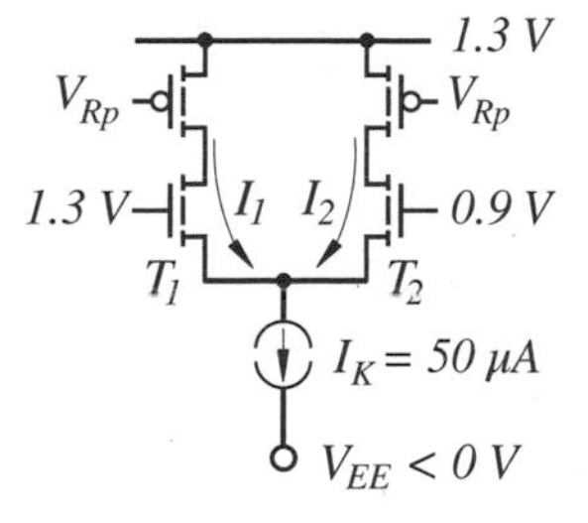

The solutions to the problems can be found under: *www.unibw-muenchen.de/campus/ET4/index.html*

REFERENCES

K.M. Chu and D.L. Pulfrey (1986) Design procedure for differential cascade voltage switch circuits. *IEEE J. Solid-State Circuits*, **SC-21**(6), 1082–7.

K.M. Chu and D.L. Pulfrey (1987) A comparison of CMOS circuit techniques: differential cascade voltage switch logic versus conventional logic. *IEEE J. Solid-State Circuits*, **SC-22**(4), 528–32.

N.F. Goncalves and H.J. DeMan (1983) NORA: a race free dynamic CMOS technique for pipelined logic structures. *IEEE J. Solid-State Circuits*, **SC-18**(3), 261–6.

R.H. Krambeck, C.M. Lee and H.F.S. Law (1982) High speed compact circuits with CMOS. *IEEE J. Solid-State Circuits*, **DC-17**(3), 614–19.

C.M. Lee and E.W. Szeto (1986) Zipper CMOS. *IEEE Circuits Devices Mag.*, May 10–16.

M. Mizuno, M. Yamashina, K. Furuta, H. Igura, H. Abiko, K. Okabe, A. Ono and H. Yamada (1996) A GHz MOS adaptive pipeline technique using MOS current-mode logic. *IEEE J. Solid-State Circuits*, **31**(6), 784–91.

P. Ng, P.T. Balsara and D. Steiss (1996) Performance of CMOS differential circuits. *IEEE J. Solid-State Circuits*, **31**(6), 841–6.

D. Radhakrishnan, S. Whitaker and G. Maki (1985) Formal design procedures for pass transistor switching circuits. *IEEE J. Solid-State Circuits*, **SC-20**(2), 531–6.

M. Shoji (1985) FET scaling in domino CMOS gates. *IEEE J. Solid-State Circuits*, **SC-20**(5), 1067–71.

Y. Suzuki, K. Odagawa and T. Abe (1973) Clocked CMOS calculator circuitry. *IEEE J. Solid-State Circuits*, **SC-8**(6), 462–9.

A. Tanabe, M. Umetani, I. Fujiwara, T. Ogura, K. Kataoka, M. Okihara, H. Sakuraba, T. Endoh and F. Masuoka (2001) 0.18-μm CMOS 10-Gb/s multiplexer/demultiplexer ICs using current mode logic with tolerance to threshold voltage fluctuation. *IEEE J. Solid-State Circuits*, **36**(6), 988–96.

T. Uehara and W.M. vanCleemput (1981) Optimal layout of CMOS functional arrays. *IEEE Trans. Comput.*, **C-30**(5), 305–12.

S. Whitaker (1983) Pass transistor networks optimize n-MOS logic. *Electronics*, September, 144–48.

M. Yamashina and H. Yamada (1992) An MOS current mode logic (MCML) circuit for low-power sub-GHz processors. *IEICE Trans. Electron.*, **E75C**(10), 1181–7.

J. Yuan and C. Svensson (1989) High-speed CMOS circuit technique. *IEEE J. Solid-State Circuits*, **24**(1), 62–70.

FURTHER READING

J.M. Rabaey (1996) *Digital Integrated Circuits*. Upper Saddle River, NJ: Prentice-Hall.

N. Weste and K. Eshraghian (1994) *Principles of CMOS VLSI Design*, 2nd edn. New York: Addison Wesley.

7
MOS Memories

In the preceding chapter, static and dynamic flip-flops are used for the implementation of storage devices. In these applications, the silicon area consumption is not that important compared to the situation in semiconductor memories, where the memory density is an outstanding parameter. An overview of existing memory devices is shown in Figure 7.1, classified in the way the information is stored. If the information is stored nonvolatile, then the power-supply can be turned off and the information remains stored. Unfortunately, this is not the case for two remaining memory groups. These can be distinguished by their clock frequency. The static memories have no requirement on the slowest clock rate, whereas the dynamic ones require a periodic clock to refresh the stored information. The memories, shown in Figure 7.1, may be used as stand-alones or in embedded form in all kinds of controllers and microprocessors. The meaning of the individual memory names is summarized in Table 7.1.

In this chapter the most important memory cells, circuit implementations, and, in conjunction, memory architectures of the MOS memories shown in Figure 7.1 are presented. It is not the intent to explain specifications of existing products that are available on the market.

7.1 READ ONLY MEMORY

The easiest way to store information in a nonvolatile way is to permanently connect or disconnect transistors in an array similar to that discussed in Section 6.4.2 for the programmable logic array (PLA). Instead of a PLA one speaks of a read only memory (ROM) when all possible input combinations lead to defined output words. In Figure 7.2 a block diagram of a ROM is shown.

The $N + M$ input words – in this case the addresses – lead to 2^{N+M} output words. With the aid of the row decoder, one word line is selected in the memory array, activating all memory cells on this line. Usually, the geometry of the memory array is such that many more bit lines (columns) than data outputs exist. This requires a column decoder for selecting the addressed bit lines. Via a sense amplifier the selected data are then fed to the outputs.

The discussed memory organization is good only up to a particular memory array size. Otherwise, the propagation delay within the memory array increases too much or

System Integration: From Transistor Design to Large Scale Integrated Circuits. Kurt Hoffmann.
 ISBN: 0-470-85407-3

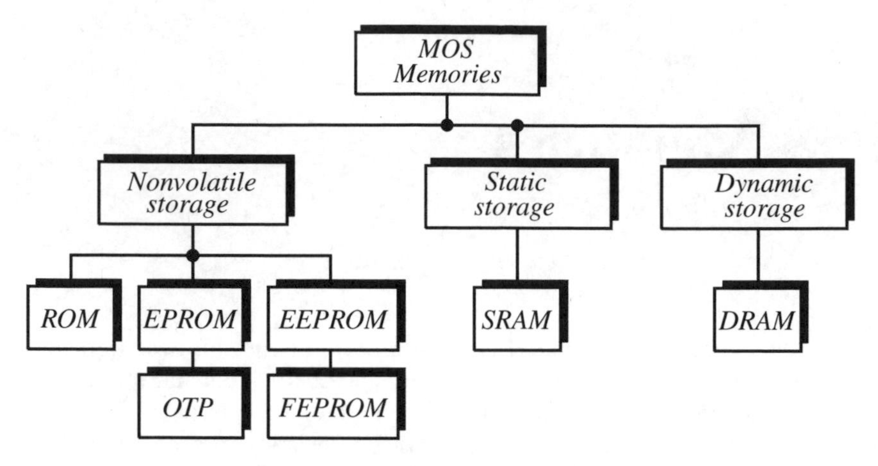

Figure 7.1 MOS memory classification

Table 7.1 Overview of commonly used memory names

Name	Memory
ROM	Read only memory
EPROM	Electrically programmable ROM
OTP	One time programmable EPROM
EEPROM	Electrically erasable programmable ROM
FEPROM	Flash erasable PROM
SRAM	Static random access memory
DRAM	Dynamic random access memory

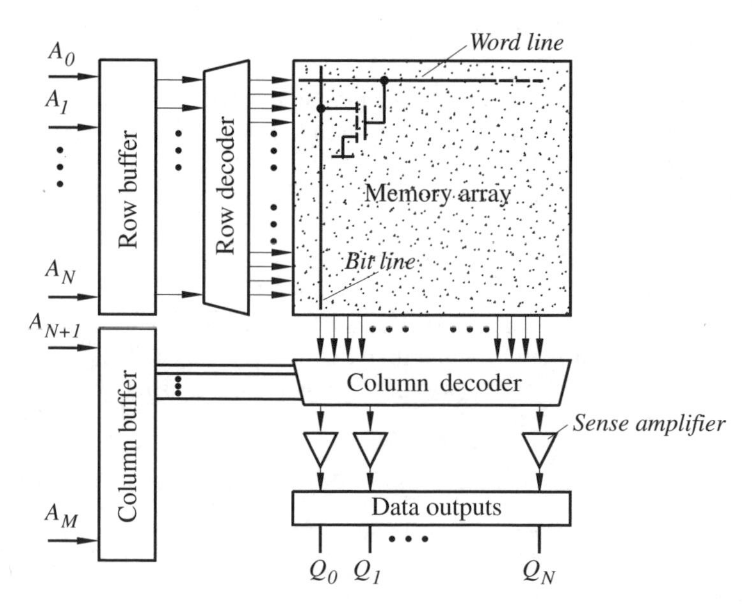

Figure 7.2 Block diagram of a ROM

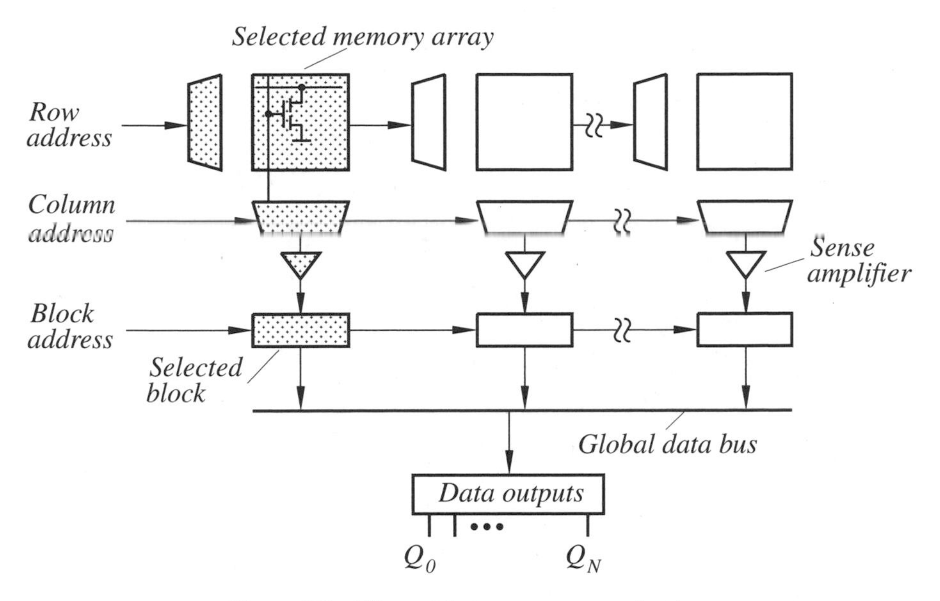

Figure 7.3 Hierarchical memory organization

the sensing of the cell signals may pose a problem. For these reasons, larger memory sizes use a hierarchical architecture (Figure 7.3).

The individual memory arrays are then selected by a block address and connected via a global data bus to the outputs. For the user it is absolutely unimportant how the internal architecture of the memory is arranged. The memory is programmed by the manufacturer with a special lithographic step, e.g. by making connections between the drains of the transistors and the bit lines – shown in the PLA of Figure 6.42(b) – as required.

If the user prefers more flexibility and wants to do the programming himself, he may use one of the following memories.

7.2 ELECTRICALLY PROGRAMMABLE AND OPTICALLY ERASABLE MEMORY

The first medium-scale integrated programmable and erasable memories were developed by Frohman-Bentchkowsky (1971) and called Floating-gate-Avalanche-injection MOS (FAMOS). In modified form this method is still used.

In this section the memory cells are considered first, then follows a discussion about resulting memory architectures.

An electrically programmable ROM cell (EPROM) is shown in Figure 7.4. The cell, which is sometimes called Stacked gate Injection MOS (SIMOS), consists of a MOS transistor with a control gate (CG) and a floating gate (FG) not connected to the outside. Instead, the gate is completely insulated by, e.g., silicon dioxide (SiO_2). The nonvolatile storage is based on the fact that electrons once placed on the floating gate are trapped and remain there for more than ten years even when the supply

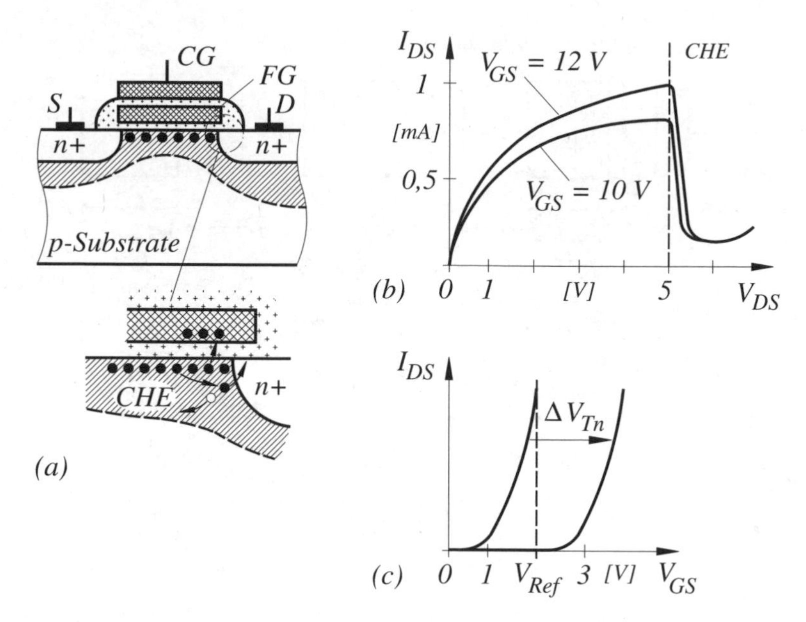

Figure 7.4 (a) EPROM cell, (b) $I_{DS}(V_{DS})$ characteristic at different V_{GS} voltages and (c) $I_{DS}(V_{GS})$ before and after programming

voltage is removed. In Figure 7.4(b) the I_{DS} current of the cell is plotted as function of the V_{DS} voltage at two different V_{GS} voltages. The resulting effective gate voltage at the floating gate influencing the semiconductor surface is given by the ratio of the capacitances between the control and floating gate and between the floating gate and inversion layer. Since a MOS transistor with short channel length is used, a large field at the drain side of the channel develops with increasing V_{DS} voltage, causing electrons to assume saturation velocity (Section 4.5.4). These so-called channel hot electrons (CHE) are so energetic that they are able to create electron–hole pairs by impact ionization, also called the avalanche process, in the pinch-off region of the transistor. The holes generated by this process are collected at the substrate and the electrons at the drain. But some electrons in this avalanche process – in the order of 10^{-5}% – can gain sufficient kinetic energy to surmount the Si–SiO_2 energy barrier (inset in Figure 7.4(a)) and charge the floating gate. Once this happens, the I_{DS} current starts to reduce (Figure 7.4(b)) due to the negative charge collected at the floating gate. This process is self-limiting since the negative charge accumulated at the floating gate reduces the electrical field at the drain side of the channel. A further consequence of the accumulated negative charge is that the threshold voltage (Equation 4.36) shifts to higher values. The time required to charge the floating gate, the so-called programming time, is about 1–10 µs at a drain current of 500 µA. For reading the information in the cell, a reference voltage of, e.g., $V_{REF} = 2\,V$ is applied to the gate and the I_{DS} current determined (Figure 7.4(c)). How the band diagram is affected during and after programming is shown in Figure 7.5.

The charge from the floating gates is removed or erased by shining ultraviolet light for about 20 minutes on the cells through a transparent window in the package. The electrons gain enough energy during this process to surmount the SiO_2 barriers toward

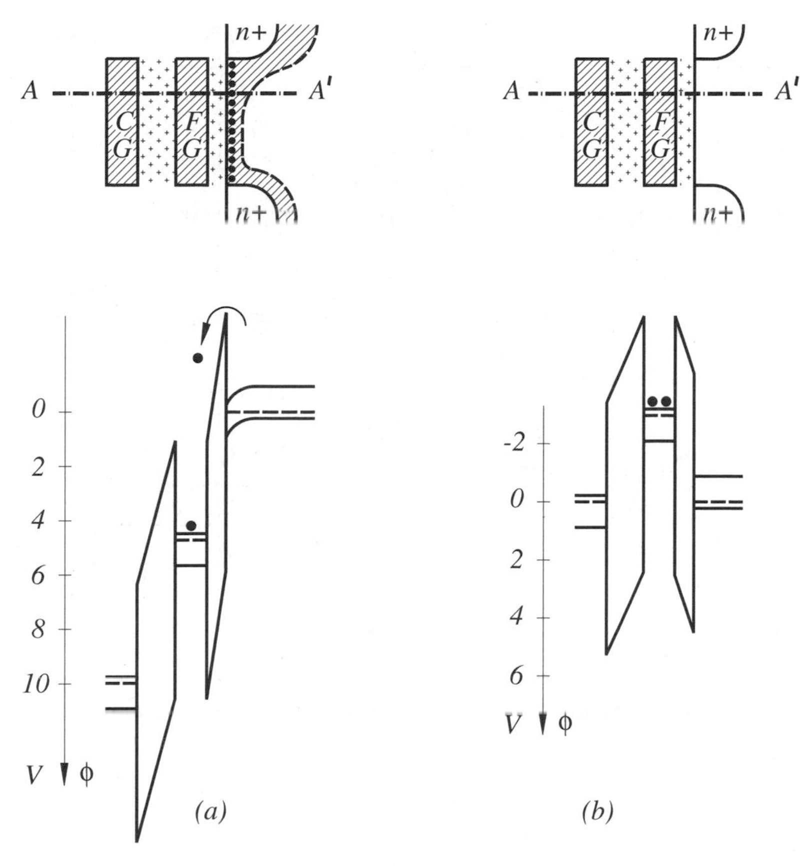

Figure 7.5 Band diagram of an EPROM cell (cross-section *A–A′*): (a) during programming and (b) after programming with 0 V at the control gate

the substrate and the control gate. An EPROM memory can be programmed and erased approximately 100 times.

The package with the transparent window is relatively expensive. To reduce the cost, the memory is placed in an inexpensive plastic package. The memory can thus be programmed only once. This device is therefore called one time programmable (OTP) EPROM.

7.2.1 EPROM Memory Architecture

The architecture of this memory (Figure 7.6) is very similar to that of the ROM.

Reading ($V_{PP} = V_{CC}$)

With the aid of the row decoder, a word line driver, e.g. WD_1 with word line WL_1, is selected. This causes the information of all the cells addressed by this word line to be

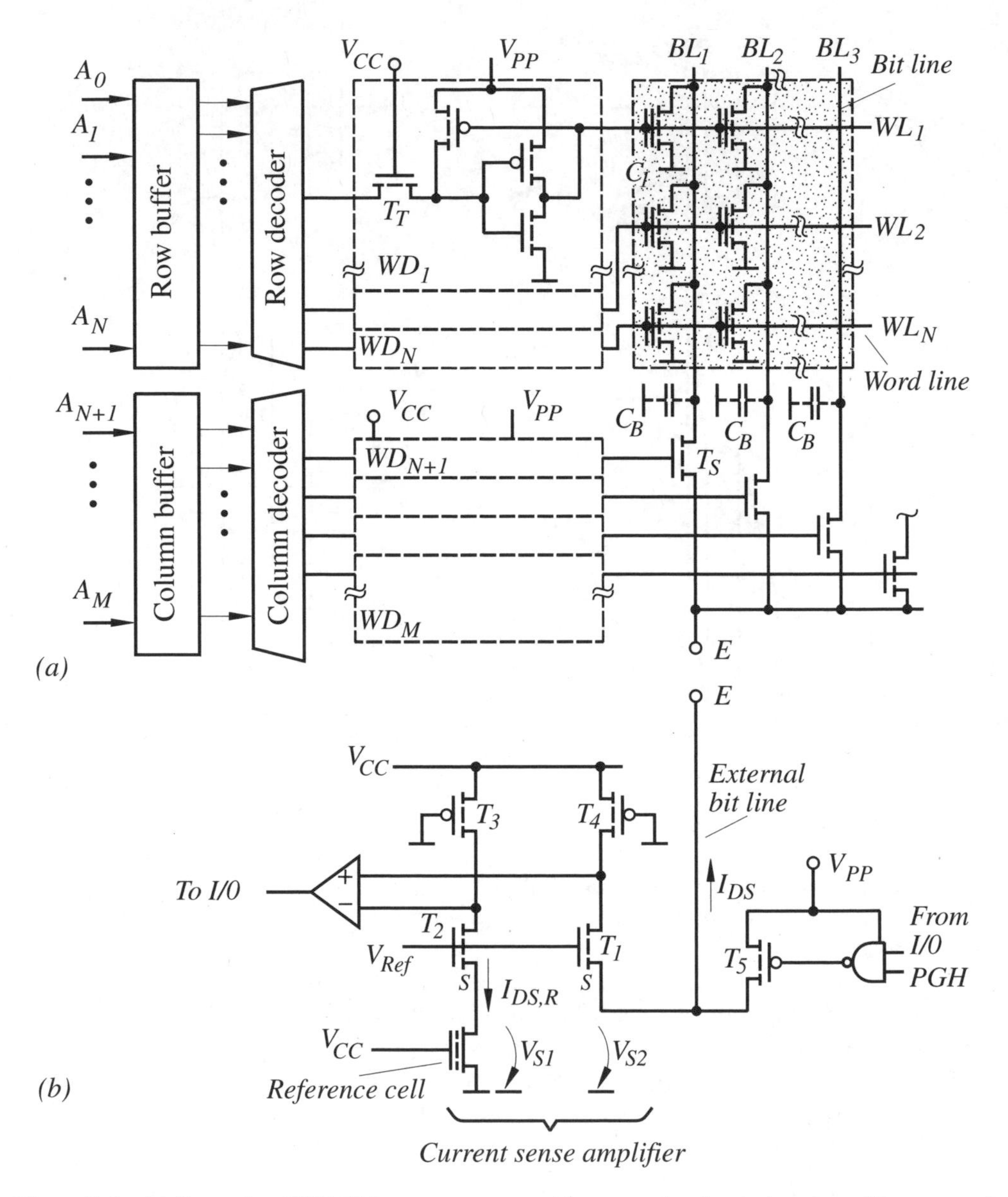

Figure 7.6 (a) Part of an EPROM memory array with row and column decoders, (b) sense and programming circuit

available on the bit lines. A column decoder then selects one or more bit lines out of the available ones, causing, e.g., the information of the cell C_1 to appear at the terminal E of the current sense amplifier (Figure 7.6(b)). This sense amplifier compares the current of the selected cell I_{DS} with a reference current $I_{DS,R}$ of a reference cell. The voltage difference between the transistors T_3 and T_4 is fed via a differential amplifier to the common input/output (I/O). The differential amplifier is treated in detail in Chapter 8.

The maximum cell current in high density memories is usually below 50 μA. In order to discharge with this small cell current the relatively large bit line capacitance C_B as fast as possible, the biased transistors T_1 and T_2 are added to the circuit. This results in voltages at the sources of these transistors of $V_{S1} = V_{Ref} - V_{Tn} - \Delta V_1$ and $V_{S2} = V_{Ref} - V_{Tn} - \Delta V_2$, respectively. The voltages ΔV_1 and ΔV_2 are the voltage changes caused by the current of the reference cell and the current of the cell, respectively. Since these voltage changes are, at 200–300 mV, relatively small, the bit line capacitance is readily discharged ($\Delta t \approx C_B \cdot \Delta V_1 / I_{DS}$). A further advantage of the reduced bit line voltage is that an unintentional programming of the cells during a read operation is less probable.

Programming ($V_{PP} \gg V_{CC}$)

The programming starts by switching the common data input/output terminal I/O into an active input (not shown) and applying a programming voltage of, e.g., $V_{PP} = 12\,\text{V}$ to the memory. This causes the word line voltage via the word line driver to be raised to 12 V. The transistor T_T in the word line driver circuit has the task of separating the high word line voltage from the row decoder. This has the advantage that only the driver circuit has to be designed for the high voltage requirement. Simultaneously, the information reaches via the *NAND* gate and transistor T_5 the bit line and thus the drain of a selected cell.

In Figure 7.6(b) a relatively simple current sense amplifier is used. With additional components this circuit can be improved substantially.

7.2.2 Current Sense Amplifier

In this circuit the voltage change at the bit line is reduced even further, and thus the time for discharging the parasitic bit line capacitance accordingly. In order to explain the circuit it is advantageous to revisit the presented simple current sense amplifier (Figure 7.6(b)) and compare the result with the improved version (Figure 7.7(b)).

Depending on the stored information – that is, the charge – the cell currents (Figure 7.7(a)) are

$$I_{DS,0} = \frac{\beta_n}{2}(V_{Ref} - V_{S,0} - V_{Tn})^2 \quad \text{or}$$
$$I_{DS,1} = \frac{\beta_n}{2}(V_{Ref} - V_{S,1} - V_{Tn})^2 \tag{7.1}$$

These currents cause the following voltage changes at the source and drain of transistor T_1 (Figure 7.7(a))

$$\Delta V_S = V_{S,0} - V_{S,1} = \sqrt{\frac{2}{\beta_n}}\left(\sqrt{I_{DS,1}} - \sqrt{I_{DS,0}}\right) \tag{7.2}$$

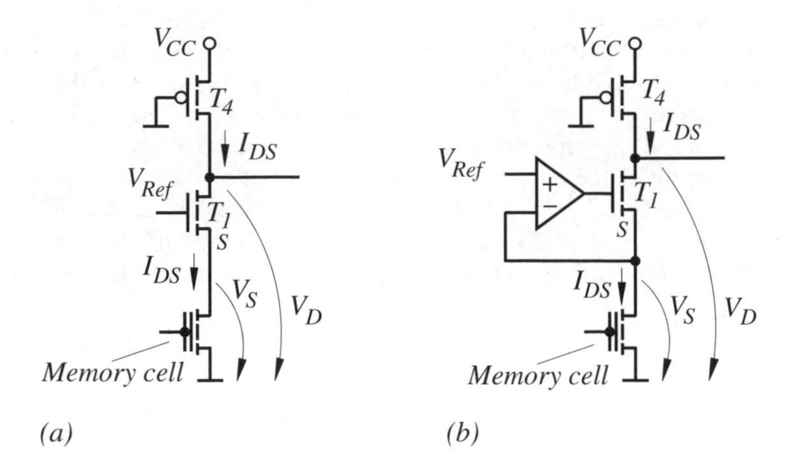

Figure 7.7 Current sense amplifier: (a) according to Figure 7.6(b) and (b) with amplifier

and

$$\Delta V_D = (I_{DS,0} - I_{DS,1}) \cdot R = \Delta I_{DS} \cdot R \tag{7.3}$$

R represents the resistance of the p-channel transistor load. As the following example demonstrates, ΔV_D is much larger than ΔV_S.

Example

The EPROM cells are able to draw a current of either 50 μA or 2 μA, depending on the information stored. How large are the voltage changes at source and drain of T_1 when $\beta_n = 1000\,\mu A/V^2$ and $R = 50\,k\Omega$?

According to the above equations the voltage changes are: at the source $\Delta V_S = 0.25$ V and at the drain $\Delta V_D = 2.4$ V.

The voltage at the source and thus bit line can be reduced even further and the switching performance improved when a circuit as shown in Figure 7.7(b) is employed (Seevinck *et al.* 1991). If, e.g., the current I_{DS} of the cell increases, this leads to a reduced V_S voltage and an increased output voltage at the differential amplifier. Since this voltage controls the gate of T_1, this results in an increase in the V_S voltage, counter-balancing the initial voltage reduction. If the differential amplifier has a very large gain, this causes the voltage between the input terminals of the amplifier to approach zero and therefore $V_S \approx V_{Ref}$. From the description it is obvious that in the ideal case the source voltage remains constant, even when the cell current changes by ΔI_{DS}. Independent of this fact, the changing current at the drain of T_1 causes a voltage change of $\Delta V_D = \Delta I_{DS} \cdot R$, which is identical to that of the preceding circuit.

The final current sense amplifier (Mills *et al.* 1995) with a reference cell is shown in (Figure 7.8). The current of the reference cell is set to a middle value between the memory cell currents of $I_{DS,1}$ and $I_{DS,0}$. To compensate for possible asymmetries in the circuit at node (1) and (2), an equalize signal *EQ* is applied prior to the read mode.

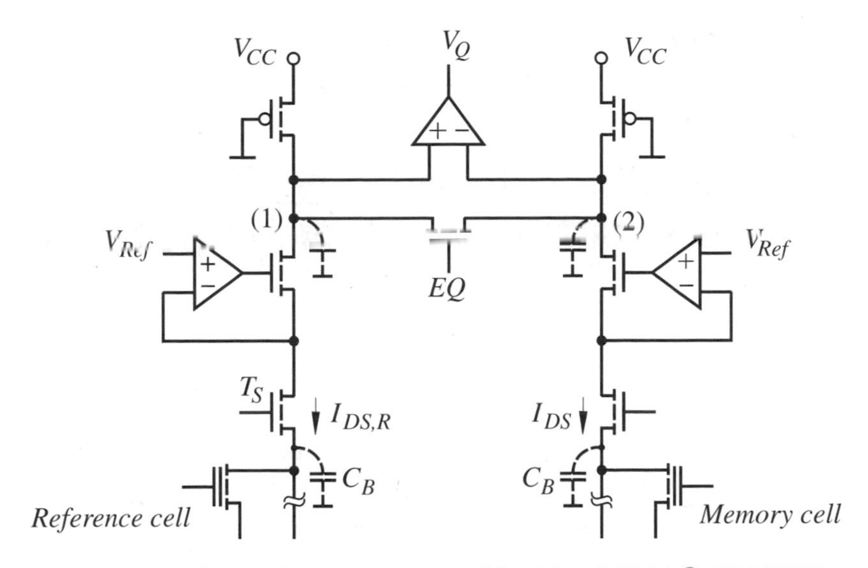

Figure 7.8 Differential current sense amplifier (after |MILL| © 1995 IEEE)

7.3 ELECTRICALLY ERASABLE AND PROGRAMMABLE READ ONLY MEMORIES

The mayor disadvantage of the EPROM is that the program and erasure procedure has to occur off-system. Therefore, the memory must be removed from the printed circuit board and placed in a special EPROM programmer. To avoid this annoying procedure and achieve more system flexibility, electrically erasable and programming read only memories (EEPROM or E^2PROM) are used. The most important ones are considered next.

7.3.1 EEPROM Memory Cells

An overview of these cells is shown in Figure 7.9. According to the general definition, programming means the injection of electrons onto the floating gate, whereas erase means the extraction of electrons from the floating gate. But one has to be aware that in some publications the opposite nomenclature is used.

ETOX cell

The programming of the EPROM tunnel oxide cell (ETOX) is achieved, as in the case before, with channel hot electrons (CHE). But for erasing the cell contents, a tunnel effect is employed. For this purpose, a thin gate oxide window of about 5 nm thickness between the overlapping gate and source is provided. With the voltages of 10 V applied to the source and 0 V to the gate, electrons are able to tunnel from the floating gate via the thin oxide to the source. This mechanism is called the Fowler-Nordheim (FN) tunnel effect and is illustrated in Figure 7.10 at the band diagram.

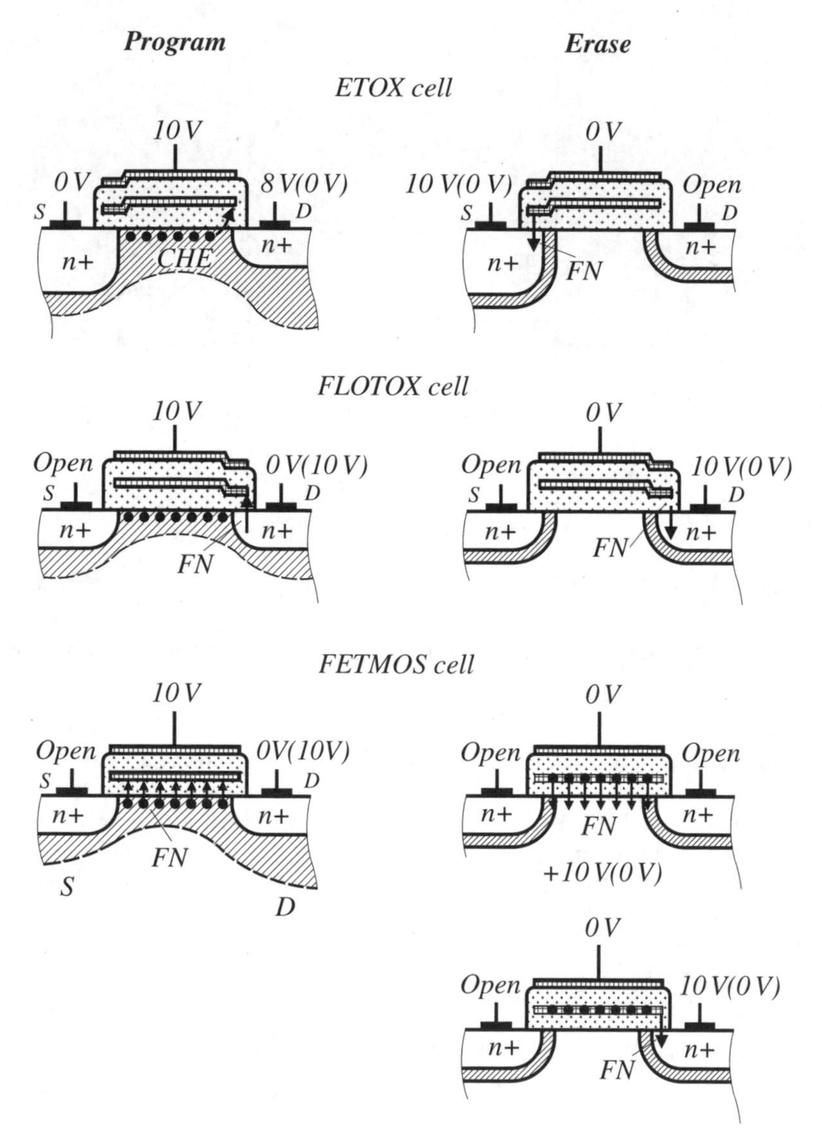

Figure 7.9 Overview of EEPROM memory cells with typical voltage conditions (inhibit function in parentheses) (substrate is connected to 0 V when not indicated otherwise)

The tunnel current is, at approximately 10^{-11} A, very small and is described by the relationship (Lenzlinger and Snow 1969)

$$I = A\mathcal{E}_{ox}^2 e^{-B/\mathcal{E}_{ox}} \tag{7.4}$$

where A and B are process-dependent constants and $\mathcal{E}_{ox}$ is the electrical field between floating gate and source. The small tunnel current is a substantial design advantage, compared to the CHE current of about 0.5 mA. But a disadvantage is the relatively

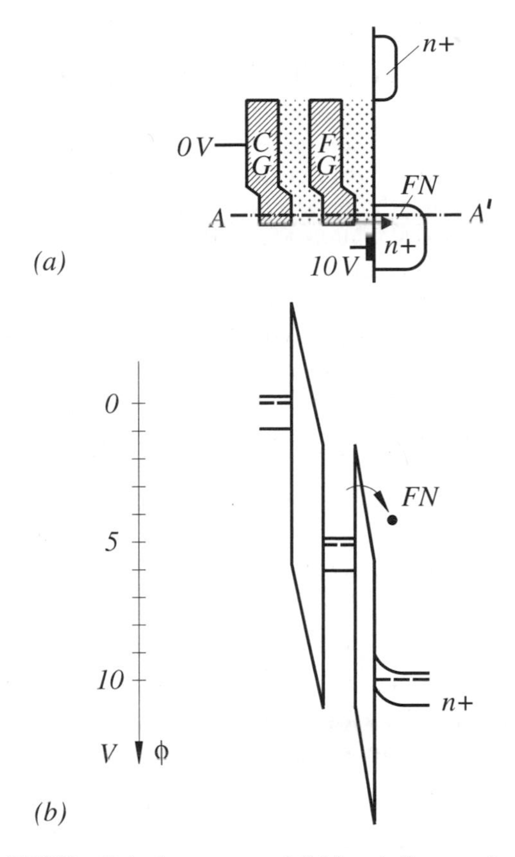

Figure 7.10 (a) ETOX cell during erase and (b) band diagram (cross-section A–A')

long erasing time of about 1 ms compared to the programming time with CHE of a few μs.

For the source region sometimes a deeper and graded junction is provided with an additional phosphorus implantation in order to increase the breakdown voltage of the junction. Not completely avoidable is the GIDL effect (Section 4.5.5), which may result in a leakage current from source to substrate.

FLOTOX cell

In the case of the FLOating gate Thin OXide (FLOTOX) cell (Figure 7.9), a thin oxide is provided between the overlapping gate and the drain diffusion. This region can be used to erase the cell or, when the voltages are reversed, to program the cell with the FN-tunnel mechanisms. The advantage is that in both cases only very small tunnel currents occur but, unfortunately, long program and erase times of about 1 ms have to be accepted.

FETMOS cell

The Floating gate Electron Tunneling MOS (FETMOS) cell uses a thin oxide in the entire gate region. Thus, the FN-tunnel mechanism occurs over the total gate area during programming and also during erasing, when the voltages are reversed. In case voltage changes at the substrate are not desirable, erasing can be done via the drain region. The advantage of using the entire gate area for the FN-tunnel mechanism is that degradation effects – which are treated next – do not affect the device as much as if this happens in a small oxide window only. Due to the thin gate oxide in the entire gate area, the gate capacitance is relatively large (Figure 7.11). This leads to the requirement that the coupling capacitance C_C between the control gate CG and floating gate FG has to be relatively large too, in order to guarantee a sufficiently large voltage V_{FG} between the floating gate and substrate during programming and erasing. In the case of programming, the floating gate voltage has a value of

$$V_{FG} = V_{GB} \frac{C_C}{C_C + C_G} \tag{7.5}$$

where the total gate capacitance C_G is given by the sum of C_S, C_B, and C_D. Since the applied V_{GB} voltage is fixed by an external or internal power-supply voltage, a large floating gate voltage can only be achieved when the coupling capacitance C_C is increased, e.g. by using as insulator material one with a high dielectric constant, such as an oxide-nitride-oxide (ONO) sandwich.

In summary, it is worth remembering that the programming by CHE is, at approximately 10 μs relatively fast, but the current of about 0.5 mA is large. This current reduces to about 10^{-11} A when the FN-tunnel mechanism is used, but the time for programming and erasing increases to about 1 ms.

Degradation mechanisms

With respect to the reliability of the presented devices, two particular aspects have to be considered. These are endurance and data retention failures. Each programming and erase event causes permanent damage to the cell and limits the number of program and erase cycles. This is shown in Figure 7.12 for a FLOTOX cell.

The observed shift in threshold voltage is caused by electron trapping within the tunneling oxide (Mielke *et al.* 1987). After approximately 10^6 program/erase cycles,

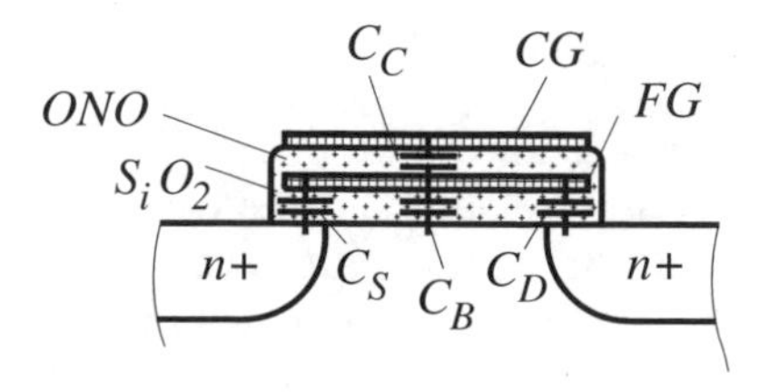

Figure 7.11 Equivalent capacitance model of the FETMOS cell

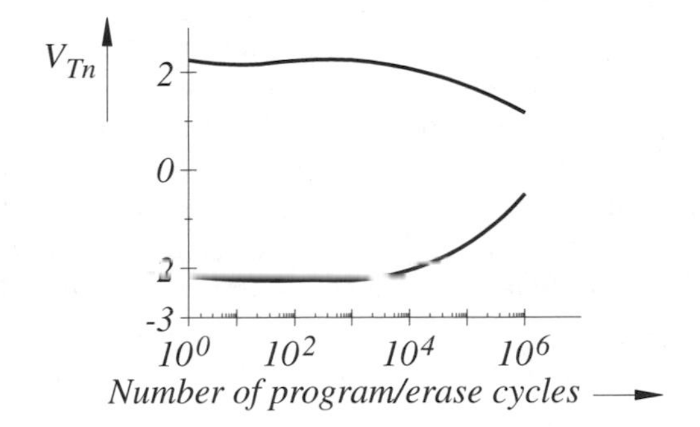

Figure 7.12 Typical threshold voltage shift as a function of program/erase cycles

the window between the H and L threshold voltages reduces to such a value that the information cannot be sensed reliably anymore and so-called endurance failures occur.

The second important reliability aspect is data retention failures. The manufacturers usually guarantee a data retention of 10 years, independent of whether a power supply is connected to the memory or not and independent of the number of read cycles performed. Data retention failures occur due to the loss of cell charge, usually at accelerated temperature. These failures can be observed after only some program/erase cycles, or sporadically where a cell loses charge at first and is afterwards fully functional again. One effective way to cope with this problem is to use an error detection and correction code.

Program disturb and over erase

In the preceding section, the reliability aspects of memory cells are discussed. Next, effects like programming disturb and over erase are considered. These effects may cause failures, if not properly taken care of in the design phase of the memory.

As an example for the program disturb problem, FOTOX cells connected in a *NOR* matrix are used and programming voltages are applied (Figure 7.13). The cells at the

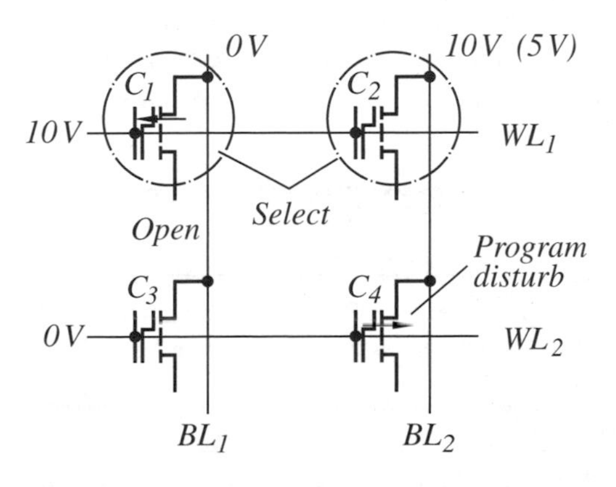

Figure 7.13 *NOR* matrix with FLOTOX cells during programming

10 V word line WL_1 are selected, and at the bit lines BL_1 and BL_2 voltages of, e.g., 0 V and 10 V are applied. Under these conditions, only electrons at cell C_1 are injected, whereas the charge at C_2 remains unaltered. The word line WL_2 with 0 V is not selected. But since 10 V exists at BL_2, electrons tunnel to the drain and erase cell C_4. This unintentional program disturb can be reduced, but not completely avoided, by, e.g., applying 5 V instead of 10 V to the bit line. A further reduction of the voltage is critical since this would cause an unintentional injection of charge at cell C_2.

During programming, electrons are injected onto the floating gate and are extracted throughout the erasing process. In the case that more electrons are extracted than injected, the floating gate is charged positively. This causes a condition called over erase, where the cells cannot be switched off anymore. The described problems can be solved by either one of the options shown in Figure 7.14, which are discussed in the following.

Two-transistor cell

This cell consists of, e.g., a FLOTOX cell and select transistor (Figure 7.15). Only where a voltage of 10 V is applied to the addressed select line *SL* it is possible to program or erase the cells. Since all transistors T_S are cut off, no I_{DS} current occurs in a cell. The sensitivity to a program disturb problem is substantially reduced, since only the addressed cells are connected to the bit lines. An over erase is of no importance since the cells are turned on or off by the select transistors. The information of the addressed cells is read by turning on the transistors T_S of the respective row and by evaluating the cell current – similar to the illustration in Figure 7.8 – when, e.g., 3 V are applied to the word and select lines.

Split-gate cells

A disadvantage of the two-transistor cell is its large area consumption. Thus, this cell type is only used in small memory arrays. As a compromise, so-called split-gate cells are employed (Figure 7.16). The control gate *CG* is used for the capacitance coupling in the memory cell and acts simultaneously as select transistor in a region *TR*, which is not controlled by the floating gate. The cell is programmed with hot electrons (CHE) at the drain side and erased by using the FN-tunnel mechanism. The cell information is read by evaluating the drain

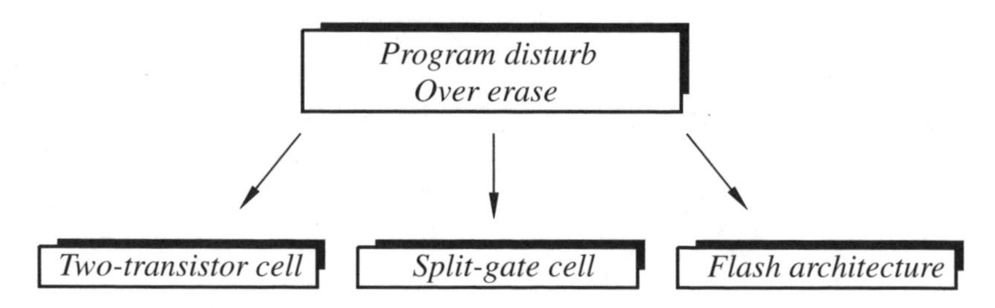

Figure 7.14 Options for solving program disturb and over erase problems

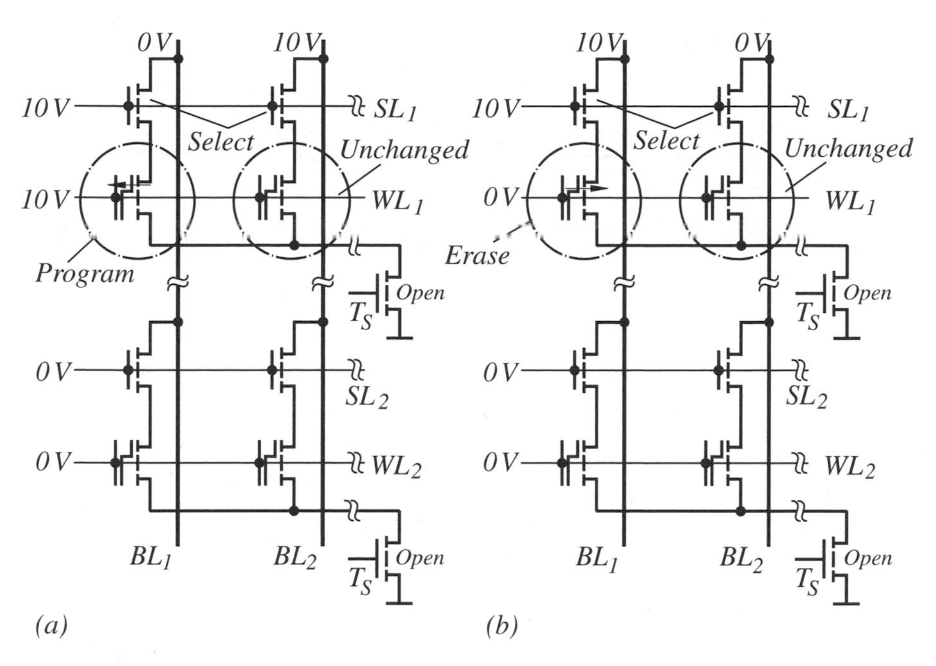

Figure 7.15 Two-transistor FLOTOX cell array: (a) programming and (b) erasing

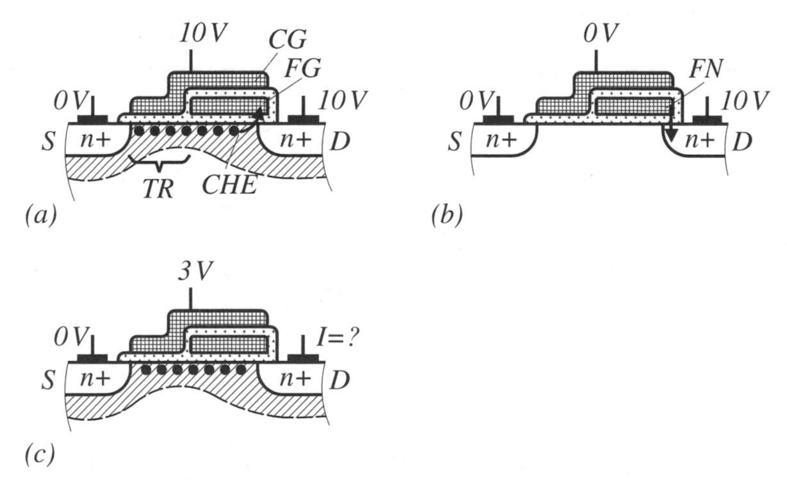

Figure 7.16 Split-gate cell: (a) programming CHE with drain side injection, (b) erasing with FN-tunnel mechanism between *FG* and drain and (c) reading cell contents

current just as described before. A disadvantage of the cell is the relatively large current of about 0.5 mA during the programming with CHE. This current can be reduced with a modified split-gate cell (Huang *et al.* 2000), shown in Figure 7.17.

The major difference to the preceding split-gate cell is the relatively large oxide gap (*G*) of about 40 nm between the floating and control gate. This causes a small energy

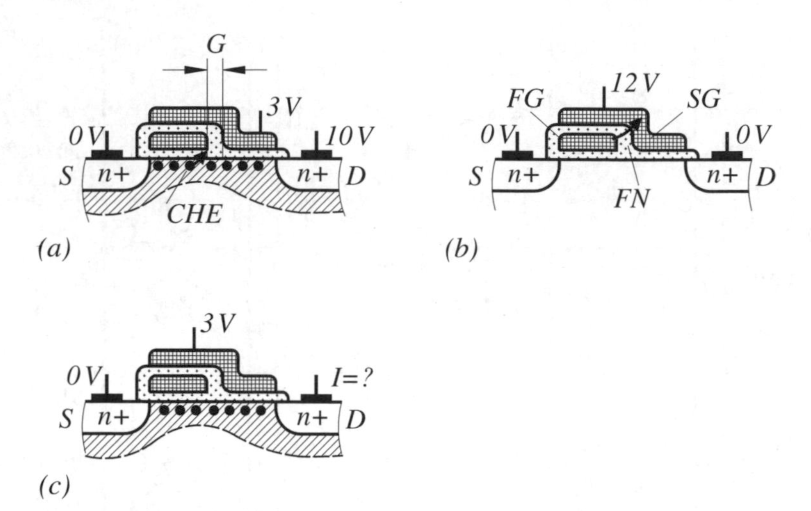

Figure 7.17 Modified split-gate cell: (a) programming CHE with source side injection, (b) erasing with FN-tunnel mechanism between *CG* and *FG* and (c) reading cell contents

barrier at the silicon surface. Electrons with sufficient energy are able to surmount this barrier and continue to move to the drain. But a fraction of the electrons is able to move across the floating gate barrier and charge the floating gate. The drain current of this so-called source side injection (SSI) scheme is reduced to about 1 μA (Silicon Storage Technology, Inc. 2000). The erasing of the cell charge is caused by the FN-tunnel mechanism, but this time between floating gate and control gate.

7.3.2 Flash Memory Architectures

If a bitwise program and erase is required, two-transistor or split-gate cells are needed for the memory core. This leads to a relatively large chip area, even with the split-gate approach. Thus, these concepts are not too attractive for ultra large scale integrated memories, required in applications such as digital cameras, digital voice recorders, and music players. One way around this problem is to use one-transistor cells – as shown in Figure 7.9 – in a flash architecture. Flash means a complete memory block or memory is erased simultaneously. Substantial chip area savings result.

Basically, the flash memories array can be built by *NOR* or *NAND* structures, which are analyzed in the following.

NOR architecture

This architecture, using for example ETOX cells (Figure 7.9), is shown in Figure 7.18. All cells are simultaneously erased with the FN-tunneling mechanism, by applying to the common source *SL*-line a voltage of, e.g., 10 V and at all word lines a voltage of 0 V. The programming of an addressed cell is caused by connecting the common source line

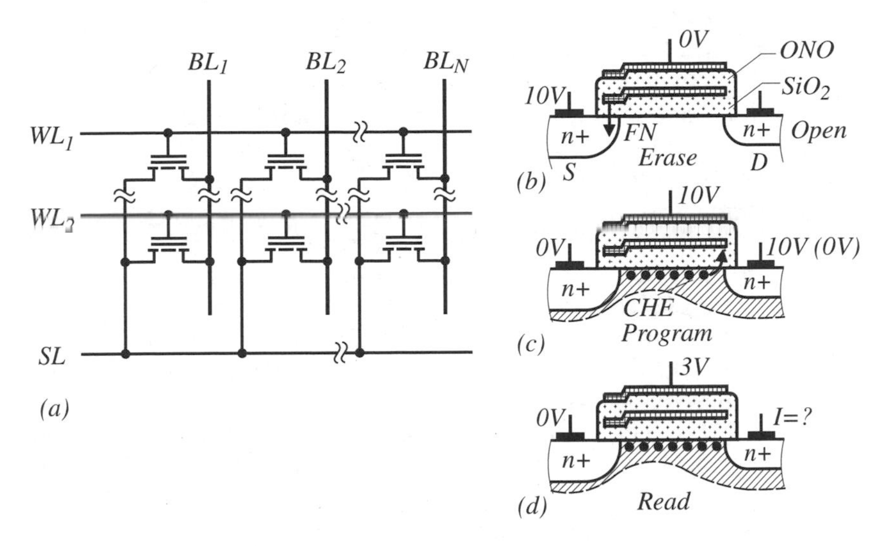

Figure 7.18 (a) NOR architecture with ETOX cells, (b) erase, (c) program and (d) read (inhibit function in parentheses)

SL to 0 V and applying at the addressed word line a voltage of, e.g., 10 V. Depending on the data at the bit lines (10 V or 0 V) CHE are, or are not, injected to the floating gate. A program disturb like in the FLOTOX cell does not occur, since in this cell programming and erasing are executed at different terminals. But the erasing process may lead to an over erase situation, when cells cannot be switched off anymore. This problem can be coped with by using an intelligent erase and modify method (Intel 1998; Tanaka *et al.* 1994) (Figure 7.19).

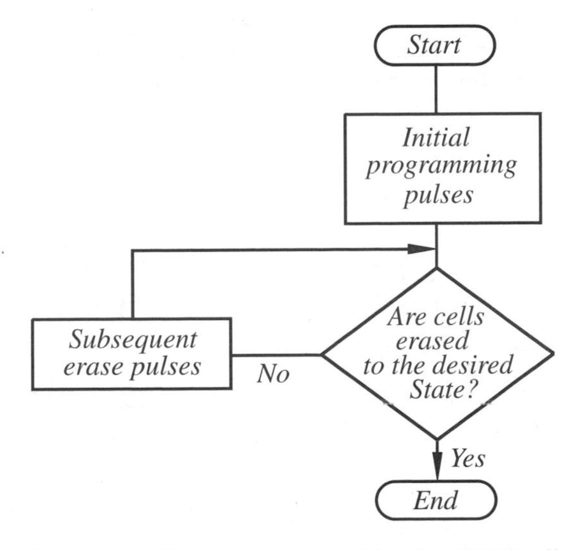

Figure 7.19 Programming algorithm for ETOX cells

To start the erasing of the cells, they are programmed first into a common condition. After this initializing, all word lines are connected to a voltage of 0 V and erase pulses of, e.g., 10 V are applied to the common source line *SL* (Figure 7.18). Next, the condition of the cells is determined and, when required, the erase procedure is repeated. The programming of the cells can be performed in a similar way. Figure 7.20(a) shows measured threshold distributions using this scheme.

The programming algorithm is so effective that even two bits per cell can be stored (Atwood *et al.* 1997). To read the different cell information, a current/voltage transformer, similar to the one shown in Figure 7.8, can be used, except that for each stored binary information a reference cell is needed (Bauer *et al.* 1995).

In the preceding example with ETOX cells, CHE are used for programming and the FN-tunnel mechanism for erasing. In the following example, an architecture using the FETMOS cells employing the FN-mechanism for programming and erasing is considered (Nozoe *et al.* 1995). The cells are connected in a *NOR* configuration between a common source line *S* and bit line *BL* (Figure 7.21). If diffusion layers are used for these lines, the cells can be placed in an area effective manner in the array without contact zones. The program and erase operations are performed sector-wise. If 0 V is applied to the *S*-line and *BL*-line, then the programming occurs when a voltage of 12 V is applied to the addressed word line (Figure 7.21(b)). An erase results when a voltage of −9 V is applied to the word line and a voltage of 3 V to the bit line. The erasing is inhibited with a voltage of 0 V (Figure 7.21(c)). This small voltage change is sufficient, since the tunnel current (Equation 7.4) depends exponentially on the electrical field (Kume *et al.* 1992). The memory architecture shown is used for file applications, where data are written or read via a register into or out of the memory array. For the control of the threshold voltage distribution, a similar programming algorithm, as described in the preceding example, is used.

NAND architecture

The *NAND* architecture offers an even higher area efficiency than the discussed *NOR* architecture. An example using FETMOS cells (Kirisawa et al. 1990) is considered. The FETMOS cell of Figure 7.9 can be erased when a positive voltage is applied to the

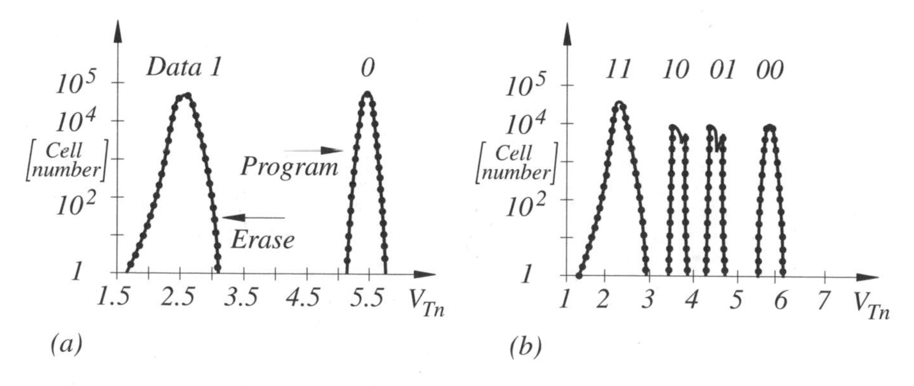

Figure 7.20 Threshold voltage distributions: (a) one bit per cell, (b) two bits per cell (reprinted by permission of Intel Corporation, © Intel Corporation 2003)

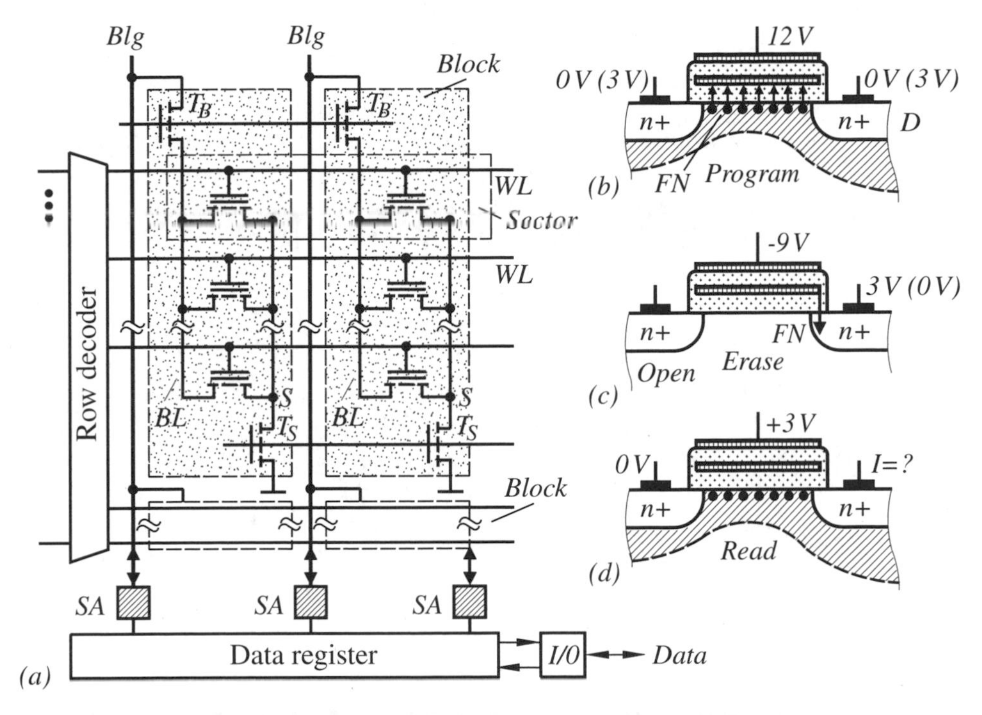

Figure 7.21 (a) *NOR* architecture with FETMOS cells, (b) program, (c) erase and (d) read (inhibit function in parentheses) (after Kume *et al.* 1992; © 1992 IEEE)

substrate of the cell. The advantage of using the entire gate area for the FN-tunnel mechanism is that degradation effects do not influence the device as much as when this happens in a small oxide window only. A slight disadvantage of this scheme is that a triple-well process is required (Figure 7.22). This is necessary in order to switch the

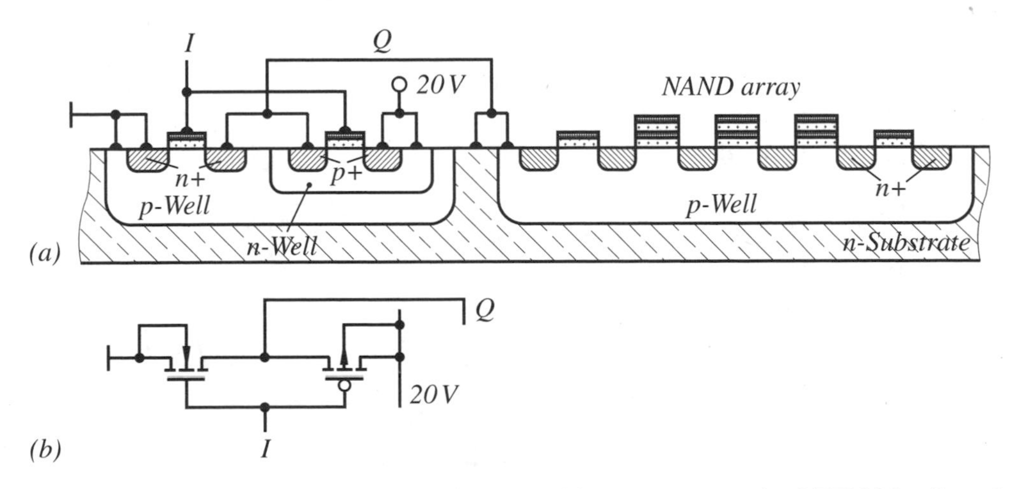

Figure 7.22 (a) Sketch of a three-well technology with *NAND* array using FETMOS cells and (b) CMOS inverter connection

substrate of the cell array, that is the p-well and the n-substrate of the chip, between the erase and program voltages by a CMOS inverter.

All cells are erased by turning off the transistors T_N and T_M and applying a voltage of 20 V to the p-well and the n-substrate and 0 V to the gates (Figure 7.23).

But there exists a substantial difference in comparison to the preceding cases. An over erase is required in order to shift the cells into a normally-on operation (Section 5.3.1, Figure 5.11). This is a must for reading the devices in a *NAND* configuration, as is discussed after considering the programming of the cells (Figure 7.24).

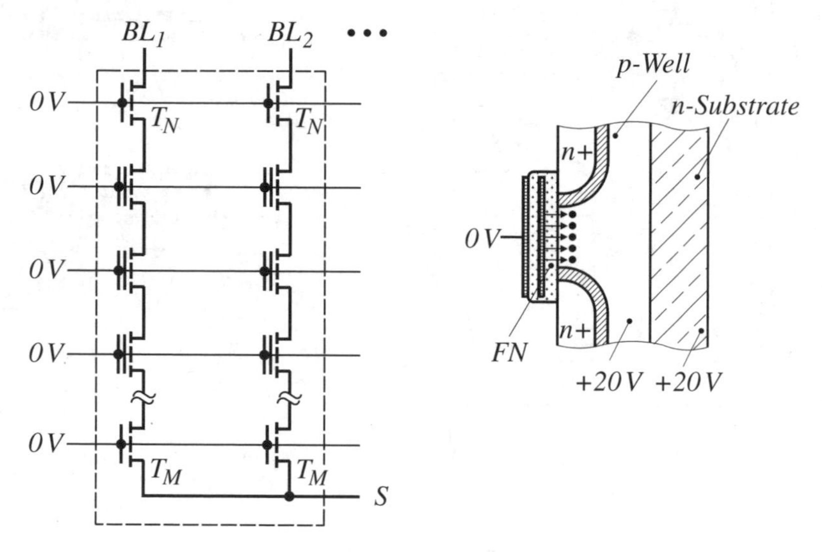

Figure 7.23 *NAND* array of Figure 7.22 under erase condition

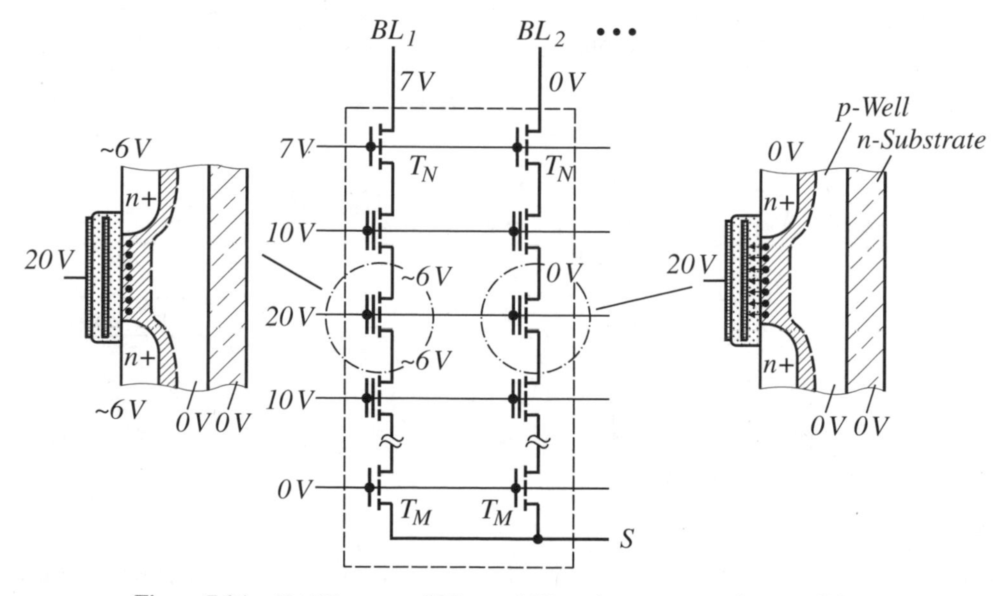

Figure 7.24 *NAND* array of Figure 7.22 under programming condition

The select transistors T_N are turned on and transistors T_M are turned off. Charge is injected by the FN-tunnel mechanism, where the bit line (BL_2) has a voltage of 0 V and the word line one of 20 V. In all other cases the FN-tunnel mechanism is suppressed. Reading the memory is comparable to the situation encountered in the PLA of Section 6.4.2, Figure 6.43. In analogy a voltage of 0 V is applied to the word line where the contents of the cells are to be determined. All other cells are turned on with a voltage, e.g., of 5 V at the word lines (Figure 7.25).

If the cell is erased and thus normally-on, a current flows through the cell chain. Otherwise the current is negligible. The series connection of the cells causes a relatively slow reading process. Usually, not more than 16 cells are therefore connected in a string (Imamiya *et al.* 1995). To increase the bit density of the *NAND* array a multi-level storage scheme can be used by applying a programming algorithm, similar to the one discussed in conjunction with the *NOR* architecture (Jung *et al.* 1996). This approach comes closer to the goal of a mass storage memory for file applications.

The presented nonvolatile memories require relatively large positive and negative voltages. These voltages should be generated on-chip in order to keep the memory user-friendly. For this purpose charge pumps are employed. Since these chip internal voltage generators are usually not able to supply a large current, only memory cells using the FN-tunnel mechanism for program and erase are suitable.

7.3.3 On-Chip Voltage Generators

In conjunction with the bootstrap buffer of Section 5.5.2 an on-chip voltage multiplication, sometimes called charge pump, is described. The starting point is the circuit of Figure 5.35, which is repeated for convenience in Figure 7.26.

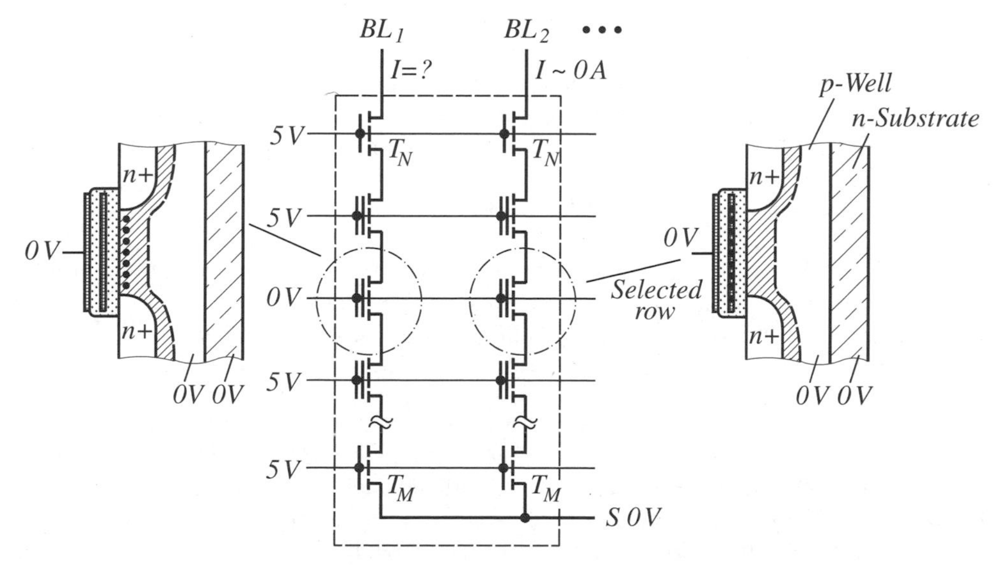

Figure 7.25 *NAND* array of Figure 7.22 during reading

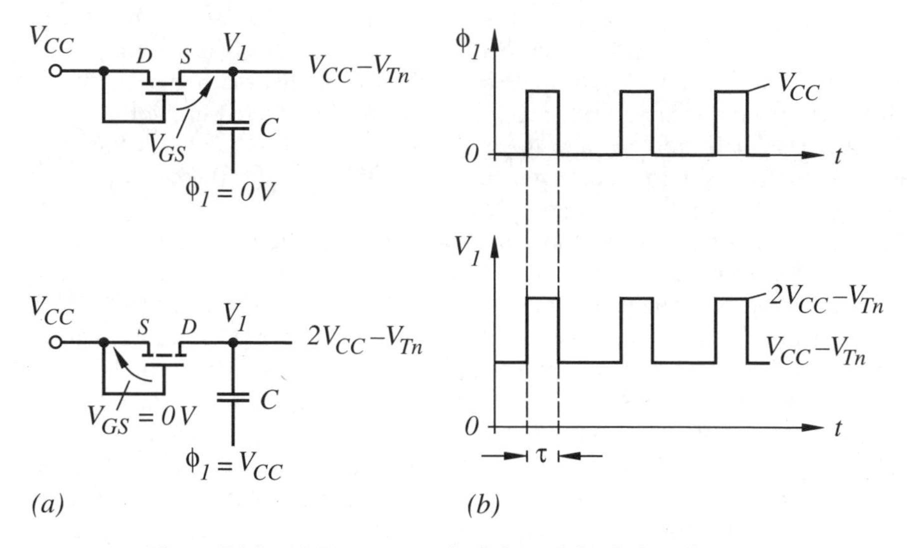

Figure 7.26 (a) Bootstrap principle and (b) timing diagram

If the clock voltage has a value of $\phi_1 = 0\,\text{V}$, then the capacitor is charged to a voltage of $V_1 = V_{CC} - V_{Tn}$. If the clock voltage is raised to $\phi_1 = V_{CC}$, the output voltage increases to $V_1 = 2V_{CC} - V_{Tn}$, when parasitic capacitances are neglected (Equation 5.65). By considering the circuit it is obvious that the voltage of the charged capacitor is added to that of the clock signal. Since in the time interval τ the voltage V_1 is larger than V_{CC}, the functions of source and drain of the transistor are reversed. With $V_{GS} = 0\,\text{V}$ the transistor is basically cut off and the charge at the capacitor remains unaltered. Since the transistor in this configuration behaves like a diode, one speaks sometimes of a MOS diode. If a further stage is added (Figure 7.27) and controlled by clock ϕ_2 an additional voltage increase to $V_2 = 3V_{CC} - 2V_{TN}$ results.

It is assumed for simplicity that the body effect can be neglected. Furthermore, it is assumed that a charge sharing effect between the two capacitors can be disregarded since $C_1 \gg C_2$. With the two bootstrap stages a maximum voltage of

$$V_n = (n+1)V_{CC} - nV_{Tn} \tag{7.6}$$

can be generated where $n = 2$. If n stages are added (Figure 7.28), the maximum voltage can be increased accordingly. In order to transform the changing voltage at node n into a DC one, an additional transistor T_D acting as a diode is included. Since a threshold voltage drop occurs across this transistor, the DC output voltage is reduced to

$$V_{DC} = (n+1)(V_{CC} - V_{Tn}) \tag{7.7}$$

In real circuits – as shown in Figure 7.28 – usually capacitors of the same size are used. This causes a charge sharing between the capacitors (Equation 5.56). But this does not mean that the DC voltage does not reach its final value; it only means that several clock cycles are required, so that the DC voltage approaches its final value.

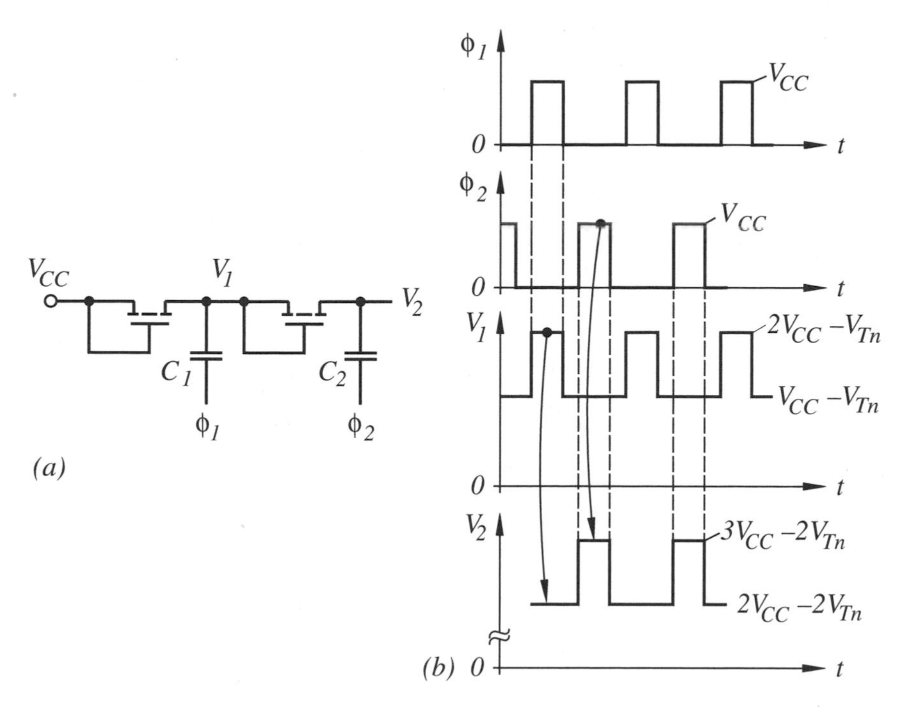

Figure 7.27 (a) Voltage generator circuit with two stages $C_1 \gg C_2$ and (b) timing diagram

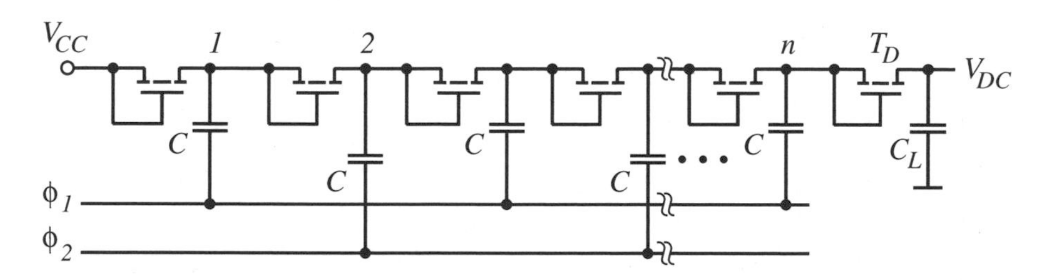

Figure 7.28 DC voltage generator circuit with n stages

The analyzed circuit, known as a Dickson charge pump, has the disadvantage that the output voltage V_{DC} is reduced by the threshold voltage (Equation 7.7). This is a particular handicap when a very small power-supply voltage exists. This can be circumvented with the circuit shown in Figure 7.29 (Wu *et al.* 1996), which requires two different threshold voltages for proper functioning.

After the initialization phase – that is, after a couple of clock cycles – the voltages at the individual nodes are multiplied, and the feedback transistors T_F are active, since they are controlled by the increased voltage of the following stage and thereby in effect short-circuiting the MOS diodes. This leads to a maximum DC voltage of

$$V_{DC} = n \cdot V_{CC} - 2V_{Tn} \tag{7.8}$$

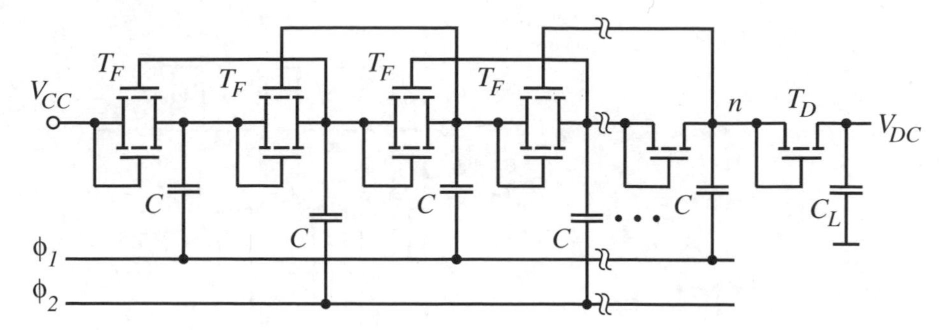

Figure 7.29 DC-voltage generator circuit with n stages for small power-supply voltages

In order that the feedback transistors T_F are active, the threshold voltage V_{Tnf} of these transistors has to meet the requirement $V_{Tnf} < 2V_{CC} - V_{Tn}$, and in order that the transistors T_F cut off properly, the threshold voltage has to be $V_{Tnf} > V_{CC}$.

Negative voltage generation

Some circuit applications require negative voltages for back-biasing a substrate or for cell erasing. How a negative voltage can be generated on-chip is the topic of the next section. For this purpose the bootstrap principle is revisited in Figure 7.30.

If the clock signal has a value of $\phi_1 = V_{CC}$ then the capacitor is charged to $V_1 = |V_{Tp}|$. If the clock signal changes from V_{CC} to 0 V, then the voltage at the output of the stage reduces to $V_1 = -V_{CC} + |V_{Tp}|$, when parasitic capacitances are ignored. All

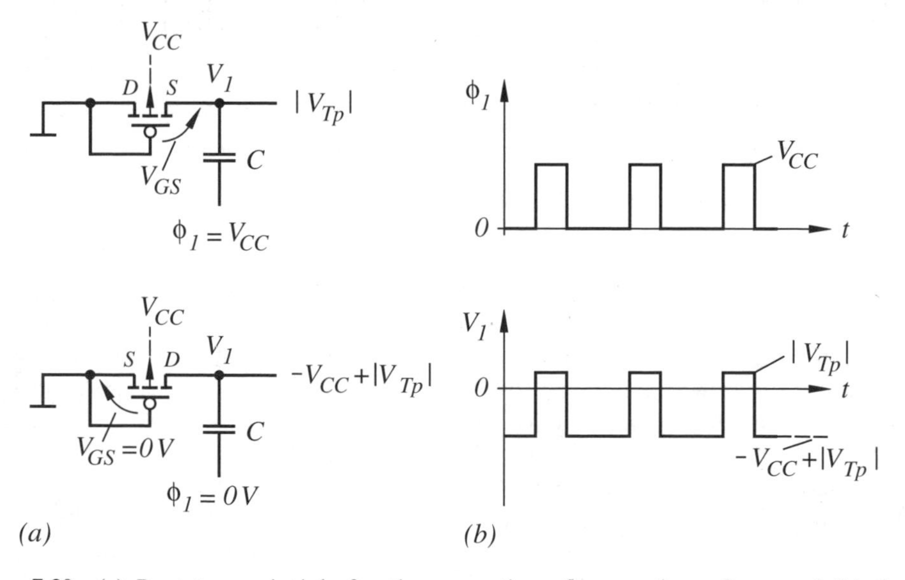

Figure 7.30 (a) Bootstrap principle for the generation of a negative voltage and (b) timing diagram

pn-junctions remain back-biased since the n-well (Section 5.3.3, Figure 5.14) is tied to V_{CC}. The functions of source and drain of the p-channel transistor are reversed (reminder: the source of a p-channel transistor is the most positive terminal) and the transistor with $V_{GS} = 0\,\mathrm{V}$ is cut off. Adding a second stage to the bootstrap circuit (Figure 7.31), a voltage of

$$V_n = n(-V_{CC} + |V_{Tp}|) \tag{7.9}$$

is generated where $n = 2$. If n stages are connected to a chain (Figure 7.32) a negative DC voltage of

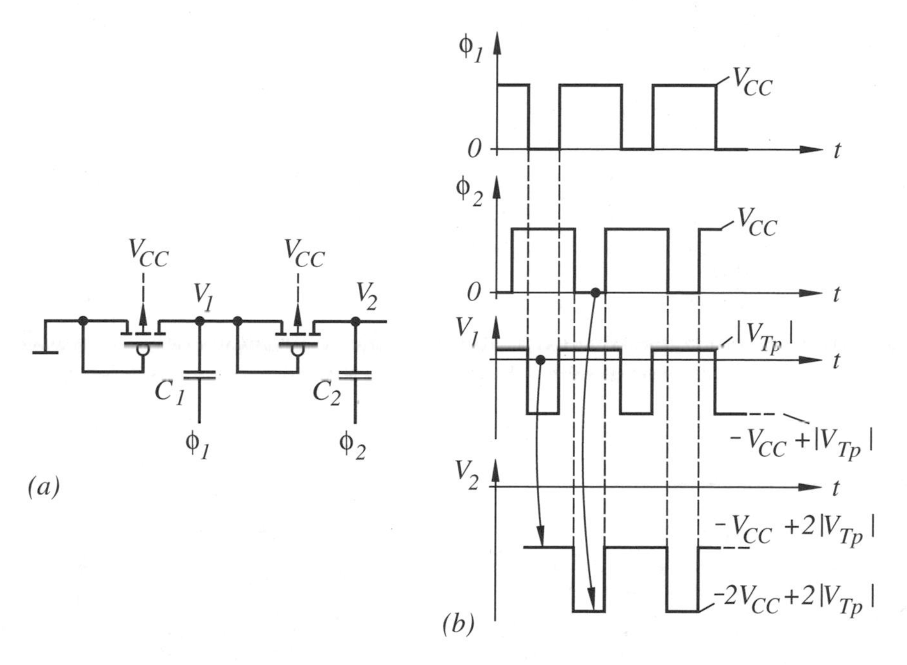

Figure 7.31 (a) Negative voltage generator circuit with two stages $C_1 \gg C_2$ and (b) timing diagram

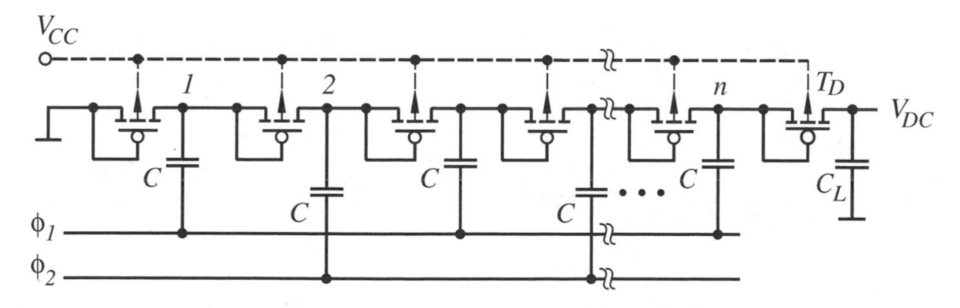

Figure 7.32 Negative DC-voltage generator circuit with n stages

$$V_{DC} = n(-V_{CC} + |V_{Tp}|) + |V_{Tp}| \tag{7.10}$$

results after some clocking cycles.

7.4 STATIC MEMORIES

These memories store the information statically. That is, the information remains stored as long as the power supply is turned on. They are usually organized in such a way that a random read/write access to the memory cells is possible. That is why they are called static random access memories (SRAM). In this section the memory cells are examined first. Then a typical architecture of an asynchronous memory is considered.

7.4.1 Static Memory Cells

The memory cell is basically a static flip-flop or register cell, introduced in Section 6.5.1. The cell, shown in Figure 7.33(a), is called a six-transistor cell. Information is written into the cell by applying, e.g., an H signal to the bit line BL and the complementary signal, an L signal to the bit line $\overline{BL}$. Next, the word line WL is activated and the select transistors T_S are turned on. This causes the flip-flop to switch into the required condition. Then the transistors T_S are turned off and the information is stored. The cell is read by first precharging both bit lines to an intermediate voltage level and then

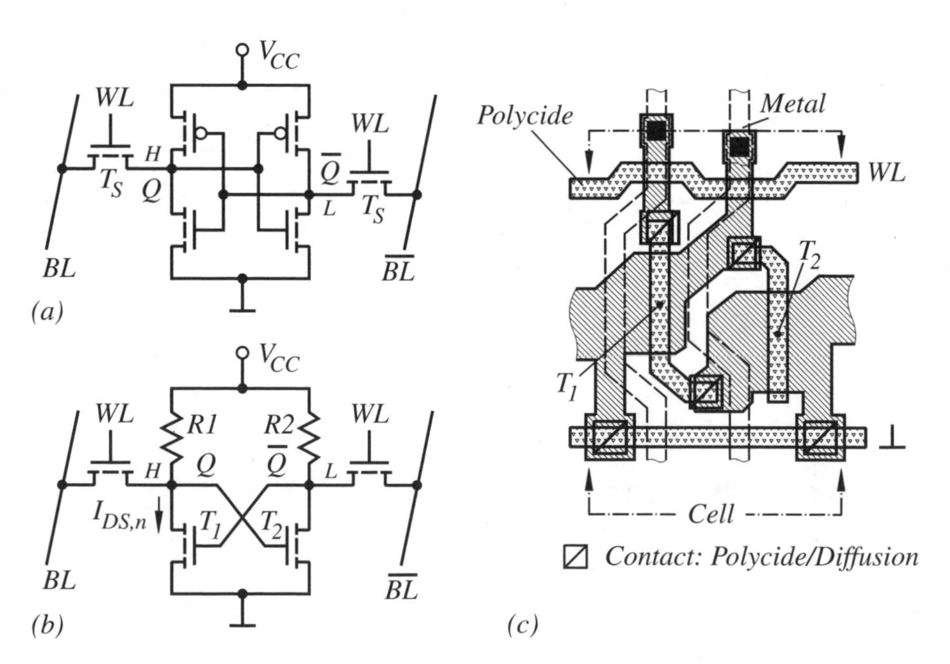

Figure 7.33 SRAM cells: (a) six-transistor cell, (b) four-transistor cell and (c) layout of four-transistor cell (second polysilicon layer not shown)

activating the select transistors. This causes the cell to generate a voltage change on the bit lines, which can be sensed by a sense amplifier.

Another static memory cell is shown in Figure 7.33(b). It is called a four-transistor cell and uses resistors as pull-up devices. In this cell one transistor e.g. T_2 is conducting whereas the other one T_1 is cut off. This means that through one resistor, in this case R_2, a current always flows. Since in a memory many cells exist, the current has to be reduced to a minimum. This is achieved by using undoped polysilicon stripes as resistors with a sheet resistance of several hundred $G\Omega/\square$. The use of such high value resistors is possible for two reasons:

(a) The leakage current $I_{DS,n}$, which is usually the sub-threshold current (Section 4.4.3) of the cut-off transistor, in this case T_1, is very small and therefore the voltage drop across the resistor is also. As long as this voltage drop is negligible on the H level, the storing property of the cell is not impaired.

(b) The extremely long charging time of the parasitic bit line capacitance by the resistors and a possible malfunctioning of the cell during reading can be avoided, by precharging the bit lines to V_{CC} prior to the read cycle. If a cell is addressed, the conducting transistor, T_2 in this example, is responsible for causing a small high–low transition at the bit line $\overline{BL}$.

The major advantage of the four-transistor cell is the cell area reduction to about $\frac{2}{3}$, compared to that of the six-transistor cell. To achieve this, a two-layer polysilicon process is required. The first one, the polycide layer, is used for the word and ground lines and the transistors (Figure 7.33(c)), whereas the second undoped one is used for the implementation of the resistors. Since the two layers are independent of each other, the second one can partly be placed upon the first one.

TFT SRAM cell

A comparison of the static current consumption reveals that the four-transistor cell has a higher power consumption than the six-transistor cell. This is important only when, e.g., 16 Mb or larger memory arrays are required. Figure 7.34 is used to clarify the situation.

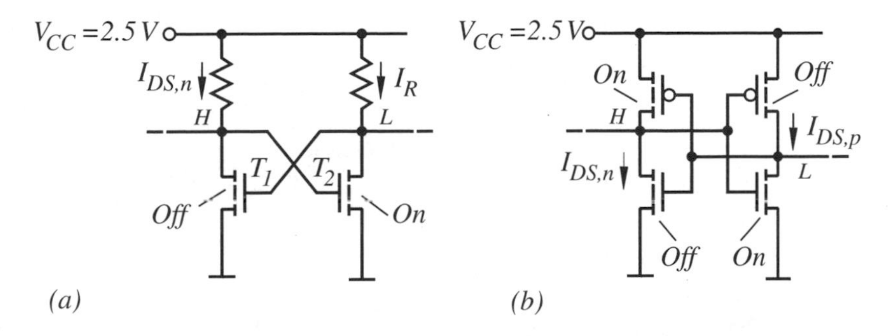

Figure 7.34 Partial SRAM cell: (a) four-transistor cell and (b) six-transistor cell

In the case of the four-transistor cell the current is

$$I = I_{DS,n} + V_{CC}/R \tag{7.11}$$

and in the case of the six-transistor cell the current has a value of

$$I = I_{DS,n} + I_{DS,p} \tag{7.12}$$

In both cases, the currents are independent of the stored binary information, due to the symmetrical nature of the cell (Section 5.3.4, Figure 5.18). The sub-threshold currents $I_{DS,n}$ and $I_{DS,p}$ can be reduced by increasing the threshold voltages (Section 4.4.3, Equation 4.65). But the resistor values cannot be increased beyond a particular value. In order to develop a feeling for the resistor values involved, the following example is used as an illustration.

Example

The stand-by current of a 16 Mb SRAM should be below 1 μA in order to facilitate a battery operation or a battery back-up. This results in a current flow per cell of about $6.3 \cdot 10^{-14}$ A. The required resistor value is thus $R = 2.5\,\text{V}/6.3 \cdot 10^{-14}\,\text{A} \approx 40 \cdot 10^{12}\,\Omega$. Hereby is assumed that the sub-threshold current is negligible.

Since the six-transistor cell requires a relatively large silicon area and the four-transistor cell may have a problem with the stand-by current, thin film transistors (TFT) are a solution.

A thin film transistor in a recrystallized polysilicon film or layer can be controlled from the back or top side of the device. The recrystallization can be achieved with the aid of a scanning laser beam. In the case where one excepts a deterioration in the transistor's cut-off behavior, the expensive recrystallization process can be omitted. In Figure 7.35 a four-transistor cell, using p-channel TFT, is shown.

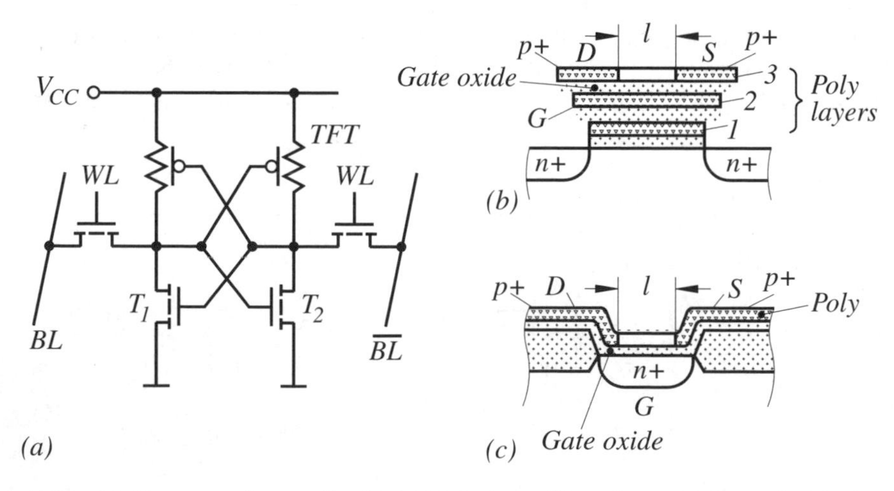

Figure 7.35 (a) Four-transistor cell with TFT devices, (b) p-channel TFT poly 2/poly 3 and (c) p-channel TFT diffusion/poly

The advantage of the TFT approach is the area saving arrangement. In Figure 7.35(b), the transistor is controlled from the back side. The gate consists of the polysilicon layer (2), separated by a gate oxide from the n-doped polysilicon layer (3) on top. This layer is implanted to form p^+-source and p^+-drain regions (Uemoto *et al.* 1992). In the realization of Figure 7.35(c) the gate is built up by the n^+-diffusion area (Ootani *et al.* 1990). A typical current–voltage relationship of a p-channel TFT transistor is shown in Figure 7.36.

With a current of $I \approx 10^{-15}$ A in cut-off, the thin film transistor complies with the requirement of the preceding example (Yamanaka *et al.* 1995). A further advantage is that the cell is less likely to produce soft errors due to alpha-particle hits (Section 7.5.4).

7.4.2 SRAM Memory Architecture

Usually these memories operate asynchronously and thus do not require an external clock. This means that no clocked circuit technique (C^2MOS) can be used, e.g. to cope with the power dissipation and propagation delay of decoders and sense circuits. To get around this problem, a chip-internal clock is derived, whenever an address or chip select changes. Using this so-called address transition detection (ATD) technique, a fast access time is possible (Sasaki *et al.* 1988).

In the following the general operation of a SRAM and then the generation of the clock signal via the ATD technique are considered. Part of a SRAM architecture is shown in Figure 7.37.

The read cycle starts at time t_0 with an address or chip select change which generates the ATD pulse. The transistors T_7 and T_8 are turned off via the signal W'. This disconnects the bus lines BS and $\overline{BS}$ from the common data I/O. The clock signal ϕ_P has a voltage of 0 V up to time t_1, causing the so-called precharge transistors T_1 and T_2 to be conducting. The purpose of these transistors is to equalize the voltage between the bit lines and bus lines. This voltage difference may either be caused by a prior read or

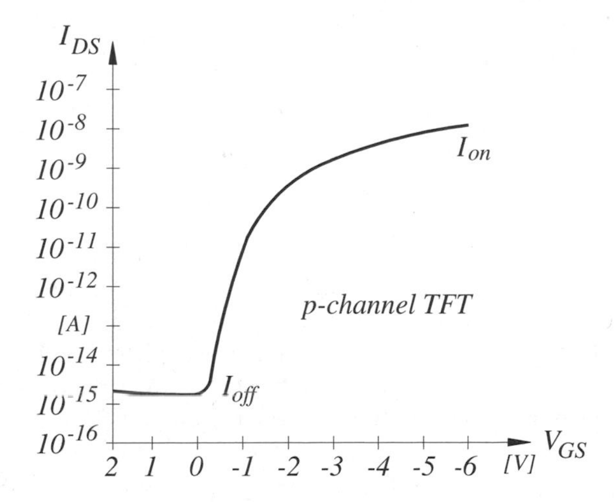

Figure 7.36 Typical p-channel TFT current voltage characteristic $w/l = 0.4\,\mu m/0.8\,\mu m$ at $V_{DS} = -3.3$ V (after Yamanaka *et al.* 1995; © 1995 IEEE)

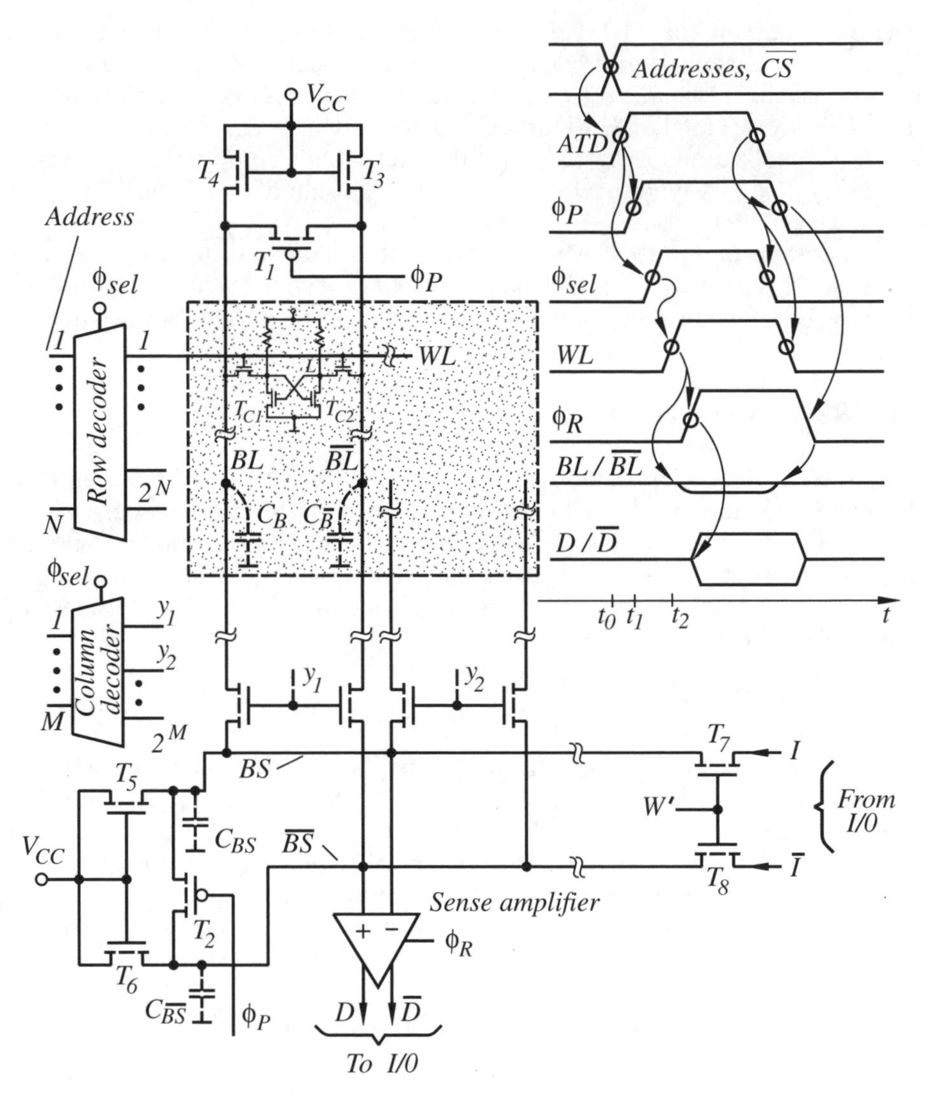

Figure 7.37 Part of a SRAM architecture with timing diagram

write cycle or by a possible difference in threshold voltage between the transistors T_3 to T_6. In this precharge phase, these lines are charged to a value of $V_{CC} - V_{Tn}$, which is about $3\,\text{V} - 0.5\,\text{V} = 2.5\,\text{V}$. The delayed clock signal ϕ_{sel} activates the decoders, which causes a low–high transition of a selected word line WL at time t_2. The select transistors of the cells are turned on. This causes the transistor in the cell with an L signal at the cross-coupled node, in this case T_{C2}, to discharge the bit line $\overline{BL}$ slightly. The generated differential signal between the bit lines is fed via the column decoder to the input of a differential sense amplifier (Section 8.3.2, Figure 8.21), which is activated by a delayed clock signal ϕ_R. The amplified data $D/\overline{D}$ are then applied and transmitted to the common I/O terminal (not shown).

Data are written into a cell by turning transistors T_7 and T_8 on and the precharge transistors T_1 and T_2 off. Then the input data are able to change via the pass transistors of the column decoder, the states of the bit lines, and thus the memory content of a selected cell.

7.4.3 Address Transition Detection

In the introduction to the preceding section it is mentioned that no external clock signal is available at an asynchronous SRAM. Therefore, a chip-internal clock is derived, whenever an address or chip select signal changes. A circuit which performs the address transition detection (ATD) is shown in Figure 7.38.

At time t_1 the signal at input A changes from L to H. This causes the signal B at the output of the inverter to change delayed at time t_2 from H to L. Due to the delay caused by the added capacitor, the *NAND* gate is activated briefly (cross-hatched area) and generates an L signal at the output C. The capacitor is implemented by two symmetrically arranged MOS transistors (Section 4.6.2, Figure 4.55). In comparison to the L to H change, an input signal variation from H to L has no influence on the output signal. Thus the logic or the circuit implementation is only good for the detection of a signal transition from L to H. In case one wants to detect the opposite signal change, an inverter in front of the input A has to be provided.

All state transitions occurring at the inputs of a SRAM are collected at a *NOR* gate and used to generate a central *ATD* clock (Figure 7.39). Transistor T_1 is controlled in such a way that a fast rise and fall time at the central *ATD* clock results. A short positive pulse at one of the local *ATD* outputs $\overline{C}$ causes an L state at the $\overline{ATD}$ signal at time t_1.

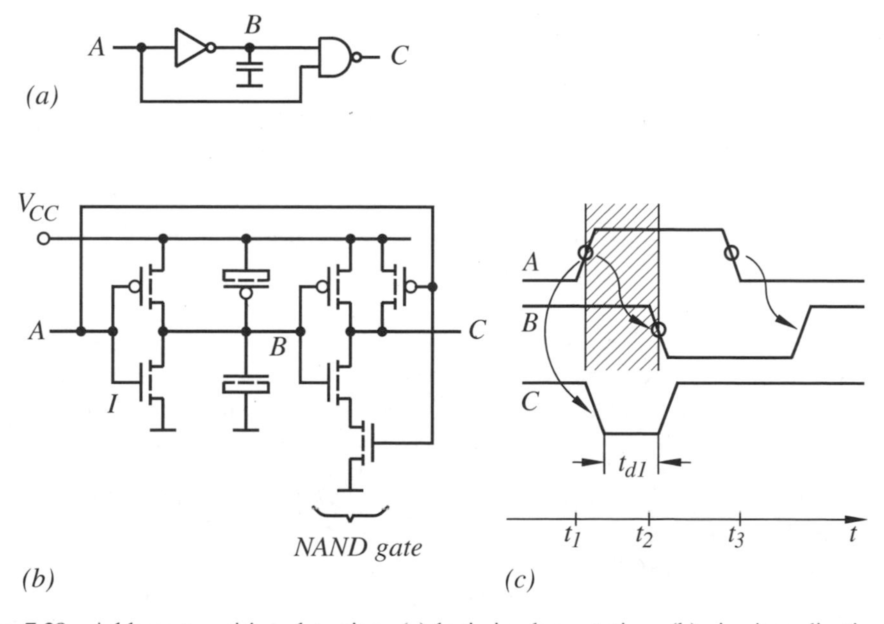

Figure 7.38 Address transition detection: (a) logic implementation, (b) circuit realization and (c) timing diagram

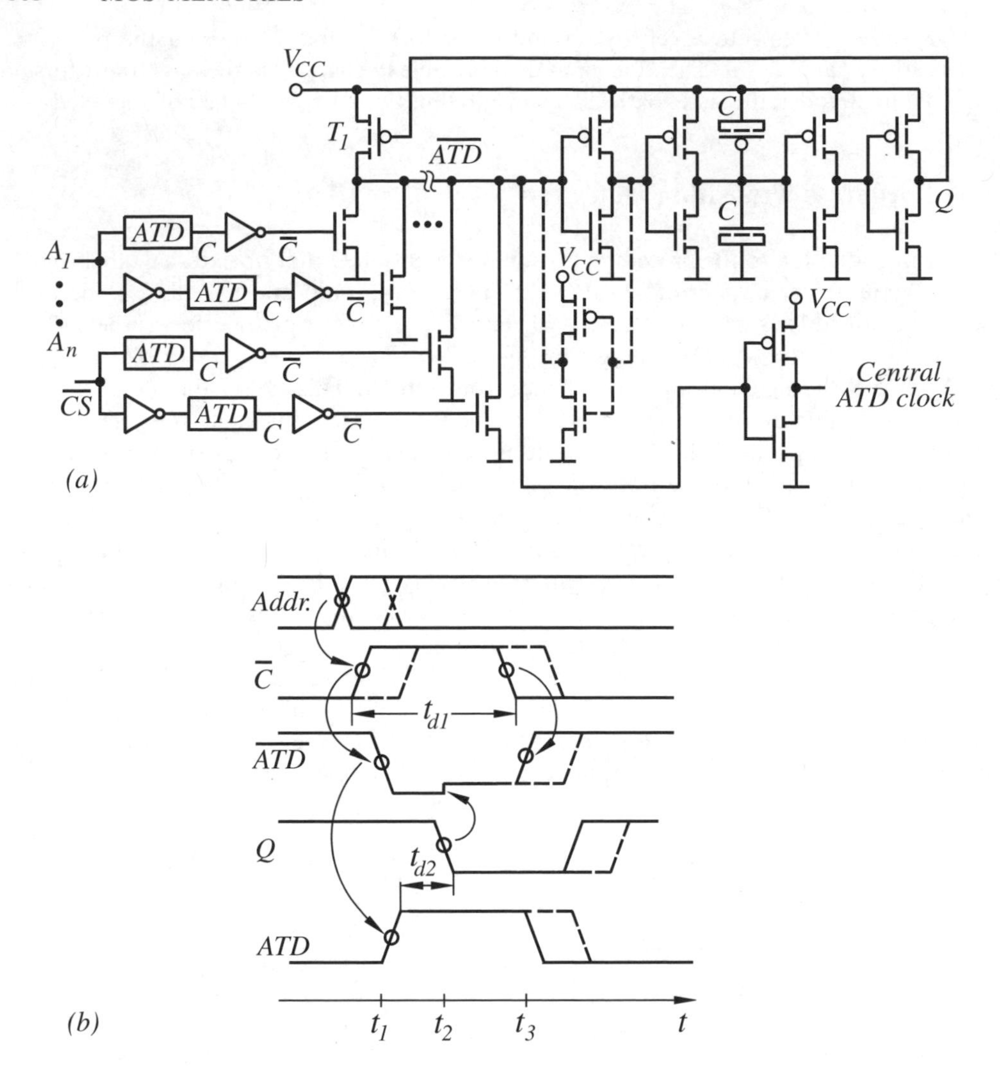

Figure 7.39 (a) ATD circuit of a SRAM and (b) timing diagram (after Kayano *et al.* 1986; © 1986 IEEE)

Transistor T_1 is at this time non-conducting since the output Q of the inverter chain has an H state. A fast fall time at the *NOR* gate results. The H state at the output Q changes after a delay time of t_{d2}, causing transistor T_1 to conduct at time t_2. This happens before the pulse $\overline{C}$ shifts into the L state at t_3. Since transistor T_1 is conducting, a fast rise time results when the $\overline{ATD}$ signal changes from the L state to the H state. The width of the central *ATD* pulse is given by the delay time t_{d1} (Figure 7.38(c)). This has the advantage that the time, even after a delayed address transition (dotted in Figure 7.39(b)), is always constant. The dotted CMOS inverter has the purpose of guaranteeing an H state at the *NOR* gate when T_1 is cut off. With the *ATD* technique, fast access times at low power dissipation are achievable. This technique is thus not only used for SRAMS but for many other circuits also, particularly nonvolatile memories.

7.5 DYNAMIC MEMORIES

These are memories where the information is stored as different charge levels at a capacitor. Due to diverse leakage currents the charge can only be guaranteed for a particular time at the capacitor. This requires a periodic refresh – that is, read and write operations – to maintain the cell's charge integrity.

7.5.1 One-Transistor Cell

Such a cell is shown in Figure 7.40. The cell consists of a cell capacitor C_S and a select transistor used for connecting the capacitor with the bit line *BL*. As capacitor a MOS structure (Section 4.2) is used. The surface of this structure can assume two conditions or states. If 0 V is applied to the bit line and the word line *WL* is activated, electrons are able to move via the n^+-reservoir and the select transistor to the surface of the MOS structure. A voltage between the semiconductor surface and the substrate ($\phi_S \approx 2\phi_F$) of about 0.8 V appears and represents the *L* state (Figure 7.40(b)). An *H* state is written into the cell by applying, e.g., 3 V to the bit line and activating the select transistor. If charge in the cell from a prior write operation exists, then the charge moves out of the cell to the n^+-region. The MOS structure is depleted of electrons and thus in a condition of deep depletion with a surface voltage of about 2.5 V, representing the *H* state (Figure 7.40(c)). Thus an *H* state or *L* state is stored in the cell after the select transistor is cut off. The deep depletion condition – that is, the *H* state – is not stable. Due to the generation of electron–hole pairs at the semiconductor surface, an inversion layer is

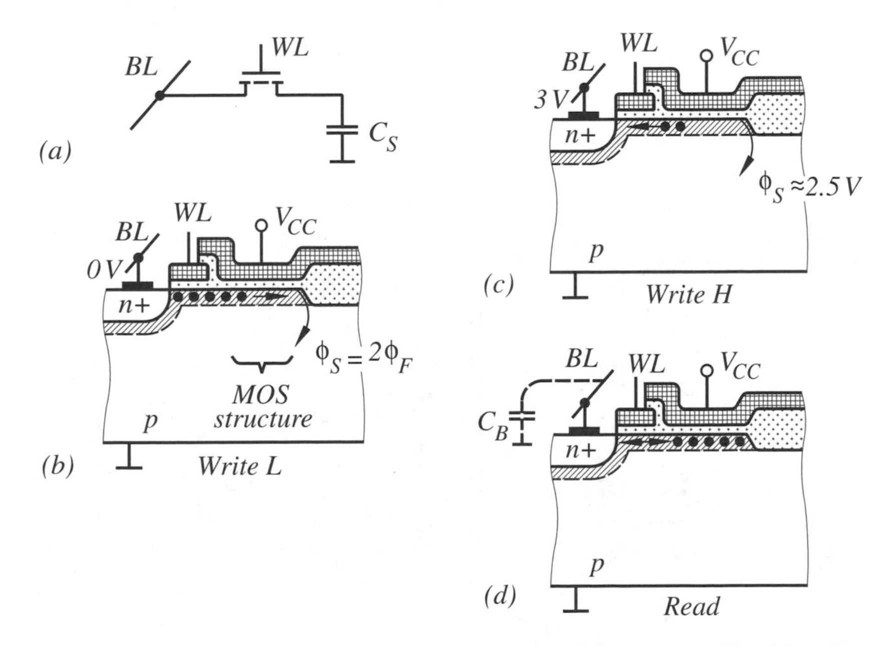

Figure 7.40 One-transistor cell: (a) equivalent circuit, (b) writing *L* state, (c) writing *H* state and (d) reading cell content

built-up (Section 4.2.1, Figures 4.7 and 4.8) and the H state changes to an L state. This takes some time. Before that happens, the information has to be read and rewritten back into the cell in at least every, e.g., 64 ms. This is the so-called refresh time of the memory. Since the refresh operation is performed simultaneously on many cells, the memory is not available to the user for less than 2%.

To read the cell information, the bit line is disconnected from the voltage supply and the parasitic capacitance of the bit line C_B is charged to, e.g., 1.5 V. If a word line is activated, electrons are able to move out of the cell to the bit line or from the bit line into the cell. This causes a voltage change at the bit line named the read signal in the order of $\pm$ 50 mV which is processed by a sense amplifier, discussed in Section 7.5.2.

In order to implement a high cell density, the cell geometry has to be scaled. Unfortunately the cell capacitor of approximately $C_S \approx 35\,\text{fF}$, and thus the capacitor area, can barely be reduced anymore, since the read signal is very small already. To circumvent this problem, trench or stacked one-transistor cells are used.

Trench cell and inverted trench cell

Trenches are formed in the substrate by an anisotropic etching process in which the MOS structure is then embedded (Figure 7.41). Depending on the depth of the trench, a relatively large capacitor in a small area can be implemented. With this technique, cell sizes of 4.8 μm² and trench depths of 4μm are realized (Fujii *et al.* 1989).

A further scaling of the cell is limited by the depletion regions' width (Equation 4.12)

$$x_d = \sqrt{\frac{2\varepsilon_0\varepsilon_{Si}}{qN_A}\phi_S} \tag{7.13}$$

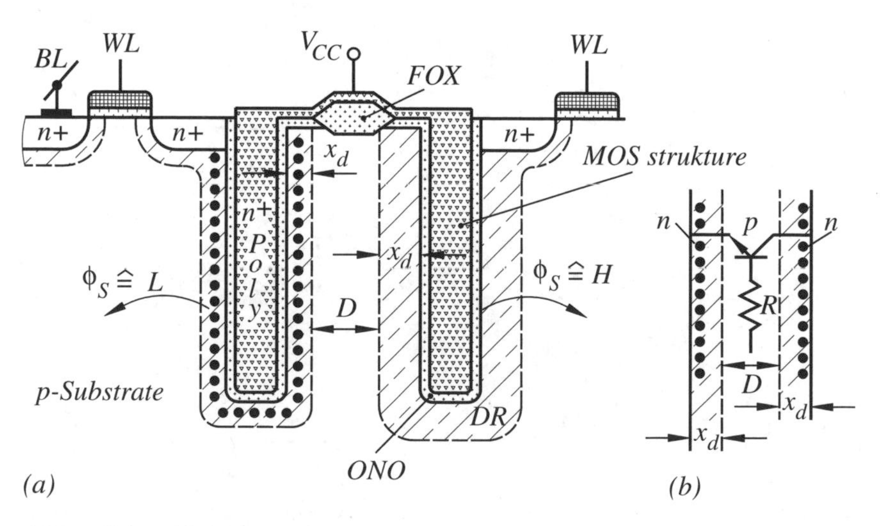

Figure 7.41 (a) One-transistor trench cell and (b) cross-section between trenches

and their distance D. This distance has to be so large that a punch-through (Section 4.5.6) between the cells is avoided. The region between the cells acts as a field-induced npn bipolar transistor, when both cells are in inversion (Figure 7.41(b)). If this transistor conducts, e.g. caused by coupling in the substrate or by alpha-particles – treated in Section 7.5.4 – charge can be lost. A remedy is the inverted trench cell, in which the n^+-terminal of the transistor is not connected to the inversion layer but instead to the n^+-poly region (gate) of the cell (Figure 7.42).

In order to reduce the distance between the cells, a sallow trench insulation (STI) is used instead of the field oxide (FOX) (Section 4.1, Figure 4.3). With this approach, a capacitor between n^+-poly and the p-substrate is built, using as insulating material an oxide-nitride-oxide (ONO) sandwich with a high dielectric constant. Unfortunately, a depletion region (DR) at the semiconductor surface exists, when the n^+-poly side of the capacitor is charged with a positive voltage. This, in effect, reduces the capacitance value of the cell. To eliminate the depletion regions around the trenches, an n-well is incorporated in the manufacturing process (Figure 7.43).

If in this case the n^+-poly side of the capacitor is charged to, e.g., 1.8 V, an accumulation layer of electrons occurs instead of a depletion region at the n-well. This has the additional advantage that the n-well resistance reduces slightly. The use of the n-well is not without a problem. A parasitic n-channel transistor, where the n^+-region of the select transistor and the n-well act as drain and source and the n^+-poly of the trench as gate (Figure 7.43(b)), exists. Since the gate is connected to the drain terminal of this transistor, charge is able to leak off. To avoid this, a thicker oxide – a so-called oxide collar – is built in the gate region of this transistor. This shifts the threshold voltage to such high values, that the transistor is basically cut off (Problem 7.4). The cell sizes are in the area of 0.6 μm² when design rules of 0.25μm are used (Nesbit *et al.* 1993) or 0.01 μm² when 0.1 μm design rules are employed.

As mentioned in the introduction to this chapter, the integrity of the cell charge can only be guaranteed for a particular time, e.g. 64 ms. The currents, responsible for the discharging of the capacitor, are manifold (Figure 7.44).

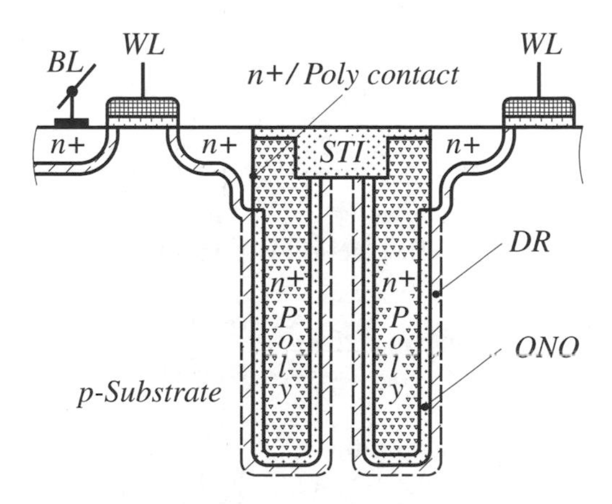

Figure 7.42 One-transistor inverted trench cell

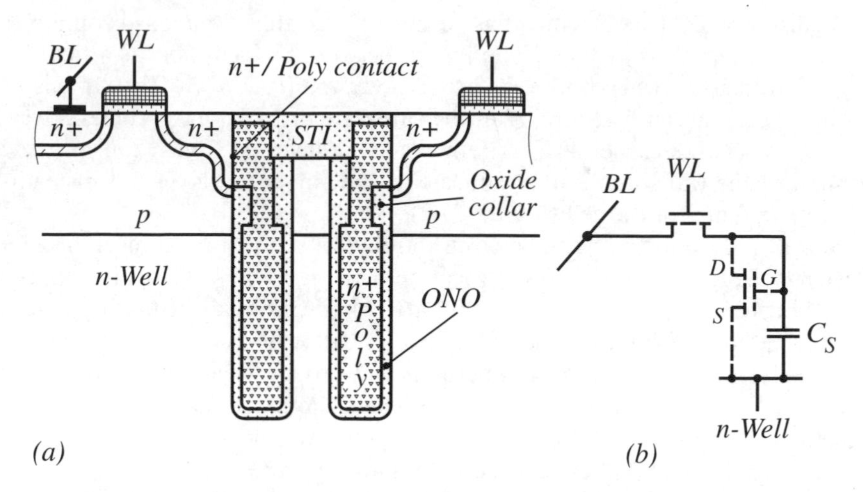

Figure 7.43 One-transistor inverted trench cell with n-well and (b) equivalent circuit (after Nesbit *et al.* 1993; © 1993 IEEE)

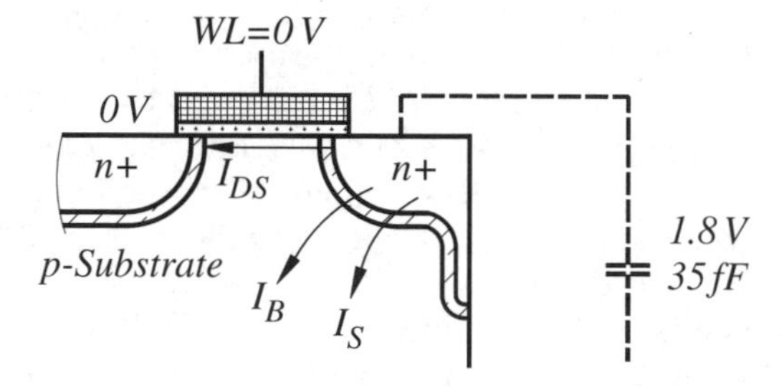

Figure 7.44 Currents responsible for discharging the capacitor of a one-transistor cell

This is the sub-threshold current of the select transistor (Equation 4.65)

$$I_{DS} = \beta_n(n-1)\phi_t^2 e^{(V_{GS}-V_{Tn})/\phi_t n}\left(1 - e^{-V_{DS}/\phi_t}\right) \tag{7.14}$$

the gate induced drain leakage current (Equation 4.93)

$$I_B = A\mathcal{E}_S e^{-B/\mathcal{E}_S} \text{ with } \mathcal{E}_S = \frac{V_{DG} - qWg}{3d_{ox}} \tag{7.15}$$

and the saturation current of the pn-junction (Equation 2.29)

$$I_S = qA\left[\frac{D_p}{w_n'}\frac{1}{N_D} + \frac{D_n}{w_p'}\frac{1}{N_A}\right]n_i^2 \tag{7.16}$$

Considering these equations, it is obvious that the worst case exists at elevated temperature. How large the sum of these currents under worst case conditions can be is illustrated with the following example.

Example

It is assumed that a voltage change of 0.2 V at the capacitor is still acceptable with respect to the sense amplifier, and that the capacitor has a value of 35 fF. With a refresh time requirement of 64 ms, this leads to a total leakage current of

$$\begin{aligned} I &\approx C_S \frac{\Delta V}{\Delta t} = 35 \cdot 10^{-15} \frac{As}{V} \frac{0.2\,\mathrm{V}}{64 \cdot 10^{-3}\,\mathrm{s}} \\ &\approx 1.1 \cdot 10^{-13}\,\mathrm{A} \end{aligned}$$

Usually, the sub-threshold current is the major contributor to the leakage current. A reduction can be achieved by increasing the threshold voltage of the select transistor (Section 4.4.3). This can also be done circuit-wise, either by applying a negative voltage of e.g. – 1 V to the p-substrate, which shifts the threshold to a higher value via the body effect (Equation 4.36), or by applying instead of 0 V a negative voltage of e.g. – 0.5 V to the word line during cut-off operation.

Stacked cell

A different implementation of a one-transistor cell is shown in Figure 7.45. In this case, the capacitor is not implemented in a trench, but instead stacked on top of the select transistors and bit line. In order to achieve a large capacitor value, insulating materials like $(Ba,Sr)TiO_3$ (called BST) with a high dielectric constant ε, with values around 50, are used (Eimori *et al.* 1993; Park and Kin 2001).

A comparison between the trench and the stacked cell reveals a more complex process for the case of the trench cell, but with the advantage that the cell capacitance, e.g., can be increased by extending the depth of the trench. Contrary is the situation for the stacked solution, where the cell capacitance is basically dependent on the kind of insulation material and its integrity.

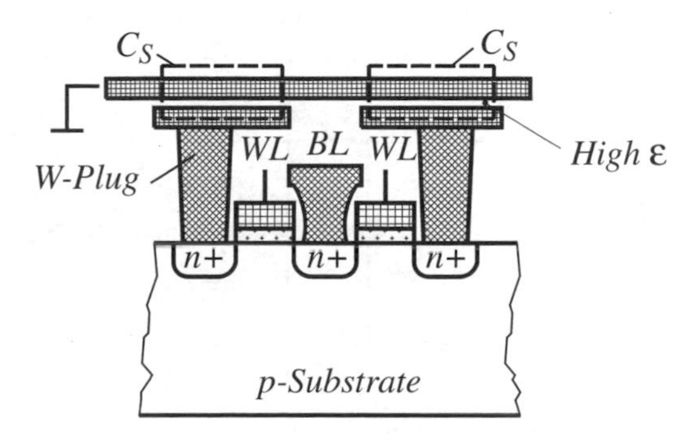

Figure 7.45 Stacked one-transistor cell

7.5.2 Basic DRAM Memory Circuits

One of the most important basic DRAM circuits are the sense amplifiers, and to some extent the word line drivers, which are considered in detail in the following section.

Sense amplifier

The basic functions such an amplifier must fulfill are demonstrated with the aid of Figure 7.46. The activation of a word line causes a charge sharing between the charge of the memory capacitor C_S and the charge of the parasitic bit line capacitance C_B. This results in a voltage of

$$V_L(BL) = \frac{Q_S + Q_B}{C_S + C_B} = \frac{C_S \cdot V_S + C_B V_B}{C_S + C_B} \tag{7.17}$$

at the bit line. In this equation V_S and V_B are the voltages at the cell capacitor and bit line capacitance, respectively, prior to the activation of the word line transistor.

In order to design a memory with a minimum of overhead, the number of cells at a bit line should be as large as possible, e.g. 512 cells. This results in a relatively large bit line capacitance C_B and thus a small read signal in the order of, e.g., $\pm 50\,\text{mV}$. This signal is amplified to the full rail to tail swing and transmitted to the output O. The reading of the cell is destructive since the initial cell charge does not exist anymore. Therefore the amplified signal is written back into the cell via transistor T_R.

A write cycle starts by turning on transistor T_S and an addressed select transistor. Data at the I-input are then able to charge the cell capacitor to the appropriate value.

The cell information – that is, the charge – must be read and written back into the cell periodically due to charge losses caused by leakage currents. The so-called refresh operates like the described read operation, except that the cells are serially addressed e.g. from WL_1 to WL_n.

The presented circuit arrangement of Figure 7.46 cannot be implemented in this form since in general, amplifiers are not precise enough to detect the absolute value and,

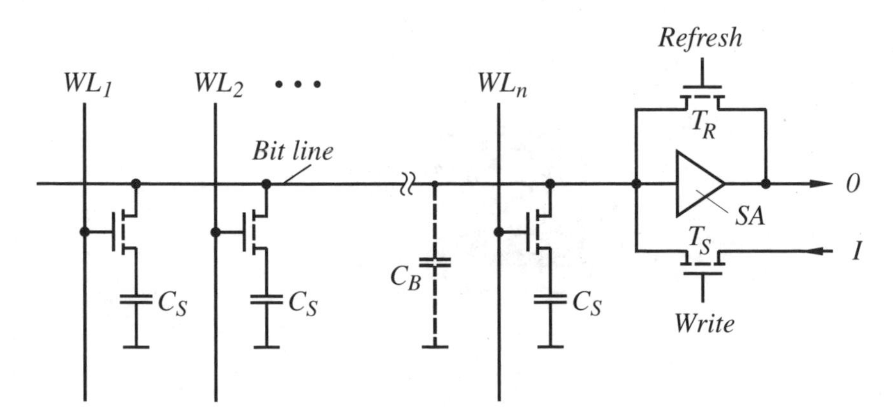

Figure 7.46 Principle functions of a sense amplifier

furthermore, use far too much silicon area for application in a memory with its tight bit line pitch. These are some reasons why differential schemes are used (Figure 7.47).

If, e.g., the word line WL_2 is activated, so are all cells of this word line connected to a respective bit line BL (Figure 7.47(a)). Simultaneously, not selected bit lines BLR are used as reference lines. The evaluation of the voltage difference between BL and BLR can then be achieved with a differential sense amplifier SA. If, e.g., the word line WL_{512} is activated (Figure 7.47(b)) the functions of BL and BLR are reversed. This scheme guarantees that a differential signal is always available and no absolute value has to be evaluated. This differential read signal has, according to Equation (7.17), a value of

$$\Delta V_L = V_L(BL) - V_B(BLR)$$
$$\Delta V_L = \frac{C_S}{C_S + C_B}[V_S - V_B(BLR)] \tag{7.18}$$

In the layout, the sense amplifiers are interleaved in such a way that no unused silicon area exists between adjacent bit lines, and one cell only connects to a sense amplifier.

The sense amplifier, fitting into such a small bit line pitch, consists of cross-coupled transistor pairs, operating similarly to a flip-flop (Figure 7.48).

The circuit function is explained by reading the cell information of cell Z and writing the refreshed information back into the cell. To start this operation, the bit line BL and BLR are precharged to a common voltage of, e.g., 0.9 V. If the word line is turned on, this causes a charge sharing between the charge of the cell capacitor C_S and the charge of the parasitic bit line capacitance C_B. If, e.g., an H level of 1.8 V is stored in the cell the bit line voltage at BL changes to, e.g., 1.0 V, whereas that at BLR remains unchanged at 0.9 V. The read signal is thus 100 mV. Next, the cross-coupled n-channel transistors are

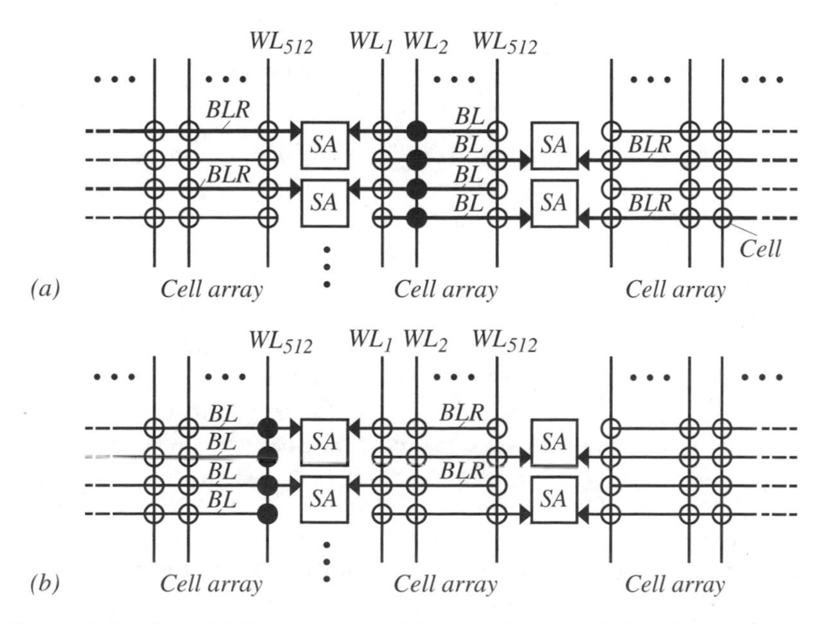

Figure 7.47 Open bit line concept: (a) WL_2 activated and (b) WL_{512} activated

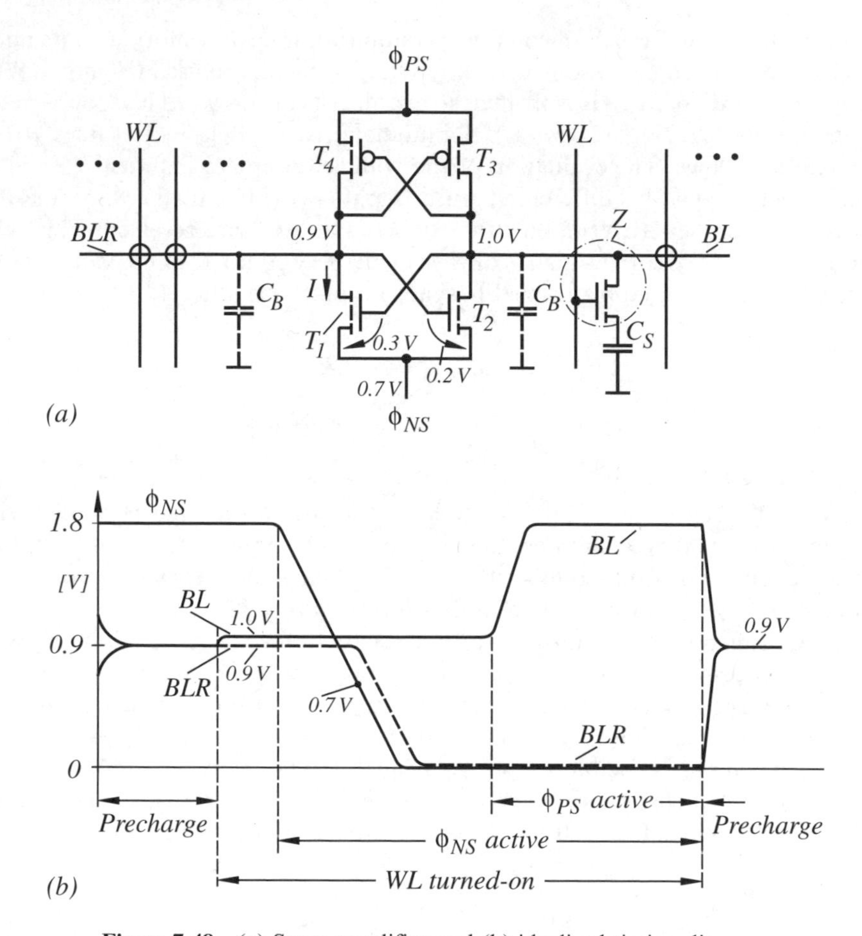

Figure 7.48 (a) Sense amplifier and (b) idealized timing diagram

activated by changing the pulse ϕ_{NS} from 1.8 V to 0 V. During this transition ϕ_{NS} assumes a value of 0.7 V. At this condition a V_{GS} voltage of 0.3 V exists at T_1, whereas at T_2 the voltage is 0.2 V. With a threshold voltage of 0.3 V for both transistors T_1 starts to conduct whereas T_2 remains in the sub-threshold region and is basically cut off. The current in T_1 begins to discharge C_B, causing a voltage reduction at bit line *BLR*. Since the transistors are cross-coupled, this leads to a reduction of the V_{GS} voltage at T_2, forcing the transistor deeper into the sub-threshold region. If the ϕ_{NS} voltage is further reduced, this causes the voltage at *BLR* to reduce also, whereas the voltage at *BL* remains basically unchanged. When the ϕ_{NS} pulse finally reaches 0 V, the bit line *BLR* has a value of 0 V and *BL* one of about 1 V. The read signal of 100 mV is thus ampified to 1 V.

Prior to the activation of the sense amplifier, an *H* level of 1.8 V was stored in the cell. This level has to be written back into the cell. It is performed by activating the p-channel transistors T_3 and T_4 with the ϕ_{PS} pulse (not shown). When this pulse changes from 0.9 V to 1.8 V, this causes T_3 to conduct whereas T_4 is cut off. If ϕ_{PS} approaches a value of 1.8 V – which corresponds to the *H* level of the cell – then the bit line *BL* is charged to

1.8 V and BLR remains at 0 V. Since the word line is still activated, the cell capacitor is charged to 1.8 V. Thus the cell signal is not only amplified but is also written refreshed back into the cell. With 0 V at the word line, the read cycle is completed.

If an L signal instead of an H signal is stored in the cell Z, then the voltage at BLR is larger than the one at BL. The reverse process takes place: BL is charged to 0 V and BLR to 1.8 V. As is obvious from the discussion, one bit line is always charged to H level, in this example to 1.8 V, whereas the other one is charged to 0 V. If both bit lines are short-circuited during precharge, a voltage of 0.9 V results at both bit lines due to charge sharing. A new read cycle can start.

In Figures 7.47 and 7.48 the bit lines are placed on each side of the sense amplifier. This is the so-called open bit line concept. This concept has the disadvantage that coupling noise influences the sense amplifier one-sided and deteriorates the read signal. This is different with the folded bit line concept (Takashima *et al.* 1994) shown in Figure 7.49

The addressed bit line BL and the one used as reference bit line BLR are geometrically placed next to each other; that is, they are folded compared to the open bit line concept. To save silicon area, the sense amplifiers are connected via the signal ϕ_L and ϕ_R either to a left cell array or to a right cell array. Capacitance couplings, e.g. shown at WL_1, are suppressed by the sense amplifier since the noise acts on both inputs simultaneously (for common mode rejection see Section 8.3.2). The arrangement of the sense amplifier – comparable to that of Figure 7.48 – is shown in Figure 7.50.

Transistors T_P are responsible for precharging the bit lines. The column decoder controls with the signal COL the transistors T_5 and T_6 which connect the bit lines via external bit lines to the I/O circuit.

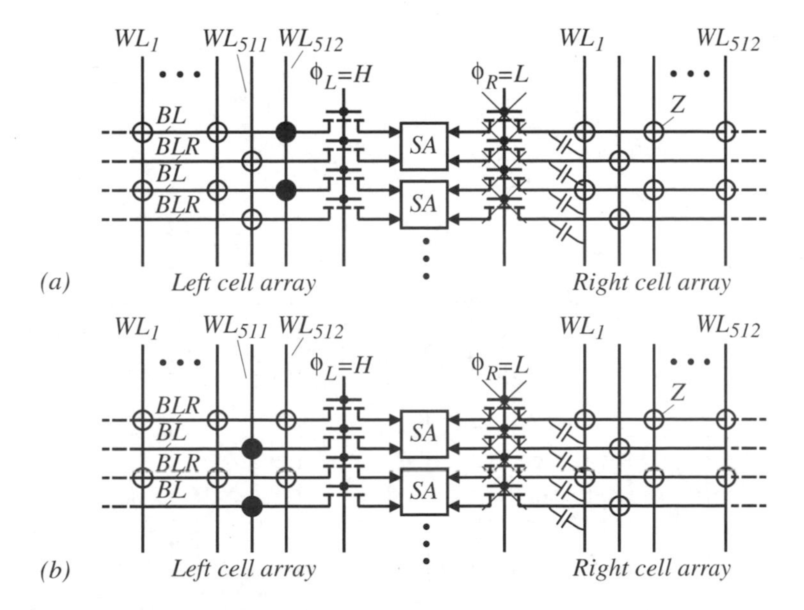

Figure 7.49 Folded bit line concept: (a) word line WL_{512} active and (b) word line WL_{511} active

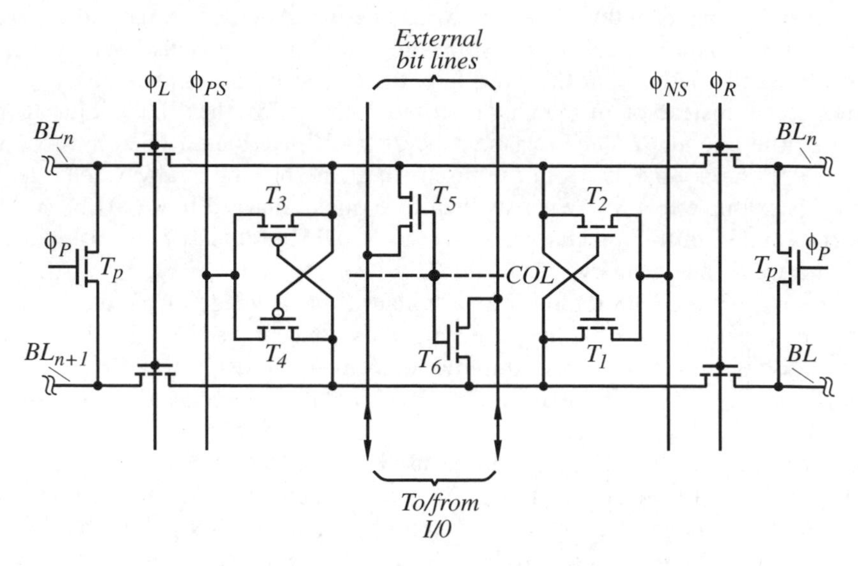

Figure 7.50 Sense amplifier arrangement in a folded bit line concept

Word line driver

In order to achieve a high silicon area efficiency, as many cells as possible, e.g. 2048, are connected to a word line. This results in a relatively large word line capacitance. Besides the requirement to charge and discharge this word line capacitance fast, the word line voltage has to be elevated beyond the V_{CC} level (Figure 7.51).

If an H level with a value of V_H is written into the cell, then the word line must have a voltage of (Equation 4.36)

$$V_{WL} = V_H + V_{Ton} + \gamma\left(\sqrt{2\phi_F + V_{SB}} - \sqrt{2\phi_F}\right) \tag{7.19}$$

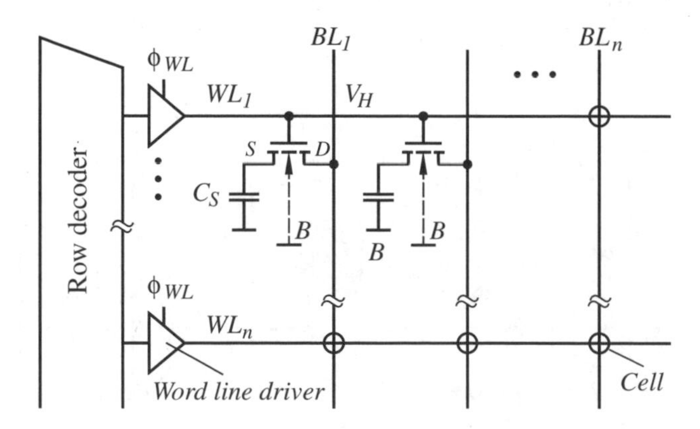

Figure 7.51 Cell array with row decoders and word line drivers

which is larger then the H-level voltage in order to overcome the threshold voltage of the select transistor including the body effect (Section 4.3.3). If an H-level voltage of V_{CC} is chosen, then the word line voltage must be larger than the power-supply voltage. This is usually achieved by employing an on-chip voltage generator (Section 7.3.3).

Another equally important requirement is that the word line driver can be fitted into the word line pitch of the memory array. A word line driver coping with the described requirements is shown in Figure 7.52. The ϕ_{WL} voltage is 0 V and the row decoder supplies H signals to all T_T transistors. This causes all driver transistors T_W to conduct and all word line capacitances C_{wn} to be charged to 0 V. Next one H state out of $n-1$ remaining L states is selected. At node (1) of the selected word line driver a voltage of $V_{CC} - V_{Tn}$ of about 2.5 V exists, whereas all other nodes (1) have a voltage of 0 V. If ϕ_{WL} changes from 0 V to V_{WL} this causes the voltage at the word line WL_n to increase and simultaneously the one at node (1) due to the bootstrap effect via C_B (Section 5.5.2). Assuming that no parasitic capacitance at node (1) exists, then the voltage at node (1) approaches a value of $V_{WL} + 2.5$ V and the word line WL_n one of V_{WL}. All other word lines remain at 0 V. The advantage of this driver schema is that the loading of the word line pulse ϕ_{WL} is given by the capacitance of the selected word line transistor T_W and its word line capacitance C_{wn} only. All remaining driver transistors T_W are cut off and contribute therefore with their small gate drain overlap capacitance only to the total loading of the ϕ_{WL} pulse.

A disadvantage of the word line driver is that the 0 V level of the unselected word lines may increase slightly due to capacitance coupling in the memory array. This could inadvertently lead to an increased sub-threshold current at the cut-off select transistors. This can be avoided by adding a CMOS inverter to the word line driver (Figure 7.53).

As mentioned before, the word line driver has to fit into the word line pitch of the cell array. To achieve this with the added inverter, several output transistors share parts of the driver components. To make this scheme work, decoded driver pulses $\phi_{WL,1}$ to $\phi_{WL,n}$ are required.

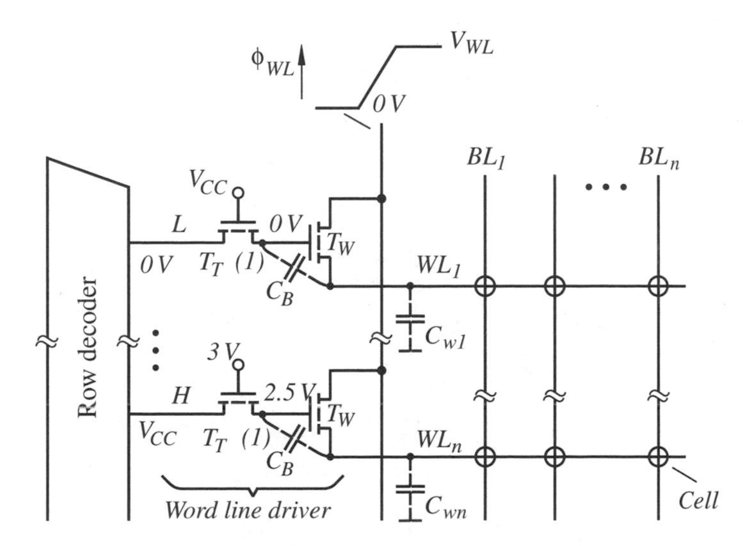

Figure 7.52 Word line drivers

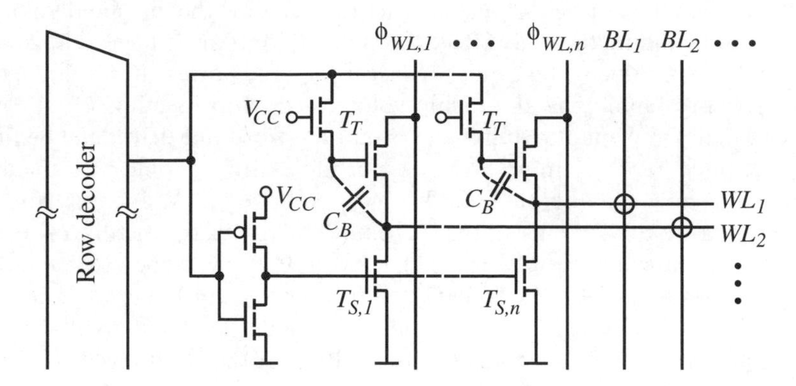

Figure 7.53 Improved word line drivers

7.5.3 DRAM Architecture

In the past the dominating characteristic factor of these memories was the memory size, as shown in Figure 7.54 (Shinozaki 1996). This is still the case but added is the requirement for a very high data rate for multimedia and microprocessor applications. To cope with this requirement so-called synchronous DRAMs (SDRAMs) play a dominating role. In order to discuss the architecture of such memories it is advantageous to start with a brief review of a conventional DRAM architecture (Figure 7.55).

In order to reduce the number of package pins, the addresses are multiplexed in two groups. The addresses A_0 to A_N are loaded into a row address buffer using an external row address strobe ($\overline{RAS}$) and the addresses A_{N+1} to A_M are loaded with an external column address strobe ($\overline{CAS}$) into the column address buffer. All cells activated by the

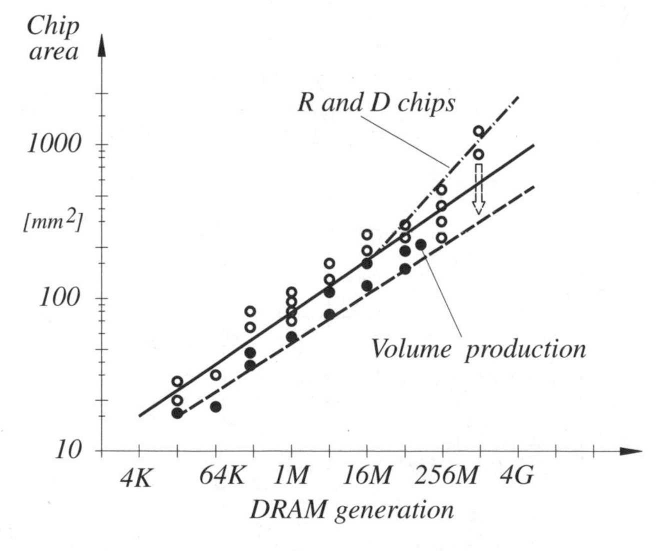

Figure 7.54 Memory generations as a function of chip area (reprinted by permission, from Shinozaki, 1996; © 1996 Infineon)

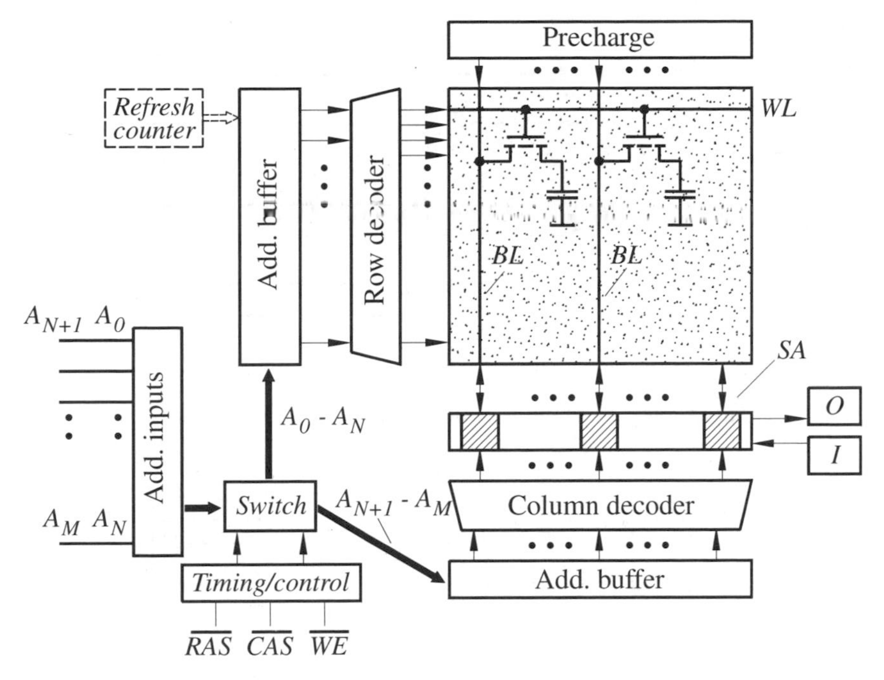

Figure 7.55 Conventional DRAM architecture

row decoder via a word line WL supply the information to sense amplifiers SA. One or more bits are selected from there by the column decoder and fed to the output O. The signal $\overline{WE}$ is used to tell the DRAM when a read or write operation is required. During refresh it is sufficient to cycle all row addresses since with each read the information is refreshed and written back into the memory cells. The refresh can be performed automatically by using an on-chip refresh counter.

If a word line is addressed, the information of all selected cells is available at the sense amplifiers. This situation can be used to feed the information sequentially to the output by using the $\overline{CAS}$ signal and the column addresses. This is shown in the timing diagram of Figure 7.56 with a four-bit word which can be extended to a hole page; that is, all addressed cells of a row.

This operation is called burst mode. Typical data are: access time starting at $\overline{RAS}$ $t_{RAS} = 50$ ns, access time starting at y-address $t_{AA} = 25$ ns, and minimum cycle time $t_{PC} = 20$ ns. This corresponds to a frequency of 50 MHz and relates to a data rate of 100 MB/s, when a memory organization with 16 data outputs is assumed.

In order to achieve an even higher data rate, clocked synchronous DRAM (SDRAM) architectures are used. All data and control signals are synchronized to the rising clock edge (Takai *et al.* 1994). This eliminates the complex and sometimes troublesome timing control by the $\overline{RAS}$ and $\overline{CAS}$ signals. As an example, a 265 Mb SDRAM architecture is shown in Figure 7.57.

The read and write operation are implemented with the $\overline{RAS}$,$\overline{CAS}$, and $\overline{WE}$ signals. As in the conventional DRAM architecture, the row and column addresses are multiplexed into the respective buffers. The memory is organized into four independent memory banks

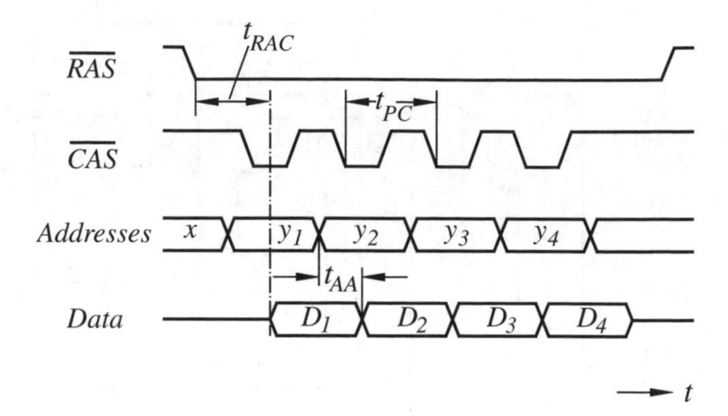

Figure 7.56 Timing diagram of a DRAM with extended data output (EDO)

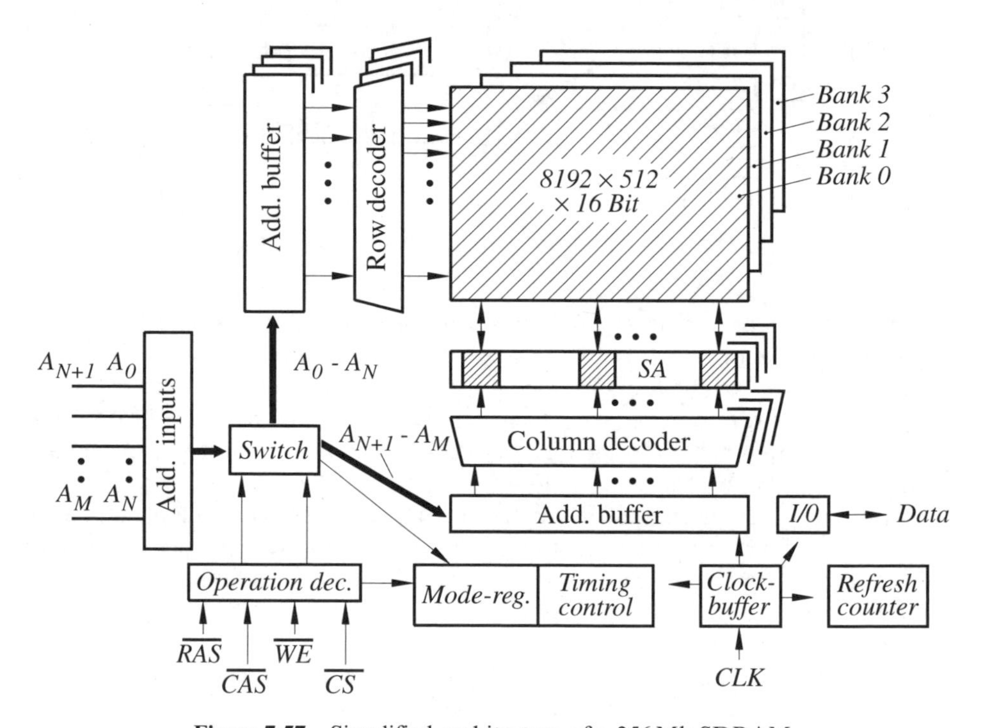

Figure 7.57 Simplified architecture of a 256 Mb SDRAM

of 64 Mb each. The operation the SDRAM performs depends on the input signals at the operation decoder and the mode register. This register has to be set at least once after the power supply is turned on. Some of the most important parameters stored are:

(a) The burst length, that is the number of data, following a read or write instruction.

(b) The burst type, that is the order in which the data appear, e.g. sequentially or interleaved between different memory banks.

(c) The $\overline{\text{CAS}}$ latency (CL), that is the number of clock cycles elapsed, after the $\overline{\text{CAS}}$-address y, before the information appears at the output. The latency time cannot be selected freely and is dependent on the chosen clock frequency. This behavior results from the pipeline structure (Section 6.5.3) of the memory which has to be adjusted for the required clock frequency.

An example of a typical SDRAM timing diagram operating in a burst read mode is shown in Figure 7.58. The burst length (BL) chosen is $BL = 4$ in a sequential order and the $\overline{CAS}$ latency is $CL = 3$.

The access of the memory starts by applying an activation command and the row addresses. Between this command and the following read or write command, at least a time t_{RCD} of three clock cycles has to elapse. After applying the read command and a column address, four data bits are automatically supplied to the output. The first bit appears at the output after the latency of $CL = 3$. Then the subsequent data appear at each clock cycle of about 8 ns at the output, which leads to a substantially improved data rate compared to the preceding case.

7.5.4 Radiation Effects in Memories

All integrated circuits, but particularly memories, are affected by soft errors. These errors are called soft since they result in a temporary loss of data only, but do not cause permanent damage to the circuit. Such errors are caused by α-particles. These particles originate primarily from the decay of ratioactive materials, such as uranium and

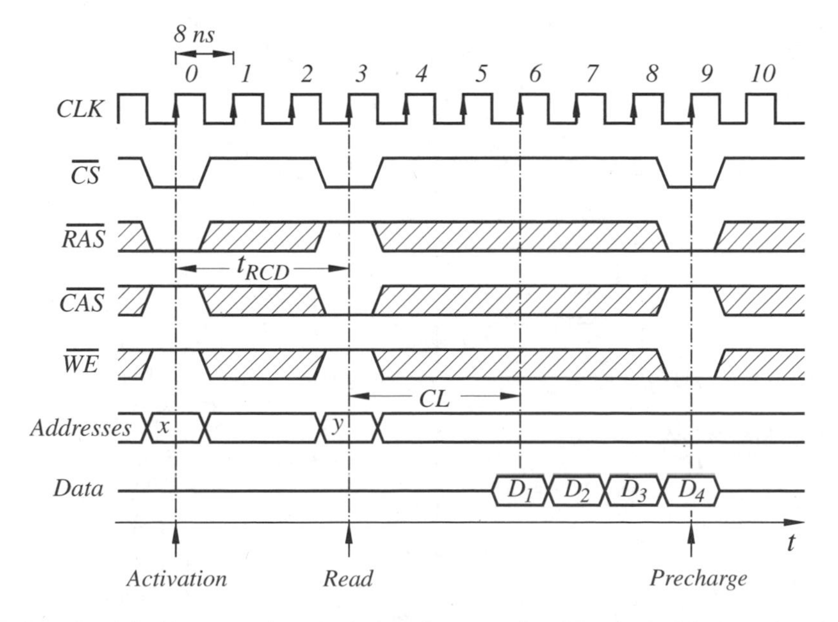

Figure 7.58 Simplified burst read mode timing diagram of an SDRAM with $CL = 3$ and $BL = 4$

thorium, found in small amounts in almost all materials, e.g. ceramic or plastic packages, metal or silicon. The α-particles consist of two protons and two neutrons and are thus identical to a helium atom. The α-particle penetrates the silicon surface and creates a cloud of electron–hole pairs along the track of the α-particle. The length of the track depends on the energy of the α-particle. If an α-particle has an energy of 5 MeV then $1.4 \cdot 10^6$ electron–hole pairs are created along a track of 23 μm within a couple of 10^{-12} seconds (Yaney *et al.* 1979). Contrary to this fast process is the lifetime of the generated electron–hole pairs, in the order of 10^{-3} seconds at room temperature, which is relatively long.

If electron–hole pairs are generated close to or in an electrical field, then the particles separate, whereby the electrons move to the positive n^+-region and the holes into the substrate (Figure 7.59). This carrier movement is amplified by the so-called charge funneling effect (Hsich 1983). This is in some respects comparable to the described Kirk effect of Section 3.3.1, where the field is pushed by the charge to the buried collector. The integral over the electrical field of the pn-junction is, according to Poisson's law proportional to the voltage across it. Therefore the electrical field, initially confined to the pn-junction, is reduced in magnitude and pushed far into the lightly doped substrate when an α-particle penetrates the junction. The penetration of the electrical field into the substrate causes additional electrons and holes to diffuse into the substrate and thus support the effect of the α-particle.

If an α-particle hits a highly resistive node in a circuit, such as the one of the cross-coupled transistors in the static four-transistor cell, this may lead to a soft error. Or if the storage capacitor in a DRAM is hit or the bit line during reading, this may lead to a soft error also. The number of soft errors, the so-called soft error rate (SER), is quoted in failure in time (FIT). One FIT corresponds to one failure in 10^9 operation hours of an integrated circuit (IC). The experimentally determined FIT number of an IC is part of the qualification process of a product and should be below 500 FIT for a memory. As a reminder, the FIT number does not say anything about cosmic ray effects which may interact with the semiconductor (Ziegler and Lanford 1981).

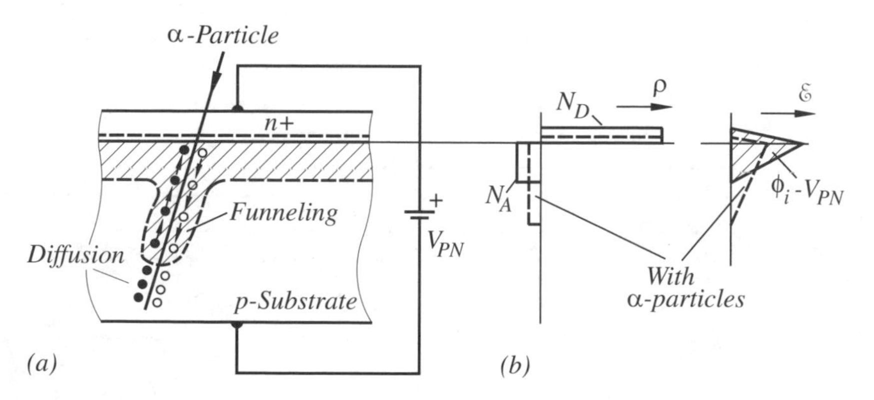

Figure 7.59 (a) Charge funneling effect at an n^+p-junction and (b) sketch of charge and field distribution

Summary of the most important results

Nonvolatile memories

The electrically programmable and erasable memory cells use either channel hot electrons or the Fowler–Nordheim tunnel mechanism for programming. For erasing the cell charge, the tunnel mechanism is used exclusively. A disadvantage of programming with channel hot electrons is the relatively large cell current, e.g. of 0.5 mA. But a definite advantage is the relatively fast programming time, of e.g. 10 μs. Contrary is the situation when the Fowler–Nordheim tunnel mechanism is used. The current is reduced to about 10^{-11} A, but the programming or erase time increases to, e.g., 1 ms. All cells have in common that the endurance – that is, the number of program/erase cycles – is limited to about 10^6 due to degradation effects. When cells are simply connected in a *NOR* matrix, over erase and program disturb problems hinder a useful programming and erase operation. Solutions are two-transistor and split-gate cells or flash architectures.

Static memories

The six-transistor cell consists of two cross-coupled CMOS inverters with select transistors. This cell type can be implemented in any CMOS manufacturing process but with the drawback of a relatively large chip area consumption. This area can be reduced to approximately $\frac{2}{3}$ when a four-transistor cell with GΩ resistors is used as pull-ups. Additional processing steps are needed. In case battery operation or an extremely low stand-by current is required, thin film transistors are a solution. Stand-by currents $<10^{-14}$ A per cell are possible.

Dynamic memories

In these memories the information is stored as different charge levels at a capacitor. Due to leakage currents, the charge can only be guaranteed for a particular time, namely the refresh time of, e.g., 64 ms. After this time the information has to be read and refreshed, that is, written back into the cell. This happens to all cells of a word line simultaneously. In order to improve the memory density, trench and stacked cells are used. These cells, together with clever sense amplifiers and word line drivers, are responsible for these memories having a very high bit density.

Problems

7.1 The ETOX cell shown is used for the storage of two bits. In order that this scheme works effectively, the threshold voltage change per logic state should be 0.8 V. How many electrons are required per state?
   Data: $l = 0.6$ μm, $w = 1.0$ μm, ε(ONO) = 5.5.

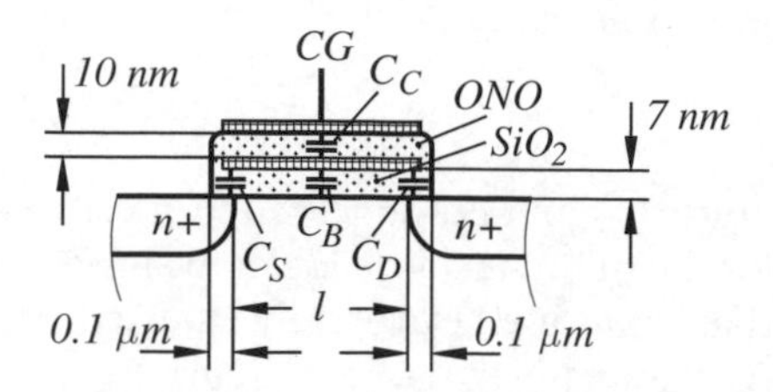

7.2 Determine the maximum resistors value of the four-transistor cell shown below. The maximum voltage drop ΔV across the resistor R is not allowed to be larger than 1 V over a temperature range from 0 °C to 90 °C at the H level. The dominating leakage current is the sub-threshold current of the transistor.

The data are: V_{Ton} (0 °C) = 0.6 V, V_{Ton} (90 °C) = 0.51 V, β_n (0 °C) = 150 $\mu A/V^2$, β_n (90 °C) = 110 $\mu A/V^2$.

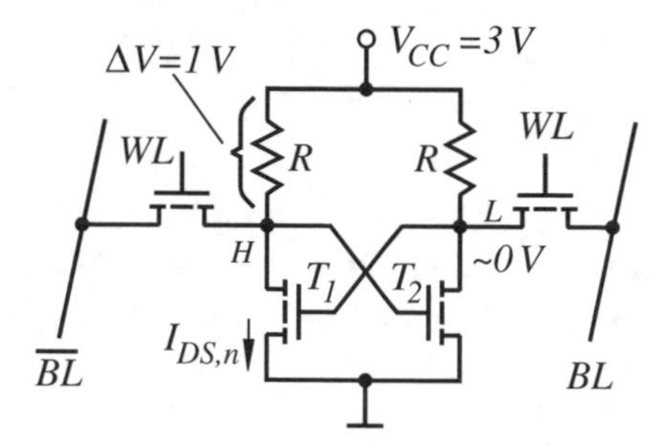

7.3 The one-transistor trench cell of a DRAM can be represented by the electrical equivalent circuit shown below.

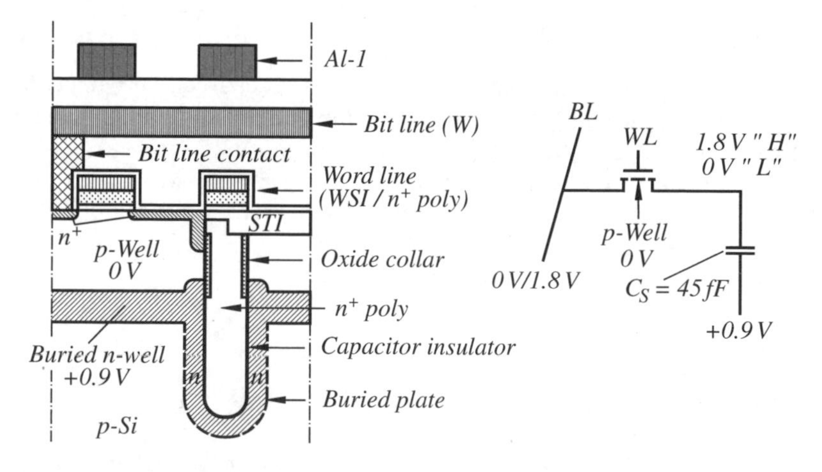

(a) Determine the sub-threshold current of the turned off n-channel transistor under worst case condition.

(b) Use the worst case condition for the determination of the refresh time. It can be assumed that a reduced H level from 1.8 V to 1.5 V and an increased L level from 0 V to 0.3 V can still be evaluated by the sense amplifier.

(c) How does the refresh time improve when a voltage of -1 V is applied to the p-well?

The data of the n-channel transistor are:

	27 °C	90 °C
I_{DS} (measured at $V_{GS} = 0.6$ V)	250 pA	4.8 nA
V_{Ton} (p-Well at 0 V)	1.0 V	0.87 V
γ	$0.3\sqrt{V}$	$0.3\sqrt{V}$
S	120 mV/decade	145 mV/decade
$2\phi_F$	0.82 V	0.76 V
$n = 1 + \frac{C'_j}{C'_{ox}}$	2	2

7.4 A trench DRAM cell is shown below. The n^+-poly side of the capacitor can be charged to 1.8 V or 0 V. This causes an unexpected large current through the parasitic n-channel transistor. Determine:

(a) The worst case voltage constellation.

(b) The threshold voltage of the parasitic n-channel transistor with a gate oxide thickness (oxide collar) of $d_{ox} = 7$ nm.

(c) The threshold voltage when the gate oxide thickness is increased to 30 nm.

(d) The threshold voltage when, additional to the 30 nm oxide thickness, a substrate voltage of -1 V is applied to the p-well.

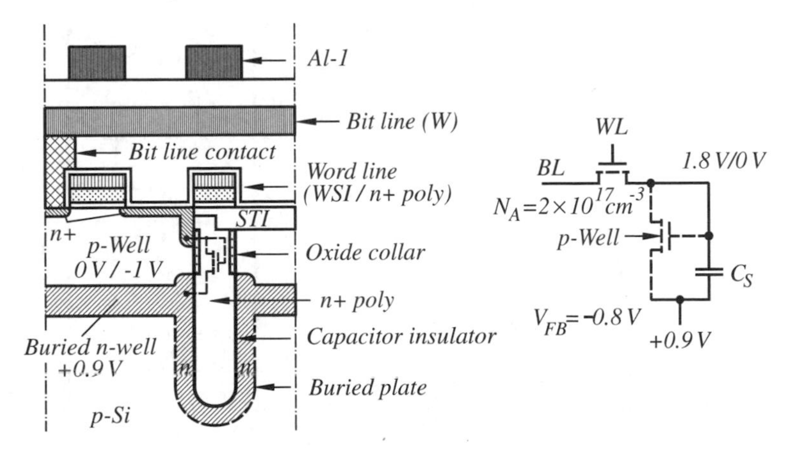

7.5 Determine the read signals ΔV_L of the open bit line scheme shown below when the capacitor of the cell is charged to an H level of 1.8 V and an L level of 0 V. Data: $C_S = 35$ fF, $C_B = 200$ fF, precharge level of the bit lines is $1.8\text{ V}/2 = 0.9$ V.

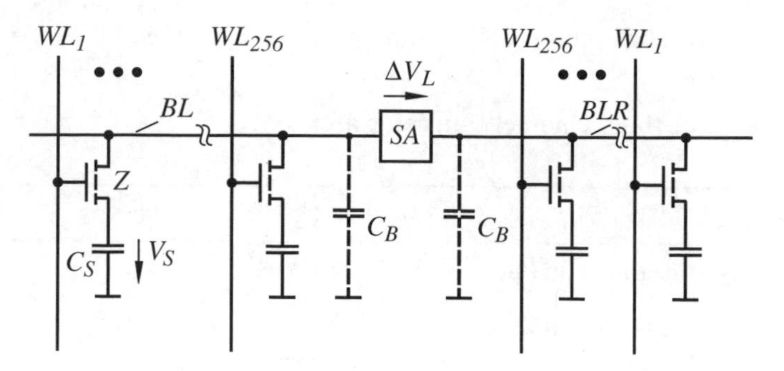

7.6 A detail of a one-transistor memory array is shown below. Determine the minimum word line voltage V_{WL} which guarantees that an H level of 1.8 V can be written into the cell.

Data: $V_{Ton} = 1.1$ V, $\gamma = 0.8$ V$^{1/2}$, $2\phi_F = 0.93$ V.

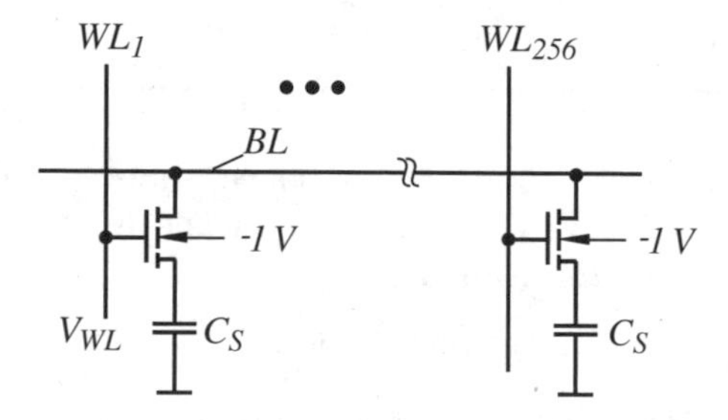

7.7 Shown below is a typical word line driver of a DRAM. Determine the voltages at the gates of the transistors T_{W0} and T_{W1} and the voltages at the word lines WL_0 and WL_1 before and after ϕ_{WL} changes from 0 V to 3.8 V.

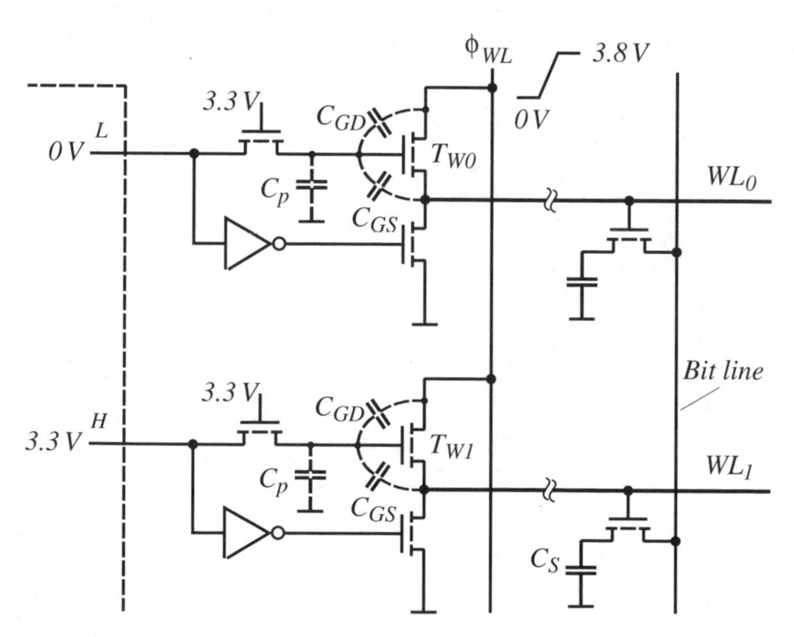

Data: $C_p = 9\,\text{fF}$, $C_{GS} = C_{GD} = 26\,\text{fF}$, $V_{Tn} = 0.6\,\text{V}$, body effect can be neglected.

The solutions to the problems can be found under: *www.unibw-muenchen.de/campus/ET4/index.html*

REFERENCES

G. Atwood, A. Fazio, D. Mills and B. Reaves (1997) Intel StrataFlash memory technology overview. *Intel Technol J.*, 4th quarter. http://www.intel.com/technology/itj/q41997/overview.pdf.

M. Bauer, R. Alexis, G. Ahwood and B. Baltar (1995) A multilevel-cell 32 Mb flash memory. *Solid State Circuits Conference, Digest of Technical Papers*, pp. 132–3.

T. Eimori, Y. Ohno, H. Kinura, J. Matsufusa *et al.* (1993) A newly designed planar stacked capacitor cell with high dielectric constant film for 256 Mbit DRAM. *IEEE International Electron Devices Meeting, Digest of Technical Papers*, pp. 631–2.

D. Frohman-Bentchkowsky (1971) A fully decoded 2048-bit electrically programmable FAMOS read-only-memory. *IEEE J. Solid-State Circuits*, **SC-6**, 301–6.

S. Fujii, M. Ogihara, M. Shinizu, M. Yoshida *et al.* (1989) A 45 ns 16-Mbit DRAM with triple-well structure. *IEEE J. Solid-State Circuits*, **24**(5), 1170 75.

C. Hsieh (1983) Collection of charge from alpha-particle tracks in silicon devices. *IEEE Trans. Electron Devices*, **30**(6), 689–93.

K.-C. Huang, Y.-K. Fang, D.-N. Yang, C.-W. Chen, H.-C. Sung, D.-S. Kuo, C.S. Wang and M.-S. Liang (2000) The impacts of control gate voltage on the cycling endurance of split gate flash memory. *IEEE Electron Device Lett.*, **21**(7), 359–61.

K. Imamiya, Y. Iwata, Y. Sugiura and H. Nakamura (1995) A 35 ns-cycle-time 3.3 V-only 32 Mb NAND flash EEPROM. *Solid-State Circuits Conference, Digest of Technical Papers*, pp. 120–31.

Intel (1998) Intel StrataFlash memory technology, application note AP-677.

T.-S. Jung, Y.-J. Choi, K.-D. Suh and B.H. Suh (1996) A 3.3 V 128 Mb multi-level NAND flash memory for mass storage applications. *Solid-State Circuits Conference, Digest of Technical Papers*, pp. 32–3.

S. Kayano, K. Ichinose, Y. Kohno, H. Shinohara, K. Anami, S. Murakami, T. Wada, Y. Kawai and Y. Akasaka (1986) 25-ns 256K × 1/64K × 4 CMOS SRAM's. *IEEE J. Solid-State Circuits*, **SC-21**(5), 686–91.

R. Kirisawa, S. Aritome, R. Nakayama, T. Endoh, R. Shirota and F. Masuoka (1990) A NAND structured cell with a new programming technology for highly reliable 5 V-only Flash EEPROM. *Symposium on VLSI Technology, 1990, Digest of Technical Papers*, pp. 129–30.

H. Kume, M. Kato, T. Adachi, T. Tanaka, T. Sasaki, T. Okazaki, N. Miyamoto, S. Saeki, Y. Ohji, M. Ushiyama, J. Yugami, T. Morimoto and T. Nishida (1992) A 1.28 μm^2 contactless memory cell technology for a 3 V only 64 Mb EEPROM. *IEEE International Electron Devices Meeting, Technical Digest*, pp. 991–2.

M. Lenzlinger and E.H. Snow (1969) Fowler–Nordheim tunneling in thermally grown SiO_2. *J. Appl. Phys.*, **40**(1), 278–83.

N. Mielke, A. Fazio and H.-C. Liou, (1987) Reliability comparison of FLOTOX and textured polysilicon EEPROM's. *Proceedings of the International Reliability Physics Symposium (IRPS)*, pp. 85–6.

D. Mills, M. Bauer, A. Bashir and R. Fackenthal (1995) A 3.3 V 50 MHz synchronous 16 Mb flash memory. *Solid-State Circuits Conference, Digest of Technical Papers*, pp. 120–21.

L. Nesbit, J. Alsmeier, B. Chen, J. DeBrosse *et al.* (1993) A 0.6 μm^2 256 Mb Trench Cell with self aligned buried strap (BEST). *IEEE International Electron Devices Meeting, Digest of Technical Papers*, pp. 627–29.

A. Nozoe, T. Yamazaki, H. Sato and H. Kotani (1995) A 3.3 V high-density AND flash memory with 1 ms/512B erase and program time. *International Solid-State Circuits Conference, Digest of Technical Papers*, pp. 124–5.

T. Ootani, S. Hayakawa, M. Kakumu, A. Aona, M. Kinugawa, H. Takeuchi, K. Noguchi, T. Yabe, K. Sato, K. Maeguchi and K. Ochii (1990) A 4-Mb CMOS SRAM with a PMOS thin-film-transistor load cell. *IEEE J. Solid-State Circuits*, **25**(5), 1082–91.

Y. Park and K. Kim (2001) COB stack DRAM cell technology beyond 100 nm technology node. *IEEE, International Electron Devices Meeting, Digest of Technical Papers*, pp. 18.1.1–4.

K. Sasaki, S. Hanamura, K. Ueda, T. Oono, O. Minato, K. Nishimura, Y. Sakai, S. Meguro, M. Tsunematsu, T. Masuhara, M. Kubotera and H. Toyoshima (1988) A 15 ns 1 Mb CMOS SRAM. *Solid-State Circuits Conference, Digest of Technical Papers*, pp. 174–5.

E. Seevinck, P.J. van Beers and H. Ontrop (1991) Current-mode techniques for high-speed VLSI circuits with application to current sense amplifier for CMOS SRAM's. *IEEE J. Solid-State Circuits*, **26**(4), 525–35.

S. Shinozaki (1996) DRAM's in the 21st century. IEEE IEDM, DRAM Short Course.

Silicon Storage Technology, Inc. (2000) Technical comparison of floating gate reprogrammable nonvolatile memories, technical paper. http://www.sst.com/downloads/tech_papers/702.pdf.

Y. Takai, M. Nagase, M. Kitamura, Y. Koshikawa, N. Yoshida, Y. Kobayashi, T. Obara, Y. Fukuzo and H. Watanabe (1994) 250 Mbyte's synchronous DRAM using a 3-stage-pipelined architecture. *IEEE J. Solid-State Circuits*, **29**(4), 426–31.

D. Takashima, S. Watanabe, H. Nakano, Y. Oowaki and K. Ohuchi (1994) Open/folded bit-line arrangement for ultra-high-density DRAM's. *IEEE J. Solid-State Circuits*, **29**(4), 539–42.

T. Tanaka, M. Kato, T. Adachi, K. Ogura, K. Kimura and H. Kume (1994) High-speed programming and program-verity methods suitable for low-voltage flash memories. *IEEE Symposium on VLSI Circuits Digest of Technical Papers*, pp. 61–2.

Y. Uemoto, E. Fujii, A. Nakamura, K. Senda and H. Takagi (1992) A stacked-CMOS cell technology for high-density SRAM's. *IEEE Trans. Electron Devices*, **39**(10), 2359–63.

J.-T. Wu, Y.-H. Chang and K.-L. Chang (1996) 1.2 V CMOS switched-capacitor circuits. *Solid-State Circuits Conference, Digest of Technical Papers*, pp. 388–9.

T. Yamanaka, T. Hashimoto, N. Hasegawa, T. Tanaka, N. Hashimoto, A. Shimuza, N. Ohki, K. Ishibashi, K. Sasaki, T. Nishida, T. Mine, E. Takeda and T. Nagano Advanced TFT SRAM cell technology using a phase-shift lithography. *IEEE Trans. Electron Devices*, **42**(7), 1305–13.

D.S. Yaney, J.T. Nelson and L.L. Vanskike (1979) Alpha-particle tracks in silicon and their effect on dynamic MOS RAM reliability. *IEEE Trans. Electron Devices*, **26**(1), 10–16.

J.F. Ziegler and W.A. Lanford (1981) The effect of sea level cosmic rays on electronic devices, *J. Appl. Phys.*, pp. 4305–4312.

FURTHER READING

B. Prince (1991) *Semiconductor Memories*, 2nd edn. New York: John Wiley & Sons.

A.K. Sharma (2003) *Advanced Semiconductor Memories, Architectures, Designs, and Applications*. New York: John Wiley & Sons.

8
Basic Analog CMOS Circuits

So far only digital circuits are considered. In this chapter basic analog circuits are analyzed. To this category belong current sources, current mirrors, and simple amplifier stages as well as differential ones. The knowledge gained is prerequisite to more elaborate amplifiers, treated in Chapter 9.

In this chapter it is assumed that the solution of transfer functions with the aid of Laplace and Fourier transforms is known. This is of the utmost importance when, e.g., a pole-zero compensation has to be performed or the stability of an amplifier has to be determined by using a Bode diagram. Since these topics are prerequisites for the understanding of analog circuits, a brief review is included in Appendix A at the end of this chapter. In this appendix as a practical example, a simplified two-stage amplifier is used to demonstrate different analyzing concepts. This example has the advantage that the results can be transferred directly to the circuits of this and the following chapters.

Before starting with the circuit discussion it is useful to revisit the small-signal equivalent circuit of the MOS transistor (Section 4.7.3), shown for convenience in Figure 8.1 again. Since the transistor is assumed to operate in current saturation, the small-signal capacitances C_{gd} and C_{gb} (Figure 4.64) are neglected.

Instead of current and voltage changes, *small-signal time-variant changes for current and voltages are assumed and indicated by small letters. Furthermore, small-signal components are indicated by small letters or indices.*

In Table 8.1 the small-signal parameters for an n-channel or a p-channel transistor are summarized.

In cases where the source terminal is not connected to the bulk, the voltage v_{sb} acts via the transconductance of the substrate on the drain current (Section 4.7.3). In order to simplify the derivation and the discussion, this second-order effect is neglected, except for cases of particular importance. Furthermore, it is assumed that the gate source overlap capacitance C_{gso} is negligible compared to the gate source capacitance C_{gs}. The simplified small-signal equivalent circuit, which is used whenever possible, is shown in Figure 8.1(b). Why this equivalent circuit is applicable to n-channel and p-channel transistors is discussed in Section 8.3.

As a reminder: the following derived equations are necessary and useful for the understanding of the circuit function, but the equations are not precise. This is not a drawback, since accurate results can only be achieved with the aid of an appropriate circuit simulation tool.

System Integration: From Transistor Design to Large Scale Integrated Circuits. Kurt Hoffmann.
 ISBN: 0-470-85407-3

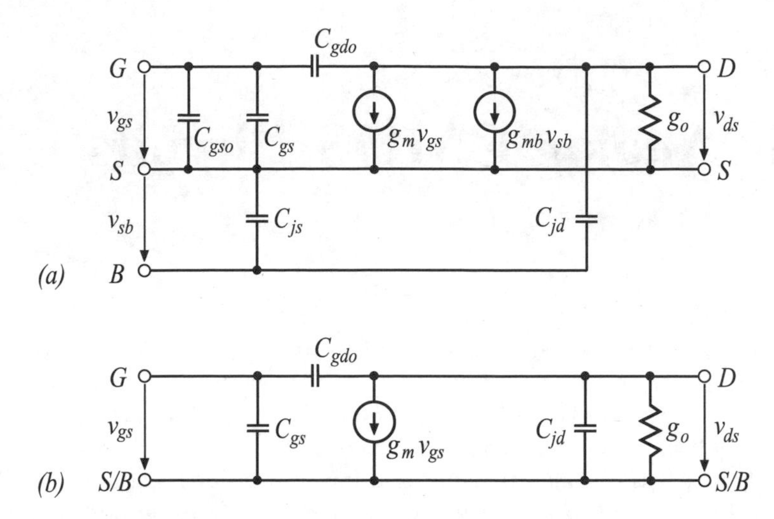

Figure 8.1 (a) Small-signal equivalent circuit of an n-channel or p-channel MOS transistor and (b) simplified version

Table 8.1 Small-signal parameters for n-channel and p-channel transistors in current saturation (for LD see Figure 5.3)

Transconductance (gate)	$g_m = \sqrt{2\lvert I_{DS}\rvert\beta(1+\lambda\lvert V_{DS}\rvert)}$
Transconductance (substrate)	$g_{mb} = \dfrac{-g_m\gamma}{2\sqrt{2\lvert\phi_F\rvert + \lvert V_{SB}\rvert}}$
Output conductance	$g_o = \dfrac{\lvert I_{DS}\rvert\lambda}{1+\lambda\lvert V_{DS}\rvert} \approx \lvert I_{DS}\rvert\lambda$
Small-signal capacitances	$C_{gs} = 2/3C_{ox}$
	$C_{gdo} = C'_{ox}wLD$
	$C_{js} = C_{jos}\left(1+\dfrac{\lvert V_{SB}\rvert}{\phi_i}\right)^{-M}$
	$C_{jd} = C_{jod}\left(1+\dfrac{\lvert V_{DB}\rvert}{\phi_i}\right)^{-M}$

8.1 CURRENT MIRROR

One very important circuit component of an amplifier is the current mirror. It is a circuit where all current sources are mirror images of a defined current I_B (Figure 8.2). This guarantees that the operation points of the individual circuit stages of an amplifier are related to each other and therefore power-supply and temperature variations do not influence the amplifier performance significantly.

The reference current I_B is realized in the simplest case by an external or internal resistor connected to V_{DD} or, in a more elaborate circuit, by a band-gap current generator

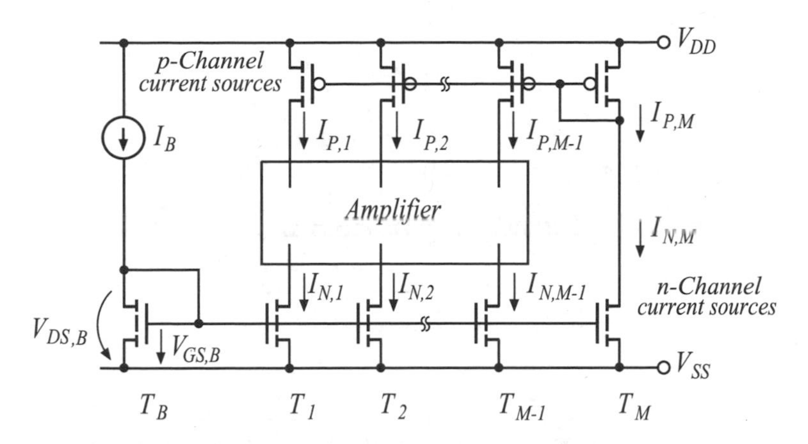

Figure 8.2 Current mirrors with n-channel and p-channel transistors

(Section 10.3). Transistor T_B operates in current saturation since gate and drain are connected as MOS diode. According to Equation (4.82), the current through the device has a value of

$$I_B = \frac{\beta_{n,B}}{2}(V_{GS,B} - V_{Tn,B})^2 \ (1 + \lambda_B V_{DS,B}) \tag{8.1}$$

The transistors T_1 to T_M, which act as current sources, are parallel connected to T_B and thus the current through these transistors has a value of

$$I_N = \frac{\beta_{n,N}}{2}(V_{GS,N} - V_{Tn,N})^2 \ (1 + \lambda_N V_{DS,N}) \tag{8.2}$$

Since the $V_{DS,B}$ voltage of T_B is also the $V_{GS,B}$ voltage of the remaining current sources, a current ratio between the transistors or a so-called current mirroring of

$$\frac{I_N}{I_B} = \frac{\beta_{n,N}(V_{GS,N} - V_{Tn,N})^2 \ (1 + \lambda_N V_{DS,N})}{\beta_{n,B}(V_{GS,B} - V_{Tn,B})^2 \ (1 + \lambda_B V_{DS,B})} \tag{8.3}$$

exists. If the threshold voltages V_{Tn} and the gain factors k_n have the same values for all transistors and the channel length modulation is negligible, then a current ratio or current mirroring of

$$\boxed{\frac{I_N}{I_B} = \frac{(w/l)_N}{(w/l)_B}} \tag{8.4}$$

results. When, e.g., the transistors T_B and T_M have the same (w/l) ratio, then the currents I_B and $I_{N,M}$ are equal and the I_B-current is mirrored to the p-channel current sources. Thus, all n-channel and p-channel current sources are related to the reference current I_B. As a result the amplifier is insensitive to voltage variations.

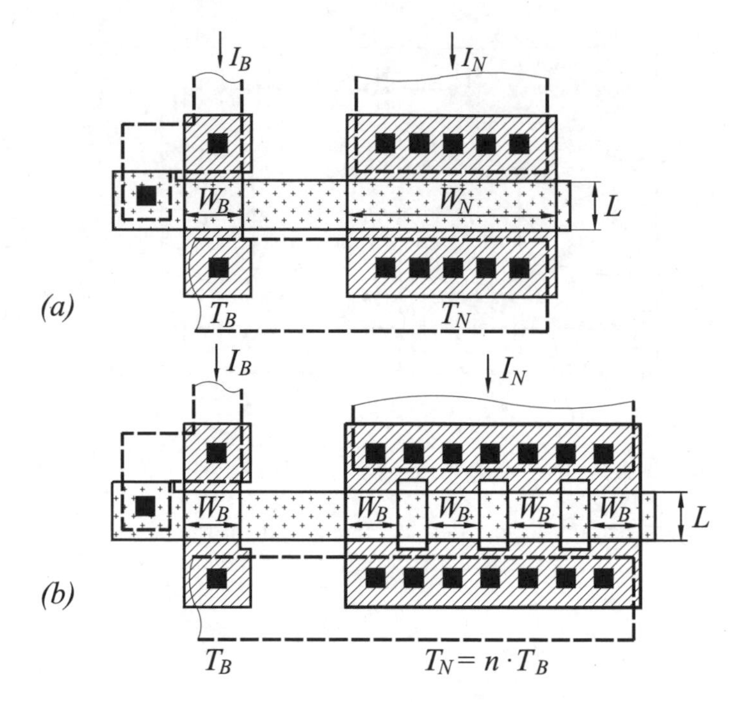

Figure 8.3 Current mirror: (a) with channel width effect and (b) without channel width effect

A layout of an n-channel current mirror is shown in Figure 8.3. Supposing that one considers the geometry in more detail, the following situation can be observed. A difference between the drawn width W and the drawn length L of the transistor and the final geometries implemented on the wafer exists (Figure 5.3). This leads to a current ratio of

$$\frac{I_N}{I_B} = \frac{(W_N - 2\Delta W)/(L_N - 2LD)}{(W_B - 2\Delta W)/(L_B - 2LD)} = \frac{(W_N - 2\Delta W)}{(W_B - 2\Delta W)} \tag{8.5}$$

The channel length variation is negligible when the same gate length is chosen for all transistors. But this is not the case for the transistor width, which is usually different for the individual current sources. To eliminate the width variation influence, e.g., on T_N, transistor T_B is duplicated n-times instead (Figure 8.3(b)).

A measure for the current change dI_{DS} caused by a drain voltage change dV_{DS} is the small-signal output conductance g_o (Table 8.1) or small-signal output resistance r_o at a defined operation point. Assuming a λ_n of $0.03\,\text{V}^{-1}$ and an operation point of $100\,\mu\text{A}$, an output resistance of $330\,\text{k}\Omega$ results (Figure 8.4).

8.1.1 Improved Current Sources

The preceding example of a current source shows a relatively large output resistance. But amplifiers need an even higher value in order to achieve a large voltage amplification.

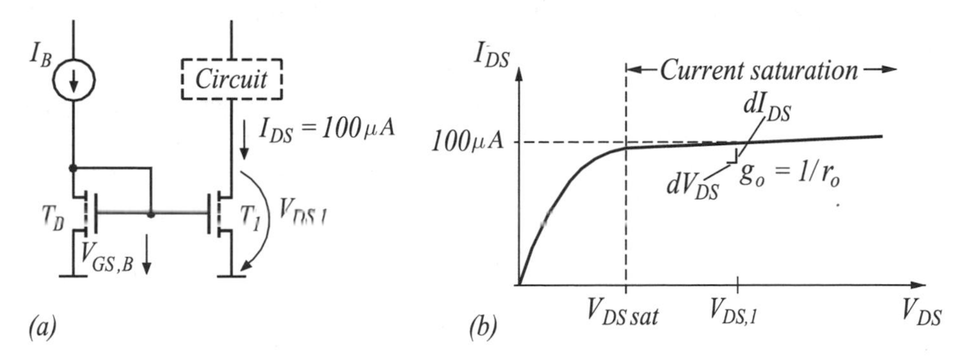

Figure 8.4 (a) Simple n-channel current mirror and (b) output characteristic

Thus, one important topic in amplifier design is the question of how to achieve this. One way is to increase the channel length and accordingly the width and thus reduce percent-wise the channel length modulation and therefore the factor λ_n (Section 4.5.2). But there is a limit to this approach due to the increase of the parasitic capacitances of the transistor. Another way is to use circuit techniques to achieve this goal. The principle is shown in Figure 8.5(a).

If the current I_1 increases this causes an increase of the voltage V_R across the resistor also. This leads to a reduction of the V_{GS} voltage of the transistor which opposes the increase of the current. A disadvantage of this approach is that in order to be effective, the feedback configuration requires a relatively large resistor and thus a large DC voltage drop across it. This can be avoided with the transistor solutions shown. In Figure 8.5(b) the resistor is replaced by transistor T_2. The biasing is achieved by two in-series-connected MOS diodes.

Of interest is the improved small-signal output resistance of the circuit. To find this resistance value one assumes that the voltage at node (1) changes by an infinitesimal amount and calculates the caused current change. The ratio of the two delivers the

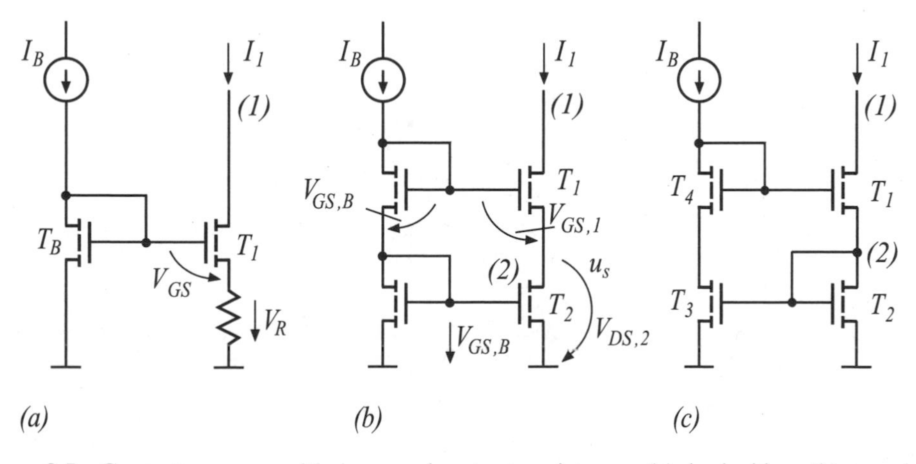

Figure 8.5 Current sources with increased output resistance: (a) basic idea, (b) transistor realization and (c) modification of implementation (b)

output resistance. For the calculation of the current the small-signal equivalent circuit is used. This circuit is then excited by a slow varying small-signal test voltage v_t and the current i_t is found by applying Kirchoff's law (Figure 8.6).

In the equivalent circuit the transistors are of interest only where voltage changes occur. These are the transistors T_1 and T_2. The test signal voltage causes no signal change at the gates of the two transistors. Thus v_1 and v_2 are zero. This leads at node (1) to the relationship

$$i_t = g_{m,1}(-v_s) + g_{o,1}(v_t - v_s) \tag{8.6}$$

and at node (2) to

$$i_t = v_s\, g_{o,2} \tag{8.7}$$

Combining the two equations leads to a small-signal output resistance of

$$r_{out} \approx r_{o,1}\, g_{m,1}\, r_{o,2} \tag{8.8}$$

Hereby it is assumed that $g_{o,1}$ and $g_{o,2}$ are much smaller than $g_{m,1}$, which is true for almost all practical cases. Supposing that one compares this result with that of the simple current source of Figure 8.4(a), with an output resistance of $r_{out} = r_{o,1}$, then one recognizes an improvement by a factor of $g_{m,1}r_{o,2}$. When the same parameters $I_1 = 100\,\mu\text{A}$ and $\lambda_1 = \lambda_2 = 0.03\,\text{V}^{-1}$ are used, as in the previous case, and furthermore a transistor gain of $\beta_{n,1} = 100\,\mu\text{A/V}^2$ is assumed, then an output resistance of 15.3 MΩ results. This is a 46.5 times improvement compared to the simple current source.

Instead of the discussed current source, the one of Figure 8.5(c) can be used. In this case the output resistance has a value of

$$r_{out} \approx r_{o,1}\, g_{m,1}\, r_{o,2} \tag{8.9}$$

which is comparable to that of Figure 8.5(b). In the derivation it is assumed that the transistors T_3 and T_2 are identical.

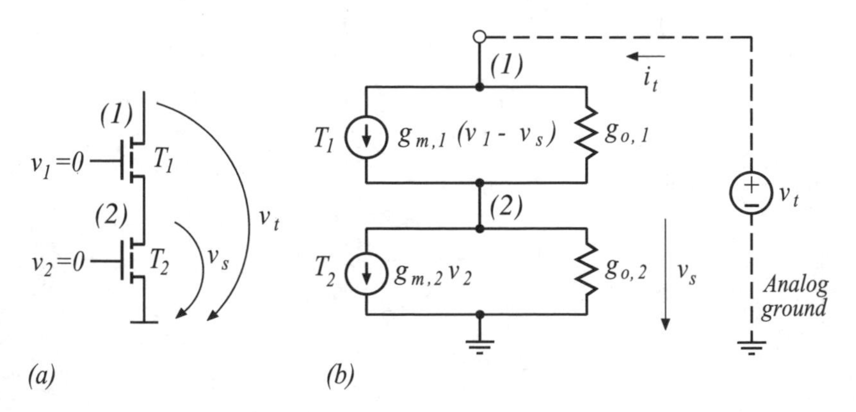

Figure 8.6 (a) Current source transistors T_1 and T_2 of Figure 8.5(b) and (b) small-signal equivalent circuit

Compared to the simple current source, the improved circuits have a higher output voltage where the transistors change from the current saturation region into the resistive region. In this region the current source can not be used due to its low resistance value. In order to analyze the cross-over voltage between the two regions the current sources are revisited (Figure 8.7).

A transistor is in current saturation when $V_{DS} \geq V_{GS} - V_{Tn}$. If a gate voltage of $V_{GS} = V_{Tn} + \delta$ is used, then the transistor remains in saturation as long as $V_{DS} \geq \delta$. With this condition the resulting cross-over voltages $V_{0\,\text{min}}$ are indicated in Figure 8.7 for the different current sources.

With the trend to smaller power-supply voltages, the cross-over voltages of versions (b) and (c) may be too large. A reduction of the output voltage without deteriorating the output resistance is possible with the circuit implementation of Figure 8.8. This circuit is a modification of the version shown in Figure 8.7(b), which yields a minimum output voltage of

$$V_{0\,\text{min}} = \delta_1 + (\delta_2 - \delta_1) = \delta_2 \tag{8.10}$$

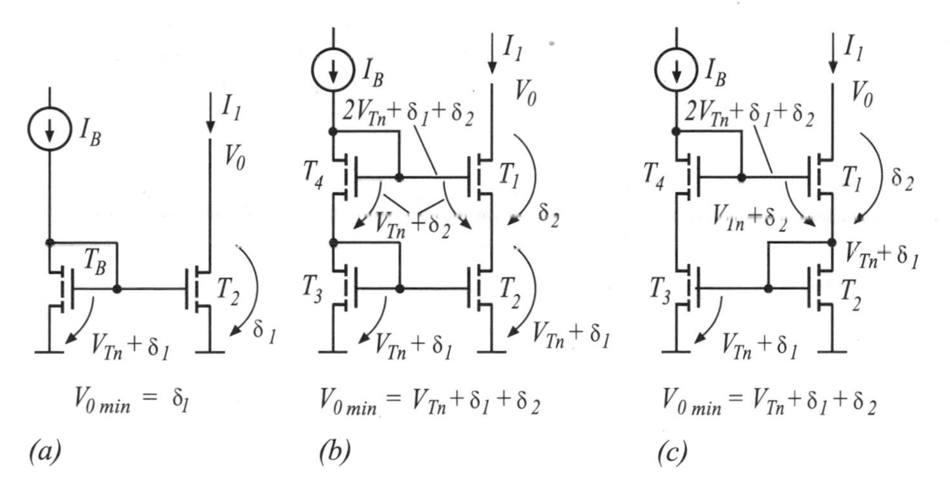

Figure 8.7 Current sources with cross-over voltages $V_{0\,\text{min}}$ $(\delta_2 > \delta_1)$

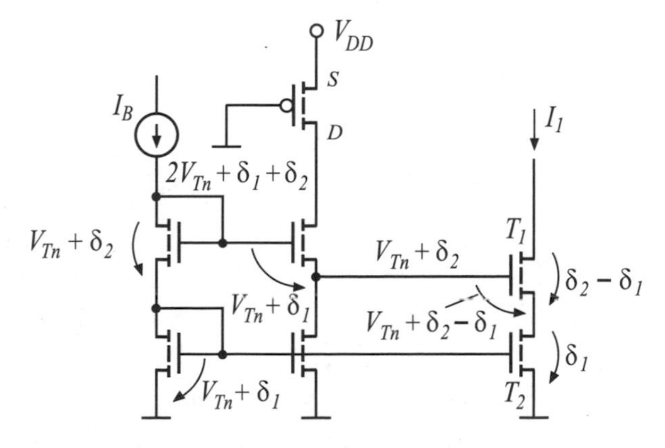

Figure 8.8 Current source with reduced cross-over voltage $(\delta_2 > \delta_1)$

and where ($\delta_2 > \delta_1$). This is a substantial improvement compared to the previous cases but with the drawback of an increased power dissipation.

8.2 SOURCE FOLLOWER

This circuit is used as a level shifter and, what is more important, as a low resistive output driver (Figure 8.9). In this circuit transistor T_1 is the source follower and T_2 implements a current source by current mirroring. The level shifting of the circuit leads to a DC voltage difference between V_B and V_o of

$$\begin{aligned} V_o &= V_B - V_{GS,1} \\ &= V_B - \left(V_{Tn} + \sqrt{\frac{2I_{DS}}{\beta_{n,1}}} \right) \end{aligned} \tag{8.11}$$

For the small-signal analyses the equivalent circuit of Figure 8.10 is used.

As mentioned at the beginning of this chapter, for simplicity reasons it is assumed that source and bulk terminals of the transistors are connected, which eliminates consideration of the body effect.

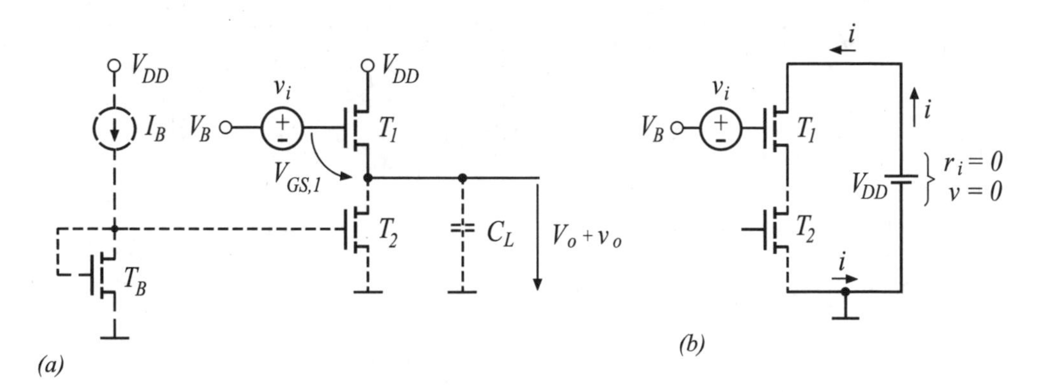

Figure 8.9 (a) Source follower and (b) small-signal behavior of a power supply

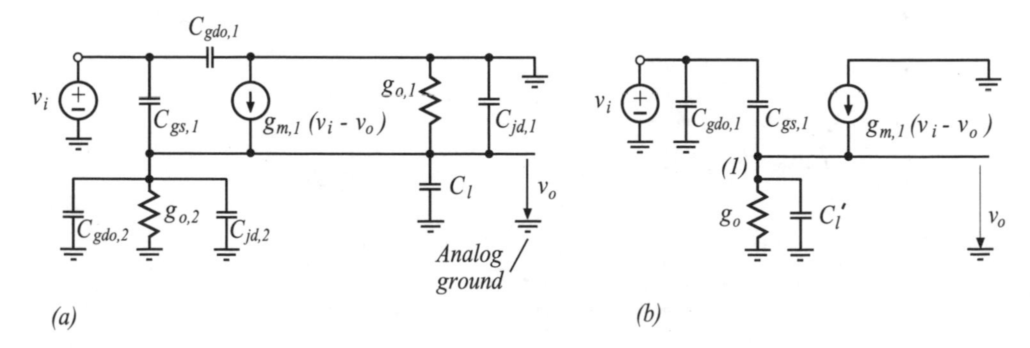

Figure 8.10 Source follower: (a) small-signal equivalent circuit and (b) simplified version

Transistor T_2 is represented by the small-signal output conductance $g_{o,2}$, the overlap capacitance $C_{gdo,2}$, and the drain junction capacitance $C_{jd,2}$. The small-signal current $g_m v_{gs}$ of T_2 is zero since no gate excitation exists. Transistor T_1 is controlled by the small-signal current generator with $g_m v_{gs}$ where $v_{gs} = v_i - v_o$. All other small-signal components used are according to the small-signal model of Figure 8.1(b).

Remark: Ground connections are shown in the figures by a small horizontal line, or if ground is not used, as most negative voltages, a connection to a V_{SS} terminal is indicated. In the small-signal circuit, e.g. of Figure 8.10, symbols are shown which can be called analog ground. These are reference points where no voltage change occurs. The DC voltage of these points is of no interest since only small-signal changes are considered. Why this is the case is shown in Figure 8.9(b). The power supply or in this case the battery can be considered to have a small-signal resistance of zero where no voltage change or drop occurs. The battery thus has no influence on the small-signal behavior of the circuit.

A simplified circuit of the source follower is shown in Figure 8.10(b). In this case the output conductances are joined together, $g_o = g_{o,1} + g_{o,2}$, and the parasitic capacitances are added into a common load capacitance C_l'. Since it is assumed that the small-signal voltage source v_i has a negligible small-signal input resistance, the capacitance $C_{gdo,1}$ does not have to be considered. Applying Kirchhoff's law to node (1) results in a transfer function in the frequency domain of

$$a(j\omega) = \frac{v_o(j\omega)}{v_i(j\omega)} = a_o(0)\frac{1 + j\dfrac{\omega}{\omega_z}}{1 + j\dfrac{\omega}{\omega_p}} \tag{8.12}$$

where the low-frequencies voltage gain is given by

$$\boxed{a_o(0) = \frac{g_{m,1}}{g_{m,1} + g_o}} \tag{8.13}$$

and the zero-frequency and pole-frequency by

$$\omega_z = \frac{g_{m,1}}{C_{gs,1}} \quad \text{and} \quad \omega_p = \frac{g_{m,1} + g_o}{C_{gs,1} + C_l'} \tag{8.14}$$

(Appendix A). Since $g_{m,1} \gg g_o$ the voltage gain of the source follower is $a_o \approx 1$. The zero-frequency is caused by the feed-forward of the signal via the capacitance $C_{gs,1}$ and the pole-frequency, called -3 dB bandwidth, is mainly caused by the load capacitance C_l'.

Usually the word frequency, expressed in Hz, is used in this book, though angular frequency $\omega = 2\pi f$, expressed in rps, is meant.

The source follower can be designed to a large extent independently of the frequency, when one implements a pole-zero compensation, as presented in Appendix A of this chapter. In order that $\omega_z = \omega_p$ the following component ratios

$$\frac{g_o}{g_{m,1}} = \frac{C_l'}{C_{gs,1}} \tag{8.15}$$

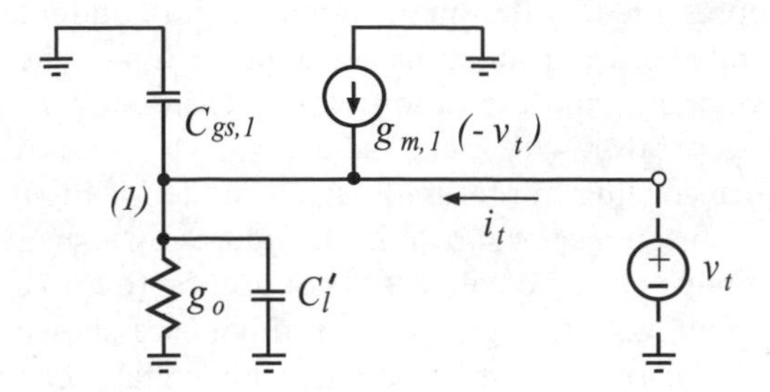

Figure 8.11 Small-signal equivalent circuit of the source follower with test voltage v_t

have to be maintained. Implementing this relationship, it may be necessary to add an additional capacitor parallel to $C_{gs,1}$. This leads to a source follower where the voltage gain (Equation 8.12) is independent of the frequency up to a very high value, where second-order effects have to be considered.

The output resistance of the circuit can be determined in a similar way as is done by the current sources of Figure 8.6. A small-signal test voltage v_t is applied to the output of the small-signal circuit, and the input is connected to analog ground, which means v_i is zero (Figure 8.11).

Solving the current equation at node (1) results in a test current of

$$i_t = \frac{v_t}{r_{out}} = v_t[g_o + g_{m,1} + j\omega(C_l' + C_{gs,1})] \tag{8.16}$$

Since $g_{m,1} \gg g_o$ and even at higher frequencies $g_{m,1} \gg \omega(C_l' + C_{gs,1})$, a small-signal output resistance of

$$\boxed{r_{out} \approx \frac{1}{g_{m,1}}} \tag{8.17}$$

results. Depending on the chosen (w/l) ratio and current, small-signal output resistance values of several Ω to kΩ can be implemented. This is important for the realization of low resistance amplifier output stages, as is discussed in Section 9.1.

8.3 BASIC AMPLIFIER PERFORMANCE

In this section a simple amplifier stage is considered in order to answer the following basic questions:

(a) Which parameters are responsible for the voltage amplification, the −3 dB bandwidth, and the unity gain frequency?

(b) How does the Miller effect influence the behavior of the amplifier stage?

To answer these questions the amplifier stage of Figure 8.12(a) is analyzed, where T_N acts as gain element and T_P as current source. For this purpose the small-signal equivalent circuit of this amplifier is drawn (Figure 8.13). It consists of a p-channel and n-channel small-signal equivalent circuit, according to Figure 8.1(b). This circuit can be simplified as shown in Figure 8.14. For this purpose all capacitances referenced to analog ground are combined in a single load capacitance C_l'. Since $v_{gs,p}$ is zero, the small-signal current $g_{m,p} \cdot v_{gs,p}$ of the p-channel transistor is zero also. The small-signal output conductances act in parallel and can thus be combined into one conductance $g_o = g_{o,p} + g_{o,n}$. The gate source capacitances of the transistors do not have to be considered since $v_{g,p}$ is zero and the voltage source v_i is assumed to have an internal resistance of zero.

The voltage gain of the circuit can be found by solving the current equation at node (1)

$$i_1 - i_2 - i_3 - i_4 = 0$$

$$j\omega C_{gdo,n}\,(v_i - v_o) - g_{m,n}v_i - g_o v_o - j\omega C_l' v_o = 0 \tag{8.18}$$

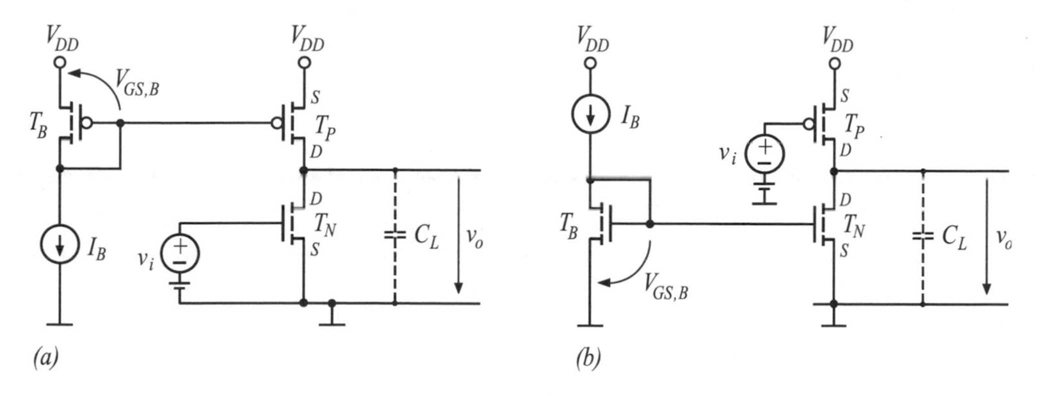

Figure 8.12 (a) Amplifier stage and (b) amplifier stage with exchanged transistor functions

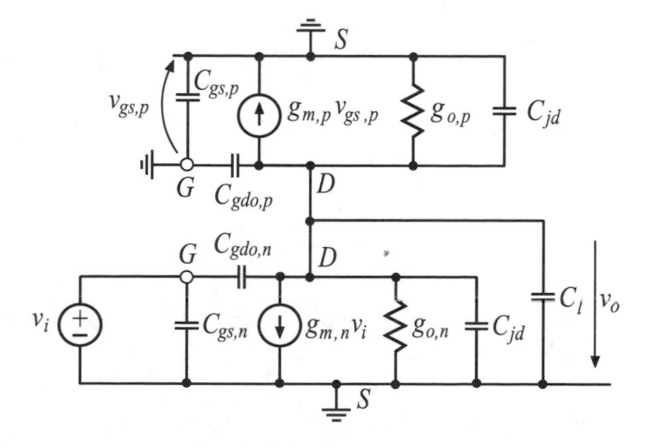

Figure 8.13 Small-signal equivalent circuit of the amplifier stage of Figure 8.12(a)

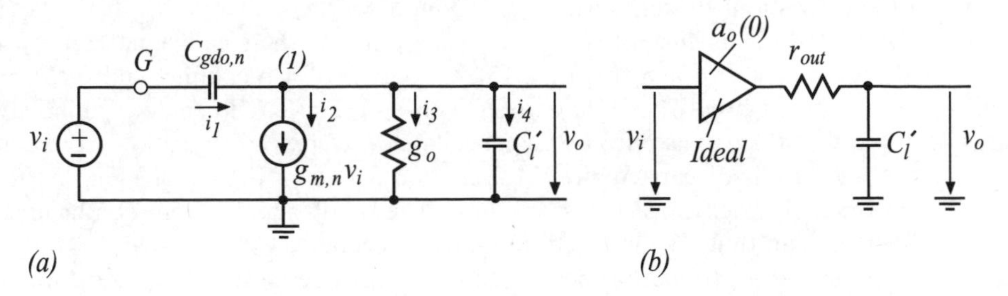

Figure 8.14 (a) Simplified equivalent circuit of Figure 8.13(a) and (b) small-signal macro model

which results in a transfer function of

$$a(j\omega) = \frac{v_o(j\omega)}{v_i(j\omega)} = a_o(0)\frac{1 - j\dfrac{\omega}{\omega_z}}{1 + j\dfrac{\omega}{\omega_p}} \tag{8.19}$$

where

$$a_o(0) = -\frac{g_{m,n}}{g_o} = -g_{m,n}r_{out} \tag{8.20}$$

is the low-frequency gain and $r_{out} = 1/g_o$ the small-signal output resistance. The minus sign of this equation indicates that a phase shift of 180° exists between output and input signal. The pole-frequency or $-3\,\text{dB}$ bandwidth (Appendix A, Equation A.11) is given by

$$\omega_p = \frac{g_o}{C_l'} \tag{8.21}$$

and the zero-frequency by

$$\omega_z = \frac{g_m}{C_{gdo,n}} \tag{8.22}$$

With the assumption, which is applicable to all practical cases, that $\omega_z \gg \omega_p$ a transfer function of

$$\boxed{a(j\omega) = \frac{a_o(0)}{1 + j\dfrac{\omega}{\omega_p}}} \tag{8.23}$$

results. A characteristic frequency of the circuit is the unity-gain frequency, sometimes called gain-bandwidth product (GBW). That is the frequency where the gain of the circuit has a value of 1 or 0 dB. With $\omega/\omega_p \gg 1$ Equation (8.23) yields a unity-gain frequency of

$$\boxed{\omega_T = |a_o|\omega_p} \tag{8.24}$$

Three characteristic parameters, namely the low frequency gain, the pole-frequency, and the unity-gain frequency, determine the behavior of the circuit. In order to gain more insight into how these parameters correlate to each other, the knowledge of their current dependence is of the utmost importance.

Substituting of equations from Table 8.1 into Equations (8.20), (8.21), and (8.24) reveals the following drain current dependence

$$a_o(0) = -\frac{g_{m,n}}{g_o} \approx -\frac{\sqrt{2I_{DS}\beta_n(1+\lambda_n V_{DS})}}{I_{DS}(\lambda_n+\lambda_p)} \sim (I_{DS})^{-1/2} \tag{8.25}$$

$$\omega_p = \frac{g_o}{C_l'} = \frac{I_{DS}(\lambda_n+\lambda_p)}{C_l'} \sim (I_{DS}) \tag{8.26}$$

$$\omega_T = |a_o|\omega_p \approx \frac{\sqrt{2I_{DS}\beta_n(1+\lambda_n V_{DS})}}{C_l'} \sim (I_{DS})^{1/2} \tag{8.27}$$

The gain a_o of the circuit is given by the ratio between the transconductance and the output conductance and is largest at small drain currents (Figure 8.15). The reason for this behavior is that the output conductance increases with the current to a much larger extent than the transconductance of the transistor (Figure 8.16). This leads to the demand of a small current for a high gain, which is contrary to the requirement of a large current for a high −3 dB bandwidth and unity-gain frequency. Therefore in an amplifier design a compromise between the two contradicting requirements has to be

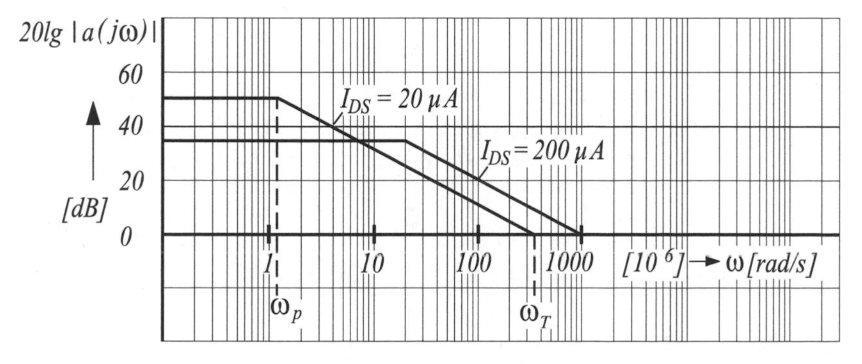

Figure 8.15 Bode plot of the simple MOS amplifier of Figure 8.12(a) ($C'_l = 1$ pF, $\beta_n = 2000\,\mu A/V^2$, $\lambda_n = \lambda_p = 0.03\,V^{-1}$, $V_{DD} = 5$ V)

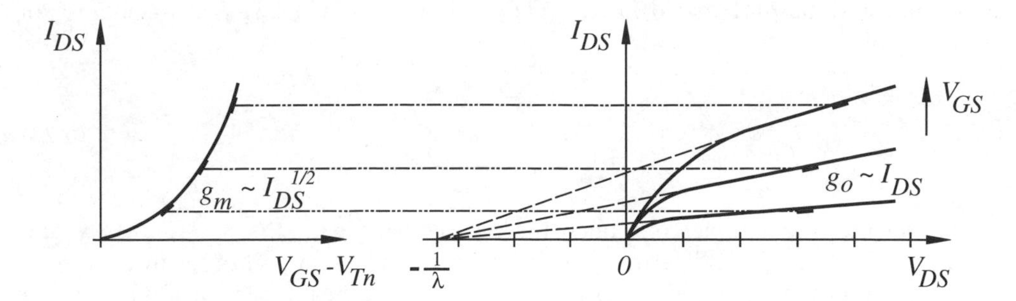

Figure 8.16 Transconductance and output conductance as function of drain current

made. This is a substantial drawback of the MOS transistor in comparison to the bipolar transistor (Section 10.4.2).

In Figure 8.12 two simple amplifiers are shown. The amplifier discussed so far (Figure 8.12(a)) uses an n-channel transistor as gain element and a p-channel transistor as current source. In the case of the implementation of Figure 8.12(b), the transistor tasks are exchanged. But the small-signal equivalent circuits of Figures 8.13 and 8.14 are valid for both amplifiers with the only exception that in the last case the small letter indexes n and p have to be exchanged. That the small-signal equivalent circuits are applicable to both amplifiers can also be seen by considering the fact that an increased input signal, e.g. by ΔV_i, leads to a reduced output signal by ΔV_o in both amplifiers. From the discussion it is obvious that therefore one small-signal equivalent circuit, as shown in Figure 8.1, is applicable to both n-channel and p-channel transistors.

The purpose of the detailed discussion of the small-signal equivalent circuit is to ease the understanding and use of more elaborate small-signal models in the following sections.

Small-signal macro model

A further point has to be made. The pole-frequency can be found by simply inspecting the small-signal equivalent circuit of Figure 8.13 or Figure 8.14. With $v_i = 0$ the output resistance can be seen to be $r_{out} = 1/g_o = 1/(g_{o,n} + g_{o,p})$. The amplifier is loaded with C_l'. This results in a pole-frequency of $\omega_p = 1/r_{out} \cdot C_l'$, which is in accordance with Equation (8.21). The amplifier has a gain (Equation 8.20) of $a_o(0) = -g_{m,n}\, r_{out}$. With these parameters, found by inspection, a macro model (Figure 8.14(b)) can be drawn. This model is comparable to the one used in Appendix A. The nice feature about this macro model is that it can be applied in many cases to much more complex circuits, without going through the laborious task of solving the current–voltage equations (Section 8.3.3).

8.3.1 Miller Effect

In the preceding derivation of the transfer function it is assumed that the input resistance of the voltage source v_i at the input of the amplifier is zero. If this is not the case (Figure 8.17) then the small-signal input current i, which is larger than expected,

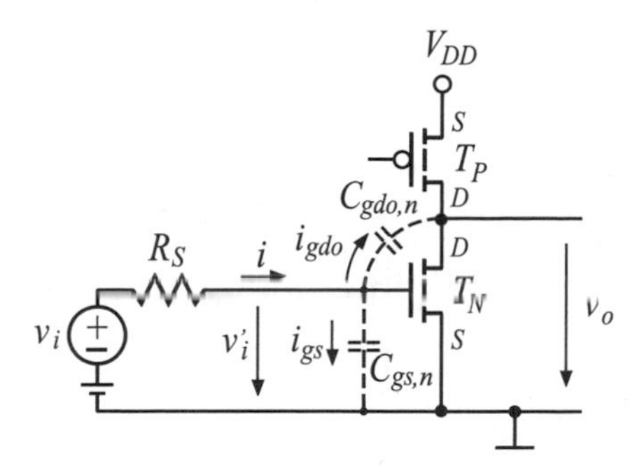

Figure 8.17 Simple amplifier stage without biasing circuit

causes a voltage drop at the source resistor R_S, reducing the input voltage to v'_i. The current at the input has a value of

$$i = i_{gs} + i_{gdo}$$
$$i = j\omega C_{gs,n} v'_i + j\omega C_{gdo,n}(v'_i - v_o) \tag{8.28}$$

This leads with the low-frequency gain of $a_o = v_o/v'_i$ to a small-signal input current of

$$i = j\omega C_{gs,n} v'_i + j\omega C_{gdo,n} v'_i(1 + |a_o|) \tag{8.29}$$

Supposing that one disregards the different capacitance values, then the i_{gdo} current is larger by the small-signal gain a_o compared to the i_{gs} current. This becomes clear, when one considers the voltages across the capacitances. In the case of $C_{gs,n}$ a voltage of v'_i exists, whereas in the case of $C_{gdo,n}$ the large voltage difference $(v'_i - v_o)$ is responsible for the current. This is the so-called Miller effect and is considered in the equivalent circuit of Figure 8.18 by increasing the input capacitance to a value of

$$C_{in} = C_{gs,n} + C_{gdo,n}(1 + |a_o|) \tag{8.30}$$

Cascode amplifier stage

The Miller effect can be reduced substantially when a transistor T_2 is added to the simple amplifier stage (Figure 8.19). The purpose of the transistor is to reduce the small-signal voltage v_k at node (k). One way to consider the effect of T_2 is to compare it

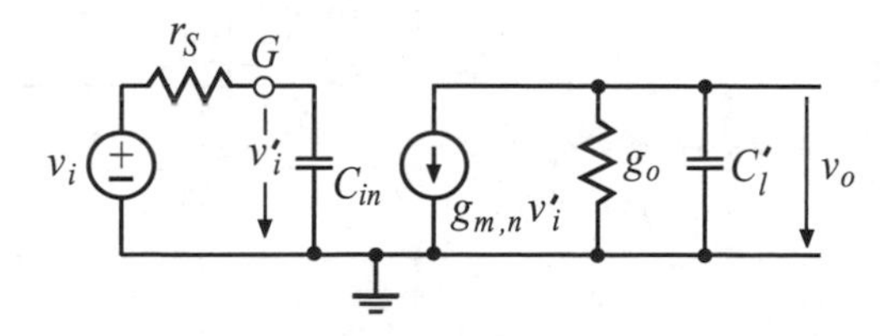

Figure 8.18 Simplified equivalent circuit of Figure 8.14 including the Miller effect

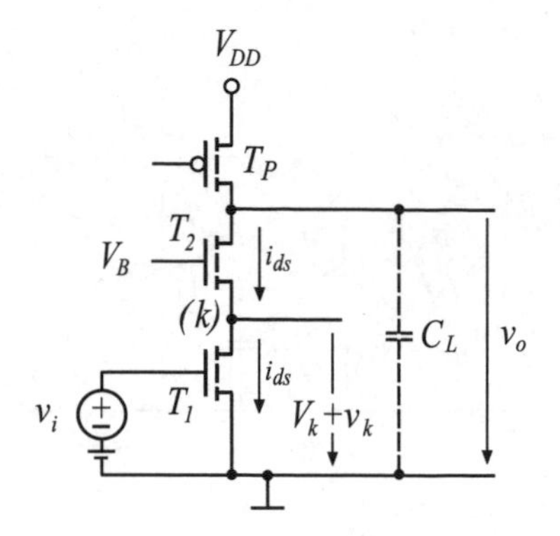

Figure 8.19 Cascode amplifier stage

with the source follower of Figure 8.9. According to Equation (8.11) the voltage at node (k) has a value of

$$V_k = V_B - \left(V_{Tn} + \sqrt{\frac{2I_{DS}}{\beta_{n,2}}} \right)$$

In case one assumes a very high transistor gain $\beta_{n,2}$, then the voltage at node (k) approaches a value of $V_B - V_{Tn}$ and does not change even when the current changes. In order to analyze the cascode amplifier stage, the equivalent low-frequency (capacitances negligible) small-signal circuit is considered (Figure 8.20).

For simplicity reasons it is assumed that the output resistances $r_{o,1} = 1/g_{o,1}$ and $r_{o,2} = 1/g_{o,2}$ are so large that the currents passing through are negligible. With $v_k = -v_{gs,2}$ this leads to

$$g_{m,1}v_i = -g_{m,2}v_k = -g_{o,p}v_o \tag{8.31}$$

This means that the voltage gain with respect to node (k) is

$$a_k(0) = \frac{v_k}{v_i} = -g_{m,1}/g_{m,2} \tag{8.32}$$

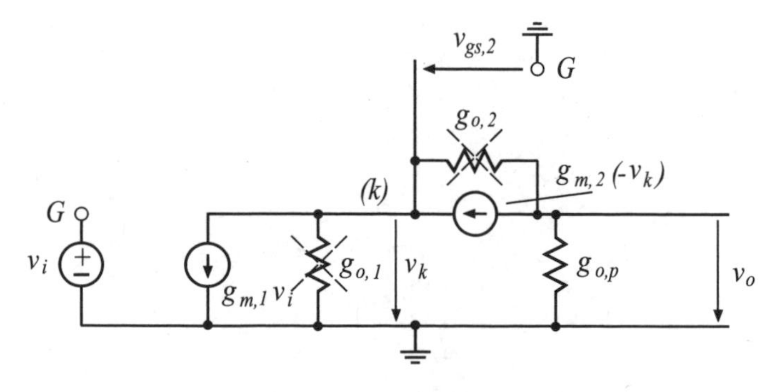

Figure 8.20 Low-frequency equivalent circuit of the cascode amplifier stage of Figure 8.19

The gain approaches zero when a very high transistor gain $\beta_{n,2}$ and thus $g_{m,2}$ is assumed for T_2. A more practical case exists when $g_{m,2}$ is chosen to be equal to $g_{m,1}$. Then the effective overlap capacitance value is reduced to $C_{gdo,1}(1 + |a_k|) = 2C_{gdo,1}$. In comparison the low-frequency small-signal gain (Equation 8.31) of the stage

$$a_o(0) = \frac{v_o}{v_i} = -g_{m,1}/g_{o,p} \tag{8.33}$$

is not affected by transistor T_2. This is also true for the -3 dB bandwidth since C_l is connected in parallel to $g_{o,p}$.

8.3.2 Differential Stage with Symmetrical Output

Nearly all amplifiers use at the input a differential stage with a symmetrical or single-ended output. In the following, these stages are analyzed with respect to the differential gain, common-mode gain, and the -3 dB bandwidth. In Figure 8.21 a differential stage with symmetrical output is shown.

In this circuit transistors T_S and T_N act as n-channel current source and the transistors T_P as p-channel sources. All currents are related to a reference current I_B by current mirroring, exactly as described in Section 8.1. If the small-signal voltages are zero then, under symmetrical biasing condition, the currents $I_{N,1}$ and $I_{N,2}$ are equal. But if $v_{i,1}$ is positive and $v_{i,2}$ negative, then $I_{N,1}$ increases slightly whereas $I_{N,2}$ decreases by the same small amount. The opposite situation exists when $v_{i,1}$ is negative and $v_{i,2}$ is positive. The change in currents leads to a small-signal output voltage v_o. In all cases, the current of the n-channel current source $I_S = I_{N,1} + I_{N,2}$ remains constant.

The input voltage v_i is divided into the two voltages $v_{i,1} = v_i/2$ and $v_{i,2} = -v_i/2$, which are referenced to analog ground. This has the advantage that the node equations of the small-signal equivalent circuit can be solved more easily.

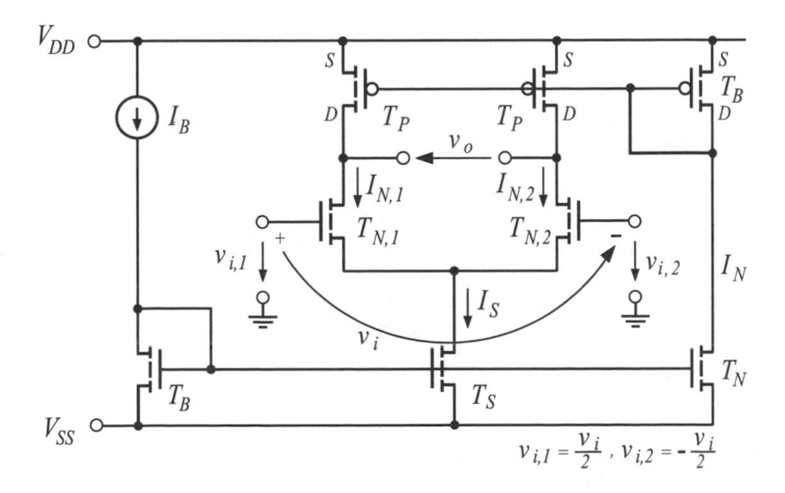

Figure 8.21 Differential input stage with symmetrical output

The differential input transistors $T_{N,1}$ and $T_{N,2}$ have to be very symmetrical in order that an offset voltage (Section 10.4.1) between the two is as small as possible, even when technology variations and mask misalignments occur. For this purpose, each transistor is divided into two parts and connected as shown in Figure 8.22. The layout with the ring structure is the preferred one since the transistors are the most symmetrically arranged ones.

Differential-mode gain

The input stage is analyzed by considering the small-signal equivalent circuit of Figure 8.23. It is assumed that the differential input transistors T_N are identical as

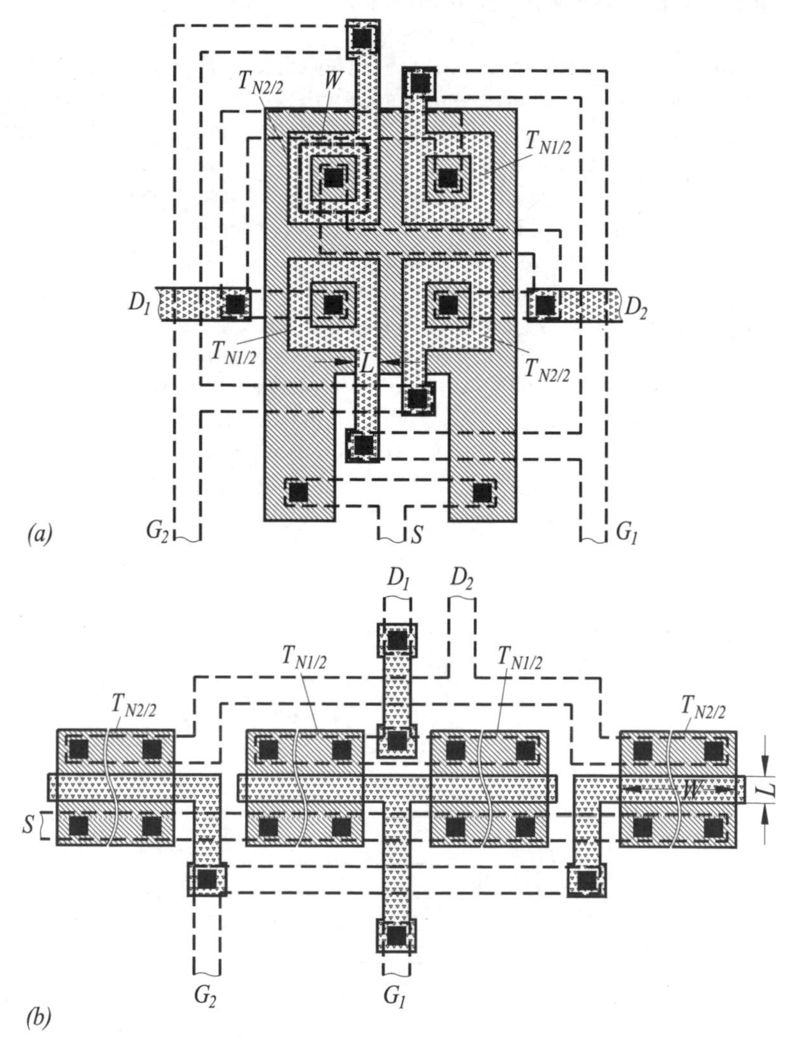

Figure 8.22 Layout of symmetrical input transistors: (a) ring structure and (b) side by side structure

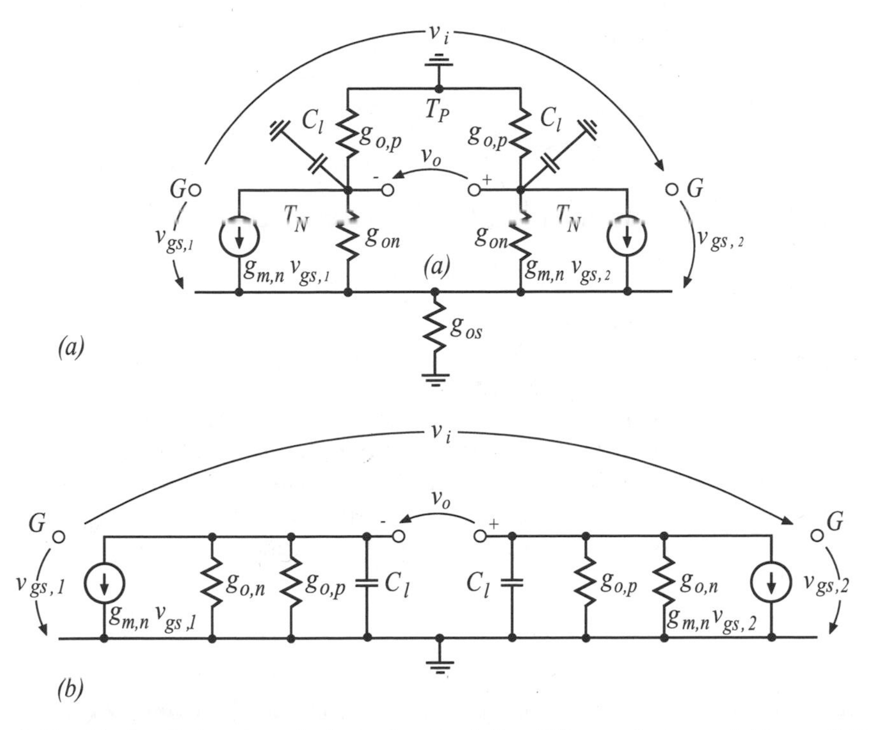

Figure 8.23 (a) Small-signal equivalent circuit of the differential stage of Figure 8.21 and (b) simplified version in differential mode

well as the transistors T_P and that the load capacitances C_l are the dominating ones in the circuit.

If $v_{i,1}$ increases and $v_{i,2}$ decreases (Figure 8.21), this is the same as if $v_{gs,1}$ and $v_{gs,2}$ increases and decreases accordingly. These voltages cause an increase of the current $g_{m,n}\, v_{gs,1}$ and a decrease of the current $g_{m,n}\, v_{gs,2}$. If the input voltages are reversed, the opposite situation occurs. The small-signal voltage at node (*a*) is zero since the two voltage-controlled currents compensate each other. Thus, this node can be considered as analog ground. This leads to the simplified circuit, as shown in Figure 8.23(b).

Solving the current voltage equations for this circuit results in a low-frequency differential gain of

$$a_{dm}(0) = \frac{v_o}{v_i} = \frac{g_{m,n}}{g_{o,n} + g_{o,p}} = g_{m,n}\, r_{out} \tag{8.34}$$

This result is identical to that of the simple amplifier stage (Equation 8.20).

The transfer function of the circuit can be derived easily, since C_l is parallel to $g_{o,p}$ and $g_{o,n}$ which results in

$$a_{dm}(j\omega) = \frac{g_{m,n}}{g_{o,n} + g_{o,p} + j\omega C_l} = a_{dm}(0)\frac{1}{1 + j\dfrac{\omega}{\omega_p}} \tag{8.35}$$

where

$$\omega_p = \frac{g_{o,n} + g_{o,p}}{C_l} \tag{8.36}$$

is the -3 dB bandwidth.

Common-mode gain

Most applications require a high differential gain often in the environment of fluctuating common-mode voltages, e.g. caused by coupling noise. To minimize the effect, the common-mode gain should be as low as possible. Using the low-frequency small-signal equivalent circuit of Figure 8.24, the common-mode gain can be found.

The left and the right side of the circuit are identical with respect to the input signal v_i. This allows to split the output conductance $g_{o,s}$ of the current source into $g_{o,s}/2$ parts for the left and right sides of the circuit. Solving the current voltage equations for this situation yields a low-frequency common-mode gain of

$$a_{cm}(0) = \frac{v_o}{v_i} \approx -\frac{g_{o,s}}{2g_{o,p}} \tag{8.37}$$

It is assumed that $g_m \gg g_{o,s}$, $g_{o,p}$ and $g_{o,n}$. Increasing g_{op} in order to reduce the common-mode gain is counterproductive with respect to the differential gain (Equation 8.35). Thus, the best way to reduce the common-mode gain is to decrease the output conductance $g_{o,s}$ of the current source. For this purpose one of the improved current sources, shown in Figure 8.5, may be used.

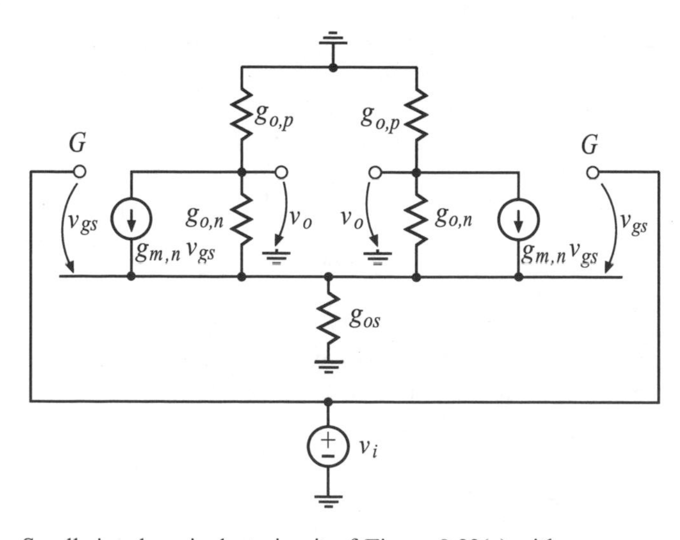

Figure 8.24 Small-signal equivalent circuit of Figure 8.23(a) with a common-mode signal

An important objective in differential amplifier design is a high common-mode rejection ratio (CMRR), which has a value for the considered circuit of

$$\mathrm{CMRR}(0) = \left|\frac{a_{dm}}{a_{cm}}\right| = 2\frac{g_{m,n}\,g_{o,p}}{g_{o,s}(g_{o,n} + g_{o,p})} \tag{8.38}$$

Essential information about the optimization of the differential stage can be gained by setting the key equations in relation to the drain current (Table 8.1). This leads to

$$\begin{aligned}
a_{dm}(0) &= \frac{g_{m,n}}{g_{o,n} + g_{o,p}} && \sim (I_{DS})^{-1/2} \\
a_{cm}(0) &\approx -\frac{g_{o,s}}{2g_{o,p}} && \neq I_{DS} \\
\mathrm{CMRR}(0) &= \frac{2g_{m,n}\,g_{o,p}}{g_{o,s}(g_{o,n} + g_{o,p})} && \sim (I_{DS})^{-1/2} \\
\omega_p &= \frac{g_{o,n} + g_{o,p}}{C_l} && \sim I_{DS}.
\end{aligned}$$

At small currents the differential gain a_{dm} and the CMRR improve, whereas the $-3\,\mathrm{dB}$ bandwidth decreases. This is the identical effect as observed by the simple amplifier stage of Section 8.3 and results from the fact that the transconductance g_m does not increase as much as the output conductance g_o of the MOS transistor when the current is increased.

The common-mode gain a_{cm} is independent of the current, since it depends on the ratio of two conductance values.

8.3.3 Differential Input Stage with Single-Ended Output

A differential output is usually only useful when the differential output signal is further processed. If this is not the case, single-ended output versions are preferred. A conversion to a single-ended output can be achieved by employing a current mirror as shown in Figure 8.25.

In this circuit the current $I_{N,1} = I_{P,1}$ is mirrored via the MOS diode $T_{P,1}$ to $T_{P,2}$. If $v_{i,1} > v_{i,2}$, this causes $I_{N,1}$ to be slightly larger than $I_{N,2}$. With $I_{N,1} = I_{P,1} = I_{P,2}$, this leads to an output current of $I_{P,2}$ which is substantially larger than $I_{N,2}$. An increased output voltage results. If $v_{i,2} > v_{i,1}$ the opposite situation exists.

MOS diode

Transistor $T_{P,1}$ is connected as a MOS diode. This transistor configuration can be represented in a small-signal equivalent circuit by the transconductance of the transistor, as Figure 8.26 illustrates.

The conductance of the MOS diode is

$$g = \frac{i}{v} = \frac{g_{m,p}v + g_{o,p}v}{v}$$

$$\boxed{g \approx g_{m,p}} \tag{8.39}$$

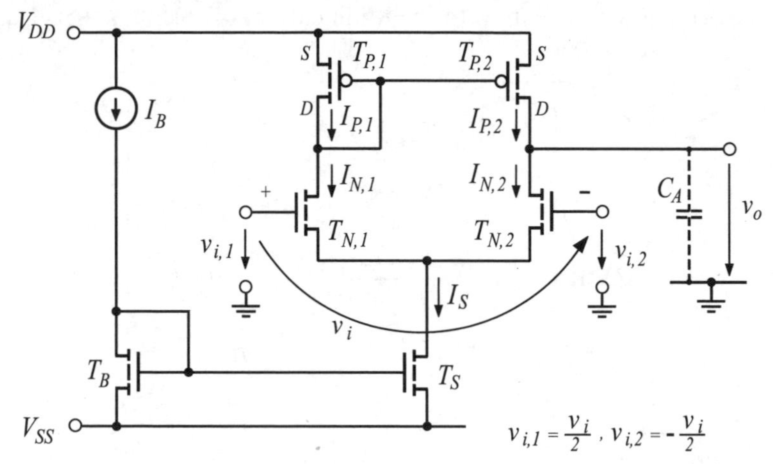

Figure 8.25 Differential input stage with single-ended output

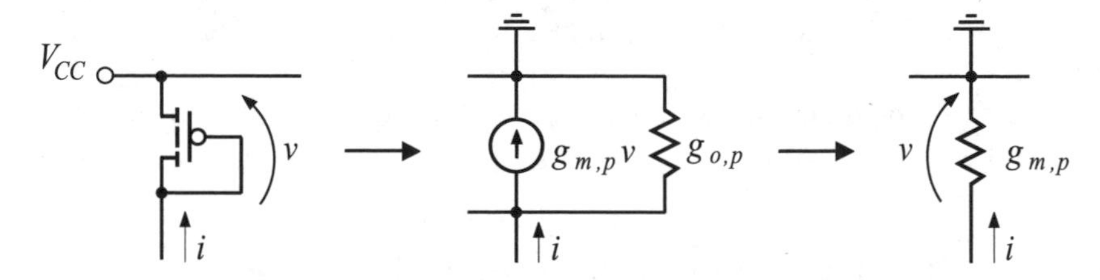

Figure 8.26 Small-signal equivalent circuit of the MOS diode

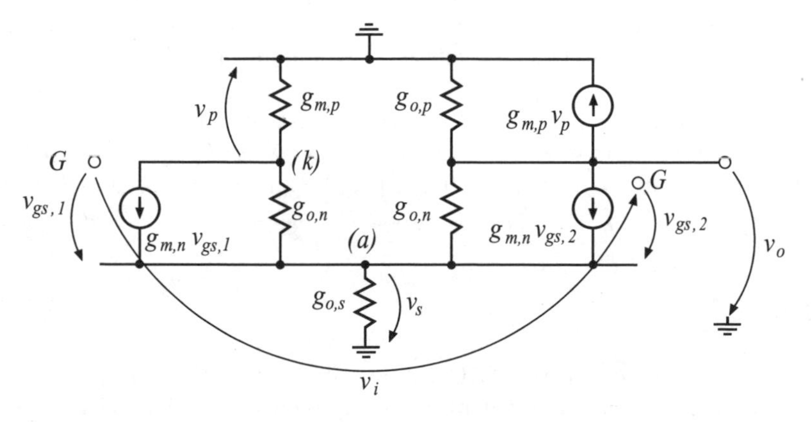

Figure 8.27 Small-signal equivalent circuit of the differential input stage with single-ended output

since in general $g_{m,p} \gg g_{o,p}$. Thus, the following small-signal equivalent circuit of Figure 8.25 results for the differential stage with single-ended output (Figure 8.27).

Solving the current voltage equations leads to the following summarized results.

$$a_{dm}(0) \approx \frac{g_{m,n}}{g_{o,n} + g_{o,p}} \sim (I_{DS})^{-1/2} \tag{8.40}$$

$$\approx g_{m,n}\, r_{out}$$

$$a_{cm}(0) \approx -\frac{g_{o,n}\, g_{o,s}}{2g_{m,p}(g_{o,p} + g_{o,n})} \sim (I_{DS})^{1/2} \tag{8.41}$$

$$\mathrm{CMRR}(0) \approx 2\frac{g_{m,n}\, g_{m,p}}{g_{o,s}\, g_{o,n}} \sim (I_{DS})^{-1} \tag{8.42}$$

In the derivation it is assumed that $g_{m,n}$, $g_{m,p} \gg g_{o,n}$, $g_{o,p}$ and $g_{o,s}$. If one compares these results with those of the stage with symmetrical output, the following observations can be made. The gain is in both stages identical. This is not surprising, since in the symmetrical case the voltage-controlled currents $g_{m,n}\, v_{gs,1}$ and $g_{m,n}\, v_{gs,2}$ act directly on the outputs, whereas in the single-ended case $g_{m,n}\, v_{gs,2}$ acts directly and $g_{m,n}\, v_{gs,1}$ as mirrored current at the output. The common-mode rejection ratio is improved, since $a_{cm}(0)$ is reduced. This is caused by the MOS diode with its relatively large transconductance value $g_{m,p}$.

Frequency behavior of the single-ended differential stage

The starting point for the derivation of the frequency behavior is the small-signal equivalent circuit of Figure 8.27. Here it is assumed that the output of the differential stage (Figure 8.25) drives a capacitance C_A which is dominating compared to all other capacitances of the circuit. Due to the current mirror the circuit can not be simplified, as is the case for the input stage with symmetrical output. Thus the derivation of the transfer

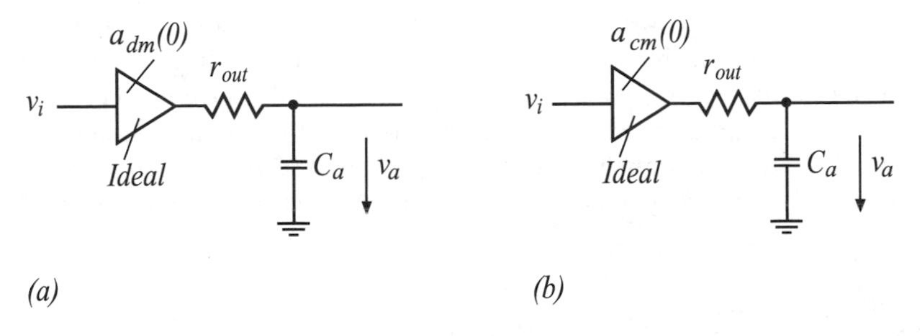

Figure 8.28 Macro model of the differential stage with single-ended output: (a) for differential mode and (b) for common mode

function is laborious. In order to avoid this, a small-signal macro model is chosen (described at the end of Section 8.3, Figure 8.14(b)) and shown in Figure 8.28(a).

The small-signal capacitance C_a is driven via the output resistance r_{out} by an ideal amplifier stage with a gain $a_{dm}(0)$. This circuit yields a transfer function of

$$a_{dm}(j\omega) = a_{dm}(0)\frac{1}{1 + j\dfrac{\omega}{\omega_p}} \tag{8.43}$$

where

$$\omega_p = \frac{g_{o,n} + g_{o,p}}{C_a} \tag{8.44}$$

is the $-3\,\text{dB}$ bandwidth of the circuit and $a_{dm}(0)$ is described by Equation (8.40). This result is identical to that of the input stage with symmetrical output. In analogy the transfer function of the common-mode operation can be found

$$a_{cm}(j\omega) = a_{cm}(0)\frac{1}{1 + j\dfrac{\omega}{\omega_p}} \tag{8.45}$$

where $a_{cm}(0)$ is given by Equation (8.41).

Extended view

In the preceding derivation it is assumed that the output capacitance C_A is dominating, compared to all other circuit capacitances. This is not necessarily the case. The so far not considered circuit capacitances may cause a severe deterioration of the common-mode rejection ratio and may generate so-called pole-zero doublets.

In order to avoid the influence of the body effect on the differential input transistors T_N the bulk terminal, that is the p-well, is connected to the source terminals (Figure 8.29). This leads to a relatively large depletion capacitance C_S of the well–substrate junction at node (1). This has no effect on the differential gain since the two voltage-controlled currents compensate each other, causing this node to behave like analog ground. This is completely different when a common-mode signal is applied to the inputs. Assuming that the capacitances C_S and C_A are much larger than C_K, then the following simplified derivation results. The common-mode gain is according to Equations (8.45), (8.41), and (8.44)

$$a_{cm}(j\omega) = a_{cm}(0)\frac{1}{1 + j\dfrac{\omega}{\omega_p}} = -\frac{g_{o,n}\,g_{o,s}}{2g_{m,p}}\frac{1}{g_{o,p} + g_{o,n} + j\omega C_a} \tag{8.46}$$

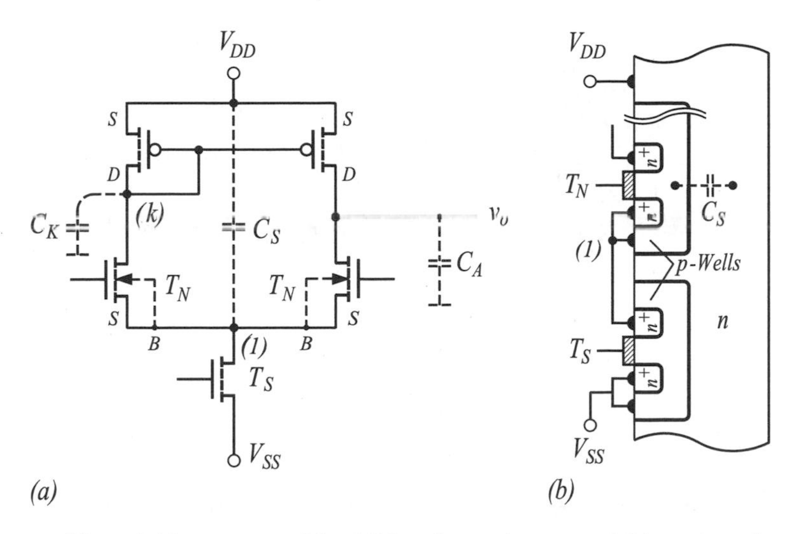

Figure 8.29 Differential input stage with additional capacitances and (b) cross-section through a p-well process with T_S and T_N

With respect to the small-signal equivalent circuit, C_S is connected parallel to the conductance $g_{o,s}$ (Figure 8.27). If this is considered a transfer function of

$$a_{cm}(j\omega) = -\frac{g_{o,n}}{2g_{m,p}} \frac{g_{o,s} \mid j\omega C_s}{g_{o,p} + g_{o,n} + j\omega C_a}$$

$$= a_{cm}(0)\frac{1 + j\omega/\omega_z}{1 + j\omega/\omega_p} \tag{8.47}$$

results, where the zero-frequency ω_z is given by

$$\omega_z \frac{g_{o,s}}{C_s} \tag{8.48}$$

This zero-frequency has the unpleasant effect that the common-mode gain increases by 20 dB/decade (Appendix A, Equation A.15) which causes of course a deterioration of the common-mode rejection ratio. This is not only the case for the differential stage with single-ended output, but also for the one with symmetrical output. In order to reduce this dependence, one could come up with the idea to reduce the effect of the depletion capacitance C_S by connecting the substrates of the T_N transistors to V_{SS} (Figure 8.30). Unfortunately, this would be contrary to the requirement of an excellent power-supply rejection ratio (PSRR) for an amplifier. For example, noise v_{ss} on the power supply (Figure 8.30) may influence the differential input transistors asymmetrically and cause, via the body effect, interference at the output.

In the discussion a p-well process is assumed. An analogous situation exists, when an n-well process is employed and p-channel transistors are used for the differential inputs.

A further interesting observation can be made when the differential mode, under the assumption that the capacitance $C_K \gg C_A$, is considered (Figure 8.30). For this

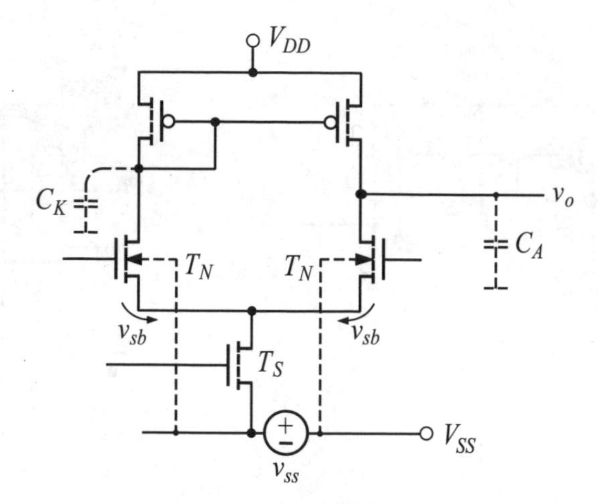

Figure 8.30 Influence of noise voltage v_{ss} on differential stage

purpose the small-signal equivalent circuit of Figure 8.27 is rearranged and analyzed (Figure 8.31).

Solving the current voltage equations results in the transfer function

$$a_{dm}(j\omega) \approx a_{dm}(0)\frac{1 + j\omega/\omega_z}{1 + j\omega/\omega_p} \tag{8.49}$$

where the pole-frequency has a value of

$$\omega_p \approx \frac{g_{m,p}}{C_k} \tag{8.50}$$

and the zero-frequency one of

$$\omega_z \approx \frac{2g_{m,p}}{C_k} \tag{8.51}$$

It is obvious from these equations that a zero-frequency ω_z exists with twice the value of the pole-frequency ω_p. The reason for this so-called pole-zero doublet is that both signals $v_{gs,1}$ and $v_{gs,2}$ are responsible for the output signal v_o but the capacitance C_k affects one input signal only, namely $v_{gs,1}$. This behavior is typical of all differential circuits where a capacitance acts on one signal path only.

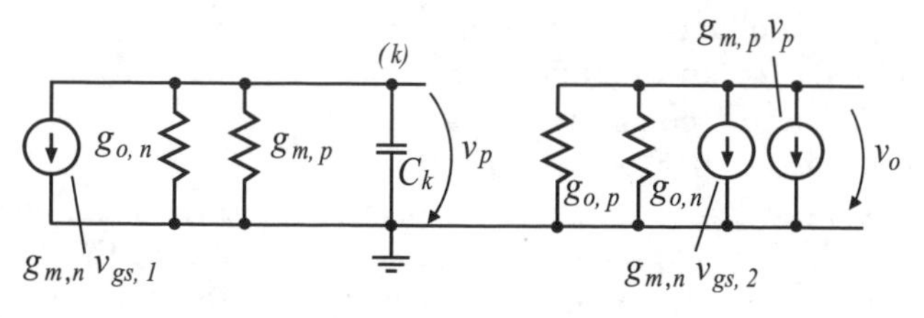

Figure 8.31 Rearranged small-signal equivalent circuit of Figure 8.27 in differential mode

Summary

A very important circuit component of an amplifier is the current mirror. This is a circuit where all current sources are mirror images of a defined current. This guarantees that the operation points of the individual circuit stages of an amplifier are related to each other, and power-supply and temperature variations do not influence the amplifier performance. With improved current sources small-signal output resistances in the MΩ region are possible. This is particularly important when high gain stages are designed.

Source followers have a voltage gain of around one and a small-signal output resistance of $1/g_m$. Depending on the chosen (w/l) ratio, small-signal resistance values of several Ω to kΩ can be implemented. This is important for the realization of low resistance amplifier stages.

An important result of the simple amplifier stage is that increasing the current reduces the voltage gain but increases the −3 dB bandwidth and the unity gain frequency of the circuit. This applies also to the differential input stages and is an unpleasant feature of the MOS transistor where $g_m \sim I_{DS}^{1/2}$ and $g_o \sim I_{DS}$.

The Miller effect is caused by the coupling of the output signal via the overlap capacitance C_{gdo} to the input. This increases the effective capacitance value by the gain of the circuit.

Problems

8.1 Determine the low-frequency output signal and voltage gain for the amplifier shown below. The data are: $I_{DS} = 100\,\mu A$, $\beta_n = 1000\,\mu A/V^2$, $\lambda_n = 0.1\,V^{-1}$, $R_L = 100\,k\Omega$, and $v_i = 50\,\mu V$. The transistor operates under saturation conditions.

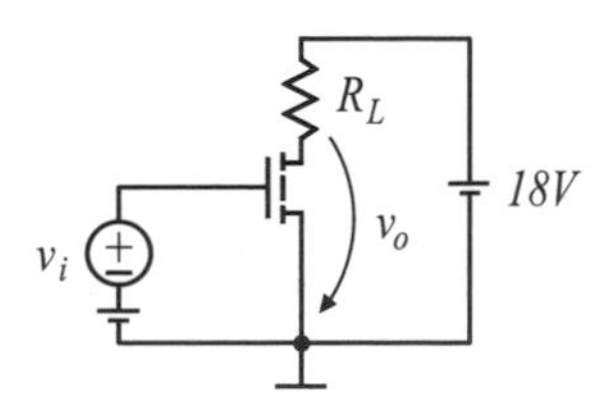

8.2 Determine the current I at room temperature, when transistor T of the current mirror circuit shifts from the saturation region into the sub-threshold region. The data are: $k_n = 120\,\mu A/V^2$, $w/l = 10$, $n = 2$, $V_{DS} > 0.1\,V$.

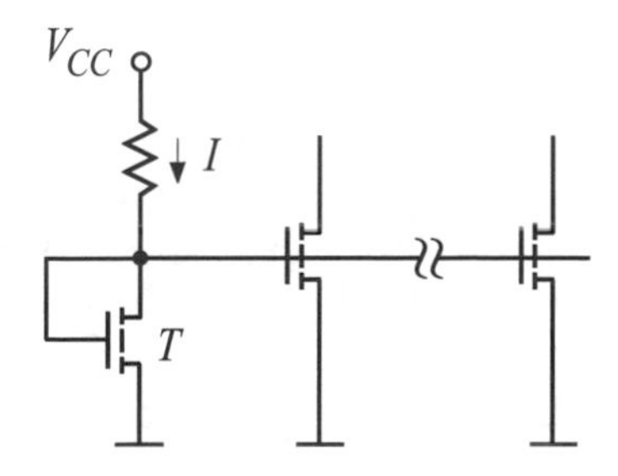

8.3 Shown below is a CMOS amplifier stage where the operation point V_P is given by a reference circuit. This makes the circuit relatively insensitive to production tolerances. Determine the low-frequency gain of the circuit when the influence of R and C is negligible. The data for the n-channel transistor are: $\beta_n = 1000\,\mu A/V^2$, $V_{Tn} = 0.5\,V$, $\lambda_n = 0.05\,V^{-1}$, and those for the p-channel transistor are: $\beta_p = 1000\,\mu A/V^2$, $V_{Tp} = -0.5\,V$, $\lambda_p = 0.05\,V^{-1}$.

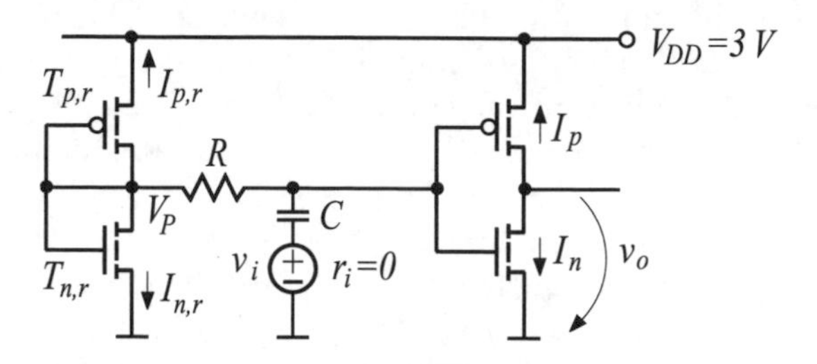

8.4 Determine the channel width of the source follower shown below when an output resistance of $r_{out} = 50\,\Omega$ is required. In order to reduce the influence of the channel length modulation, a gate length of 1.5 μm is selected. Therefore $\lambda_n V_{DS} \ll 1$.

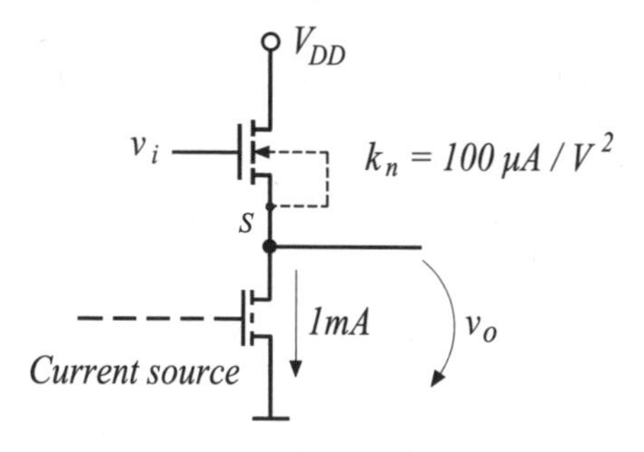

The solutions to the problems can be found under: *www.unibw-muenchen.de/campus/ET4/index.html*

APPENDIX A

Transfer Functions

The terms *transfer function* or *system function* are used in general for time-invariant linear systems described by the linear differential equation

$$a_n \frac{d^n g(t)}{dt^n} + \ldots + a_1 \frac{dg(t)}{dt} + a_o g(t) = b_m \frac{d^m f(t)}{dt^m} + \ldots + b_1 \frac{df(t)}{dt} + b_o f(t) \tag{A.1}$$

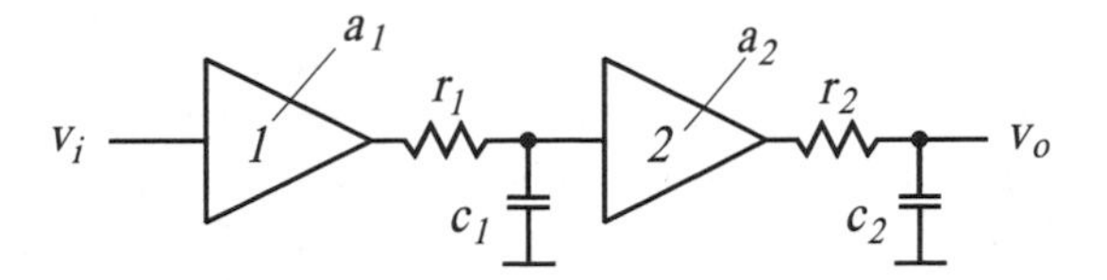

Figure A.1 Two-stage amplifier of second order

where the constant coefficients are marked by small letters. A typical example of such a system is a two-stage amplifier, shown in Figure A.1 which represents to a large extent an integrated two-stage amplifier. The coefficients a_1 and a_2 are constant gain factors of the amplifiers. The resistors r_1 and r_2 describe the output resistances of the amplifier stages and c the load capacitances. The relationship between output and input voltages can be found by using Kirchoff's current–voltage law and the relationship $i = c\ \mathrm{d}v/\mathrm{d}t$. This leads to the differential equation

$$a_1 a_2 v_i(t) = c_1 r_1 c_2 r_2 \frac{\mathrm{d}^2 v_o(t)}{\mathrm{d}t^2} + (c_1 r_1 + c_2 r_2) \frac{\mathrm{d}v_o(t)}{\mathrm{d}t} + v_o(t) \tag{A.2}$$

In order to find the transfer function $v_o(t)/v_i(t)$, the differential equation has to be solved. One way of achieving this is to use the Laplace transform. This leads to a linear equation which can be solved by linear algebra. Applying to this solution the inverse Laplace transform, the transfer function in the time domain can be found.

The Laplace transform of the function $f(t)$ is defined by

$$F(s) = \int_0^\infty f(t) e^{-st} \mathrm{d}t \tag{A.3}$$

where $s = \sigma + j\omega$ is a complex variable. It is assumed that $f(t) = 0$ for $t < 0$. Applying this transformation to the differential equation of (A.2) leads to the linear equation

$$a_1 a_2 v_i(s) = c_1 r_1 c_2 r_2 s^2 v_o(s) + (c_1 r_1 + c_2 r_2) s v_o(s) + v_o(s) \tag{A.4}$$

and to a transfer function of

$$a(s) = \frac{v_o(s)}{v_i(s)} = \frac{a_o}{\left(1 + \frac{s}{p_1}\right)\left(1 + \frac{s}{p_2}\right)} \tag{A.5}$$

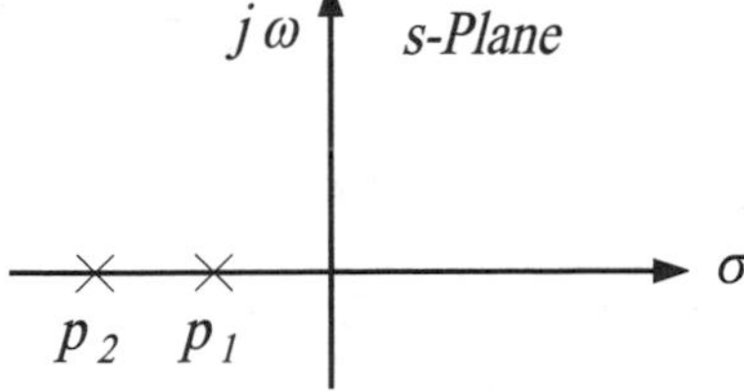

Figure A.2 Pole locations of Equation (A.5) in the s-plane

where $a_o = a_1 \cdot a_2$, $p_1 = 1/c_1 r_1$, and $p_2 = 1/c_2 r_2$. With p poles are denoted. They are a characteristic feature of the transfer function. They describe in the s-plane the value where the denominator of the transfer function is zero and thus $v_o(s)/v_i(s)$ becomes infinite (Figure A.2). In this example the poles are real. But they may be complex also, e.g. caused by a feedback configuration. The locations of the poles in the s-plane are of the utmost importance for the design of an amplifier since they lead to the time behavior of the transfer function in response to a unit-step function. This is shown for a variety of pole locations in Figure A.3

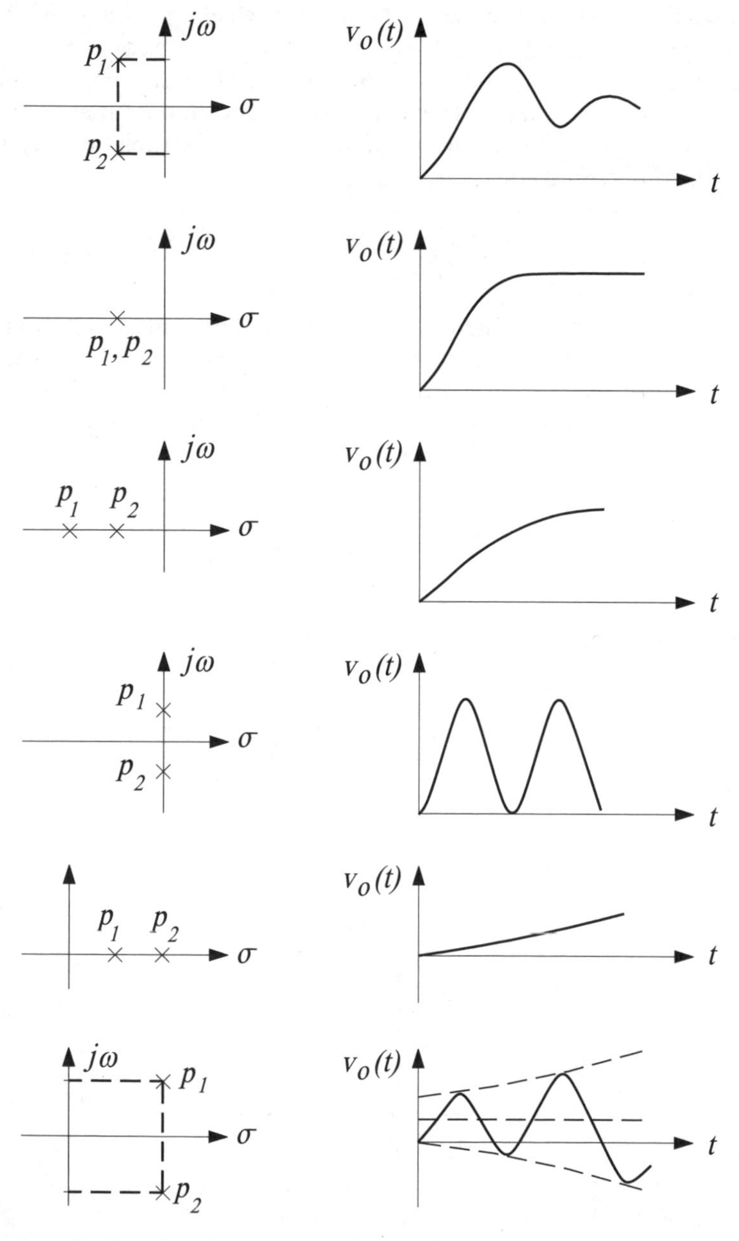

Figure A.3 Pole locations in the s-plane and its influence on a step response

Applying the Laplace transform to a step function $v_i(t)$ leads to $1/s$. Substituting this result, e.g., in Equation (A.5), solving for $v_o(s)$, and using the inverse Laplace transform results in the step response of the transfer function in the time domain. Figure A.3 indicates that a system is stable only when $v_o(t)$ does not increase exponentially or does not have a sinusoidal or even growing sinusoidal behavior. To guarantee a stable system, the poles must be in the Left Half side of the s-Plan (LHP). It is obvious that knowledge of the pole positions and their current dependence is an important feature in amplifier design. For this purpose, most circuit design tools are able to supply this information.

In the preceding description the Laplace transform is used for the solution of differential equations. The Fourier transform

$$F(j\omega) = \int_{-\infty}^{\infty} f(t)e^{-j\omega t}\mathrm{d}t \tag{A.6}$$

can be considered to be a special case of the Laplace transform, where $s = j\omega$, $\sigma = 0$, and $f(t) \neq 0$ if $t \leq 0$. Applying this transform to the presented example leads to

$$a(j\omega) = \frac{v_o(j\omega)}{v_i(j\omega)} = \frac{a_o}{\left(1 + \dfrac{j\omega}{\omega_{p,1}}\right)\left(1 + \dfrac{j\omega}{\omega_{p,2}}\right)} \tag{A.7}$$

where the operator s is replaced by $j\omega$. In this equation $\omega_{p,1}$ and $\omega_{p,2}$ are called pole-(angular) frequency to indicate that the Fourier transform is used, even if $\omega_{p,1} = p_1$ and $\omega_{p,2} = p_2$. The transfer function $a(j\omega)$ can be written in the polar form

$$a(j\omega) = \frac{a_o}{|a_{p,1}|e^{j\phi_{p,1}}|a_{p,2}|e^{j\phi_{p,2}}} \tag{A.8}$$

and analyzed using the Bode plot or diagram. This diagram consists of two parts, the magnitude and phase angle part of the transfer function displayed on logarithmic scales. Logarithmic scales are used because they simplify the construction, manipulation, and interpretation of the transfer function considerably. Separating Equation (A.8) into the two parts results in

$$[20\,\mathrm{lg}|a(j\omega)|]\mathrm{dB} = \left[20\,\mathrm{lg}a_o - 20\,\mathrm{lg}\sqrt{1 + \left(\frac{\omega}{\omega_{p,1}}\right)^2} - 20\,\mathrm{lg}\sqrt{1 + \left(\frac{\omega}{\omega_{p,2}}\right)^2}\,\right]\mathrm{dB} \tag{A.9}$$

and

$$\phi = \phi_o - \phi_{p,1} - \phi_{p,2} \tag{A.10}$$

The product of terms in the polar form is equal to the sum of the magnitudes of the individual terms in the Bode diagram presentation.

The square root terms of Equation (A.9) can be approximated by the construction of the magnitude plot of the Bode diagram. If $(\omega/\omega_{p,i})^2 > 1$, the real part can be neglected. This leads to a magnitude change of $20\,\mathrm{lg}\,[(10\omega/\omega_{p,i})/(\omega/\omega_{p,i}) = 20\,\mathrm{dB/decade}$ for each square root term. In Figure A.4 the individual contributions to the magnitude plot are shown

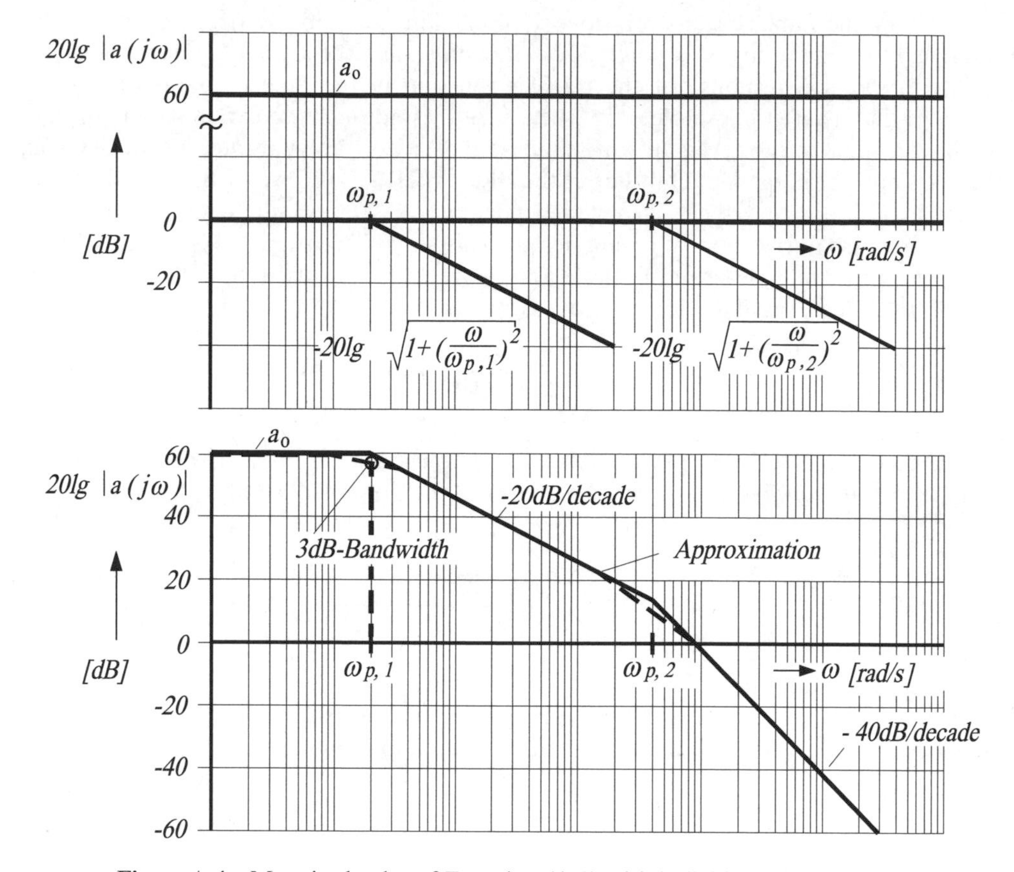

Figure A.4 Magnitude plot of Equation (A.7) with individual contributions

in asymptotic approximated form. When these terms are added or in this case subtracted from the a_o term, according to Equation (A.9), the magnitude plot of the function results.

In this plot a_o is the low-frequency gain of the circuit, e.g. 60 dB, and ω_p the -3 dB bandwidth. If $\omega_{p,1} \ll \omega_{p,2}$, then Equation (A.9) yields at $\omega_{p,1}$ a reduction in low-frequency gain a_o of

$$[20\,lg|a(j\omega_{p,1})|]\,\text{dB} \approx \left[20\,lga_o - 20\,lg\sqrt{1+\left(\frac{\omega_{p,1}}{\omega_{p,1}}\right)^2}\right]\text{dB}$$

$$\approx [20\,lga_o]\text{dB} - 3\,\text{dB}. \qquad \text{(A.11)}$$

Usually the word frequency, expressed in Hz, is used in this book, though angular frequency $\omega = 2\pi f$ expressed in rps is meant.

The phase angle plot of Equation (A.10) can be found similarly by an asymptotic approximation. The phase angle for each term of Equation (A.8) is determined by

$$\phi_{p,i} = \tan^{-1}\frac{\text{imag.part}}{\text{real part}} = \tan^{-1}\frac{\omega}{\omega_{p,i}} i = 1, 2 \ldots \qquad \text{(A.12)}$$

This equation has the following values

$$\phi_{p,i}(\omega = 0.1\omega_{p,i}) = \tan^{-1}(0.1) \approx 0^\circ$$
$$\phi_{p,i}(\omega = 10\omega_{p,i}) = \tan^{-1}(10) \approx 90^\circ \tag{A.13}$$

which can be used for the construction of the angular plot. It is obvious that at $\omega = \omega_{p,i}$ the phase has a value of

$$\phi_{p,i}(\omega = \omega_{p,i}) = \tan^{-1} \omega_{p,i}/\omega_{p,i} = 45^\circ \tag{A.14}$$

The phase angle plot with its individual contributions is shown in Figure A.5.

In the following example a transfer function of the form

$$a(j\omega) = a_o \frac{\left(1 + j\dfrac{\omega}{\omega_{z,1}}\right)}{\left(1 + j\dfrac{\omega}{\omega_{p,1}}\right)\left(1 + j\dfrac{\omega}{\omega_{p,2}}\right)} \tag{A.15}$$

is assumed. This function has additional to the pole-frequencies $\omega_{p,1}$ and $\omega_{p,2}$ a zero-frequency ω_z. The zero describes in the s-plane (Figure A.3) the value where the numerator of the transfer function is zero and thus $v_o(s)/v_i(s)$ becomes zero. In this

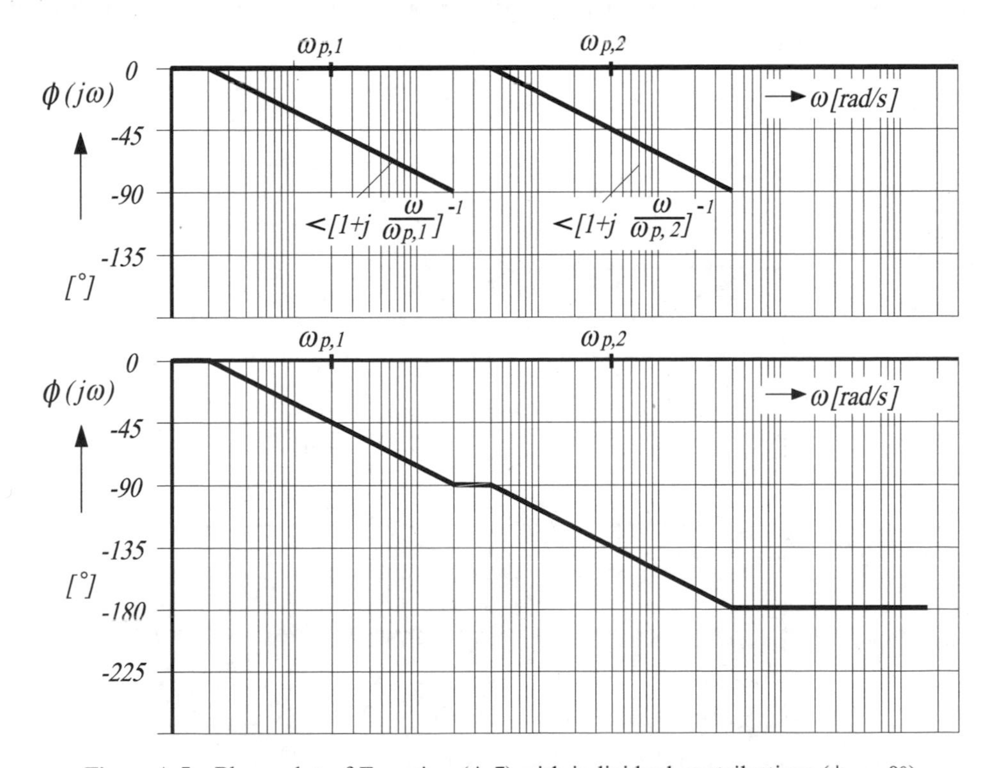

Figure A.5 Phase plot of Equation (A.7) with individual contributions ($\phi_o = 0^\circ$)

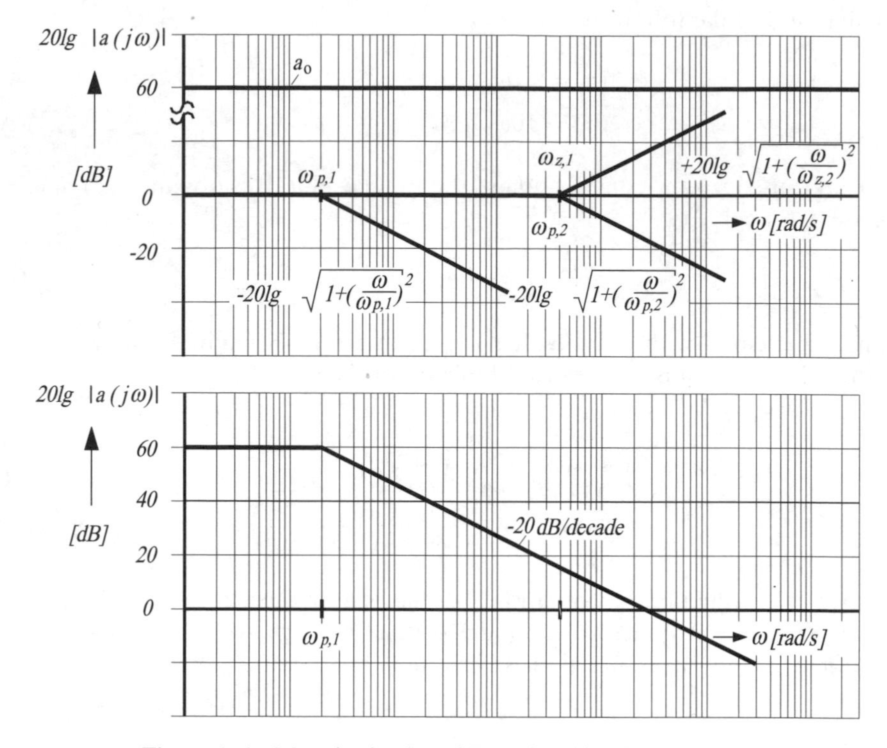

Figure A.6 Magnitude plot of Equation (A.15) $a_o = 60\,\text{dB}$

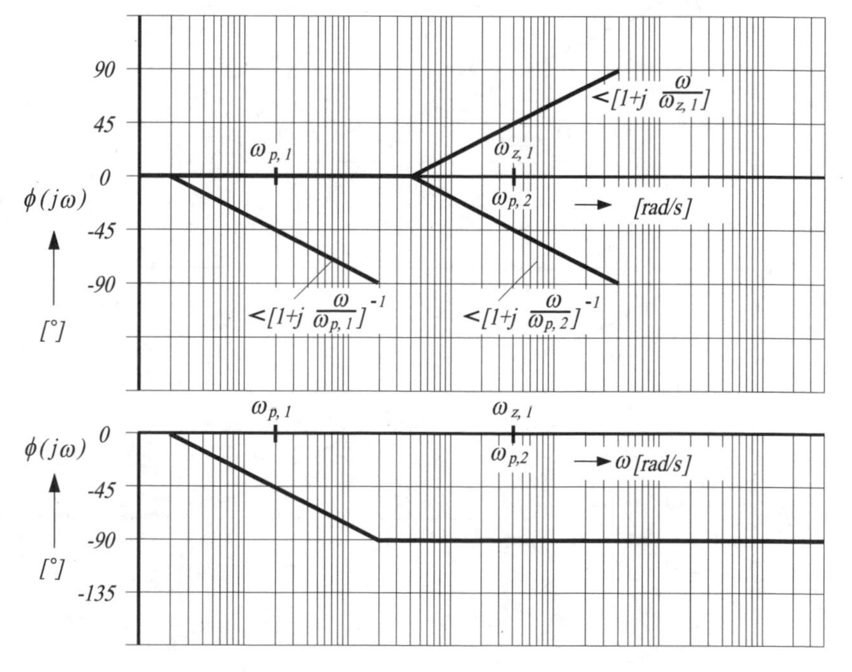

Figure A.7 Phase plot of Equation (A.15) $\phi_o = 0$

example the location of the zero is in the Left Half side of the s-Plan (LHP). Magnitude and phase plots of the function are shown in Figures A.6 and A.7. As the plots illustrate, the zero-frequency (Equation A.15) causes an increase in gain of +20 dB/decade and a phase reduction by 90°. Since the pole-frequency $\omega_{p,2}$ is chosen to have the same value as the zero-frequency ω_z, a so-called pole-zero compensation occurs.

FURTHER READING

K.R. Laker and W.M.C. Sansen (1994) *Design of Analog Integrated Circuits and Systems.* New York: McGraw-Hill.

B. Razavi (2001) *Design of Analog CMOS Integrated Circuits.* Boston, MA: McGraw-Hill.

P.R. Gray (2001) *Analysis and Design of Analog Integrated Circuits*, 4th edn. New York: John Wiley & Sons.

9

CMOS Amplifiers

In the preceding chapter analog circuits are considered. These basic building blocks are used to design two commonly used amplifiers. That is a Miller amplifier, where a stable operation is achieved by the cancellation of a zero-frequency, and a folded cascode amplifier. The latter is modified into a class *AB* operation for better current driving capability at lower power dissipation.

9.1 MILLER AMPLIFIER

The described differential input stage with single-ended output of Section 8.3.3 has a gain in the order of 100. By adding a second stage, the gain can be increased to about 10 000 or 80 dB. This leads to the Miller amplifier shown in Figure 9.1.

As second stage the amplifier of Figure 8.12(b) is used. The low frequency gain of this stage is in analogy to Equation (8.20)

$$a_2(0) = -\frac{g_{m,2}}{g_{o,1} + g_{o,2}} = -g_{m,2}\, r_{out,2} \tag{9.1}$$

and the one of the first stage (Equation 8.40) is

$$a_{dm,1}(0) \approx -\frac{g_{m,n}}{g_{o,n} + g_{o,p}} = -g_{m,n}\, r_{out,1} \tag{9.2}$$

This results in a total low-frequency gain of the Miller amplifier of

$$a_{dm}(0) \approx a_{dm,1}(0)a_2(0) \approx +\frac{g_{m,n}}{g_{o,n} + g_{o,p}}\frac{g_{m,2}}{g_{o,1} + g_{o,2}}. \tag{9.3}$$

When the amplifier is used in a feedback configuration and drives a capacitance load, the circuit is not stable. The reasons for this behavior are two pole-frequencies of the amplifier. In order to analyze this situation a small-signal macro model of the amplifier is used (Figure 9.2). The basics for this small-signal model are outlined at the end of Section 8.3 (Figure 8.14(b)). Each amplifier stage is represented by an ideal amplifier

System Integration: From Transistor Design to Large Scale Integrated Circuits. Kurt Hoffmann.
 ISBN: 0-470-85407-3

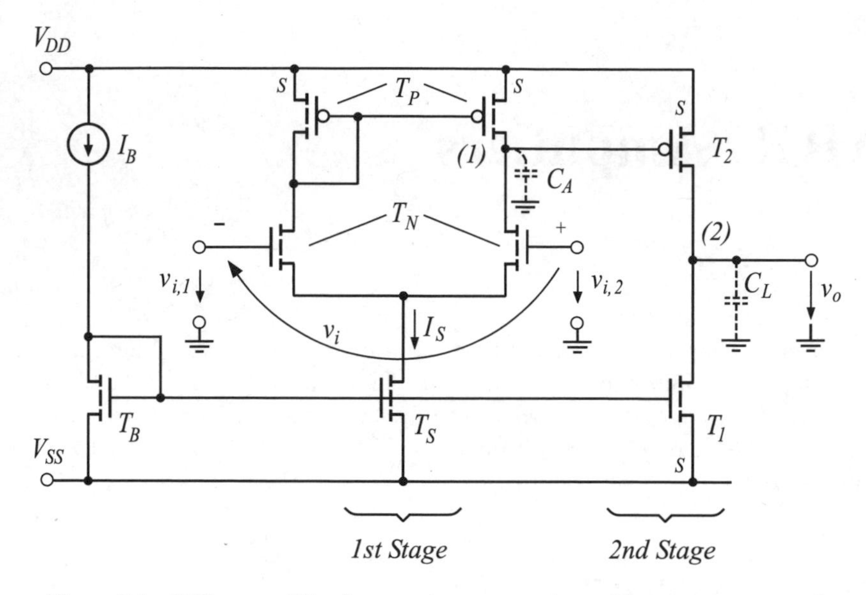

Figure 9.1 Miller amplifier (nomenclature + and − with respect to output)

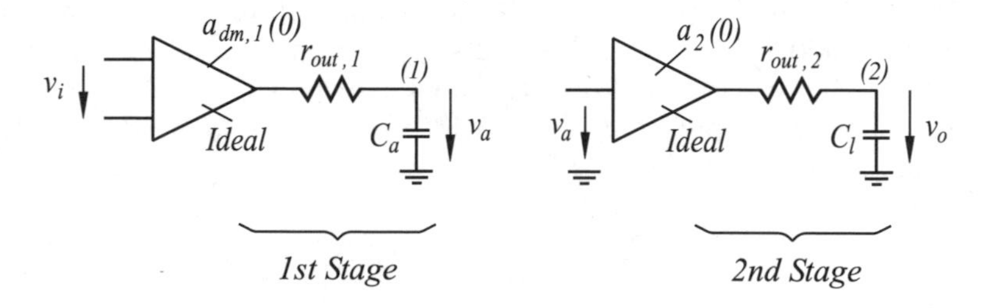

Figure 9.2 Small-signal macro model of the Miller amplifier

having an output resistance, described by Equations (9.1) and (9.2). This equivalent circuit leads to the following transfer function

$$a_{dm}(j\omega) = a_{dm}(0)\frac{1}{\left(1 + j\dfrac{\omega}{\omega_{p,1}}\right)\left(1 + j\dfrac{\omega}{\omega_{p,2}}\right)} \tag{9.4}$$

where the pole-frequencies are given by

$$\omega_{p,1} \approx \frac{g_{o,n} + g_{o,p}}{C_a} \quad \text{and} \quad \omega_{p,2} \approx \frac{g_{o,1} + g_{o,2}}{C_l} \tag{9.5}$$

and the gain $a_{dm}(0)$ by Equation (9.3). At high frequencies with $\omega \gg \omega_{p,1}$ and $\omega_{p,2}$ Equation (9.4) changes to

$$a_{dm}(j\omega) \approx -a_{dm}(0)\frac{\omega_{p,1}\,\omega_{p,2}}{\omega^2} \tag{9.6}$$

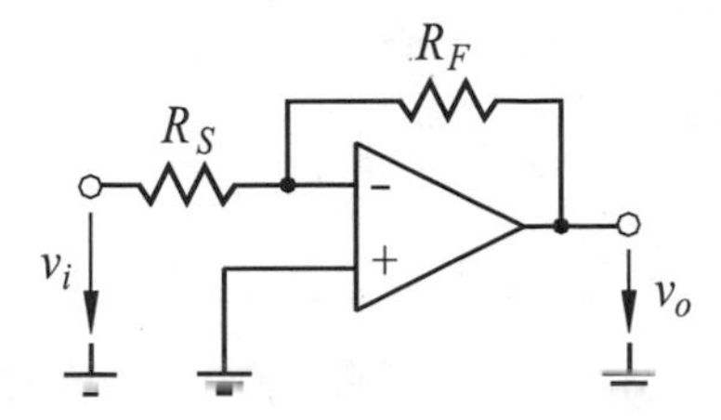

Figure 9.3 Amplifier connected as voltage follower $v_o/v_i = -R_F/R_S$

This equation demonstrates that the output signal lags the input signal by 180°. If the amplifier is connected as voltage follower (Figure 9.3) this configuration tends to excessive ringing. In order to avoid this, a frequency compensation is required.

Frequency compensation

In real amplifiers several poles exist. Fortunately, in most practical cases one can assume that one pole is dominant and one is non-dominant. Beyond the non-dominant pole there may be many other poles, but these are too far away to have an effect on the frequency performance. The circuit behaves therefore basically like a two-pole amplifier. In this case, a pole splitting compensation is possible. For this purpose a capacitor C_C is inserted between nodes (1) and (2) of the Miller amplifier (Figure 9.1). This results in the modified small-signal macro model shown in Figure 9.4. This circuit leads to a transfer function of

$$a_{dm}(j\omega) = a_{dm}(0)\frac{\left(1 - j\dfrac{\omega}{\omega_z}\right)}{\left(1 + j\dfrac{\omega}{\omega_{p,1}}\right)\left(1 + j\dfrac{\omega}{\omega_{p,2}}\right)} \tag{9.7}$$

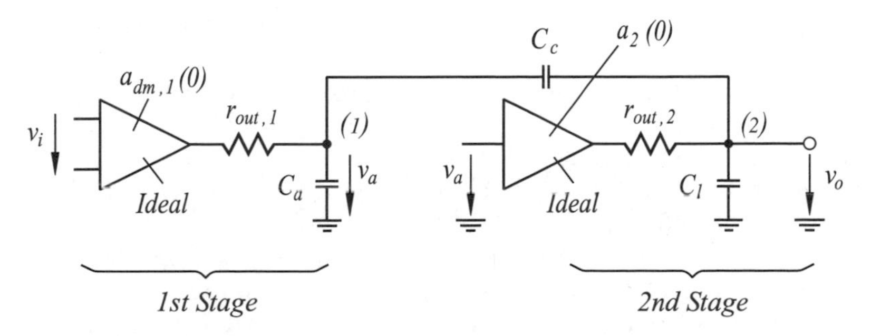

Figure 9.4 Small-signal macro model of Figure 9.2 with pole splitting capacitor C_C

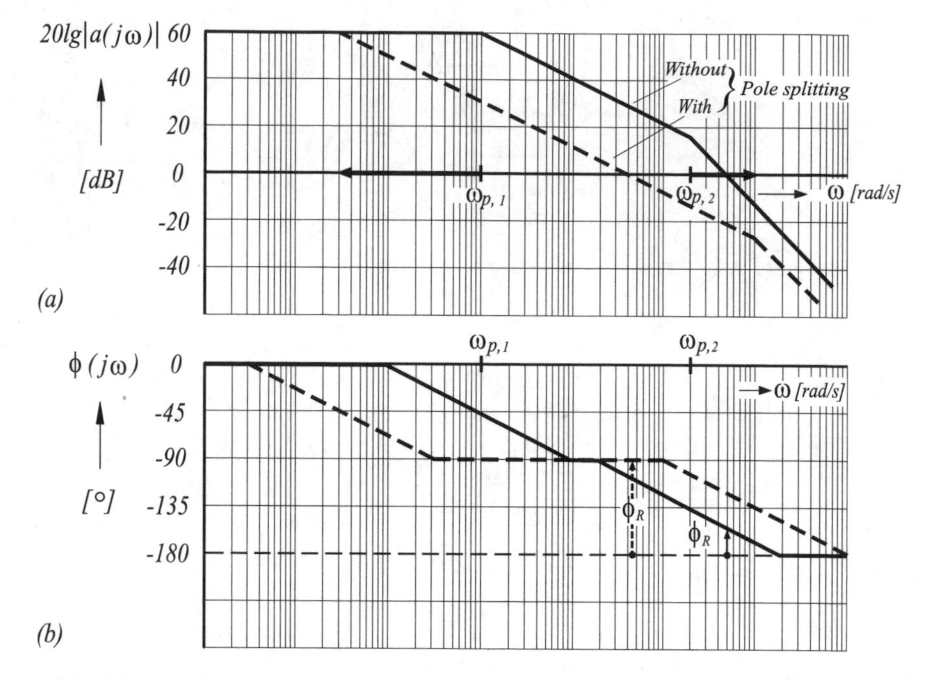

Figure 9.5 Magnitude and phase plot of Miller amplifier with and without pole splitting (zero-frequency ω_z not included)

with one zero-frequency and two pole-frequencies of

$$\omega_z = \frac{g_{m,2}}{C_c} \tag{9.8}$$

$$\omega_{p,1} \approx \frac{g_{m,n}}{a_{dm}(0)\, C_c} \tag{9.9}$$

$$\omega_{p,2} \approx \frac{g_{m,2}}{C_a C_l} \cdot \frac{1}{1/C_c + 1/C_a + 1/C_l} \tag{9.10}$$

and a differential gain $a_{dm}(0)$ given by Equation (9.3). In the derivation of this equation it is assumed that $g_m \gg g_{o,p} + g_{o,n}$ and $g_{m,2} \gg g_{o,1} + g_{o,2}$.

A zero-frequency is generated by the feed-forward coupling via C_c. The pole-frequency $\omega_{p,1}$ is influenced by the Miller effect (Section 8.3.1), which in effect increases the capacitance value C_c by a factor of $a_{dm}(0)$ and causes a reduction of the pole-frequency $\omega_{p,1}$. Simultaneously C_c causes an increase of the pole-frequency $\omega_{p,2}$. This is shown in Figure 9.5, where the influence of the zero-frequency is not included.

This compensation leads to an improved phase margin ϕ_R determined at a gain of 0 dB. The pole-splitting effect caused by C_c is also shown in the s-plane of Figure 9.6. Furthermore, a zero is included in the Right Half s-Plane (RHP). Its influence on the frequency performance is discussed next.

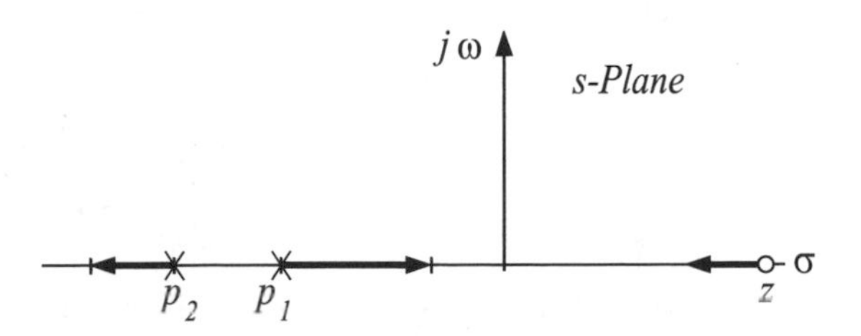

Figure 9.6 Pole-splitting effect shown in the s-plane

Zero in the right half s-plane (RHP)

When C_c is added to the circuit, the amplifier may still have excessive ringing or will oscillate when connected in a feedback configuration. This is because C_c has shifted a previously negligible zero-frequency from higher frequencies to lower frequencies. The neglected zero-frequency is caused by the signal feed-forward via the overlap capacitance C_{gdo} of transistor T_2 (Figure 9.1) to the output. Since C_c is connected in parallel to C_{gdo}, and $C_c \gg C_{gdo}$, the previously neglected zero-frequency moves from higher to lower frequencies (Figure 9.7).

The effect is that the magnitude plot of the transfer function, in the frequency range $\omega > \omega_z$, does not reduce with -20 dB/decade but is compensated instead by the zero-frequency with $20\,\mathrm{lg}[1 + (\omega/\omega_z)^2]^{1/2} = +20$ dB/decade. The zero-frequency causes further a phase shift of $\phi = -\tan \omega/\omega_z$, because of the negative sign in the numerator of Equation (9.7), which eliminates any phase margin ($\phi_R < 0°$).

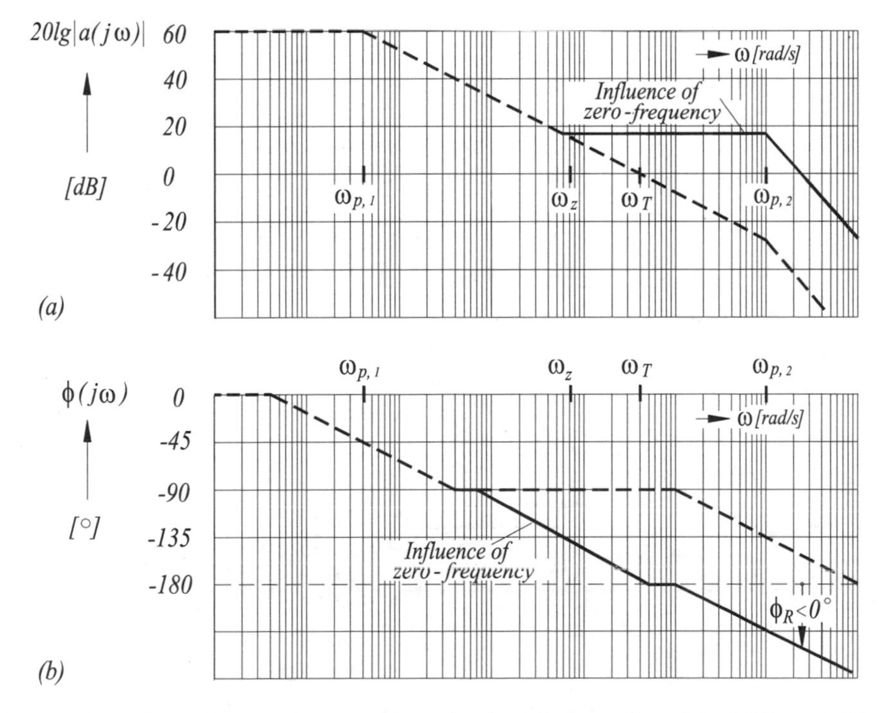

Figure 9.7 Effect of an RHP zero on magnitude and phase plot of the Miller amplifier

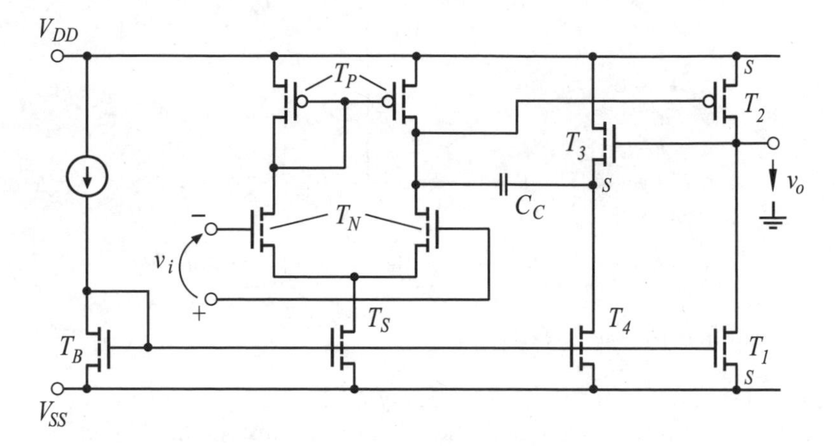

Figure 9.8 Miller amplifier with source follower to eliminate RHP zero

Eliminating the RHP zero

One way to eliminate the feed-forward effect via C_c is to add a source follower stage, consisting of the transistors T_3 and T_4, to the circuit (Figure 9.8). The pole-splitting effect is still preserved in the amplifier, provided that the gain of the source follower (Equation 8.13) is close to unity. A disadvantage of this approach is the additional power dissipation of the source follower. This can be avoided with the following solution.

Cancellation of the RHP zero with a resistor

A cancellation of the RHP zero is possible when a resistor is connected in series to the compensation capacitor (Figure 9.9). The transfer function can be found by using the

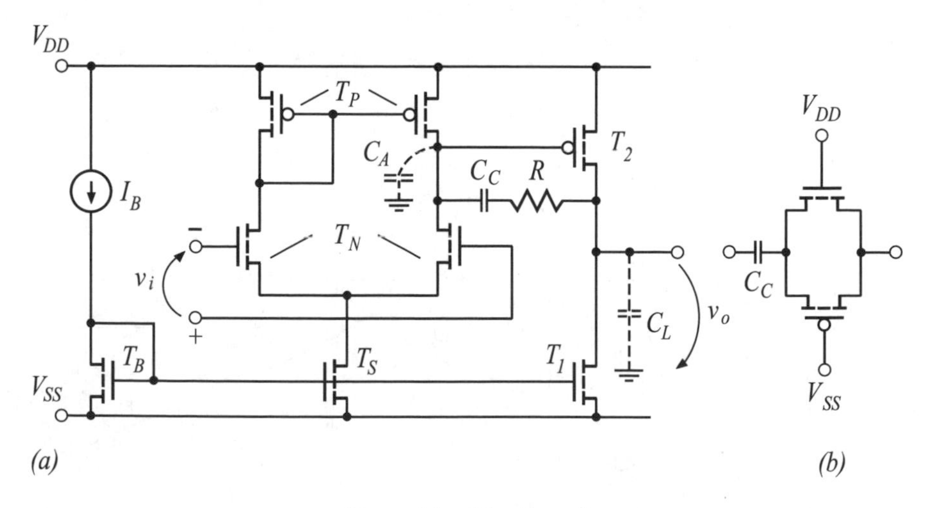

Figure 9.9 (Continued)

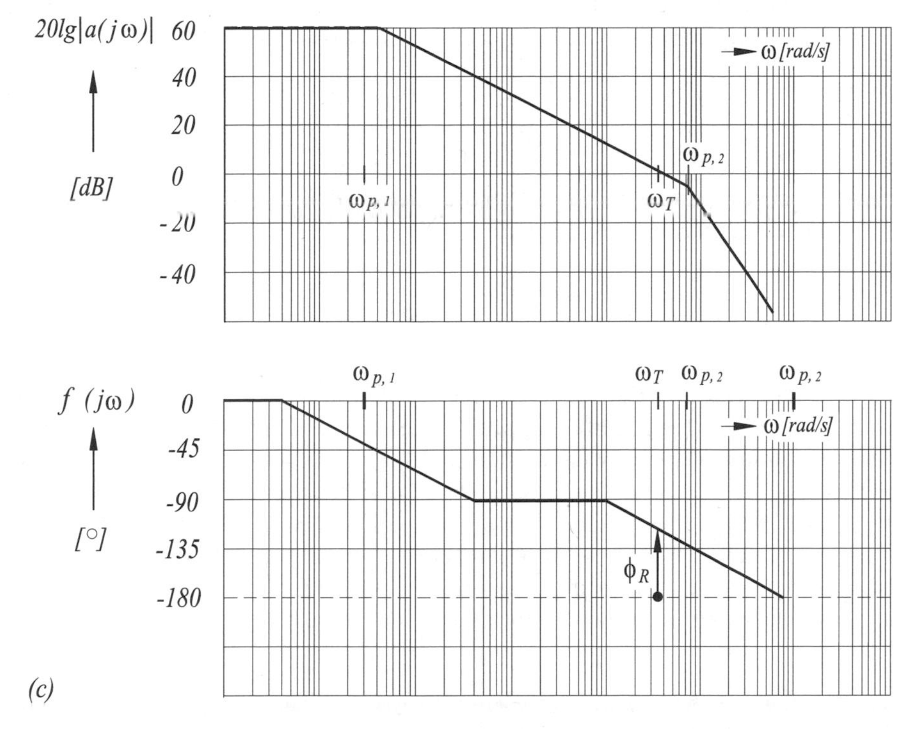

Figure 9.9 (a) RHP zero compensated Miller amplifier, (b) RHP cancellation network and (c) Bode diagram

small-signal macro model of Figure 9.4, where a small-signal resistor r is connected in series to C_c. Solving the current voltage equations, this leads to

$$a_{dm}(j\omega) = a_{dm}(0)\frac{\left(1 + j\dfrac{\omega}{\omega_z}\right)}{\left(1 + j\dfrac{\omega}{\omega_{p,1}}\right)\left(1 + j\dfrac{\omega}{\omega_{p,2}}\right)\left(1 + j\dfrac{\omega}{\omega_{p,3}}\right)} \tag{9.11}$$

where the zero-frequency has a value of

$$\omega_z = \frac{1}{C_c\left(r - \dfrac{1}{g_{m,2}}\right)} \tag{9.12}$$

This zero is now located in the LHP. If $r = 0$ the zero moves back to its original location in the RHP. The pole-frequencies, Equations (9.9) and (9.10)

$$\omega_{p,1} \approx \frac{g_{m,n}}{a_{dm}(0)C_c}$$

$$\omega_{p,2} \approx \frac{g_{m,2}}{C_a C_l} \cdot \frac{1}{1/C_c + 1/C_a + 1/C_l}$$

remain unchanged by the resistor r, as does the differential gain $a_{dm}(0)$, described by Equation (9.3). But the resistor has generated an additional pole-frequency

$$\omega_{p3} = \frac{1}{r}\left(\frac{1}{C_c} + \frac{1}{C_a} + \frac{1}{C_l}\right) \tag{9.13}$$

The zero-frequency (Equation 9.12) can be cancelled or, in other words, moved to infinity, when a small-signal resistor with a value of

$$r = \frac{1}{g_{m,2}} \tag{9.14}$$

is chosen. Independent of the resistor value the advantage of the pole-splitting effect remains preserved. In this circuit an approximate symmetrical resistor of the compensation network is implemented by parallel-connected n-channel and p-channel transistors. For the capacitor the capacitance between two polysilicon layers is used. In most designs usually a C_C value close to that of the load capacitance C_L is chosen.

An alternative cancellation of the RHP zero is possible with a pole-zero compensation, similar to the one described in Appendix A of Chapter 8. In this case $\omega_z = \omega_{p,2}$, which leads, according to Equations (9.10) and (9.12), to a resistor value of

$$r \approx \frac{1}{g_{m,2}C_c}(C_l + C_c) \tag{9.15}$$

This pole-zero compensation is critical to $g_{m,2}$ variations at low power-supply voltages and is therefore not pursued further.

As is stated in Chapter 8 at several occasions, for the design of an amplifier knowledge of the dependence of key amplifier parameters on the drain current and thus operation point of the circuit is of the utmost importance. A summary showing these relationships, based on Equations (9.3), (9.9), (9.10), and the conductances of Table 8.1, is given below.

$$a_{dm}(0) = \frac{g_{m,n}}{g_{o,n} + g_{o,p}} \frac{g_{m,2}}{g_{o,1} + g_{o,2}} \sim I_{DS}^{-1} \tag{9.16}$$

$$\omega_{p,1} \approx \frac{g_{m,n}}{a_{dm}(0)C_c} \sim I_{DS}^{3/2} \tag{9.17}$$

$$\omega_{p,2} \approx \frac{g_{m,2}}{C_l} \sim I_{DS}^{1/2} \tag{9.18}$$

$$\omega_T = a_{dm}(0)\omega_{p,1} \approx \frac{g_{m,n}}{C_c} \sim I_{DS}^{1/2} \tag{9.19}$$

Furthermore, the relationship between the unity-gain frequency and the pole-frequency, as described by Equation (8.24), is included. Additionally, it is assumed that $C_c = C_l$

and that the pole-frequency $\omega_{p,3}$ is at such a high frequency that basically no influence on the frequency performance exists.

These equations indicate the same behavior as found by the simple amplifier stage of Section 8.3, Equations (8.25) to (8.27), which is typical for MOS amplifiers. The gain is largest at small drain currents and the -3 dB bandwidth and unity-gain frequency is largest at high drain currents. Thus, a compromise between the conflicting requirements has to be made in the design phase of the amplifier.

In order to reduce damped ringing of the amplifier, the phase margin should be $\phi_R > 60°$. This leads to the requirement (Figure 9.9(c)) that

$$\omega_{p,2} > 2\omega_T \tag{9.20}$$

With $C_c = C_l$, Equations (9.18) and (9.19) yield that a transconductance of

$$g_{m,2} > 2g_{m,n} \tag{9.21}$$

is required. In real designs usually a relationship of $g_{m,2} \approx 3g_{m,n}$ is selected. Considering the equations further, it becomes obvious that the ratio of $\omega_{p,2}/\omega_T$ is independent of the current, which results in a relatively stable amplifier operation independent of parameter variations.

Slew rate

When a large input voltage step V_i is applied to the amplifier, connected as voltage follower (Figures 9.3) the output will be slew rate limited, as indicated in Figure 9.10. The reason for this behavior is that the current which charges C_C is limited by the current I_S of the differential stage. If an input voltage step is applied, $T_{N,1}$ becomes conducting and $T_{N,2}$ non-conducting. Since the current of $T_{N,1}$ is mirrored to $T_{P,2}$, the capacitor is charged by I_S. If the situation is reversed, with $T_{N,2}$ conducting and $T_{N,1}$ non-conducting, then the capacitor is discharged by I_S. If one assumes that the

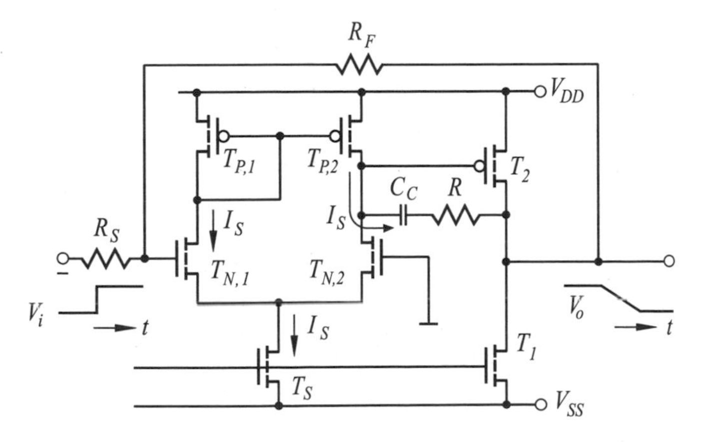

Figure 9.10 Miller amplifier in voltage follower configuration

charge or discharge is not limited by the resistor, or the output stage, then a slew rate of

$$\boxed{\mathrm{SR} = \frac{\mathrm{d}u_o}{\mathrm{d}t} \approx \frac{I_S}{C_C}} \tag{9.22}$$

results.

Design requirement

Usually, only a few performance requirements, such as minimum unity-gain frequency, slew rate, low-frequency gain, CMRR or maximum output load, dictate certain relationships in the design. This is in addition to the already fixed ones.

With a required unity-gain frequency and the known capacitance load, the following parameters (Equations 9.19, 9.21, and 9.14) are already defined

$$\begin{aligned} g_{m,n} &\approx C_c \omega_T \\ C_l &= C_c \\ g_{m,2} &\approx 3 g_{m,n} \\ r &= 1/g_{m,2} \end{aligned}$$

A slew-rate requirement (Equation 9.22) leads to the needed current I_S of the current source. A low-frequency gain specification (Equation 9.3) can be used to find via the output conductances

$$(g_{o,n} + g_{o,p})(g_{o,1} + g_{o,2}) = g_{m,n}\, g_{m,2}/a_{dm}(0) \tag{9.23}$$

the gate length of the transistors. The CMRR requirement (Equation 8.42) yields the output conductance of the current source

$$g_{o,s} \approx 2 \frac{g_{m,n}\, g_{m,p}}{g_{o,n}\mathrm{CMRR}(0)} \tag{9.24}$$

and thus the gate length of this transistor.

The most important relationships and typical performance parameters are summarized in Table 9.1.

Output stage

In order to have a phase margin of $\phi_R > 60°$, the capacitances are chosen such that $C_c = C_l$. This limits the driving capability of the amplifier to relative small capacitance values. An improvement can be achieved when a source follower (Section 8.2) is added to the Miller amplifier (Figure 9.11). T_2 acts as current source and T_3 as source follower.

Table 9.1 Typical performance parameter of a Miller amplifier ($V_{DD} = 2.5\,\text{V}$, $V_{SS} = -2.5\,\text{V}$)

Parameter/Relationship	Example
$a_{dm}(0) = \dfrac{g_{m,n}}{g_{o,n} + g_{o,p}} \dfrac{g_{m,2}}{g_{o,1} + g_{o,2}}$	75 dB
$f_T \approx g_{m,n}/2\pi C_c$	3 MHz
$\text{SR} \approx I_S/C_c$	4 V/μs
$\text{CMRR}(0) = 2\, g_{m,n}\, g_{m,p}/g_{o,s}\, g_{o,n}$	80 dB
ϕ_R	60°
Load	10 pf

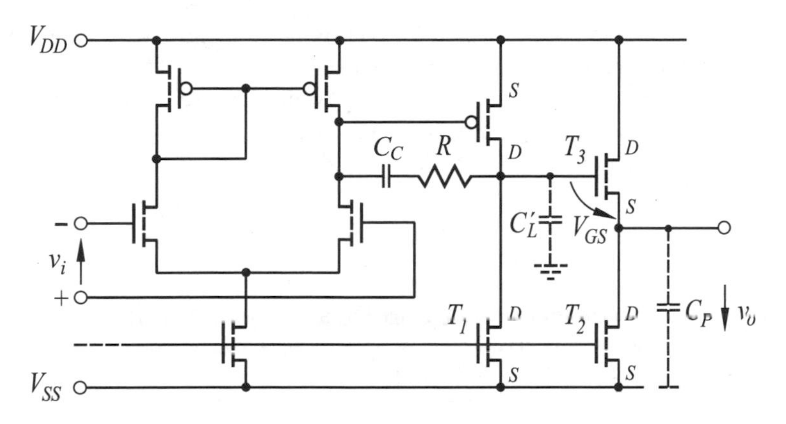

Figure 9.11 Miller amplifier with source follower output stage

The source follower has a gain (Equation 8.13) of approximately one and a low output resistance (Equation 8.17) of

$$r_{out} \approx 1/g_{m,3} \tag{9.25}$$

Since the dominant pole-frequency of the circuit $\omega_{p,1}$ (Equation 9.17) and the low-frequency gain (Equation 9.16) are unaffected by the source follower, this applies also to the unity-gain frequency (Equation 9.19). But what has changed is the pole-frequency $\omega_{p,2}$, which has moved to very high frequencies, since C_L' is assumed to be much smaller than C_L of Figure 9.9. When $\omega_{p,2}$ is much larger than ω_T, the influence of this pole-frequency is negligible. But now there exists a new pole-frequency at the output given by

$$\omega_{p,out} \approx g_{m,3}/C_p \tag{9.26}$$

which replaces the previously considered $\omega_{p,2}$ one. With the need of a large phase margin (Equations 9.20 and 9.21)

$$\begin{aligned} \omega_{p,out} &> 2\omega_T \\ g_{m,3}/C_p &> 2g_{m,n}/C_c \end{aligned} \tag{9.27}$$

this leads to the following required small-signal output resistance of the source follower stage

$$r_{out}^{-1} \approx g_{m,3} > 2g_{m,n}C_p/C_c \tag{9.28}$$

A drawback of the source follower is definitely the increased power dissipation by this stage and the reduced output swing due to the V_{GS} voltage drop at transistor T_3. A way of solving these problems is discussed in the next section.

9.2 FOLDED CASCODE AMPLIFIER

In the Miller amplifier no provision exists to reduce the Miller effect, described in Section 8.3.1. This can be accomplished by changing the input stage to a cascode one by adding the transistors T_3 and T_2 (Figure 9.12).

Additionally, an improved p-channel current source is used. This is the p-channel version of the one shown in Figure 8.7(c). As a result, the low-frequency gain of the Miller amplifier stage changes from (Equation 8.40)

$$a_{dm}(0) \approx g_{m,n}(g_{o,n} + g_{o,p})^{-1} = g_{m,n}\, r_{out}$$

to a higher one of

$$a_{dm}(0) \approx g_{m,n}\left(\frac{g_{o,6}\, g_{o,1}}{g_{m,1}} + \frac{g_{o,n}\, g_{o,2}}{g_{m,2}}\right)^{-1} = g_{m,n}\, r_{out,f} \tag{9.29}$$

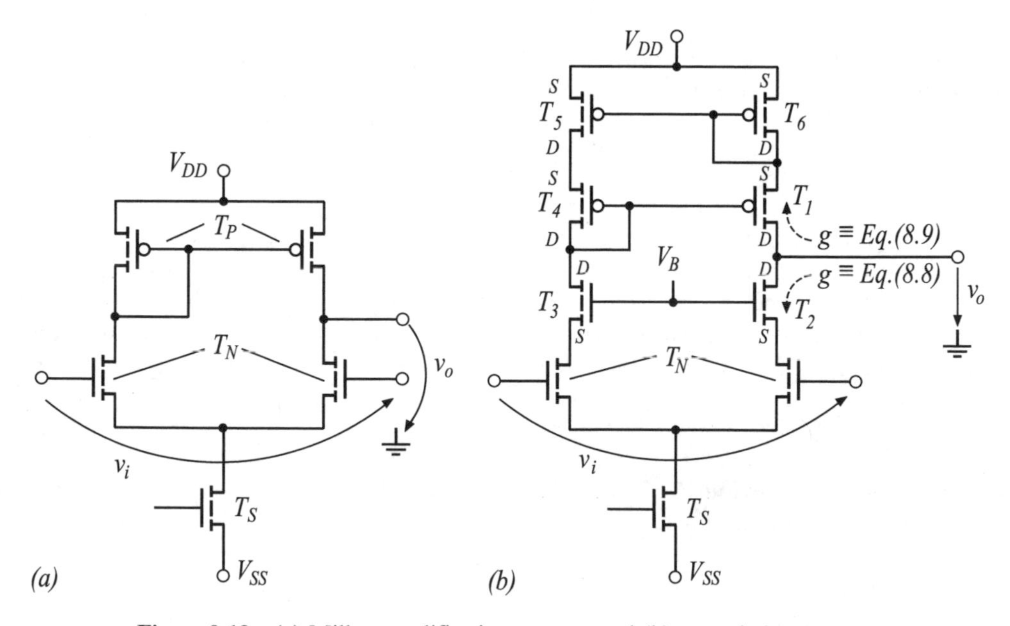

Figure 9.12 (a) Miller amplifier input stage and (b) cascode input stage

The output conductance is found by adding the conductances resulting from T_1, T_6, and from T_2, T_N in accordance and analogy to Equations (8.9) and (8.8).

A drawback of this stage is the relatively large power-supply voltage needed. This can be circumvented by folding the p-channel current source from V_{DD} to V_{SS} (Figure 9.13). For this purpose, n-channel transistors are used as current source, which is now identical to the one of Figure 8.5(c). The n-channel cascode transistors T_2 and T_3 of Figure 9.12(b) are exchanged to p-channel ones with their sources connected to the differential input transistors. The additional transistors T_P are used for current supply. If one assumes that these transistors have a relatively large channel length and thus a low output conductance, then the gain of the folded cascode amplifier is given by Equation (9.29).

The dominant pole-frequency of the circuits exists at the output

$$\omega_p = \frac{1}{C_l\, r_{aus,f}}$$
$$\omega_p = \frac{1}{C_l}\left(\frac{g_{o,6}\, g_{o,1}}{g_{m,1}} + \frac{g_{o,n}\, g_{o,2}}{g_{m,2}}\right) \tag{9.30}$$

when the loading capacitance C_l is relatively large. This leads to a transfer function of

$$a(j\omega) = a_{dm}(0)\frac{1}{1 + j\dfrac{\omega}{\omega_p}} \tag{9.31}$$

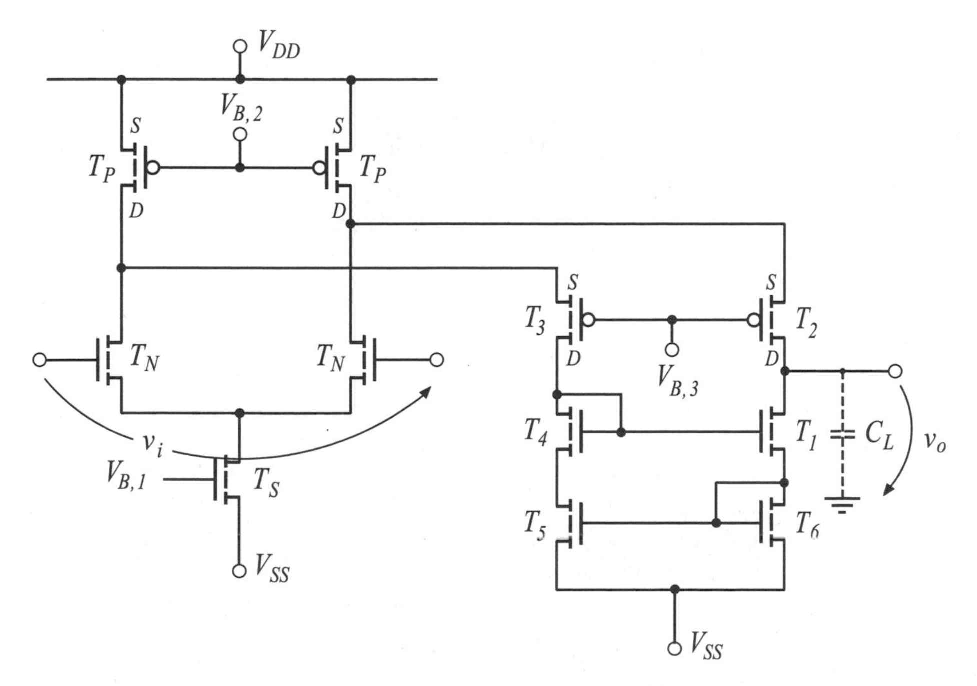

Figure 9.13 Folded cascode amplifier

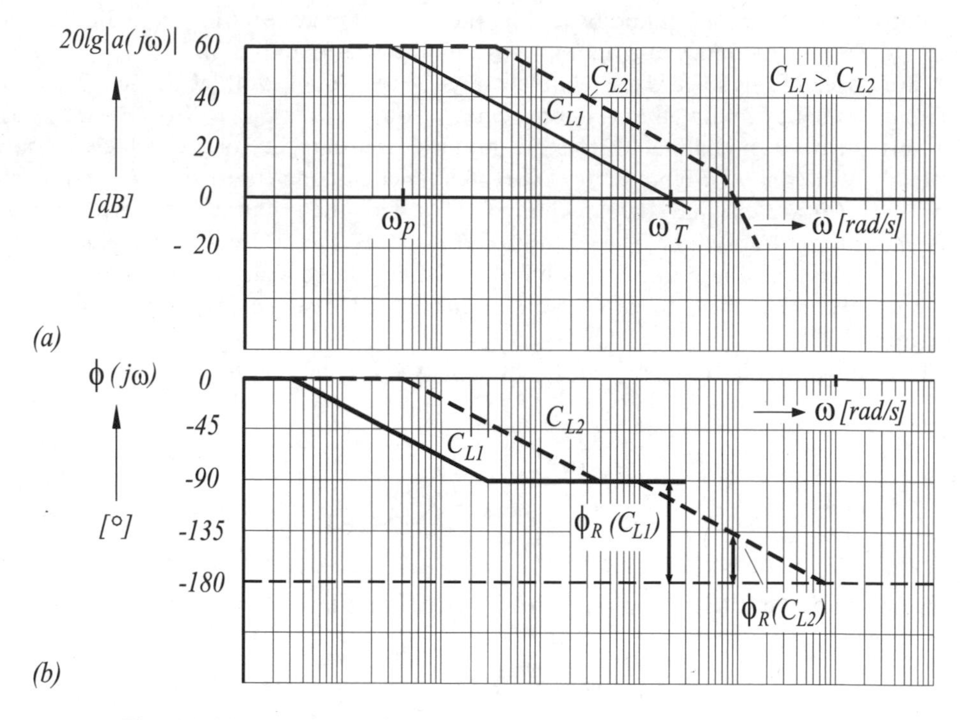

Figure 9.14 Bode plot of the folded cascode amplifier with $C_{L1} > C_{L2}$

The amplifier has one dominant pole-frequency only and is thus very stable with a phase margin $\phi_R(C_{L1})$ of 90° (Figure 9.14).

Only when the loading capacitance is small, so far neglected pole-frequencies are able to reduce the phase margin to $\phi_R(C_{L2})$. This is a considerable difference and an advantage compared to the discussed Miller amplifier.

9.3 FOLDED CASCODE AMPLIFIER WITH IMPROVED DRIVING CAPABILITY

In this section an example of a low voltage folded cascode amplifier with a class *AB* output driver is presented (Hogervorst *et al.* 1994). An overview of the schematic is shown in Figure 9.15. The input stage consists of a folded cascode circuit, using as differential inputs p-channel transistors instead of n-channel ones. This is the only way that source and bulk terminals can be connected in an n-well process (Section 5.3.3, Figure 5.14) in order to eliminate the body effect. To ease the circuit schematic representation, symbols are used for current sources. Depending on the power-supply voltage available, these current sources can be implemented by an appropriate circuit described in Section 8.1.1. A class *AB* output stage is used for the reduction of the power dissipation. To guarantee a class *AB* operation, a biasing network is required which is almost independent of the power-supply voltage and parameter variations.

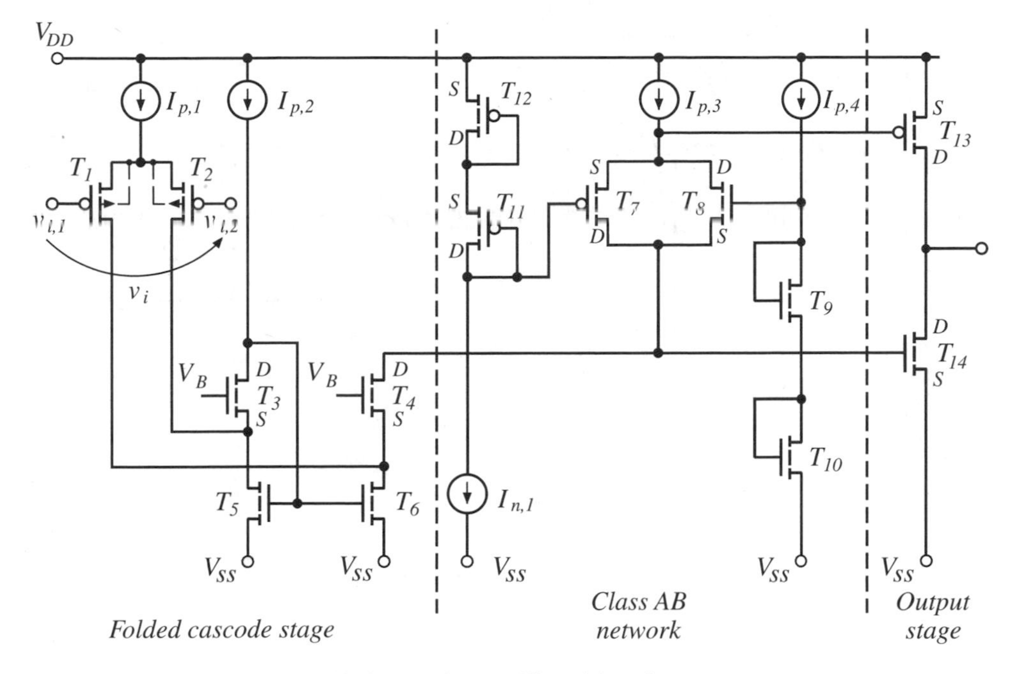

Figure 9.15 Folded cascode amplifier with a class *AB* output stage

Folded cascode stage

This circuit consists of the two input transistors T_1 and T_2 and a so-called wrap-around circuit, which is very advantageous at a low power-supply voltage (Figure 9.16).

The floating voltage source V_{AB} is used to define the operation conditions of the output transistors. This voltage source is implemented by the transistors T_7 to T_{12}, which is considered after the cascode stage is analyzed. The V_{AB} voltage source has no effect on the small-signal behavior of the circuit, since it can be assumed that its small-signal resistance is zero.

One way to explain the function of the wrap-around circuit is to assume that an input voltage v_i causes the shown current changes ΔI. The increased I_2 current causes a reduced $V_{GS,3}$ voltage, which leads to an increased $V_{GS,5}$ voltage. Since the gates of T_5 and T_6 are connected, the current of T_6 increases, which causes an increase of the $V_{GS,4}$ voltage. This increase is boosted by a reduced I_1 current. The increased $V_{GS,4}$ voltage leads to an increased $I_{4,6}$ current and a reduced output voltage v_a. When the input signal is reversed, then v_a increases.

In order to determine, e.g., the low-frequency gain of the circuit, one can use the small-signal equivalent model of the circuit and solve the current voltage equations. This is, from a particular number of transistors, an elaborate effort. To cope with these situations, CAD programs are recommended to be used (e.g., Gielen and Sansen 1991; Gielen *et al.* 1989).

Under the assumption that the following transistor pairs are implemented, $T_1 = T_2$, $T_3 = T_4$, $T_5 = T_6$, and that the currents $I_{p,2}$ and $I_{p,3}$ are equal, a low-frequency gain of the first stage

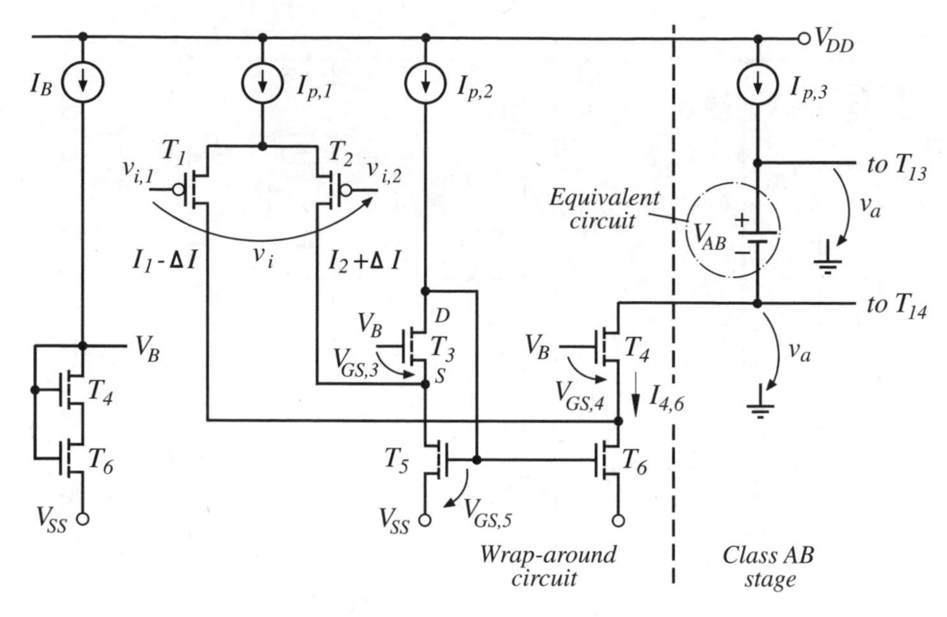

Figure 9.16 Folded cascode stage

$$a_{dm,1}(0) = v_a/v_i \approx -\frac{g_{m,1}\,g_{m,3}}{(g_{o,1}+g_{o,5})g_{o,3}} \tag{9.32}$$

can be determined.

The low-frequency gain of the output stage (Figure 9.17) is the result of the ratio of the individual transconductances to output conductances

$$a_2(0) = v_o/v_a = -\frac{g_{m,13}+g_{m,14}}{g_{o,13}+g_{o,14}} \tag{9.33}$$

The product of the low-frequency gains leads to a total amplifier gain of

$$a_{dm}(0) \approx \frac{g_{m,1}\,g_{m,3}}{(g_{o,1}+g_{o,5})g_{o,3}}\frac{g_{m,13}+g_{m,14}}{g_{o,13}+g_{o,14}} \tag{9.34}$$

AB output stage

In Figure 9.16 a floating voltage source V_{AB} is used to guarantee that the output transistors T_{13} and T_{14} operate under all circumstances in class *AB* operation. This saves a substantial amount of current compared to a class *A* operation needed, e.g., for the source follower output shown in Figure 9.11. The principle *AB* output stage is shown in Figure 9.17.

The voltage source V_{AB} is floating. That means current $I_{p,3}$ flows through the V_{AB} voltage source via T_4 and T_6 to V_{SS}. If the current through T_4 and T_6 is larger than $I_{p,3}$,

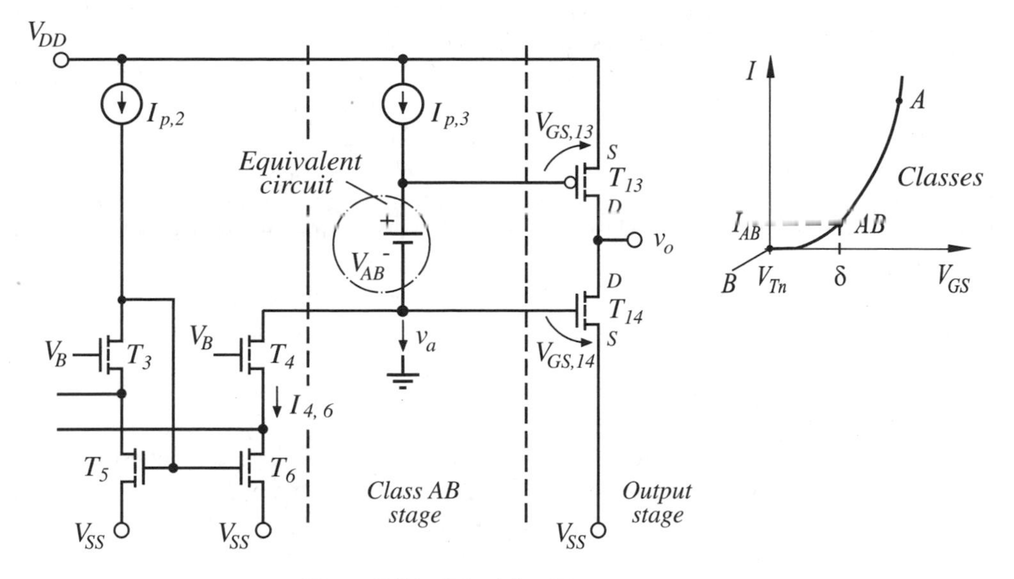

Figure 9.17 Principle *AB* output stage

the voltage source moves toward V_{SS}. This leads to a decrease of the $V_{GS,14}$ voltage and an increase of the $V_{GS,13}$ voltage, and thus to an increased current through T_{13} and an decreased one through T_{14}, whereas in the other case a reverse situation exists. The class *AB* network generating the V_{AB} voltage source is shown in Figure 9.18.

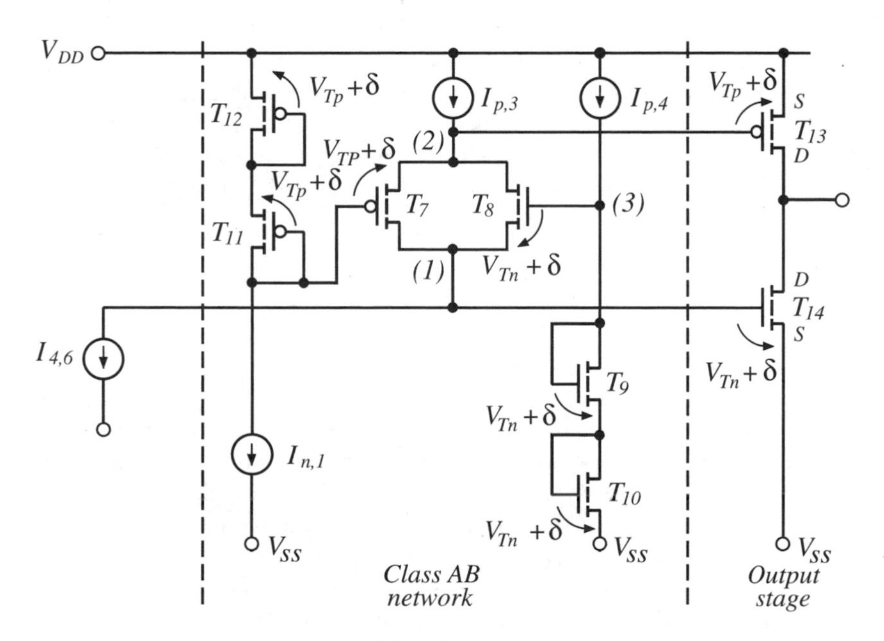

Figure 9.18 Class *AB* network

To ease the discussion it is assumed that at all transistors a gate voltage of $V_{GS} = V_{Tn(p)} + \delta$ exists. δ describes the overdrive voltage of the transistors, which is positive or negative depending on whether a n-channel or p-channel transistor is considered. With this assumption, the voltage at node (3) has a value of $2(V_{Tn} + \delta)$ and the voltage at node (1) one of $V_{Tn} + \delta$. Therefore, the output transistor T_{14} has an overdrive voltage of δ. Analogous to this situation is that of the p-channel transistors, where T_{13} also has an overdrive voltage of δ. This voltage determines therefore how far the operation point (AB) of the output transistors is located in the active region (inset of Figure 9.17). The nice feature of the AB network is that absolute parameter variations have no influence on the overdrive voltage. A further feature is that a rail to rail swing at the output is possible, which is considered next.

(a) Input voltage conditions $v_{i,1} \gg v_{i,2}$
 In this case $I_2 \gg I_1$ (Figure 9.16) (T_1 and T_2 are p-channel transistors) and therefore $I_{4,6} \gg I_{p,3}$ (Figure 9.18). At node (1) a voltage of $V_{GS,14} < V_{Tn}$ exists, causing T_{14} to cut off and simultaneously T_8 and thus T_{13} to turn on. An output voltage of V_{DD} results.

(b) Input voltage condition $v_{i,1} \ll v_{i,2}$
 In this case $I_2 \ll I_1$ and therefore $I_{4,6} \ll I_{p,3}$. At node (2) (Figure 9.18) a voltage larger than $V_{DD} - |V_{Tp}|$ exists. This causes T_{13} to be cut off and T_7 and T_{14} to be turned on. An output voltage of V_{SS} results.

Note: Studying Chapters 8 and 9 reveals that knowledge of the transfer function with the pole-frequencies and zero-frequencies is a prerequisite for the successful design of an amplifier. With the aid of a computer-aided AC analysis, the Bode plot or the location of the poles and zeros can be found but not necessarily the parameters, e.g. w/l-ratios, threshold voltages, gain factors or currents, influencing their dependence. One could start drawing the small-signal equivalent circuit and solve the current–voltage relationships in order to find this dependence, but from a particular number of transistors the complexity is too great. In this case it is recommended to use programs to find the required relationships (Gielen and Sansen 1991; Gielen *et al.* 1989).

Summary

A step-by-step Miller amplifier design is discussed. It turns out that the low-frequency gain is largest at small drain currents and the $-3\,\text{dB}$ bandwidth, and the unity-gain frequency is largest at high drain currents. This is the same behavior found by the simple amplifier stage of Chapter 8, which is typical of MOS amplifiers. Thus, a compromise between the conflicting requirements has to be made in the amplifier design. Since this amplifier leads to instabilities, a pole splitting with a capacitor is employed and an RHP zero is eliminated by adding a resistor. This in effect changes the amplifier into a two-pole version. The Miller effect can be reduced when a folded cascode amplifier is used. This version has an improved phase margin compared to the Miller amplifier. For driving large capacitance loads at a small power-supply voltage, a modified folded cascode amplifier with a rail to rail *AB* output stage is recommended.

Problems

9.1 Derive the low-frequency small-signal output resistance equation of the improved current source shown below. The transistors T_3 and T_2 are assumed to be identical.

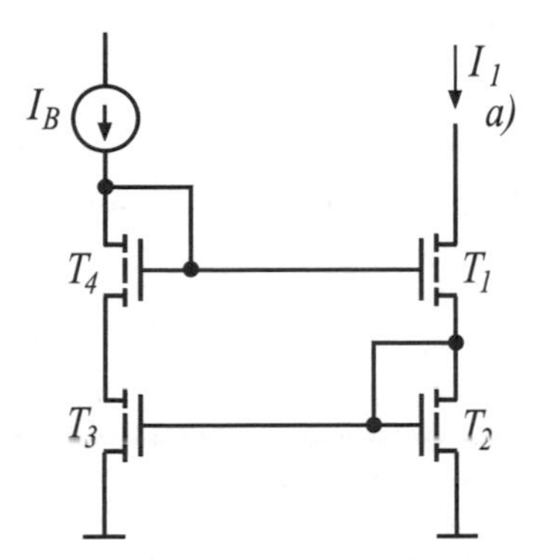

9.2 Determine the low-frequency gain of the amplifier stage shown below. Transistor T_5 operates in current saturation. The data are: β_p (T_1 to T_4) $= 800\,\mu A/V^2$, $\lambda(T_1$ to $T_5) = 0.01\,V^{-1}$, $\beta_5 = 5000\,\mu A/V^2$. It can be assumed that $V_{DS}\lambda \ll 1$.

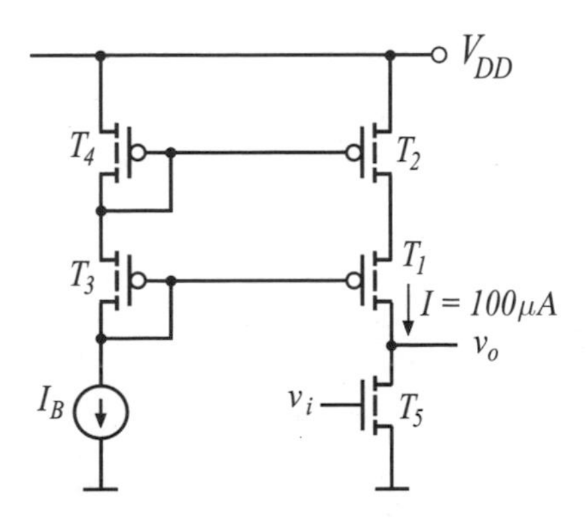

9.3 The folded cascode amplifier of Figure 9.15 uses the wrap-around stage shown below. Determine the V_B voltage range, guaranteeing that both transistors operate in current saturation.

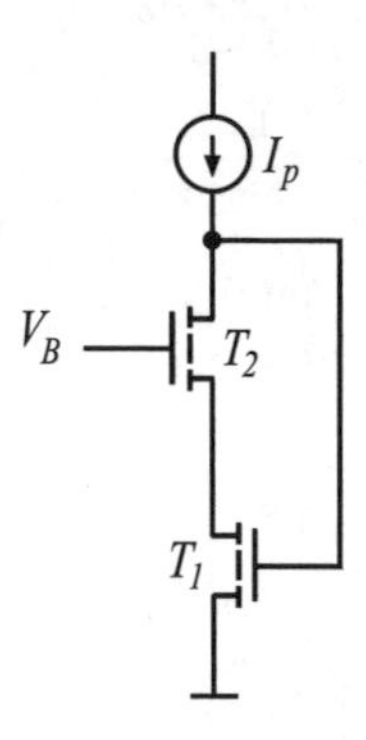

The solutions to the problems can be found under: *www.unibw-muenchen.de/campus/ET4/index.html*

REFERENCES

G. Gielen and W. Sansen (1991) *Symbolic Analysis for Automated Design of Analog Integrated Circuits*. Boston, MA: Kluwer Academic.

R. Hogervorst, J.P. Tero, R.G.H. Eschauzier and J.H. Huijsing (1994) A compact power-efficient 3V CMOS rail-to-rail input/output operational amplifier for VLSI cell libraries. *IEEE J. Solid-State Circuits*, **29**(12), 1505–12.

G. Gielen, H.C.C. Walscharts and W.M.C. Sansen (1989) ISAAC: a symbolic simulator for analog integrated circuits. *IEEE J. Solid-State Circuits*, **24**(6), 1587–97.

For further reading see Chapter 8.

10

BICMOS

Chronologically, the first integrated digital and analog circuits were bipolar ones. During the following years these circuits approached better and better performances, but simultaneously the NMOS and later the CMOS technique started to dominate. This was mainly caused by the high packaging density and the higher yield of MOS technologies and the reduced power dissipation of the circuits. Despite this fact, bipolar circuits still have an advantage in analog and high speed applications. That is the reason why the combination of both technologies in a BICMOS manufacturing process is the optimum from a design point of view (Figure 10.1).

BICMOS applications are, e.g. fast memories and microcontrollers, and, of particular importance, wire and wireless communicating systems.

Before starting with the detailed discussion it is useful to revisit the current–voltage characteristic of both transistors (Figure 10.2). The MOS transistor has in the subthreshold region $V_{GS} < V_{Tn}$ an exponential behavior which changes into a quadratic one when $V_{GS} > V_{Tn}$. Contrary is the situation of the bipolar transistor, which shows an exponential behavior in the entire V_{BE} region until the transistor approaches the knee current I_K at the onset of strong injection. How these different characteristics and other features influence the circuit performance is the topic of this chapter.

10.1 CURRENT STEERING TECHNIQUES

In Section 6.3 a current steering technique with MOS transistors is described which achieves a high switching speed by reducing the voltage swing. This is a classical approach at first implemented with bipolar transistors and called current mode logic (CML).

10.1.1 CML Circuits

The basic element is the current switch (Figure 10.3). The use of resistors instead of p-channel transistors – as is usually the case in MCML circuits (Section 6.3) – has the advantage that parasitic capacitances can be reduced and higher clock frequency be achieved.

System Integration: From Transistor Design to Large Scale Integrated Circuits. Kurt Hoffmann.
 ISBN: 0-470-85407-3

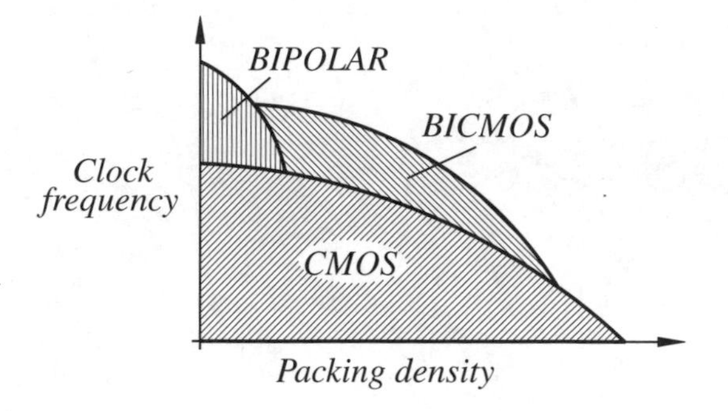

Figure 10.1 Relationship between packing density and clock frequency for different technologies

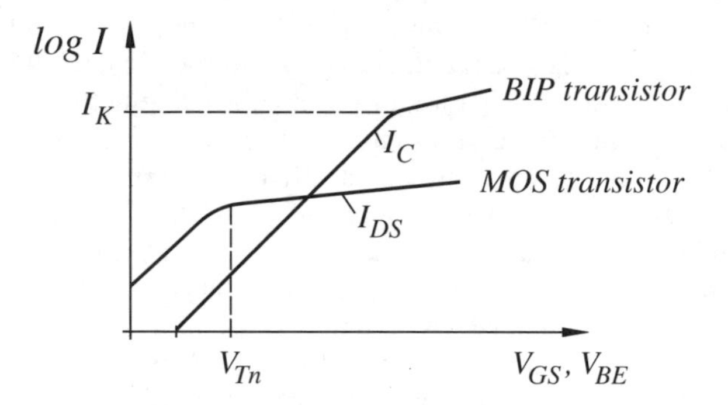

Figure 10.2 Current–voltage characteristics of BIP transistor and MOS transistor

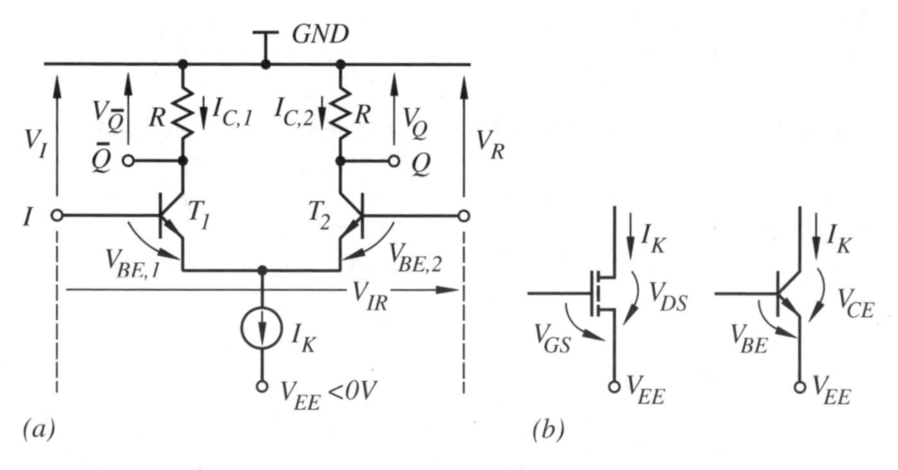

Figure 10.3 (a) Current switch and (b) current sources

In this case, the most positive terminal of the power supply, ground (*GND*), is used as reference point. This has the advantage that the negative input and output levels are dependent on currents only and not on the negative power-supply voltage V_{EE}. The current source can be implemented by one of the implementations discussed in Section 8.1

or by an equivalent bipolar configuration. Since the current sources have to operate in current saturation, the voltage requirements $V_{DS} \geq V_{GS} - V_{Tn}$ or $V_{CE} \geq V_{BE}$ have to be met. The MOS solution is preferable since V_{DS} can be designed smaller than V_{CE}, which is advantageous at a small power-supply voltage.

If a signal of $V_{IL} < V_R$, where V_R is a reference voltage, exists at the input, then in the ideal case, transistor T_1 is non-conducting and T_2 is conducting since $V_{BE,2} > V_{BE,1}$. This voltage condition steers the total current of the current source I_K through transistor T_2, when the base current is neglected. A voltage drop at the output Q of

$$V_Q = V_{QM} = -I_K R \tag{10.1}$$

occurs, whereas the voltage at output $\overline{Q}$ is zero. If $V_{IH} > V_R$ the opposite situation exists. In order to have a symmetrical voltage transfer characteristic, the reference voltage at T_2 is placed in the middle between V_{IH} and V_{IL}

$$V_R = \frac{V_{IH} + V_{IL}}{2} \tag{10.2}$$

The voltage transfer characteristic of the current switch shows some interesting relationships which are considered in the following.

At the input a differential voltage of

$$\begin{aligned} V_{IR} &= V_{BE,1} - V_{BE,2} \\ &= \phi_t \left(\ln \frac{I_{C,1}}{I_{SS}} - \ln \frac{I_{C,2}}{I_{SS}} \right) \\ &= \phi_t \ln \frac{I_{C,1}}{I_{C,2}} \end{aligned} \tag{10.3}$$

exists. In this equation symmetrical transistors with identical transport currents I_{SS} are assumed. The output voltage has a value of

$$V_Q = -I_{C,2} R \tag{10.4}$$

When the base currents are small and neglected, the relationship

$$I_{C,1} + I_{C,2} = I_K \tag{10.5}$$

is valid. These three equations together with Equation (10.1) yield a voltage transfer function of

$$\boxed{V_Q = V_{QM} \left(1 + e^{V_{IR}/\phi_t} \right)^{-1}} \tag{10.6}$$

which is plotted in normalized form for different V_{QM} values in Figure 10.4. This function demonstrates a voltage gain of

$$G = \frac{dV_Q}{dV_{IR}} = -V_{QM} \frac{e^{V_{IR}/\phi_t}}{\phi_t(1 + e^{V_{IR}/\phi_t})^2} \tag{10.7}$$

which is smallest at a low V_{QM} voltage, since the transistors operate in the flatter region of the I_C (V_{BE}) function. This limits the logic levels to about $4\phi_t = 4kT/q$ (Treadway 1989). The maximum possible voltage difference between the logic levels depends on the kind of CML cascading (Figure 10.5).

Between the outputs a voltage difference of

$$V_Q - V_{\overline{Q}} = V_{BC} \tag{10.8}$$

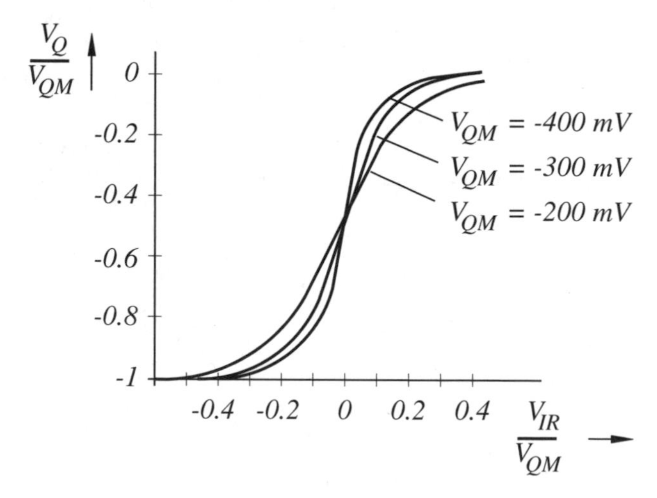

Figure 10.4 Normalized transfer function of a CML circuit with V_{QM} as parameter at room temperature

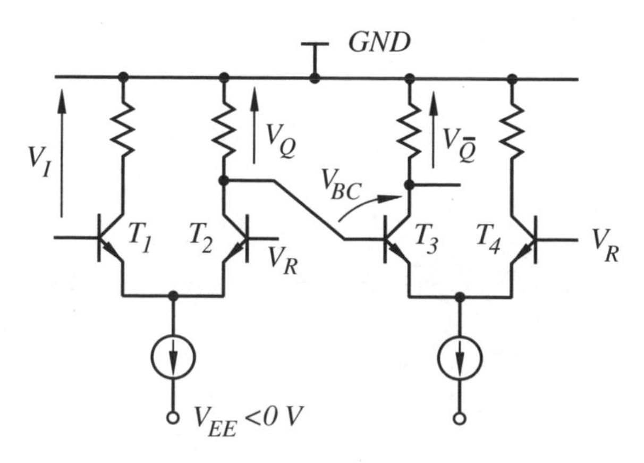

Figure 10.5 Cascaded current switches

exists, which can be positive or negative depending on the logic conditions of the input signal V_I. The positive voltage has to be limited to such a value that the transistor, in this case T_3, operates in weak voltage saturation (Section 3.2.3) in order to avoid an adverse effect on the switching performance of the CML circuit. This requirement leads to a voltage difference between the logic levels of $V_Q - V_{\overline{Q}} \approx \pm 300\,\text{mV}$, when the internal collector resistances (Figure 3.36) are negligible. An example of cascaded CML circuits with typical voltages is shown in Figure 10.6. In this case the required power-supply voltage is

$$V_{EE} = V_{SD} + V_{EB} + \Delta V \tag{10.9}$$

where V_{SD} is the saturation voltage of the MOS transistor used as current source. With the values of Figure 10.6 a power-supply voltage of $V_{EE} = -1.3\,\text{V}$ results.

The described CML circuit can be changed in a multi-input *NOR*/*OR* gate as shown in Figure 10.7. If at least one input gate has a V_{IH} signal, then the *NOR* output has an

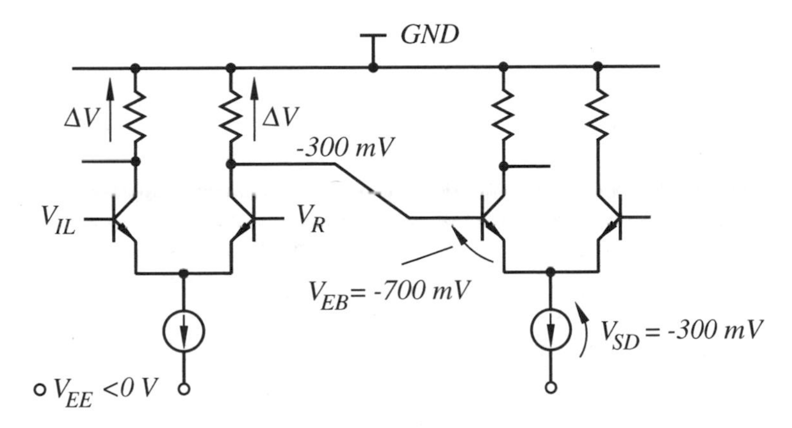

Figure 10.6 Cascaded current switches with typical voltages

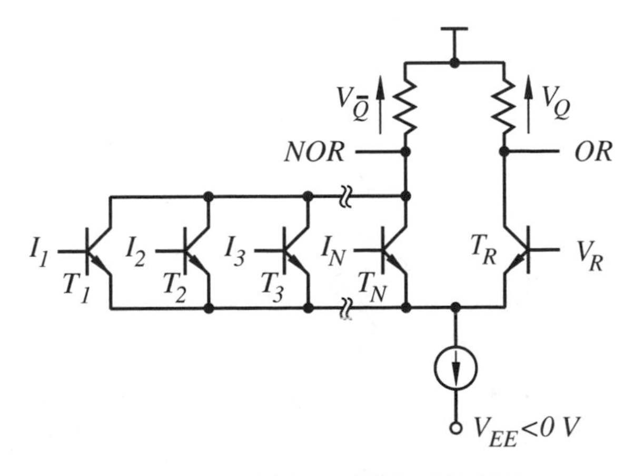

Figure 10.7 Multi-input CML *NOR*/*OR* gate

L signal and the OR output an H signal. If all transistors T_1 to T_N are conducting, this causes a shift in the output levels. This limits the number of parallel-connected transistors, as is demonstrated with the following transfer function. The N parallel-connected transistors act together as one transistor with an N times larger transport current. Since the current of T_R is not affected, this leads to an asymmetric transfer function (Equations 10.3 to 10.6) of

$$\boxed{V_Q = V_{QM}\left(1 + Ne^{V_{IR}/\phi_t}\right)^{-1}} \tag{10.10}$$

This is shown in Figure 10.8 for a V_{QM} voltage of $-300\,\text{mV}$ and for different numbers of connected transistors.

The asymmetry is at $V_Q = V_{QM}/2$ (Equation 10.10)

$$\Delta V_{IR} = \phi_t \ln \frac{1}{N} \tag{10.11}$$

With $N = 5$ one gets a value of $-42.8\,\text{mV}$. This leads to an increased V_{QH} level and a decreased V_{QL} level. The number of possible input transistors thus depends directly on the minimum acceptable V_{QL} level.

When more complex logic functions are implemented a typical CML circuit technique, called series gating, is employed (Section 6.3). A simple gate is used as an illustration (Figure 10.9). The transistors are connected in such a way that the source current I_K is either steered to the left or right branch of the gate but never into both branches simultaneously.

A result of the series gating is that each logic level must have adjusted voltages. At level 1 the voltages are unchanged whereas at level 2 the voltages are reduced by V_{BE}. This causes a more negative power-supply voltage of

$$V_{EE} = V_{SD} + EV_{EB} + \Delta V \tag{10.12}$$

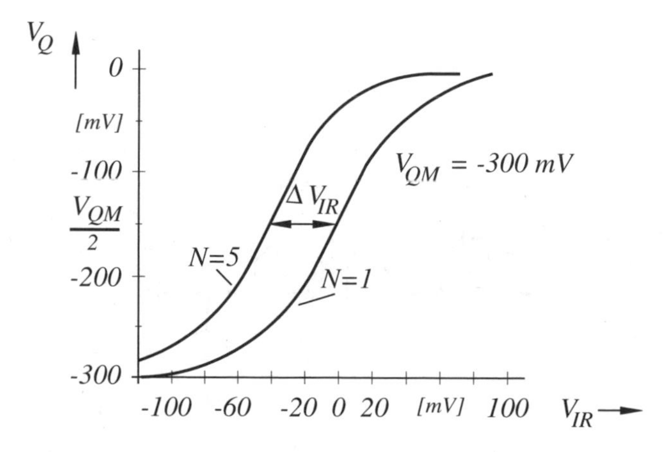

Figure 10.8 Transfer function of a NOR/OR gate with $N = 1$ and $N = 5$ at room temperature

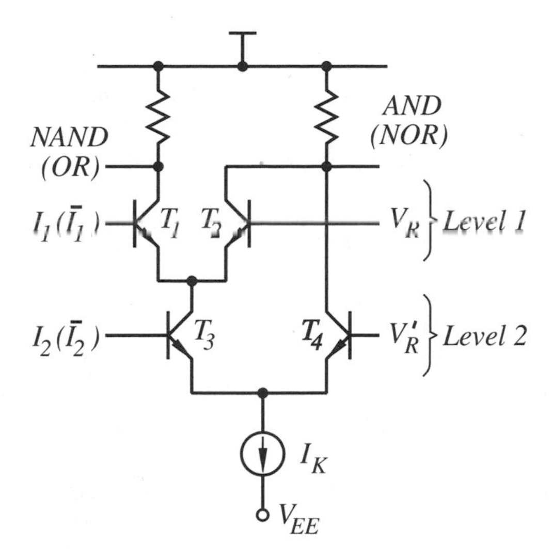

Figure 10.9 $NAND(OR)/AND(NOR)$ gate

where E determines the number of logic levels. This level shift is a definite disadvantage compared to the MOS solution of Section 6.3.

The level shift is realized by the additional transistors T_6 and T_5. The last one is connected as a diode (Section 3.4) with a voltage drop of V_{BE} (Figure 10.10).

Since a current always flows through these transistors, parasitic capacitances are charged by only the small input voltage swing ΔV_I. The level shifter thus degrades the switching performance of the CML circuit to only a small extent. But the series gating reduces the performance. That is why usually not more than three logic levels are used.

A characteristic of the CML circuits is the small voltage levels. If more robust levels are required, this leads to emitter coupled logic (ECL) circuits, discussed in the following section.

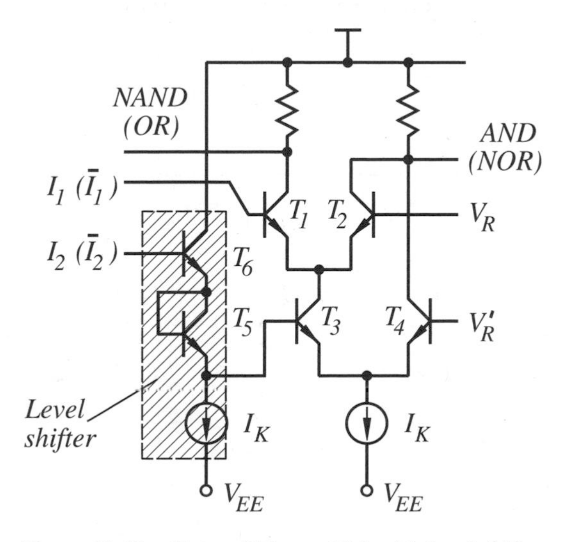

Figure 10.10 Gate of Figure 10.9 with level shifter

10.1.2 ECL Circuits

Compared to the circuit of Figure 10.5 an emitter follower *EF* is inserted between the current switches (Figure 10.11). The circuit causes a level shift of V_{BE} between the current switches. This results, in comparison to Equation (10.8), in an increased voltage difference between the outputs of

$$V_Q - V_{\overline{Q}} = V_{BC} + V_{BE} \tag{10.13}$$

The voltage is positive or negative depending on the input signal. But the value of the logic levels can now be chosen in such a way that V_{BC} is zero or negative. This avoids the steering transistors operating under weak injection conditions. The situation is illustrated in Figure 10.12 for an example where $V_{IH} = -V_{BE} \approx -0.7\,\text{V}$ and $V_{IL} = -2V_{BE} \approx -1.4\,\text{V}$. The reference voltage for this case is chosen to be

$$V_R = \frac{V_{IH} + V_{IL}}{2} = -1.5V_{BE} \tag{10.14}$$

The emitter follower not only has the advantage of level shifting but also transforms a high input impedance into a low output one, which leads to an improved switching performance. This is similar to the situation of the source follower discussed in Section 8.2.

Differential ECL circuits

Even faster switching times can be achieved when a differential ECL technique is used. The speed advantage results from the fact that each circuit node is only changed by half of the logic swing voltage (Figure 10.13).

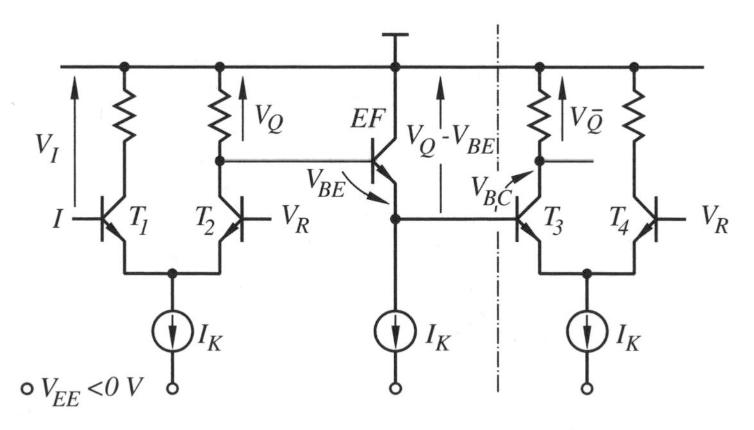

Figure 10.11 Level shift between current switches with an emitter follower *EF*

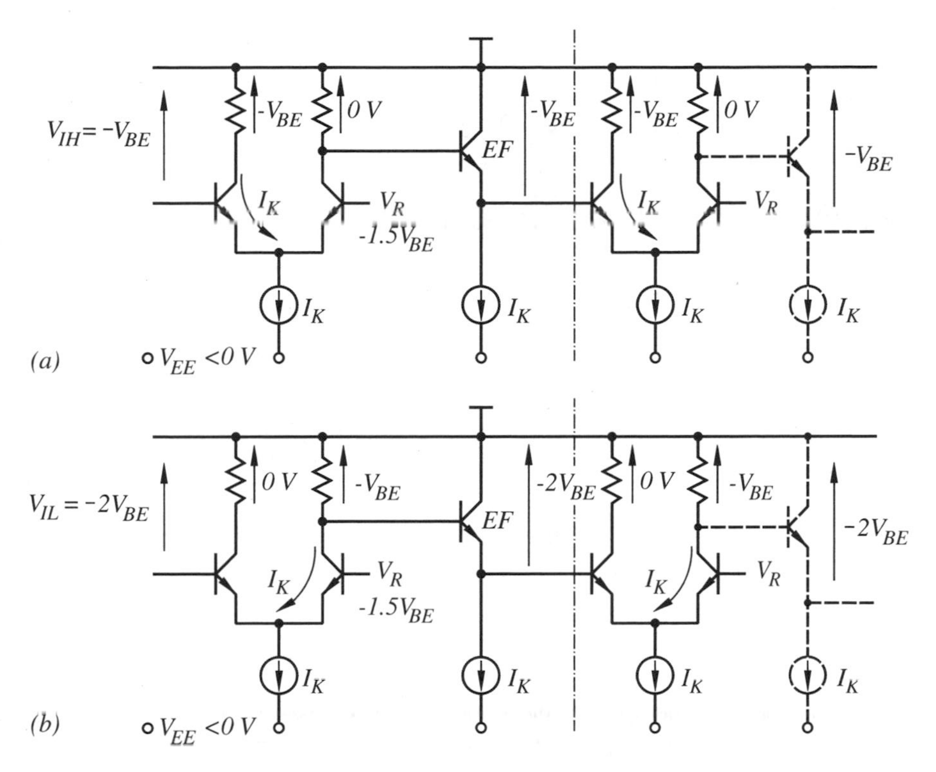

Figure 10.12 Voltage levels at cascaded current switches with emitter follower: (a) $V_{IH} = -V_{BE}$ and (b) $V_{IL} = -2V_{BE}$

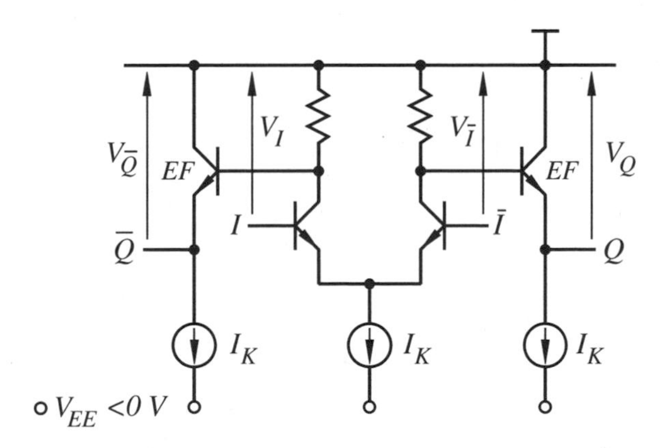

Figure 10.13 Differential current switch with emitter followers *EF*

An example of a differential gate, comparable to the MOS solution of Figure 6.30(a), is shown in Figure 10.14. The required level shift can be performed with relatively little overhead due to the already available emitter followers.

Another example is the D flip-flop of Figure 10.15, which is comparable to the MOS counterpart of Figure 6.47.

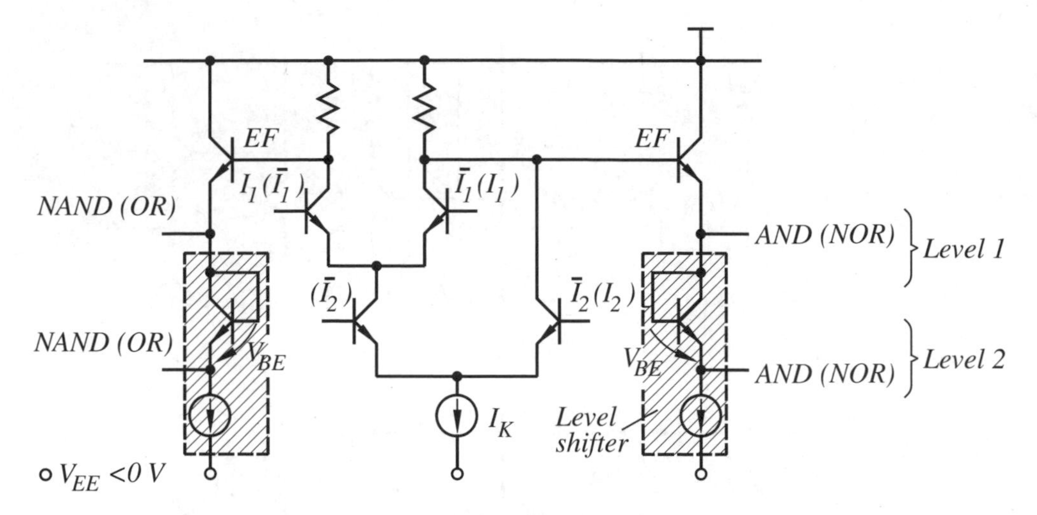

Figure 10.14 Differential *NAND*(*OR*)/*AND*(*NOR*) gate

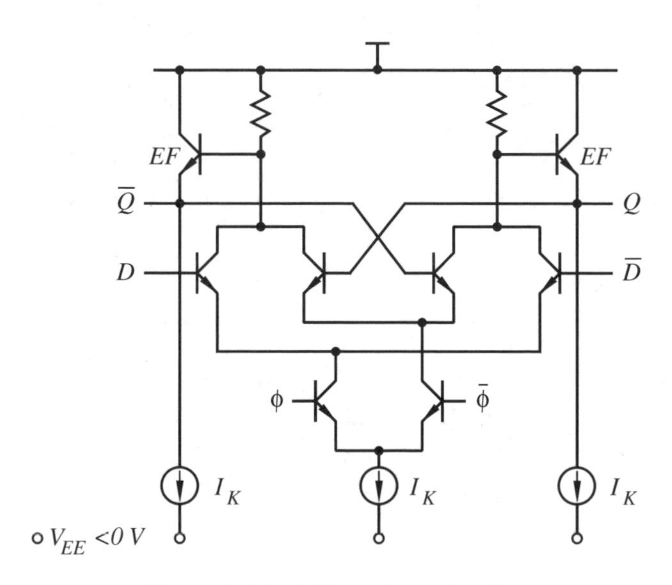

Figure 10.15 Differential D flip-flop

If one compares the MCML technique, described in Chapter 6, with the CML/ECL one, then the following advantages and disadvantages result. The bipolar solutions work with smaller signal swings since this technique is by far less sensitive to offset voltages caused by process variations (Section 10.4.1). This leads to a substantial increase in data rate. A disadvantage is that CML/ECL solutions require a higher power-supply voltage and thus have a larger power dissipation due to the level shifting. When logic gates are loaded this increases the delay time. This is far less pronounced in bipolar implementations, due to the exponential I_C (V_{BE}) behavior, than in MOS realizations (Yamashina and Yamada 1992). Due to these features the CML/ECL circuits are the fastest to date (Wurzer *et al.* 1999).

So far, circuit techniques are considered using mainly bipolar transistors. Combining bipolar transistors and MOS transistors leads to BICMOS circuits with new characteristic features. Typical examples of buffer and gates are considered in the following section.

10.2 BICMOS BUFFER AND GATES

One of the most important BICMOS circuits is the buffer. This buffer shows, in comparison to a CMOS implementation, better driving capabilities, which lead to a faster switching speed (Figure 10.16). If an L signal is applied to the input of the buffer (Figure 10.16(a)), the p-channel transistor M_2 is conducting and the n-channel transistor M_1 non-conducting. Since M_2 conducts, a base current of $I_{B,2}$ flows into the bipolar transistor T_2. This results in an amplified current, namely the emitter current $I_{E,2}$, which charges the capacitance C_L to a value of $V_{QH} = V_{CC} - V_{BE,2}$. If the input signal changes from L to H state, then M_2 is non-conducting and M_1 conducting. A base current $I_{B,1}$ flows into the bipolar transistor T_1. Amplified, this causes a collector current $I_{C,1}$ to discharge the capacitance to a value of $V_{QL} = V_{BE}$. The connection of M_1 between base and collector of T_1 is chosen in order to avoid the BC-junction becoming forward biased and deteriorating the switching performance of the transistor. Since ideally the transistor pairs T_1M_1 and T_2M_2 do not conduct simultaneously, only a dynamic power dissipation occurs. Derivations from this behavior are discussed at the end of this section.

From the previous discussion one could conclude that the BICMOS buffer has an improved charging and discharging current caused by the current gain B_F of the transistors. Unfortunately, this is not completely true, as the following discussion reveals.

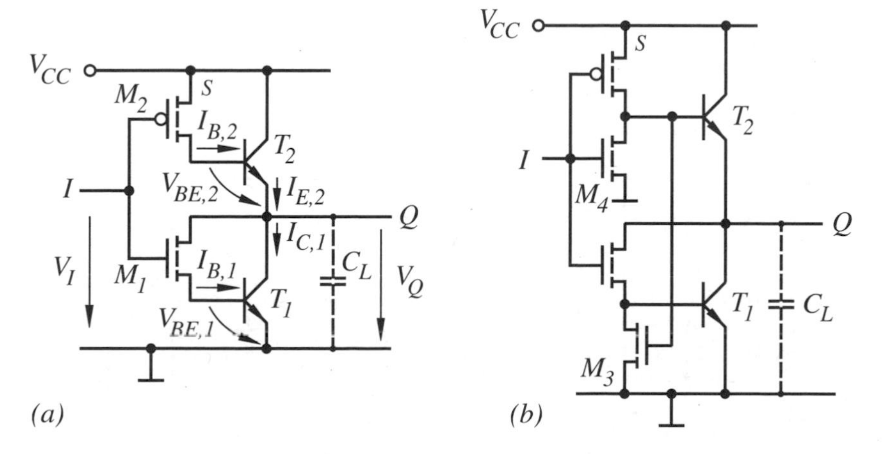

Figure 10.16 (a) Principle BICMOS buffer and (b) BICMOS buffer with improved switching characteristic

At first the discharging of C_L is considered (Figure 10.17). The MOS transistor M_1 is replaced by a current source. This is not entirely correct since the transistor changes during discharging into the resistive region, but this is negligible to the unfavorable behavior of the bipolar transistor, as will be shown. Furthermore, the capacitances of the MOS transistor are neglected to ease the discussion. Using the equivalent circuit of Figure 10.17(b), a differential equation describing the discharging can be derived.

The base current is calculated by

$$\begin{aligned} I_{B,1} &= I_Q + I_B \\ &= \frac{\mathrm{d}Q_{BE}}{\mathrm{d}t} + \frac{I_{C,1}}{B_F} \end{aligned} \tag{10.15}$$

According to Equation (3.91), the stored charge has a value of $Q_{BE} = \tau_F I_{C,1}$ which leads to the following differential equation

$$\frac{\mathrm{d}I_{C,1}}{\mathrm{d}t} + \frac{I_{C,1}}{\tau_F B_F} = \frac{I_{B,1}}{\tau_F} \tag{10.16}$$

where the BE-depletion capacitance is neglected. The solution of this equation results in a time-depending collector current of

$$\boxed{I_{C,1}(t) = I_{B,1}\, B_F\left(1 - e^{-t/\tau_F B_F}\right)} \tag{10.17}$$

Two cases are of particular interest:

(a) when $t \gg \tau_F B_F$

$$\boxed{I_{C,1} = I_{B,1} B_F} \tag{10.18}$$

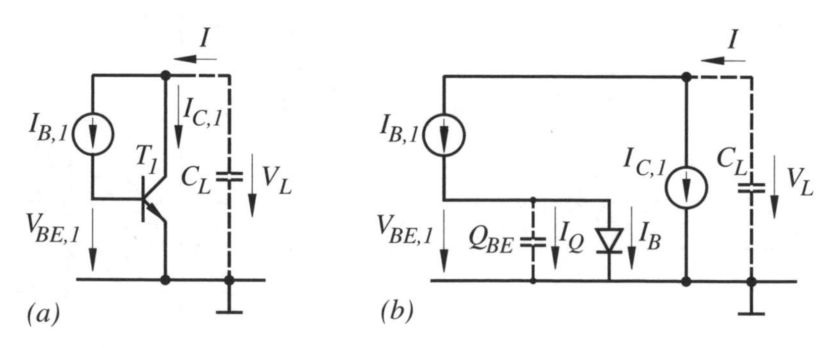

Figure 10.17 BICMOS buffer: (a) effective circuit elements during discharging and (b) simplified equivalent circuit (Figure 3.39)

(b) when $t \ll \tau_F B_F$

$$\boxed{I_{C,1}(t) = I_{B,1}\frac{t}{\tau_F}} \tag{10.19}$$

The result of case (a) is probably expected. But case (b) shows a linear increasing collector current with time, which is smaller than the current in case (a). The reason for this behavior is that during the time $t \ll \tau_F B_F$ most of the base current is needed to build up the charge Q_{BE}. Under these conditions the BICMOS buffer has no advantage compared to a standard CMOS version.

Rise and fall time of the BICMOS buffer

The result of the previous discussion is that the BICMOS buffer has a substantial advantage when a large capacitive loading exists where $t \gg \tau_F B_F$. In this case a collector current of (Equation 10.18)

$$I_{C,1} = I_{B,1}\, B_F$$

results, which leads to a fall time of

$$\begin{aligned} t_f &= C_L \frac{\Delta V}{I_{B,1} + I_{C,1}} \\ t_f &= C_L \frac{\Delta V}{I_{B,1}(B_F + 1)} \end{aligned} \tag{10.20}$$

where ΔV describes the voltage change $V_{QH} - V_{QL}$ at the output of the buffer. As one expects, the shortest fall time results at a large current gain B_F.

So far only the fall time is considered, which depends on the currents $I_{B,1}$ and $I_{C,1}$. In the case of the rise time the situation is comparable, since C_L is charged by $I_{B,2}$ and $I_{C,2}$. When one assumes the same currents for T_1 and T_2 then the rise time is also described by Equation (10.20).

In order to avoid a transient current at the BICMOS buffer during switching, transistor T_1 (Figure 10.16(a)) has to be turned off before transistor T_2 is turned on or vice versa. To guarantee this, the stored base charge of the respective transistor has to be removed rapidly. This is achieved with the additional transistors M_3 and M_4, shown in Figure 10.16(b).

The BICMOS buffer requires a relatively large breakdown voltage of the BE-junctions, as the following discussion reveals. If an L signal is applied to the input of the buffer (Figure 10.16(b)), then C_L is charged to a voltage approaching V_{CC}. If the input signal changes from the L state to the H state, this causes transistor M_4 to conduct. As a result 0 V is applied to the base of T_2. If the output capacitance is relatively large, and thus the fall time slow, a voltage of about $-V_{CC}$ across the BE-junction of T_2 exists at the beginning of the discharging phase. Thus the breakdown voltage of the BE-junctions should be larger than V_{CC}, which is not necessarily easy to achieve (Section 3.2.5).

In summary, the BICMOS buffer has, because of $t \gg \tau_F B_F$, from a particular capacitance value, a substantially reduced delay time t_d, compared to the CMOS buffer. This is shown in Figure 10.18 with a cross-over at about 1 pf.

A disadvantage of the BICMOS buffer is that the power-supply voltage cannot easily be reduced below 3 V because of the V_{BE} voltage drops at the output.

It is possible to modify the BICMOS buffer into *NAND* and *NOR* gates with improved driving capability, as shown in Figure 10.19. A drawback is the relatively large transistor count, since the *L* levels have to be guaranteed by the n-channel transistors and the *H* level by the p-channel transistors (Figure 10.20).

This is similar to the situation mentioned in Section 6.1.1, where complementary circuits are discussed. A simpler implementation results when a p-channel transistor is used instead of n-channel transistors between collector and base of T_1 (Figure 10.21).

Depending on the values of V_{Tp} and V_{BE}, different *L* signals are possible at the output. If $|V_{Tp}| > V_{BE}$, then an output voltage of $V_{QL} = |V_{Tp}|$ exists. But if $V_{BE} > |V_{Tp}|$

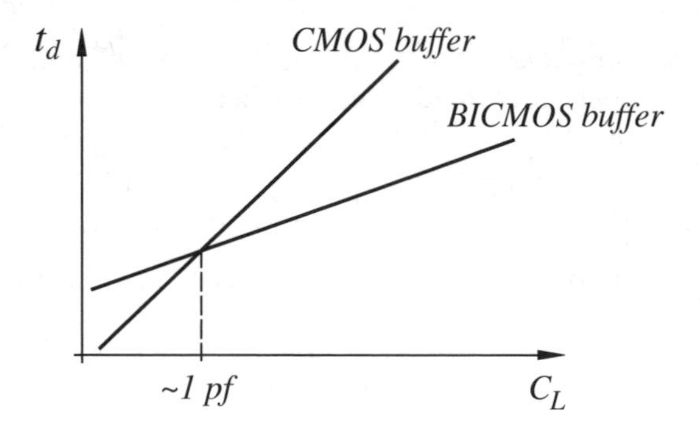

Figure 10.18 Delay time comparison between CMOS and BICMOS buffers

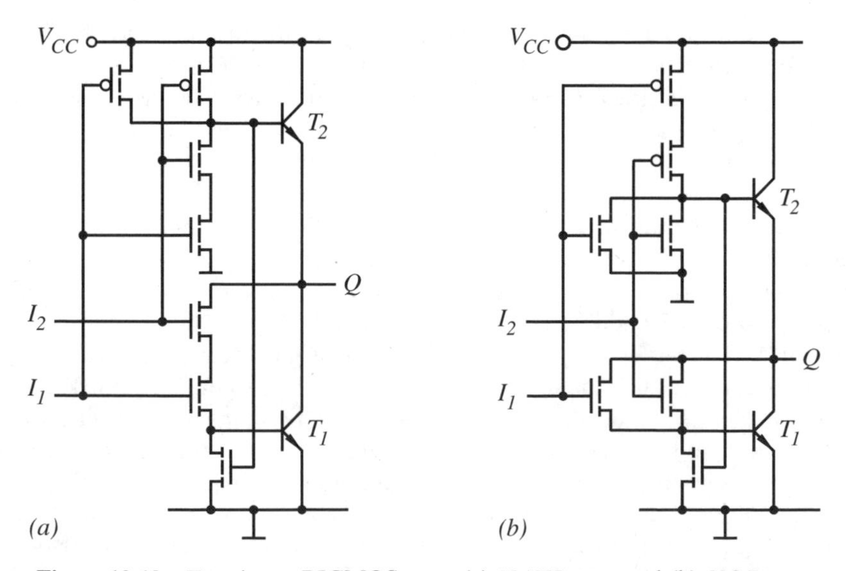

Figure 10.19 Two-input BICMOS gate: (a) *NAND* gate and (b) *NOR* gate

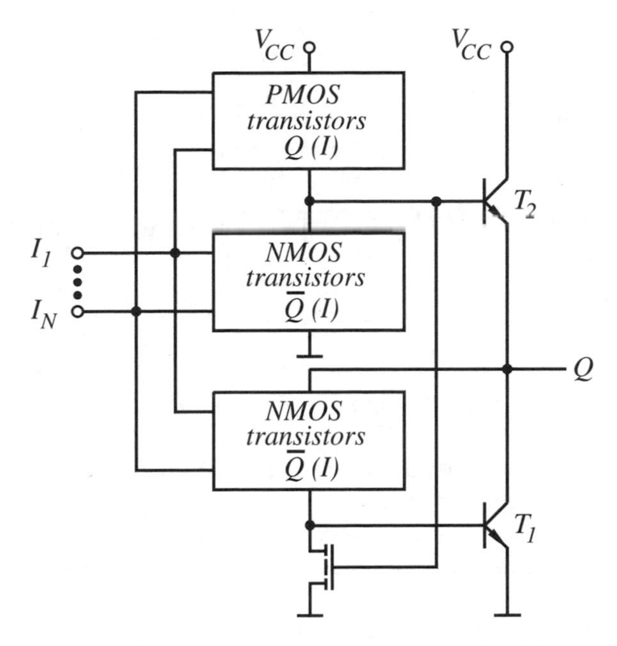

Figure 10.20 General implementation of BICMOS gates

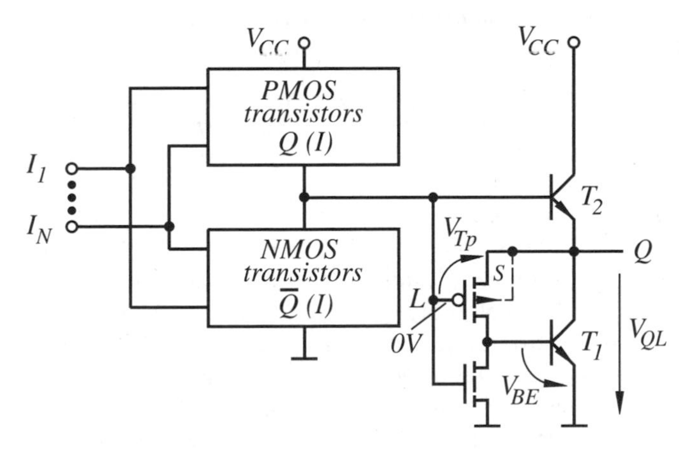

Figure 10.21 Simplified BICMOS gate

then a value of $V_{QL} = V_{BE}$ results. A disadvantage of the circuit is that due to the p-channel transistor with its reduced mobility, a slower fall time results.

10.3 BAND-GAP REFERENCE CIRCUITS

Band-gap reference circuits are a classical approach to implement precise and almost temperature-independent voltage references on a chip (Widlar 1971). The principle idea is demonstrated in Figure 10.22.

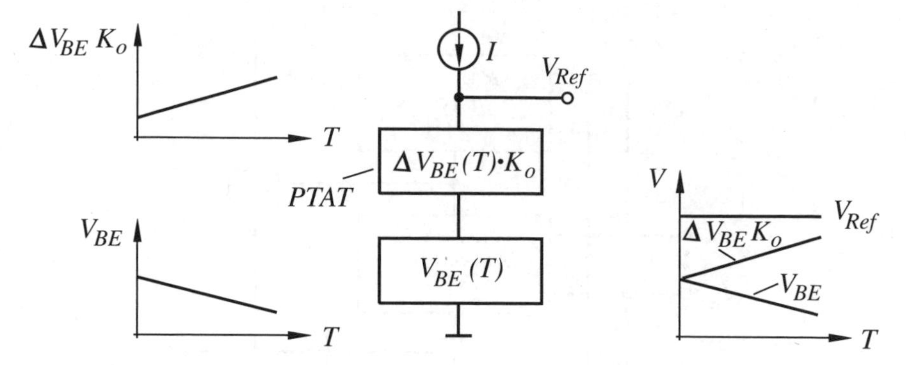

Figure 10.22 Principle of a band-gap reference circuit

The negative temperature coefficient of the V_{BE} voltage is compensated by a voltage source with a positive temperature coefficient, which is proportional to the absolute temperature (PTAT). This voltage source is generated by evaluating the difference between two V_{BE} voltages. How this leads to a PTAT behavior is considered in the following.

The collector current of the bipolar transistor (Equation 3.7) is described by

$$I_C = I_{SS}\, e^{V_{BE}/\phi_t} \tag{10.21}$$

when $V_{BE} > 100\,\text{mV}$. This results in a base emitter voltage of

$$V_{BE} = \phi_t \ln\frac{I_C}{I_{SS}} = \frac{kT}{q}\ln\frac{I_C}{I_{SS}} \tag{10.22}$$

Considering this equation, one could get the idea that this voltage increases with temperature. The opposite is true, since the temperature behavior of the transport current I_{SS} dominates. In order to eliminate the influence of the transport current, the difference between two V_{BE} voltages is generated (Figure 10.23).

The bipolar transistors are connected as pn-diodes. It has the advantage, as discussed in Section 3.4, that the base resistance can be reduced substantially. In the circuit $V_{BE,1} > V_{BE,2}$. This is achieved by increasing the current through R_1 by a factor of m, compared to the current through R_2, and by reducing $V_{BE,2}$ by duplicating transistor T

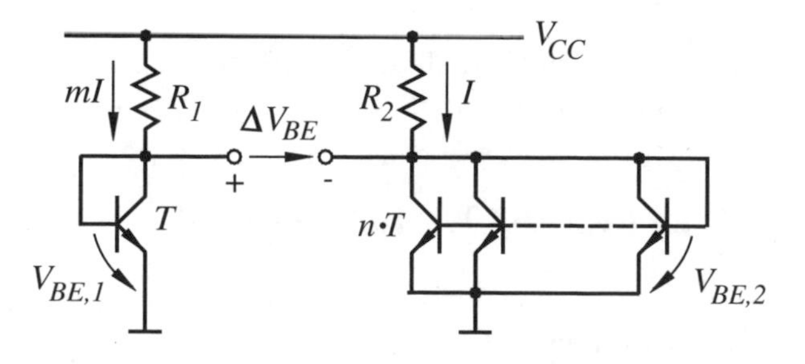

Figure 10.23 Generation of a PTAT voltage

n times. Parallel-connected transistors are used instead of only one with a larger geometry in order to have transistors with identical characteristic data. Under this prerequisite a PTAT voltage of

$$\Delta V_{BE} = V_{BE,1} - V_{BE,2}$$

$$\Delta V_{BE} = \phi_t \ln \frac{mI}{I_{SS}(1 + 1/B_F)} - \phi_t \ln \frac{I}{n \cdot I_{SS}(1 + 1/B_F)}$$

$$\boxed{\Delta V_{BE} = \frac{kT}{q} \ln mn} \tag{10.23}$$

results. This voltage has the required positive temperature coefficient of

$$\frac{d\Delta V_{BE}}{dT} = \frac{k}{q} \ln mn \tag{10.24}$$

The principle circuit of Figure 10.22 thus yields a voltage of

$$V_{Ref} - V_{BE}(T) + K_o \Delta V_{BE}(T) \tag{10.25}$$

where K_o is a temperature-independent constant.

The temperature behavior of the V_{BE} voltage can be approximated by

$$\boxed{V_{BE} = V_{Go} - NT} \tag{10.26}$$

where V_{Go} corresponds to the extrapolated band-gap E_{Go}/q voltage when the temperature approaches zero Kelvin. N is a temperature coefficient with a value of about 1.6 mV/K. A more precise description follows at the end of this section. The above equations lead to a reference voltage of

$$V_{Ref} = V_{Go} - NT + K_o \frac{kT}{q} \ln mn \tag{10.27}$$

With the parameters selected to meet the requirement

$$K_o \frac{k}{q} \ln mn = N \tag{10.28}$$

a temperature-independent reference voltage of $V_{Ref} = V_{Go}$ results. As mentioned before, this voltage corresponds to the extrapolated band-gap E_{Go}/q voltage of

1.205 V, when the temperature approaches zero Kelvin and thus explains the given name of the circuit.

How the basic idea can be used to build a real band-gap reference circuit is considered next. For this purpose the PTAT circuit of Figure 10.23 is modified (Figure 10.24). The V_o voltage of the circuit is raised until the voltages V_a and V_b at the terminals (a) and (b) are equal. This is the case at the voltage point V_x of the inset of Figure 10.24. Then the voltage V_o is somehow continuously adjusted to guarantee that the difference between the V_a and V_b voltages remains zero, even when the temperature changes. Under this condition, terminal (b) has always a voltage of $V_{BE,1}$, which leads to a PTAT voltage of ΔV_{BE} across R_3. The last one is added to the $V_{BE,2}$ voltage of the parallel-connected transistors. A temperature-independent voltage at terminal (b) exists.

A real band-gap circuit results when the V_o voltage adjustment is performed by an amplifier as shown in Figure 10.25. If the gain of the amplifiers is sufficiently large, then the voltage between the terminals (a) and (b) approaches zero. If, e.g., the voltage at terminal (b) is smaller than the voltage at terminal (a), this causes the output voltage V_{Ref}

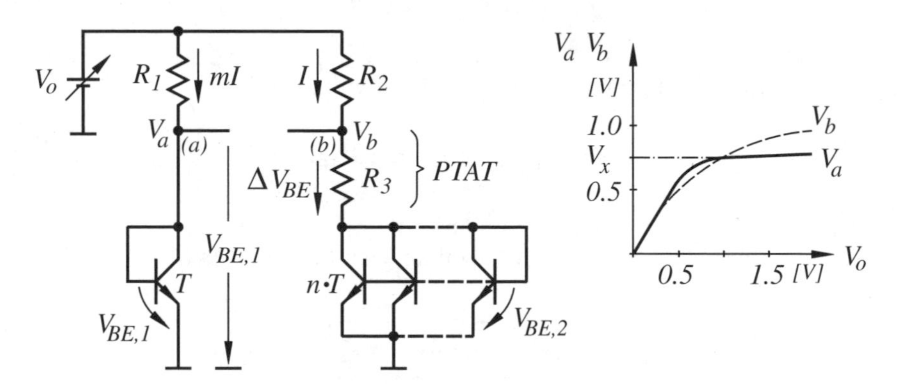

Figure 10.24 Modified PTAT circuit

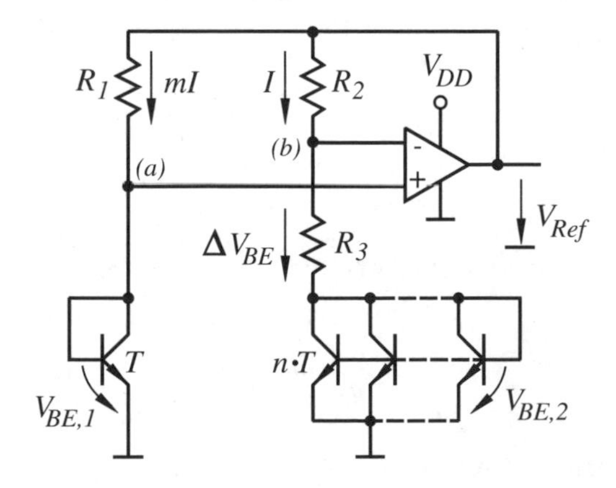

Figure 10.25 Band-gap reference circuit

to increase until the difference between the two input voltages approaches zero. If the voltage at terminal (*b*) is larger than that at terminal (*a*), a reverse situation exists. This circuit leads, according to Figure 10.25, to a reference voltage of

$$V_{\text{Ref}} = V_{BE,2} + I(R_2 + R_3) \tag{10.29}$$

Substituting Equations (10.23 and 10.26) and the one resulting directly from the circuit

$$I = \Delta V_{BE}/R_3 \tag{10.30}$$

into Equation (10.29) gives

$$\boxed{V_{\text{Ref}} = V_{Go} - NT + \frac{kT}{q}(\ln mn)\left(1 + \frac{R_2}{R_3}\right)} \tag{10.31}$$

In order to compensate for the temperature influence, the requirement

$$\frac{k}{q}(\ln mn)\left(1 + \frac{R_2}{R_3}\right) = N \tag{10.32}$$

has to be met. This leads to a resistor ratio of $R_2/R_3 = 3.7$ if $m = 5$, $n = 10$, and a temperature coefficient of $N = 1.6\,\text{mV/K}$ is assumed.

An advantage of this implementation is that the reference voltage is dependent on the ratio of the resistors and not dependent on an absolute resistor value, which is difficult to control in production.

Effect of an offset voltage

Offset voltage is the voltage applied between the two inputs of an amplifier to compensate for all asymmetries of the amplifier arising from input to output. For example, if the amplifier shown in Figure 10.25 has an offset voltage of $\pm V_{off}$, then the voltage does not approach zero between the inputs of the amplifier, but instead the value of the offset voltage. When one assumes a voltage at terminal (*a*) of $V_{BE,1} \pm V_{off}$, this leads to a ΔV_{BE} of $V_{BE,1} \pm V_{off} - V_{BE,2}$ and accordingly to a reference voltage of

$$V_{\text{Ref}} = V_{Go} - NT + \left[\frac{kT}{q}(\ln mn) \pm V_{off}\right]\left[1 + \frac{R_2}{R_3}\right] \tag{10.33}$$

The offset voltage thus causes an error at the reference voltage, which is amplified by the factor $(1 + R_2/R_3)$. The easiest way to reduce the influence of the offset voltage is to use a bipolar amplifier, since the offset voltage is substantially smaller than that of a MOS amplifier (Section 10.4.1).

Temperature coefficient of the V_{BE} voltage

In the preceding discussion the temperature behavior of the V_{BE} voltage is linearly approximate (Equation 10.26)

$$V_{BE}(T) = V_{Go} - NT$$

In reality, the temperature coefficient N is not constant but depends slightly on the temperature which is discussed in the following.

According to Equation (3.40), the collector current has a temperature behavior of

$$I_C(T) = E\left(\frac{T}{300\,\mathrm{K}}\right)^{(4-a_n)} e^{\frac{-E_G(T)}{kT}}\left(e^{\frac{q}{kT}V_{BE}} - 1\right) \tag{10.34}$$

which yields a temperature-dependent V_{BE} voltage of

$$V_{BE}(T) = \frac{kT}{q}\left[\ln\frac{I_C(T)}{[\mathrm{A}]} - \ln\frac{E}{[\mathrm{A}]} - (4-a_n)\ln\frac{T}{[300\,\mathrm{K}]} + \frac{E_G(T)}{kT}\right] \tag{10.35}$$

which has a value at $T \to 0\,\mathrm{K}$ of

$$V_{BE}(T \to 0\,\mathrm{K}) = \frac{E_G(T \to 0\,\mathrm{K})}{q} = V_{Go} \tag{10.36}$$

The band-gap variation with temperature is given by Equation (3.41)

$$E_G(T)/q = V_G(T) = V_{Go} + \varepsilon T \tag{10.37}$$

where ε has a value of $-2.8 \cdot 10^{-4}$ V/K. The required temperature dependence of the collector current in Equation (10.35) is found, e.g., for the parallel-connected transistors directly from Figure 10.25

$$I_C(T) = \Delta V_{BE}/R_3 = \frac{kT}{q}(\ln mn)/R_3 = FT \tag{10.38}$$

Substituting of the above equations into Equation (10.35) results in a V_{BE} voltage of

$$V_{BE}(T) = V_{Go} - \left[\frac{k}{q}(4-a_n)\ln\frac{T}{[300\,\mathrm{K}]} + \frac{k}{q}\ln\frac{E}{FT} - \varepsilon\right]T \tag{10.39}$$

If one compares this result with the linear approach of Equation (10.26) then it is obvious that the temperature coefficient N is not constant but instead temperature dependent. The change of the coefficient with temperature can be found from the following derivative

$$\frac{\mathrm{d}V_{BE}}{\mathrm{d}T} = -\frac{1}{T}\left[V_{Go} + \frac{kT}{q}(3-a_n) - V_{BE}(T, I_C)\right] \tag{10.40}$$

Example

The temperature coefficient of the V_{BE} voltage has to be determined at a temperature of $T = 300\,\text{K}$ (27 °C). The data are: $I_C = 10^{-5}\,\text{A}$, $I_{SS} = 10^{-18}\,\text{A}$, $V_{Go} = 1.205\,\text{V}$, and $a_n = 1.5$. These data yield a V_{BE} voltage of

$$V_{BE}(27\,°\text{C}, 10\,\mu\text{A}) = 26\,\text{mV} \ln \frac{10^{-5}}{10^{-18}} = 0.778\,\text{V}$$

and a respective temperature coefficient (Equation 10.40) of

$$\left. \frac{\mathrm{d}V_{BE}}{\mathrm{d}T} \right|_{\substack{T = 27\,°\text{C} \\ I_C = 10\,\mu\text{A}}} = -1.65\,\text{mV/K}$$

In a circuit, a curvature compensation, as described in Gunawa *et al.* (1993), can be used to minimize the influence of the changing temperature coefficient.

A modified band-gap reference circuit with fewer circuit components and less power dissipation – because of the missing amplifier – is shown in Figure 10.26. The circuit uses a feedback loop via transistor T_4. The function of the loop can be explained in the following way: a reduction, e.g. in the V_{Ref} voltage, causes also a reduction in the $V_{BE,3}$ voltage which leads to a reduced collector current $I_{C,3}$ and thus to an increased voltage at node (a). This counterbalances the reduced V_{Ref} voltage via T_4. A stable operation point exists where the reference voltage has a value of

$$V_{\text{Ref}} = V_{BE,3}(I, T) + V_T(T) \tag{10.41}$$

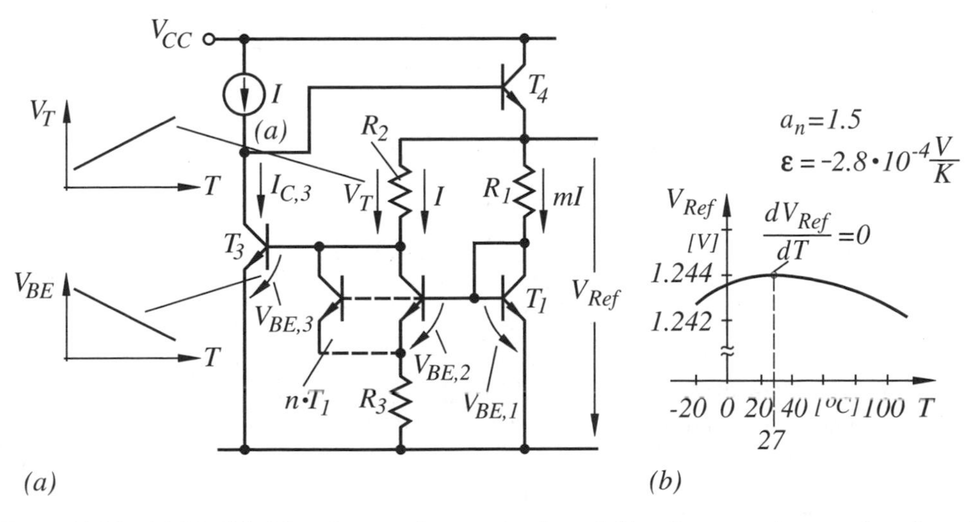

Figure 10.26 (a) Modified band-gap reference circuit and (b) reference voltage as function of temperature

With the PTAT voltage

$$
\begin{aligned}
V_T &= R_2 I \\
&= R_2 \frac{V_{BE,1} - V_{BE,2}}{R_3} \\
&= \frac{R_2}{R_3} \frac{kT}{q} \ln nm
\end{aligned}
\tag{10.42}
$$

where Equation (10.23) is used, this leads to a band-gap reference voltage of

$$
V_{\text{Ref}} = V_{BE,3}(I, T) + \frac{R_2}{R_3} \frac{kT}{q} \ln nm \tag{10.43}
$$

How the circuit has to be dimensioned is considered next. The temperature coefficient of the reference voltage can be found using Equations (10.43) and (10.40)

$$
\begin{aligned}
\frac{\mathrm{d}V_{\text{Ref}}}{\mathrm{d}T} = &-\frac{1}{T}\left[V_{Go} + \frac{kT}{q}(3 - a_n) - V_{BE,3}(I, T)\right] \\
&+ \frac{R_2}{R_3} \frac{k}{q} \ln nm
\end{aligned}
\tag{10.44}
$$

where a linear temperature dependence of the collector current is assumed.

The temperature change at T_R, e.g. room temperature, is required to be

$$
\left.\frac{\mathrm{d}V_{\text{Ref}}}{\mathrm{d}T}\right|_{T_R} = 0 \tag{10.45}
$$

In this case the individual temperature parameters have to compensate each other and Equation (10.44) has to meet the requirement

$$
\frac{1}{T_R}\left[V_{Go} + \frac{kT_R}{q}(3 - a_n) - V_{BE,3}(I, T_R)\right] = \frac{R_2}{R_3} \frac{k}{q} \ln nm \tag{10.46}
$$

This leads to a reference voltage of the circuits at room temperature, according to Equation (10.43), of

$$
V_{\text{Ref}}(T_R) = V_{Go} + \frac{kT_R}{q}(3 - a_n) \tag{10.47}
$$

Example

The circuit of Figure 10.26 has to be designed in such a way that the temperature meets the requirement of Equation (10.45) at $T_R = 300\,\text{K}$. With $V_{Go} = 1.205\,\text{V}$ and $a_n = 1.5$ a reference voltage of

$$
V_{\text{Ref}}(300\,\text{K}) = 1.244\,\text{V}
$$

results.

With these data, the circuit has a temperature dependence (Equations 10.39, 10.46, and 10.47) of

$$V_{\mathrm{Ref}}(T) = \frac{kT}{q}\left[(3 - a_n)\ln\frac{T_R}{T} + \frac{1}{k}\left(\frac{E_G(T)}{T} - \frac{E_G(T_R)}{T_R}\right)\right] + \frac{T}{T_R}V_{\mathrm{Ref}}(T_R) \qquad (10.48)$$

which is shown in Figure 10.26(b) for the case that $\varepsilon = -2.8 \cdot 10^{-4}$ V/K (Equation 10.37). The observed temperature deviation is less than 2 mV. Implemented circuits show a slightly larger temperature deviation of less than 15 mV in a temperature range between 30 °C and 150 °C (Tran *et al.* 1989).

The complete circuit, including the current sources, is shown in Figure 10.27. The current $I_{C,5}$ is determined by transistor T_5 and its operation condition given by the reference voltage V_{Ref} and the resistor R_4. This current is mirrored via $T_{P,1}$ to $T_{P,2}$ and thus responsible for the current I_P. This arrangement has the advantage that the current is not dependent on power-supply variations.

In most band-gap reference circuits the possibility exists, when V_{CC} is applied, that no current flows. In order to avoid this undesirable operation point, a start-up circuit is provided. With no current flowing $V_{\mathrm{Ref}} = 0$ V and transistor T_6 is conducting. This causes a current to be mirrored via $T_{P,1}$ and $T_{P,2}$ into the band-gap circuit, which leads to the generation of the reference voltage V_{Ref}, which turns off transistor T_6.

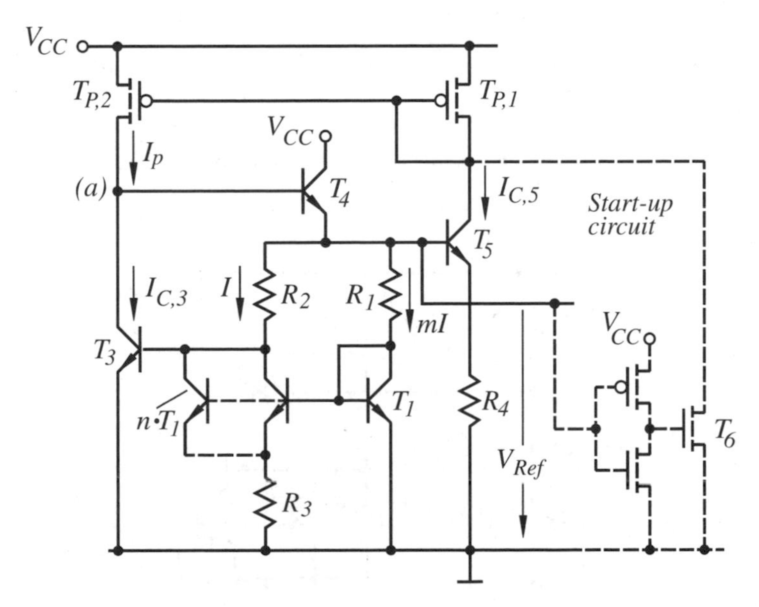

Figure 10.27 Band-gap reference circuit of Figure 10.26 with current sources (after Tran *et al.* 1989; © 1989 IEEE)

Band-gap reference circuit using a CMOS process

It is obvious that a band-gap reference circuit is an extremely useful circuit component, which designers like to use also, when instead of a BICMOS a CMOS manufacturing process is available. Figure 10.28 shows how bipolar transistors can be utilized in a CMOS process, described in Chapter 4.

A lateral and a vertical pnp transistor exist, with current gains roughly between 2 and 10. The p-channel transistor is not active since the gate is connected to V_{CC}. In order to avoid a parasitic current path through the vertical transistor, both transistors are connected in parallel and the collectors are tied to 0 V (Figure 10.29).

The bipolar transistors act as pn-diodes. This has the advantage, as discussed in Section 3.4, that the base resistance can be reduced substantially. For the realization of the amplifier the Miller one, presented in Section 9.1, may be used.

An alternative to this implementation (Razavi 1999) is shown in Figure 10.30. This solution is recommended for use when a very low power dissipation is required. In this circuit the sum of voltages $V_{EB,1} + V_{GS,1}$ is equal to that of $V_{EB,2} + I_2 R_1 + V_{GS,2}$. If one assumes that all MOS transistors have the same (w/l) ratio then in each branch of the circuit the same current flows due to the mutual current mirroring and $V_{GS,1} = V_{GS,2}$. Under these condition a current (Equation 10.23 with $m = 1$) of

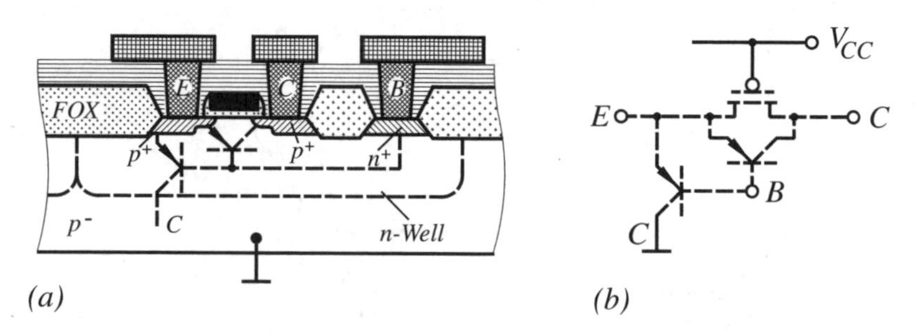

Figure 10.28 (a) Lateral and vertical pnp transistors and (b) equivalent circuit

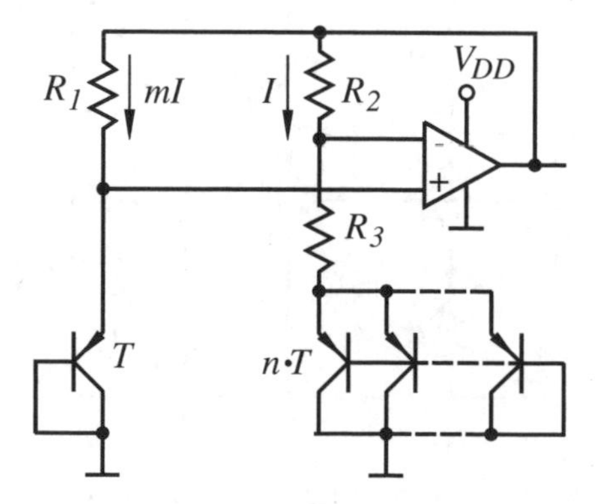

Figure 10.29 Band-gap reference circuit with pnp transistors

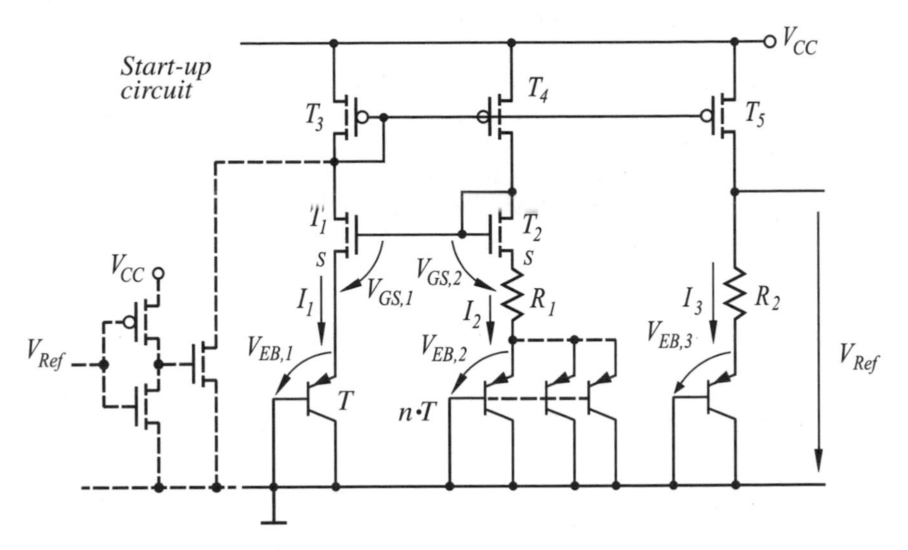

Figure 10.30 Low power band-gap reference circuit (after Razavi 1999 © 1999 IEEE)

$$I_2 = \frac{1}{R_1}(V_{EB,1} - V_{EB,2}) = \frac{1}{R_1}\frac{kT}{q}\ln n \tag{10.49}$$

flows through resistor R_1. This current has a positive temperature coefficient (PTAT). Since the current I_2 is mirrored to I_3 and thus $I_2 = I_3$, this leads to a band-gap reference voltage of

$$V_{\text{Ref}} = V_{EB,3}(I, T) + \frac{R_2}{R_1}\frac{kT}{q}\ln n \tag{10.50}$$

which is comparable to the one derived by Equation (10.43) of the preceding circuit. An advantage of the circuit is that only two resistors are required. These resistors usually need a relatively large layout area, when high resistor values for a low power application are required.

In the preceding example it is explained that when the power supply is turned on, it may happen that no current flows in the circuit. In order to avoid this situation a start-up circuit is provided.

10.4 ANALOG APPLICATIONS

So far, the pros and cons of the BICMOS technique with respect to digital circuits are discussed. What kind of advantages emerge when analog circuits are considered is the topic of this section. For this purpose, the transfer functions of bipolar transistors and MOS transistors are compared and the offset behavior of the transistors is determined.

10.4.1 Offset Voltage of Bipolar and MOS Transistors

An important parameter is the offset voltage of differential amplifiers. This voltage has a value which compensates for asymmetries caused by process variations and shortcomings in the design of the amplifier. In order to determine the offset voltage contribution, resulting from the matched input transistors only, the illustration of Figure 10.31 is used. Load elements are not considered in order to apply the analysis exclusively to the input transistors. In the case of the MOS circuit, the offset voltage can be calculated from the difference in gate voltage required to cause the same current flow in both transistors.

$$V_{off} = V_{GS,2} - V_{GS,1}$$

$$V_{off} = V_{Tn,2} + \sqrt{\frac{2I_{DS,2}}{\beta_{n,2}}} - \left(V_{Tn,1} + \sqrt{\frac{2I_{DS,1}}{\beta_{n,1}}}\right) \tag{10.51}$$

With $I_{DS,1} = I_{DS,2}$ this leads to an offset voltage of

$$V_{off} = \Delta V_{Tn} + \sqrt{2I_{DS}}\left(\sqrt{\frac{1}{\beta_n - \frac{\Delta\beta}{2}}} - \sqrt{\frac{1}{\beta_n + \frac{\Delta\beta}{2}}}\right) \tag{10.52}$$

where ΔV_{Tn} is the difference between the threshold voltages and $\Delta\beta_n$ the difference in current gain. The above equation can be simplified to

$$\boxed{V_{off} \approx \Delta V_{Tn} + \frac{V_{GS} - V_{Tn}}{2}\frac{\Delta\beta_n}{\beta_n}} \tag{10.53}$$

In most practical cases the gain deviation is given by the geometry variation between the transistors

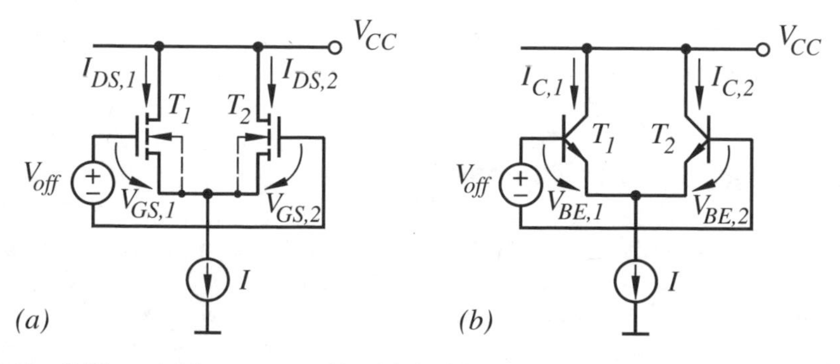

Figure 10.31 Differential input stage: (a) with MOS transistors and (b) with bipolar transistors

$$\frac{\Delta\beta_n}{\beta_n} \approx \frac{\Delta(w/l)}{w/l} \tag{10.54}$$

A similar derivation can be performed for a bipolar transistor pair. With $V_{BE} > 100\,\text{mV}$ an offset voltage of

$$V_{off} = V_{BE,2} - V_{BE,1} = \phi_t \left(\ln \frac{I_{C,2}}{I_{SS,2}} - \ln \frac{I_{C,1}}{I_{SS,1}} \right) \tag{10.55}$$

results. With $I_{C,1} = I_{C,2}$ this leads to

$$V_{off} = \phi_t \ln \frac{I_{SS,1}}{I_{SS,2}} = \phi_t \ln \frac{I_{SS} + \frac{\Delta I_{SS}}{2}}{I_{SS} - \frac{\Delta I_{SS}}{2}} \tag{10.56}$$

where ΔI_{SS} is the variation in transport current between the two transistors. This equation can be simplified to

$$\boxed{V_{off} \approx \phi_t \frac{\Delta I_{SS}}{I_{SS}}} \tag{10.57}$$

where

$$\frac{\Delta I_{SS}}{I_{SS}} \approx \frac{\Delta A_E}{A_E} \tag{10.58}$$

is given approximately by the variation between the two emitter areas.

When one compares the offset voltages between the two transistor pairs (Equations 10.53 and 10.57), then the following becomes obvious.

The geometrical variation between the MOS transistors is amplified by a factor of $(V_{GS} - V_{Tn})/2$, which is about 250 mV. In the case of the bipolar transistors the area variation is amplified by ϕ_t, which has a value of 26 mV at room temperature. Supposing one assumes the same geometrical variations between the transistor pairs, then the MOS transistor pair shows an offset voltage approximately one magnitude larger. Additionally, the variation in threshold voltage has to be added.

In most practical cases bipolar transistor pairs show an offset voltage between 1 and 2 mV, whereas that of MOS transistor pairs varies approximately between 5 and 20 mV.

10.4.2 Comparison of Small-Signal Performance

This comparison can be substantially simplified when a simple amplifier stage is used. Considered are small-signal voltages at input and output as well as small-signal currents. In the terminology of a two-port network these are the parameters h_{12} and h_{21}.

Small-signal voltage gain of the MOS transistor

In Figure 10.32 a simplified amplifier stage is shown which drives a capacitance load C_L. An ideal current source, with $g_{ol} = 0$, is assumed for current supply. To simplify things further, parasitic resistances are neglected. With these assumptions, the so-called intrinsic transistor behavior can be analyzed.

The small-signal analysis of this circuit is described in Section 8.3 and is repeated for ease of discussion. According to Equation (8.23) the transfer function of the circuit can be described by

$$a(j\omega) = \frac{v_o}{v_i}(j\omega) = \frac{a_o}{1 + j\dfrac{\omega}{\omega_p}} \tag{10.59}$$

where the characteristic parameters low-frequency gain a_o (Equation 8.25), $-3\,\text{dB}$ bandwidth ω_p (Equation 8.26), and unity-gain (angular) frequency (Equation 8.27) have the following drain current dependence

$$a_o = -\frac{g_{m,n}}{g_o} \approx -\frac{\sqrt{2I_{DS}\beta_n(1+\lambda_n V_{DS})}}{I_{DS}\lambda_n} \sim (I_{DS})^{-1/2} \tag{10.60}$$

$$\omega_p = \frac{g_o}{C_l} = \frac{I_{DS}\lambda_n}{C_l} \sim I_{DS} \tag{10.61}$$

$$\omega_T = |a_o|\omega_p \approx \frac{\sqrt{2I_{DS}\beta_n(1+\lambda_n V_{DS})}}{C_l} \sim (I_{DS})^{1/2} \tag{10.62}$$

These equations demonstrate that with an increased drain current the low-frequency gain reduces and the $-3\,\text{dB}$ bandwidth and unity-gain frequency increase (Figure 10.33).

The reason for this behavior is that the output conductance increases to a much larger extent with the current than the transconductance of the transistor (Figure 10.34).

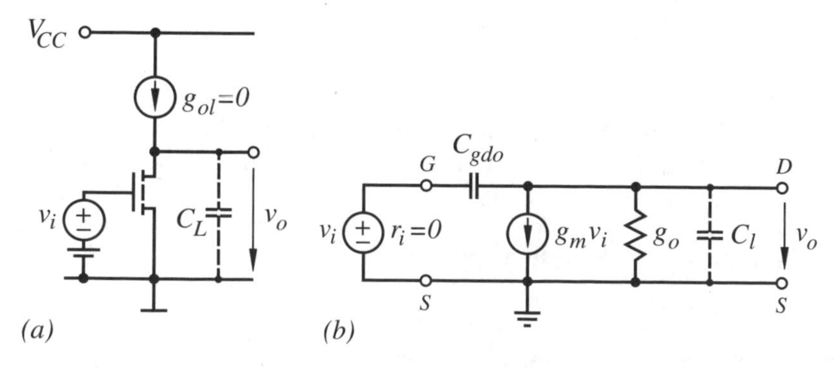

Figure 10.32 (a) Simplified MOS amplifier stage and (b) small-signal equivalent circuit

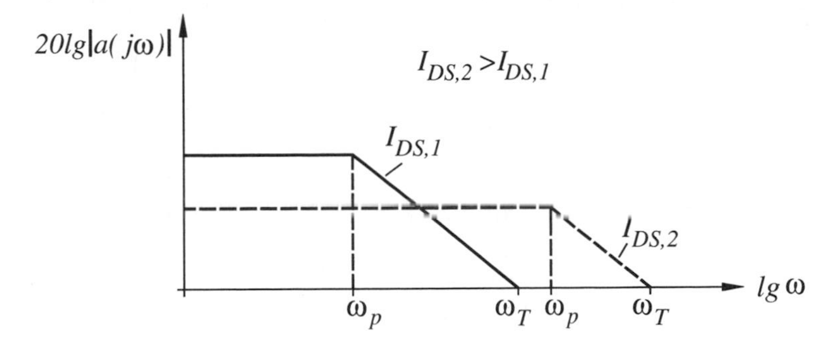

Figure 10.33 Magnitude plot of the MOS amplifier stage of Figure 10.32

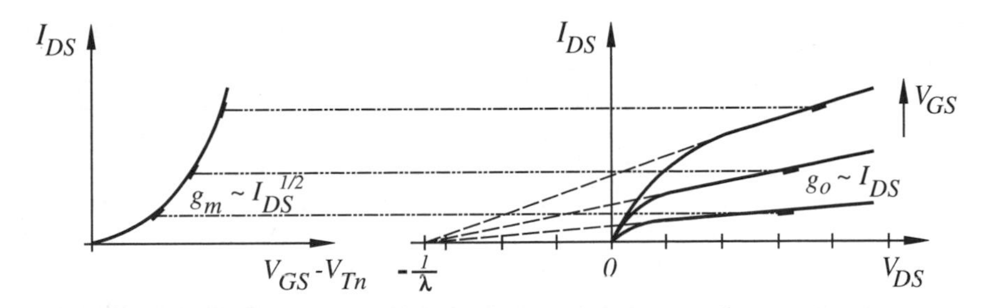

Figure 10.34 Comparison of transconductance and output conductance at different drain currents I_{DS}

Small-signal voltage gain of the bipolar transistor

A similar voltage gain stage like the one considered is shown in Figure 10.35, but with a bipolar transistor. The transfer function is the same as in the preceding case

$$a(j\omega) = \frac{v_o}{v_i}(j\omega) = \frac{a_o}{1 + j\dfrac{\omega}{\omega_p}}$$

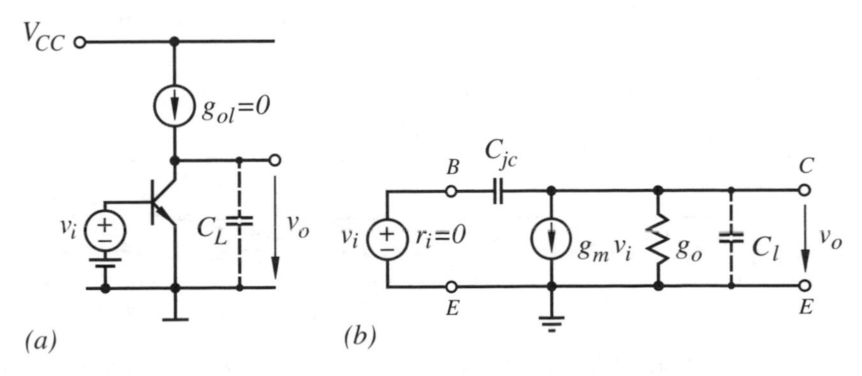

Figure 10.35 (a) Simplified bipolar amplifier stage and (b) small-signal equivalent circuit

except that the characteristic parameters display a totally different current dependence (Problem 10.5)

$$a_o = -\frac{g_m}{g_o} = -\frac{I_C/\phi_t}{I_C/V_{AF}} = -\frac{V_{AF}}{\phi_t} \neq I_C \qquad (10.63)$$

$$\omega_p = \frac{g_o}{C_l} = \frac{I_C/V_{AF}}{C_l} \qquad \sim I_C \qquad (10.64)$$

$$\omega_T = |a_o|\omega_p = \frac{I_C/\phi_t}{C_l} \qquad \sim I_C \qquad (10.65)$$

This is shown in the magnitude plot of Figure 10.36.

The most important observation is that the small-signal gain a_o is independent of the collector current. Why this is the case is illustrated in Figure 10.37. Caused by the exponential current–voltage behavior of the transistor, the transconductance g_m increases in the same manner as the output conductance g_o. The ratio of the two – that is, the small-signal voltage gain – is thus independent of the collector current.

In comparison to the MOS transistor, the bipolar transistor shows a much larger voltage gain (Figure 10.38). With, e.g., an Early voltage of 50 V, this transistor has at room temperature a gain of 2000. Assuming the same current dissipation and capacitance loading, then both transistors have the same −3 dB bandwidth but the unity-gain frequency of the bipolar transistor is much higher due to the large small-signal voltage gain.

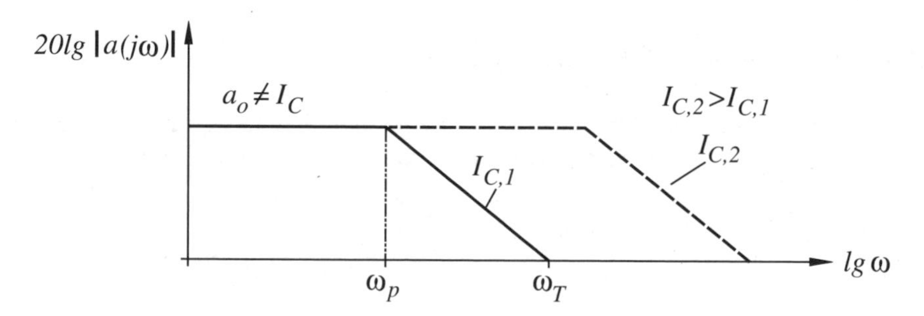

Figure 10.36 Magnitude plot of the amplifier stage of Figure 10.35

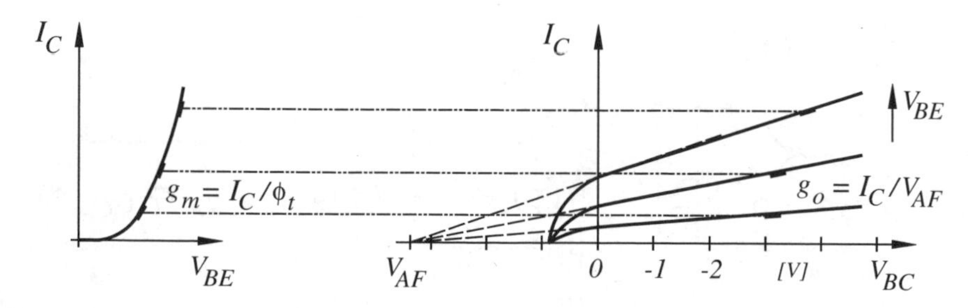

Figure 10.37 Comparison of transconductance and output conductance at different collector currents I_C

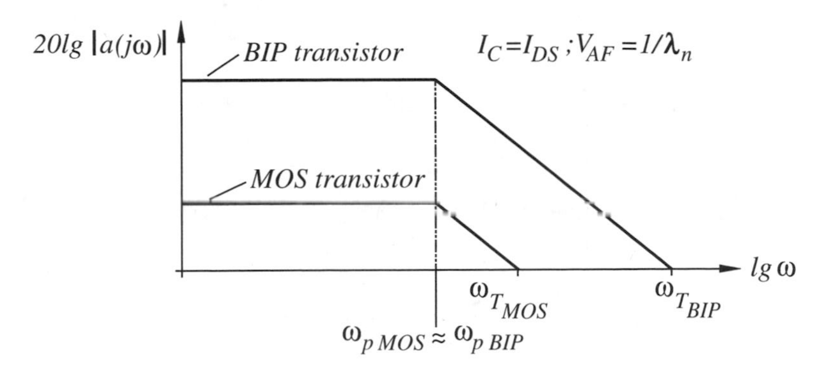

Figure 10.38 Comparison of voltage gain magnitude plots

Small-signal current gain of the MOS-transistor

So far, small-signal output and input voltages are compared. What kind of situation exists when the small-signal output and input currents are considered is the topic of this section. A simplified MOS current gain stage is shown in Figure 10.39. Output capacitances do not play a role, since the output is connected to V_{CC}. But the input impedance has to be considered. The resulting transfer function of the equivalent circuit $\beta(j\omega) = i_o/i_g$, and the unity-gain frequency $\omega_T(i_o = i_g)$ (Problem 10.4) are

$$\beta(j\omega) = \frac{\sqrt{2I_{DS}\beta_n(1+\lambda_n V_{DS})}}{j\omega(C_{gs}+C_{gdo})} \sim I_{DS}^{1/2} \tag{10.66}$$

$$\omega_T = \frac{\sqrt{2I_{DS}\beta_n(1+\lambda V_{DS})}}{C_{gs}+C_{gdo}} \sim I_{DS}^{1/2} \tag{10.67}$$

These relationships are shown in the magnitude plot in Figure 10.40.

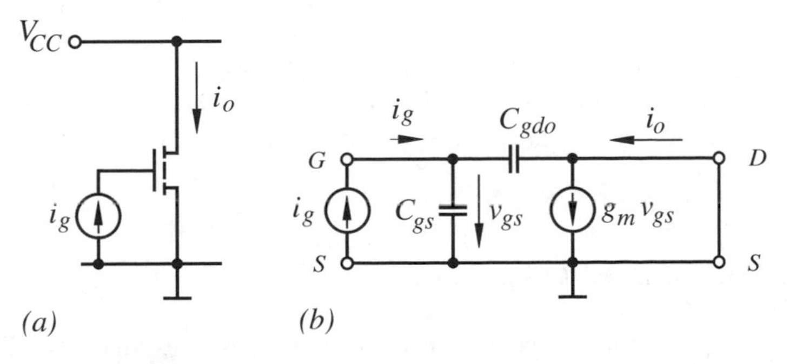

Figure 10.39 (a) Simplified MOS current gain stage and (b) small-signal equivalent circuit

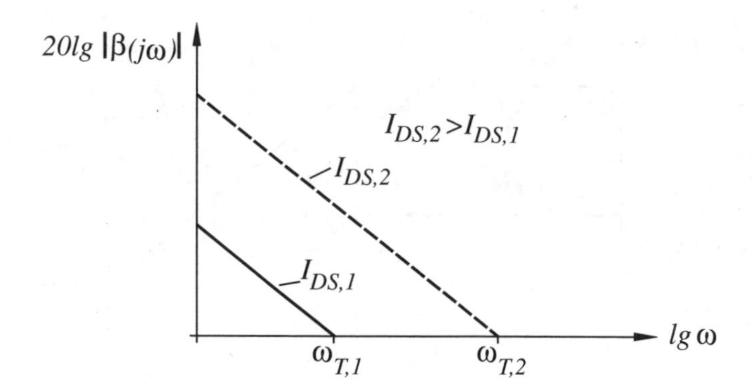

Figure 10.40 Magnitude plot of the MOS current gain stage of Figure 10.39

At $\omega \to 0$ the current gain approaches infinity since the input current

$$i_g = j\omega(C_{gs} + C_{gdo})v_{gs} \tag{10.68}$$

approaches zero. An increase in drain current causes an increase in gain and in unity-gain frequency until second-order effects, e.g. mobility degradation and internal transistor delays, limit the performance. It is interesting to find out which parameter influences the unity-gain frequency the most. For this purpose Equations (4.82) and (4.115) are substituted into Equation (10.67), which results in a unity-gain (angular) frequency of

$$\omega_T \approx \frac{(w/l)\mu_n C'_{ox}(V_{GS} - V_{Tn})}{(2/3)wlC'_{ox}}$$

$$\boxed{\omega_T \approx \frac{3}{2}\frac{\mu_n(V_{GS} - V_{Tn})}{l^2}} \tag{10.69}$$

The overlap capacitance and the channel length modulation are neglected for simplicity. As the equation demonstrates, the transistor gate width does not influence the unity-gain frequency. The reason is that an increase in gate width leads to an increased current, which is counterbalanced by an increased gate capacitance.

According to Equation (10.69), the most effective way to increase the unity-gain frequency is to reduce the channel length. Values larger than 70 GHz are reported at a gate length of 0.18 μm (Knoblinger *et al.* 2000; Mahnkopf *et al.* 1999) (Figure 10.41).

The reduction in unity-gain frequency at high currents is caused by mobility degradation and the increase in frequency as a function of the V_{DS} voltage by the channel length modulation.

Small-signal current gain of the bipolar transistor

Next, the small-signal current gain of the bipolar transistor is analyzed. This analysis is included in Section 3.5.4 and is repeated here for convenience. The transfer function

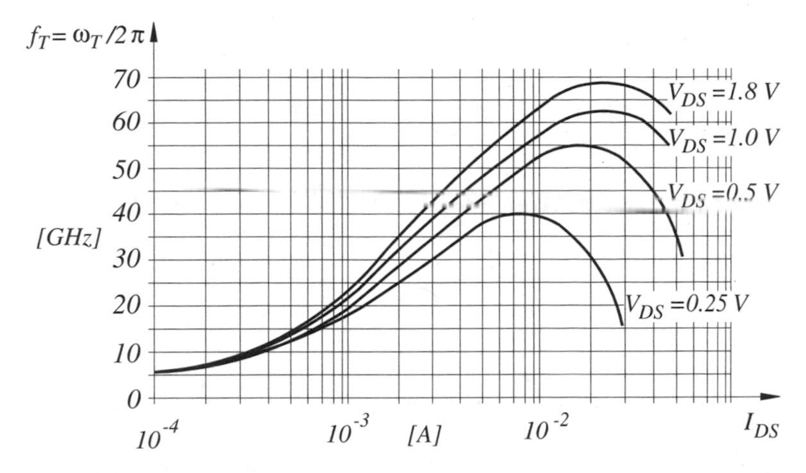

Figure 10.41 Unity-gain frequency of an n-channel transistor as a function of I_{DS}. (w/l = 108 μm/0.18 μm) (reprinted by permission, from Knoblinger *et al.* 2000; © 2000 Infineon).

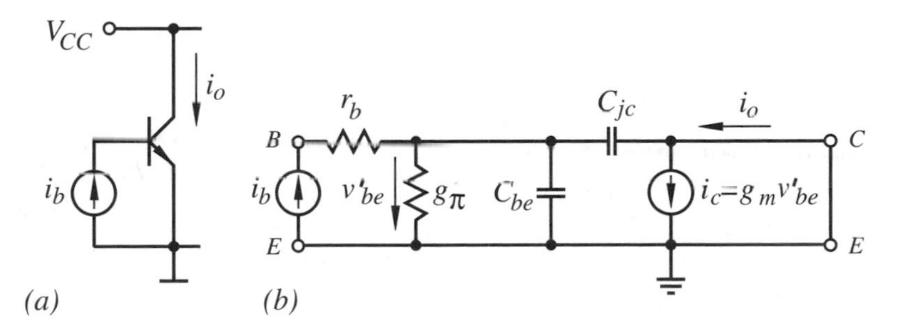

Figure 10.42 (a) Simplified bipolar current gain stage and (b) small-signal equivalent circuit

(Equation 3.114) is given by

$$\beta(j\omega) = \frac{i_o}{i_b}(j\omega) = \beta_F \frac{1}{1 + j\dfrac{\omega}{\omega_p}} \tag{10.70}$$

With respect to the collector current, the characteristic frequencies show the following dependencies (Equations 3.117 and 3.118)

$$\omega_p = \frac{1}{\beta_F[\tau_F + \frac{\phi_t}{I_C}(C_{je} + C_{jc})]} \sim I_C \tag{10.71}$$

$$\omega_T = \frac{1}{\tau_F + \frac{\phi_t}{I_C}(C_{je} + C_{jc})} \sim I_C \tag{10.72}$$

This behavior is sketched in Figure 10.43. The important result is that from a particular collector current the −3 dB bandwidth and the unity-gain frequency do not increase

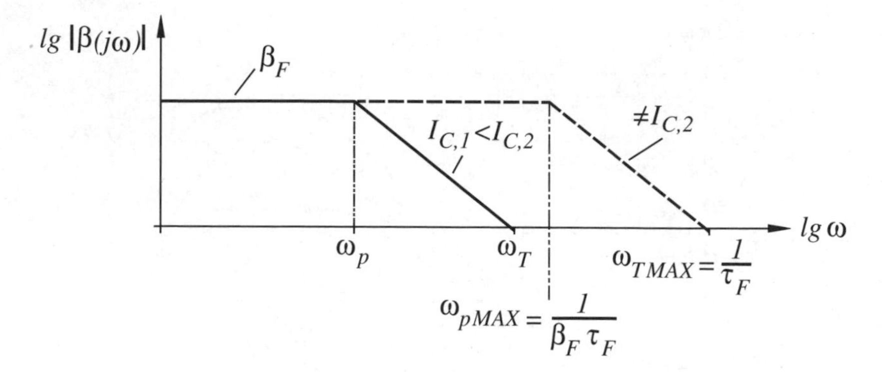

Figure 10.43 Magnitude plot of the bipolar current gain stage of Figure 10.42

anymore because of $1/I_C$. This yields a maximum unity-gain (angular) frequency (Equation 3.121) of

$$\boxed{\omega_{TMAX} = \frac{1}{\tau_F} = \frac{2D_{nB}}{x_B^2}} \tag{10.73}$$

which is proportional to the inverse of the base width square. This fact leads to transistors with unity-gain frequencies far into the GHz region. An example is shown in Figure 10.44 (Klein and Klepser 1999).

The reduction in unity-gain frequency at high currents is caused by an increased transit time τ_F at strong injection, and the increase in unity-gain frequency with V_{CE} by the base-width modulation.

A comparison of the behavior of the two transistor types reveals that the unity-gain frequencies of both transistors are inversely proportional to the square of the gate

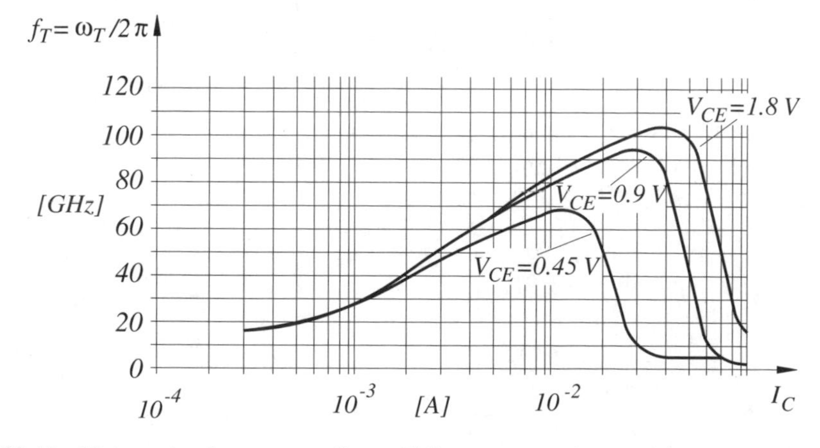

Figure 10.44 Unity-gain frequency of an SiGe npn transistor with an emitter area of 0.25 μm · 5.75 μm (reprinted by permission, from Klein and Klepser 1999; © 1999 Infineon)

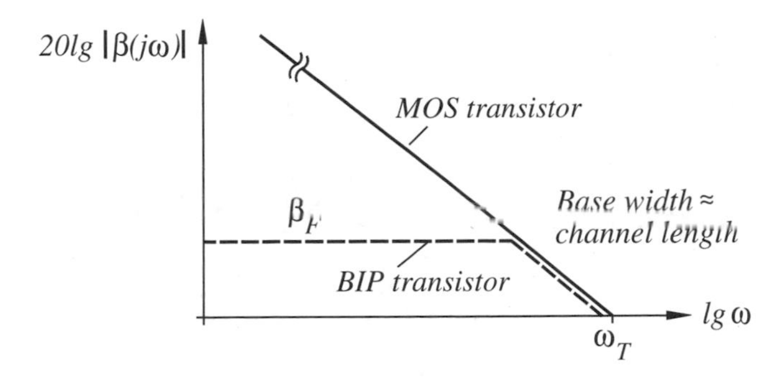

Figure 10.45 Comparison of current gain magnitude plots

length or base width, respectively. Thus, modern MOS and bipolar transistors show unity-gain frequencies of the same magnitude. This behavior is summarized in the current-gain magnitude plot of Figure 10.45.

Summary

The BICMOS technology utilizes the advantages of the bipolar technique additionally to those of the MOS technique. The highest data rates are achievable with CML and ECL circuits. They operate at reduced logic levels and are less sensitive to offset voltages. BICMOS buffers and gates show an improved driving capability as long as the switching times $t \gg \tau_F B_F$.

Band-gap reference circuits are a classical way to implement precise and almost temperature-independent reference voltages on-chip. The negative temperature coefficient of the V_{BE} voltage is compensated by a positive one, which is generated by evaluating the difference between two V_{BE} voltages. A comparison of the offset voltage between MOS and bipolar transistor pairs shows that the MOS transistors, because $\Delta V_{Tn} + (V_{GS} - V_{Tn})\Delta\beta_n/2\beta_n$, have about a 10 times larger offset voltage than bipolar transistors with $\phi_t \Delta I_{SS}/I_{SS}$.

A comparison between the transfer functions of the MOS and bipolar transistors reveals the advantages with respect to analog applications. The results are summarized in Figure 10.46. The most important results are:

(a) The voltage gain $a_o = v_o/v_i$ of the bipolar transistor is, because of V_{AF}/ϕ_t, independent of the collector current. Since the gain of a bipolar transistor is much larger than that of a MOS transistor, a much higher unity-gain frequency results. Thus, bipolar transistors are much more suitable for the design of amplifiers.

(b) The magnitude plot of the current gain shows a comparable unity-gain frequency for both transistor types. This causes MOS designs to penetrate the domain of bipolar transistors. By clever combining of MOS and bipolar transistors one can expect that novel designs will show up, particularly in communication systems and fiber optic applications.

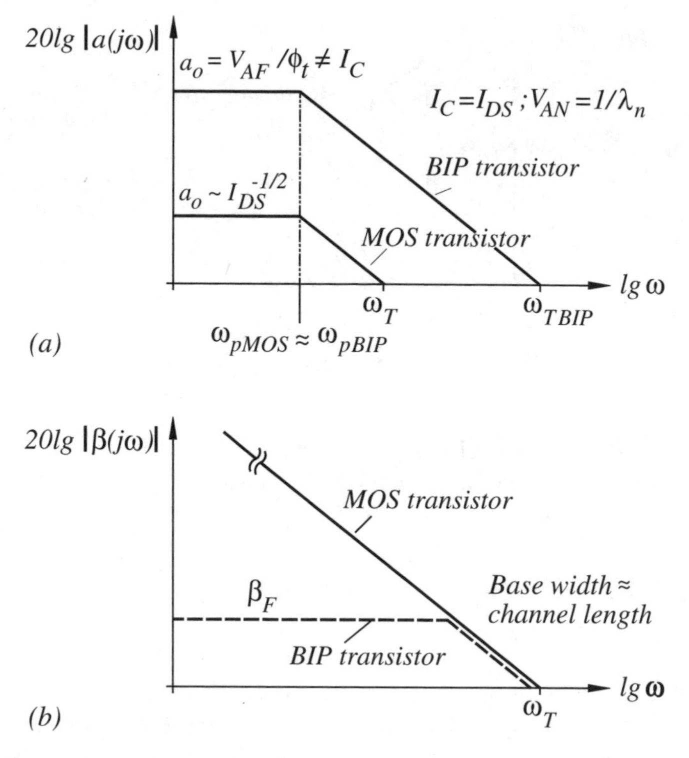

Figure 10.46 Comparison of magnitude plots: (a) small-signal voltage gain and (b) small-signal

Problems

10.1 Determine the minimum possible power-supply voltage V_{CC} of the BICMOS drivers shown below.

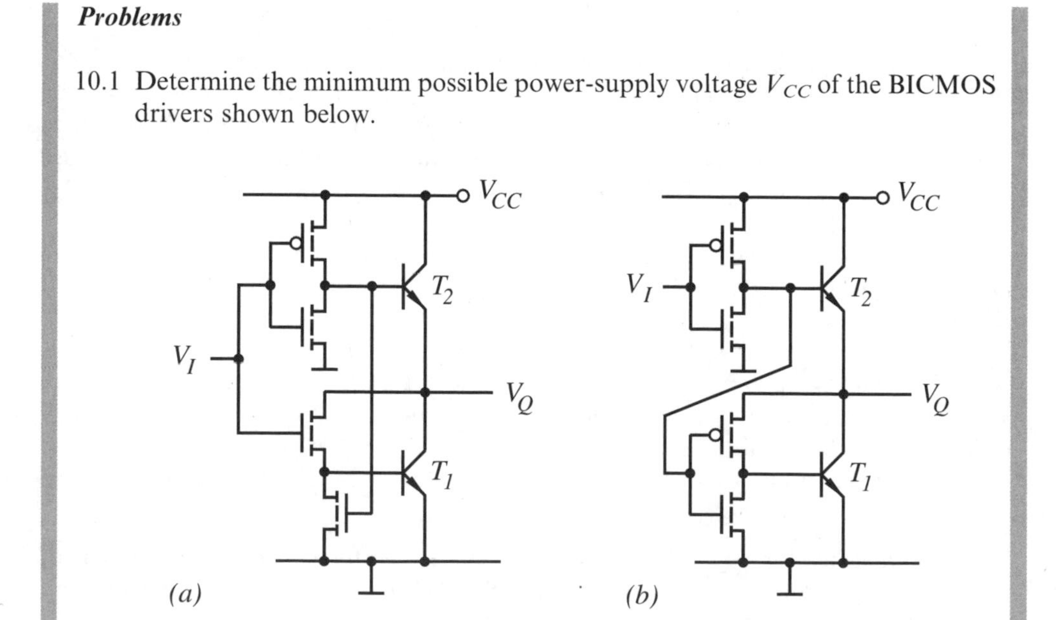

10.2 The current source shown below with the currents I_{B1} and I_{B2} is independent of the power-supply voltage when the channel length modulation is negligible. Derive the relationship for the current I_{B1} as a function of the transistor geometries when $(w/l)_3 = (w/l)_4$ and $(w/l)_2 > (w/l)_1$. It can be assumed that all transistors operate in current saturation.

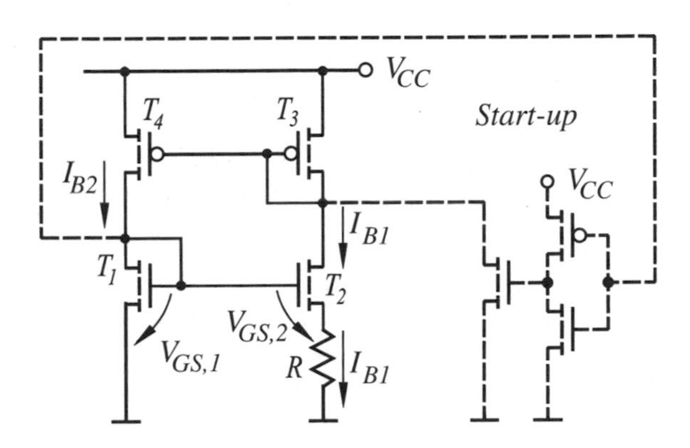

10.3 Shown below is a modified band-gap reference voltage circuit. Determine the reference voltage under the assumption of an ideal amplifier.

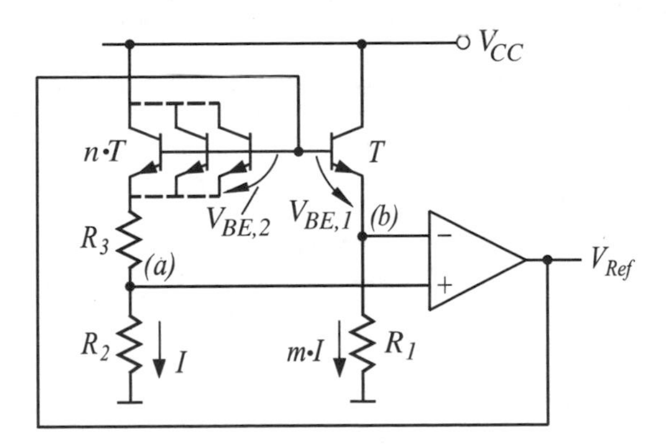

10.4 In Figure 10.39 a simple MOS amplifier stage with equivalent circuit is shown. Determine the i_o/i_g transfer function.

10.5 In Figure 10.35 a simple bipolar amplifier stage with equivalent circuit is shown. Determine the transfer function v_o/v_i when parasitic resistances are negligible.

10.6 Compare the input capacitance values of a MOS and a bipolar transistor at a current of 1 mA. The data are:

Bipolar transistor: $V_{BE} = 0.85\,\text{V}$, $f_T = 30\,\text{GHz}$ corresponds to $\tau_F = 5.3\,\text{ps}$.

MOS transistor: $V_{GS} - V_{Tn} = 1.0\,\text{V}$, $C'_g = 4\,\text{fF}/\mu\text{m}^2$, $k_n = 120\,\mu\text{A}/\text{V}^2$, $l = 0.15\,\mu\text{m}$.

The solutions to the problems can be found under: *www.unibw-muenchen.de/campus/ET4/index.html*

REFERENCES

M. Gunawa, G.C.M. Meijer, J. Fonderie and J.H. Huijsing (1993) A curvature-corrected low-voltage bandgap reference. *IEEE J. Solid-State Circuits*, **28**(6), 667–70.

W. Klein and B.U. Klepser (1999) 75 GHz bipolar-production technology for the 21st century. In H.E. Mase (ed.) *ESSDERC, '99: Proceedings of the 29th European Solid-State Device Research Conference*. Paris: Editions Frontieres, pp. 88–94.

G. Knoblinger, P. Klein and M. Tiebout (2000) A new model for thermal channel noise of deep submicron MOSFETs and its application in RF-CMOS design. *IEEE Symposium on VLSI Circuits, Digest of Technical Papers*, pp. 150–3.

R. Mahnkopf, K.-H. Allers, M. Armacost, A. Augustin *et al.* (1999) System-on-a-Chip technology platform for 0.18 μm digital, mixed signal and eDRAM application. *IEEE International Electron Devices Meeting, Digest of Technical Papers*, pp. 849–52.

B. Razavi (1999) CMOS technology characterization for analog and RF design. *IEEE J. Solid-State Circuits*, **34**(3), 268–73.

H.V. Tran, P.K. Fung and D.B. Scott (1989) BiCMOS current source reference network for uLSI BiCMOS with ECL circuitry. *IEEE International Solid-State Circuits Conference, Digest of Technical Papers*, pp. 120–1.

R.L. Treadway (1989) DC analysis of current mode logic. *IEEE Circuits Devices*, **5**(2), 21–35.

R.J. Widlar (1971) New developments in IC voltage regulators. *IEEE J. Solid-State Circuits*, **6**, 2–7.

M. Wurzer, J. Bock, H. Knapp, W. Zirwas, F. Schumann and A. Felder (1999) A 40-Gb/s integrated clock and data recovery circuit in a 50-GHz f_t silicon bipolar technology. *IEEE J. Solid-State Circuits*, **34**(9), 1320–4.

M. Yamashina and H. Yamada (1992) An MOS current mode logic (MCML) circuit for low-power sub-GHz processors. *IEICE Trans. Electron.*, E75C(10), 1181–7.

Index

System Integration: From Transistor Design to Large Scale Integrated Circuits. Kurt Hoffmann.

ISBN: 0-470-85407-3